The Lost Constellations

A History of Obsolete, Extinct, or Forgotten Star Lore

John C. Barentine

The Lost Constellations

A History of Obsolete, Extinct, or Forgotten Star Lore

Springer

Published in association with
Praxis Publishing
Chichester, UK

John C. Barentine
Tucson, AZ, USA

SPRINGER PRAXIS BOOKS IN POPULAR ASTRONOMY

Springer Praxis Books
ISBN 978-3-319-22794-8 ISBN 978-3-319-22795-5 (eBook)
DOI 10.1007/978-3-319-22795-5

Library of Congress Control Number: 2015947001

Springer Cham Heidelberg New York Dordrecht London

Cover image: "Lacerta, Cygnus, Lyra, Vulpecula and Anser", plate 14 in Urania's Mirror, a set of celestial cards accompanied by A familiar treatise on astronomy . . . by Jehoshaphat Aspin. London. Astronomical chart, 1 print on layered paper board : etching, hand-colored. Available from the United States Library of Congress's Prints and Photographs division under the digital ID cph.3g10063.

Printed on acid-free paper

Springer International Publishing AG Switzerland is part of Springer Science+Business Media (www.springer.com)

To my grandparents,
Gerald and Verne Anne Danley

Preface

I grew up at the edge of Phoenix, Arizona, just before the start of the vast, sprawling suburbs that completely surround the city. In the late 1980s, beneath a sky already so light polluted that fourth-magnitude stars had essentially vanished, I came of age as an amateur astronomer. With a small backyard telescope, my views were largely limited to the Moon, bright planets, interesting double stars, and a handful of the very brightest nebulae, galaxies, and star clusters. In high school, I upgraded to a simple but sturdily constructed 8-in. Newtonian telescope that brought some of those distant objects closer; in the very middle of the night, when the city light was at a relative minimum, I could just begin to glimpse a few of the brighter galaxies in the NGC catalog. But the bright stars, those needing no telescope for a fine view, were first and foremost among my companions on many long nights, and to this day their seasonal comings and goings take them away and bring them back regularly like old friends.

When I was in college, I met the famed amateur comet-hunter David Levy, fresh off the spectacular impact of his co-discovery with Carolyn and Gene Shoemaker, Comet Shoemaker-Levy 9, with Jupiter in the summer of 1994. David was an inspiration ever since I first read about him 10 years earlier, as a kid. I asked who inspired him in the same phase of life, and he pointed me to a very under-appreciated book by an unfamiliar author: Lesile Peltier (1900–1980). The book was *Starlight Nights*[1] (1965), in which Peltier recounts his life as, in Harlow Shapley's words, "the world's greatest nonprofessional astronomer." As a boy growing up in rural western Ohio during the earliest years of the twentieth century, he first set out to learn the names of the stars; along the way, they became as familiar as the earth beneath his feet. Peltier wrote that it took him

> about a year to become acquainted with the stars. This may have been a longer apprenticeship than some would care to serve but I have found it well worthwhile for in the end I had much more than just a mere assortment of names and places in

[1] David arranged for a reprinting of the book Sky Publishing in 2007, along with an excellent new foreword he wrote.

the sky. Each star had cost an effort. For each there had been planning, watching, and anticipation. Each one recalled to me a place, a time, a season. Each one was now a personality. The stars, in short, had now become *my* stars.

I, too, set about to learn the names of the stars—at least a few bright ones visible from my part of the world—and in time I began to assemble them into asterisms and constellations. The Big Dipper was familiar from childhood; the fainter stars comprising the rest of the Great Bear remained a challenge. Orion rose in the fall and ruled the winter sky, while an appearance by the bright orange star Arcturus, low on the northeastern horizon in late February, was an annual reminder those long winter nights were drawing to a close. Sagittarius took its place low in the southeast by midnight in May, but I could only imagine the steam issuing forth from its famous "Teapot" because the sky was far too bright to show any trace of the Milky Way. By the time the Summer Triangle marched overhead at sunset in September, the long days of school vacations were already over. These were the cycles I came to expect every year, without exception. While I knew fainter stars existed from occasional overnight trips out of town to darker locations, they were otherwise always minor players in the nightly spectacle, watching from behind the folds of dark velvet curtains in the wings.

Various books and how-to guides on astronomy also made me aware that, from our latitude, about one-third of the entire night sky was rendered permanently invisible. There were other bright stars marking exotic constellations, but they never rose above our southern horizon. For several years in my early teens, I had a pen pal in Australia with whom I exchanged letters every few weeks; I wrote asking for descriptions of his night sky and marveled at his recollection of seeing Supernova 1987A from his home in the Sydney suburbs, a bright star visible to the naked eye despite the light pollution problems of the big city. I was left to imagine what southern skies looked like.

One winter night when I was about 14 years old I ventured out into the front yard of our house to set up my telescope, a 60 mm refractor received as a gift the previous Christmas. Although it had little light-gathering power or angular resolution, it was "good enough" given the bright skies of Phoenix. My quarry that night was a very young crescent Moon that would appear in the gap between the tall trees on either side of our street as it gently curved toward the southwest. The front of our house faced south, and as I wrangled the telescope tripod through the open front door, I looked up and spotted something grazing the top of the peaked roof of the house on the other side of the street. It was a bright, steely blue-white star, twinkling fiercely in the murk only a few degrees off the southern horizon. And it wasn't supposed to be there.

I stopped and set the tripod down, leaned against one of the brick support structures holding up the roof above the front porch, and waited. *Surely it must be an airplane*, I thought as I waited for the subtle tell-tale motion, parallel to the roof line, of planes on approach from the west into Sky Harbor International Airport. I waited for the steady *blink-blink-blink* of the warning lights that would betray the interloper as a commercial airliner. I waited several minutes, noting only a slow, constant march westward at the sidereal rate. It didn't change the shape, brightness, or color. It was a star, all right, and a quick examination of my how-to books showed it had a name: *Canopus*.

Canopus is the brightest star in the constellation Carina. It is a virtual twin, in apparent brightness and color, to the star Sirius in Canis Major, the brightest star in the entire sky. From the southern hemisphere, on summer nights, the two lights follow each other, passing nearly overhead. From my location at 32° north latitude, its southerly location ensured that it never rose nearly so high. The effect of atmospheric extinction on Canopus was clear, rendering it noticeably fainter than Sirius. The experience felt like noticing an intriguing and altogether new detail in a painting hanging on the same wall for all one's life, hiding in plain sight the entire time.

Carina, I learned, represents "the Keel," as in a part of a ship's functional anatomy. Then I read something astonishing: at Carina's southerly declinations, there were also Sails (Vela), a mariner's Compass (Pyxis), and a Poop Deck (Puppis). There was an entire *ship* down there, one long hacked into bits that are now recognized as entirely separate constellation. But at one time, centuries ago, that vessel—the ship *Argo*—had ruled the starry waves. Considered a single constellation, it dwarfed every other figure in the sky from antiquity until around 1900.

Why did Argo suffer the fate of being broken up into its constituent parts, like a real boat dashed on coastal rocks? The seeds of its disassembly were sewn in the mid-eighteenth century when the French astronomer Nicolas Louis de Lacaille decided it was too unwieldy to easily yield to the designs he had on mapping the stars of the southern sky. By popular acclamation, his suggestion was gradually adopted over the next 150 years and finally formalized when the International Astronomical Union (IAU) promoted the changes as authoritative in its 1928 declaration of the modern constellation boundaries. Only in the past several decades have professional astronomers decided among themselves what (and which) constellations properly *are*, but for centuries the nomenclature of the sky was subject to a kind of Wild West mentality. The story that emerges is one of technologies, the diffusion of knowledge, voyages of discovery, the egotism of astronomers and cartographers, and nationalistic pride.

While we were at college, my friend Marci Winter gave me a book she had mistakenly bought in hopes it would be useful for her introductory astronomy course. That book, Richard Hinckley Allen's *Star Names: Their Lore And Meaning* (1899), is the other key to understanding why I wrote *this* book. Although written over a century ago, Allen's book remains the definitive written work on the history of the constellations. Allen was a gifted polymath with an encyclopedic knowledge of history. As a young man, in an era when astronomical research was driven by visual telescopic observations, his hopes to pursue professional astronomy as a career were dashed by his poor eyesight. While he later experienced reasonable success as a businessman, he never lost his passion for the night sky, and *Star Names* is the love letter expressing his devotion to understanding how the lore of the sky came to be. Structured as a series of section headings, one per constellation, Allen dug deeply into history, literature, art, and science to collect essentially everything then-known about the constellations into a single volume, now attractively priced as an inexpensive reprint. While he wrote extensively about references to the constellations in classical antiquity, he also collected the myths and legends of Medieval, Arab, Indian, Babylonian, and Chinese traditions.

Thumbing through *Star Names* for the first time, I made a second astounding discovery: Argo Navis wasn't the only "former" constellation lurking in the night sky. There were many others consigned to an even worse fate. They had vanished entirely leaving behind as the only vague indicator of their former status at the top of the astronomical food chain some quirky irregularities in the boundaries of the modern constellations. Allen is a detailed, carefully researched source whose name appears repeatedly here, even as some modern critics have dismissed some of his sources as obsolete and his star names as unreliable. Most importantly, his research included constellations already disappeared from the then-current maps in the interest of completeness.

One winter long ago, I made a new friend among the stars, one that would lead me on a journey more than a quarter-century in the making. That journey now culminates in this book. For years I kept notes on scraps of paper and photocopies of the odd article now and then on obsolete constellations from popular astronomy magazines. But I was always dismayed that no proper history of these celestial also-rans had ever been written other than the few tantalizing glimpses in Allen and a handful of other sources, such as Ian Ridpath's excellent *Star Tales* (1989). After finishing my Ph.D., I suddenly found myself with free time for the first time in several years, and writing this book helped fill that void. What has emerged is a very human story that transcends astronomy itself, somewhere at the confluence of history, mythology, folklore, exploration, and psychology.

Like many "Generation X" kids who later went into science I was profoundly influenced by Carl Sagan's 1980 television series *Cosmos*, in which Sagan described the constellations as a "set of human hopes and fears placed among the stars." The same hopes and fears informed the process by which the divisions of the night sky eventually became canonical. How the constellations won (and lost) the astronomical popularity contest, resulting in historical immortality or obsolescence, contains all the essential elements of entertainment, featuring moments of comedy, tragedy, and whimsy. This book tells how we got to where we are, and what we cast off along the way.

Many people contributed indirectly to this book through many years of unfailing support, and I am fortunate that my friends, family, and teachers number too many individuals to name them all here. A few people stand out and are worthy of special recognition for their roles in the realization of this work. I am forever indebted to the love of my mother, Delsia, and of my father, John (1952–2010) who, sadly, did not live to see it published. Their support of my hobby-turned-career set me on the path I continue to walk today. To my friend of more than two decades, Gilbert Esquerdo (Smithsonian Astrophysical Observatory/Planetary Science Institute), I owe thanks for many useful conversations in the early stages of its planning. My former research supervisor Dr. Roger Culver (Colorado State University), the best mentor a student could ask for, heard one of the earliest pitches for this story. Dr. Joel Cruz (Elmhurst College, Illinois), historian and gentleman, encouraged me to see this work to publication and shared with me his knowledge of the academic publishing world. And finally, I am grateful to Marci Winter Lister for gifting me her unneeded copy of Allen's *Star Names* and launching me on a fascinating and rewarding journey.

Lastly, it is a pleasure to acknowledge the influence of two authors whose works loomed large over the writing of this book: Richard Hinckley Allen's *Star Tales* and *Burnham's Celestial Handbook* by Robert Burnham, Jr. Both writers brought the rigor of scholarship to understanding the history of the heavens, and I have tried to emulate their respective styles—thorough but not exhaustive—in telling the story of parts of the night sky hidden in plain view.

Tucson, AZ
June 2015

JOHN C. BARENTINE

Technical Note

SOURCES

The goal of this book is to be comprehensive without being exhaustive. Original sources of the works referenced herein were preferred in every possible situation. Where the primary works were unavailable, secondary citations were used; I have endeavored to make this distinction clear.

ILLUSTRATIONS

In addition to consulting original written works, I have preferred first printed editions of various charts and atlases as the source of most illustrations in this book. This approach is taken is to show as many interesting depictions of lost constellations as practicable without reproducing every known instance. Depictions from certain seminal works, such as Johannes Hevelius' *Prodromus Astronomiae* (1690) and Johan Elert Bode's *Uranographia* (1801b), are included in every appropriate case; otherwise, the choice of illustrations is made to adequately trace the origin and evolution of constellations in as straightforward as possible a manner.

I have employed a limited amount of manipulation of images from historical atlases strictly for the purpose of improving the clarity and legibility of those images while never altering the figures therein contained. Mild enhancements, such as those undertaken to increase contrast and reduce the background "noise" of discolored or damaged paper, are not generally noted. Any instances of significant image processing that fundamentally alters the source material, such as digitally joining globe gores to produce a seamless composite map, have been noted in the text.

Photographs of non-printed works such as paintings and other illustrations have been reproduced with image density adjustments for clarity only. I have made an effort in every case to try to include in image captions information about the dimensions of the original, the medium, and current location and/or catalogue information where obtainable. For works not in the public domain, credit is given to the creator along with usage information such as Creative Commons licenses.

TRANSLATIONS

As a result of preferring original sources, I have often confronted passages in original Latin, German, and French. I render these in English as best as I can, being fluent in none of those languages; wherever possible, I have checked with native speakers or those with extensive formal training in Latin. Sometimes the renderings are imprecise, but I have tried to retain some of the flavor of the original and always the essence. In every instance I have quoted passages in their original (non-English) languages as footnotes throughout the text such that the reader can decide if my translations are good. Any deficiency in the essence of the translations will be corrected in a future edition. Otherwise, when using others' translations, I have indicated the translator's name and corresponding bibliographic information when known.

NOMENCLATURE

Since by definition the constellations described in this book had fallen into complete disuse by the time of the first General Assembly of the International Astronomical Union (IAU), where the canon of modern constellations was decided by the international governing body of professional astronomers, they were never subjected to the process by which the IAU formalized a set of genitive cases and three-letter abbreviations (see Chap. 2). There is also the issue of the names of the constellations themselves; as they passed in and out of fashion and were rendered by authors writing in, variously, Latin, English, French, German, and other languages, a variety of spellings often ensued. I describe here how I settled on a means of standardizing names, cases, and abbreviations across the chapters corresponding to individual constellations.

Constellation Names

The names of constellations adopted by the IAU are a mix of Latin and Greek words; the latter generally derive from the names in circulation at the time Ptolemy wrote the *Almagest* in the second century AD Others were Roman inventions, but the names of all constellations in the Ptolemaic canon were Latinized. Some of the first new constellations added since the time of Ptolemy referred to discoveries made by explorers to southern hemisphere destinations and the New World. Latin had no native word for the toucan, for example, so when the native name "tukana" came from the Tupi language of Brazil via Portuguese, it was appropriately Latinized as "Tucana." Petrus Plancius suggested a constellation representing the toucan in 1598, labeling according to the borrowed Latin. Constellations created in the eighteenth century to celebrate the apparatus of the arts and sciences often required contrived Latin names for concepts unknown to the ancients (e.g., "Globus Aerostaticus" for the Hot Air Balloon and "Machina Electrica" for the Electrical Generator). Sometimes they repurposed ancient words for similar devices such as "Antlia Pneumatica" (later shortened to "Antlia") to describe a mechanical air pump, whereas the word "antlia" referred to a water pump in ancient sources.[1]

I have retained the preference for Latin names in this book in all practical cases; fortunately, many of the extinct constellations here discussed were introduced by their creators with native Latin (or Latinized) names. In isolated cases, constellation names were never Latinized by their creators or involve words that have no obvious Latin equivalent. An example is the Battery of Volta, described in Volume 2. Since an electrical battery has no conceptual expression in Classical or Medieval Latin, I borrowed the Latin word "pila," meaning a pillar or column, as of stone, to indicate the original sense of a battery as a "Voltaic pile." Thus, while I use "Battery of Volta" as the formal name of the constellation, I render its genitive as "Pila Voltae" and its three-letter abbreviation as "PiV."

Genitives

The genitive grammatical case is used to indicate possession, in the sense that a particular star "belongs" to the constellation inside whose boundaries it falls. The widespread use of this convention originated in Johann Bayer's *Uranometria* (1603). Bayer devised a system of cataloging the stars in a particular constellation visible to the unaided eye by the use of Greek letters. According to this scheme, the brightest star in a constellation was labeled "α," the next brightest "β," and so forth through the 24 letters of the Greek alphabet. However, most constellations had more than 24 visible stars; when he ran out of Greek letters in a particular constellation, Bayer ran through the lowercase Roman alphabet starting at "a," among which he omitted the lowercase letters "j" and "v." That brought the total number of available letters to 48. Bayer never exceeded this number in any constellation, but later astronomers sought to extend the series using uppercase Roman letters beginning with "A" following "z" and finishing at "Q," inclusive of "J." Within any given constellation, Bayer proceeded from one whole magnitude to the next in half-magnitude intervals; for example, a "third-magntitude star" is any having a visual magnitude between +3.5 and +2.4. He further proceeded in an overall north-to-south pattern, then repeated the process for the next-faintest magnitude bin. In other cases, Bayer changed the order of the letters for historical or other reasons.

To complete the designation, Bayer added the classical constellation name in the genitive case; for example, the brightest star in the constellation Canis Major became α Canis Majoris ("alpha of Canis Major"). This convention followed the rules of Latin noun declension, which in some cases required the Latinization of constellation names

[1]For example, Martial, *Epigrammata* 9, 14, 3; C. Suetonius Tranquillus, *Tiberius* 51.

Table 1 Modern constellations whose names consist of two words: nominative case, genitive case, and meaning

Nominative	*Genitive*	*Meaning*
Canes Venatici	Canum Venaticorum	The Hunting Dogs
Canis Major	Canis Majoris	The Greater Dog
Canis Minor	Canis Minoris	The Lesser Dog
Coma Berenices	Comae Berenices	Berenice's Hair
Corona Australis	Coronae Australis	The Southern Crown
Corona Borealis	Coronae Borealis	The Northern Crown
Leo Minor	Leonis Minoris	The Lesser Lion
Piscis Austrinus	Piscis Austrini	The Southern Fish
Triangulum Australe	Trianguli Australis	The Southern Triangle
Ursa Major	Ursae Majoris	The Greater Bear
Ursa Minor	Ursae Minoris	The Lesser Bear

originally derived from ancient Greek. For instance, Orion became "Orionis" in the genitive case and the star Betelguese, which appeared to Bayer as the brightest in that constellation, became α Orionis.

Among extinct constellations, a variety of names occur in more than one language. For the purposes of standardizing their genitives as closely as possible to the convention implicitly adopted by the IAU, I constructed genitives based on Latinized forms of the constellation names as described above. Naturally, there are some special cases. Modern constellations whose names contain a noun and a modifier, like Ursa Major and Corona Borealis, appear in Table 1. In the "Name/Modifier" paradigm, both words take the genitive case, so the above examples become Ursae Majoris and Coronae Borealis, respectively. Often the modifier is itself already rendered in the genitive, such as in the case of Caput Medusae (see Volume 2). In these situations, the modifier remains unchanged in the genitive, while the name changes case, so "Capitis Medusae" ("*of* the Head *of* Medusa"). There are a few instances of unusual Latin declensions, such as Argo Navis, which becomes Argūs Navis in the genitive. In that singular example I have kept the macron over the "u" in order to specify the genitive as completely as possible.

Abbreviations

At the first IAU General Assembly at Rome in 1922, the Union approved a list of 88 constellations which remain with us today as a modern canon used by professional and amateur astronomers alike. To simplify written references to the constellations, a three-letter abbreviation was devised for each after delegates expressed dislike of a proposed four-letter scheme. The convention determining how a three-letter abbreviation is rendered depends on whether the approved constellation name consists of one or two words. For one-word constellations (e.g., Orion, Taurus, Sagittarius), the word is contracted so as to make each distinct from any other with a similar spelling. So Orion becomes "Ori" and Taurus becomes "Tau," but Sagittarius becomes "Sgr" rather than "Sag," as the latter was reserved for the constellation Sagitta. Therefore, the preferred rendering is the first three letters of the Latinized name unless some other constellation exists for which those initial three letters are the same. In all cases, the first letter only of the three is capitalized.

If a constellation consists of two words, the format for the three-letter abbreviation is capital-lowercase-capital, where the first capital and lowercase letters refer to the first word in the name (typically Name in the "Name Modifier" paradigm), and the final capital letter refers to the second word. So Corona Borealis becomes "CrB" and Piscis Austrinus becomes "PsA." However, there are irregularities. In some cases, the abbreviation is formulated capital-capital-lowercase, as "CVn" for Canes Venatici and "UMa" for Ursa Major. Then there is the completely inexplicable "Com" for Coma Berenices, ignoring entirely the second word. Rather than explicitly naming a formula for making these determinations, the IAU simply published the abbreviations as a list.

I have imposed the following set of rules in creating abbreviations for extinct constellations:

- For three-letter abbreviations of constellations containing two words, I followed the predominant IAU convention of capital-lowercase-capital. When "Major" or "Minor" is the modifier, I follow the form capital-capital-lowercase *except* if the result is identical to an existing IAU abbreviation. So, e.g., Cancer Minor cannot be shortened as "CMi" because Canis Minor already holds that abbreviation. So, Cancer Minor becomes CnM.
- In a few situations there are totally unique combinations such as "Cerberus et Ramus Pomifer." This I rendered as "CeR" ("Cerberus et Ramus").
- For names consisting of three words, each word receives one capital letter. So, e.g., Gladii Electorales Saxonici becomes GES.
- There are two names consisting of three words where the last word is a modifier and the names differ *only* by that modifier: Telescopium Herschelii Major and Telescopium Herschelii Minor. Since there is no proper way to reduce these names to three-letter abbreviations without losing essential elements, I opted for four letters: THMa and THMi.

Contents

Part I
Toward the Modern Night Sky

1

What Is a Constellation?

> [A]ncient customs are difficult to overcome, and it is very probable, that, except the recently-named groups, which we may now suppress, the venerable constellations will always reign.
>
> – Camille Flammarion, *Astronomie Populaire* (1880)

From a dark location on Earth, far from sources of artificial light pollution, a few thousand stars are sufficiently bright to be seen by the unaided human eye. They are spread across the night sky in a seemingly random way, although the keen observer will note a few structural consistencies. The sky is bisected by the path of the Milky Way, the faint band of light that represents the plane of the galaxy in which we live, its innumerable faint stars blurred together into softly glowing clouds as seen by the human eye. There are generally more visible stars in the direction of the Milky Way's center, in the constellation Sagittarius, than its "anticenter" (the point opposite the center on the sky) in the constellation Auriga. Otherwise, the distribution of stars visible to the naked eye does not yield many clues as to the Earth's position in the universe.

One popular misconception about the stars is that brightness indicates distance, fainter stars being located further away from Earth than the bright ones; were it the case that all stars had identical *intrinsic* brightnesses, this would be true. But by the twentieth century, astronomers realized that the luminosities of stars spanned an enormous range of values, from those many thousands of times intrinsically brighter than the Sun to those just a fraction of a percent of our own star's luminosity. The brightness of stars, then, does not tell us much about the immediate volume of space we inhabit. That some stars are brighter and others are fainter is also only part of the story by which certain ones come to form recognizable patterns to humans.

The main influence on the distribution of stars in the night sky has to do with our location in (and the structure of) the Milky Way. Our home galaxy is a common type known a "disk spiral," consisting of a relatively flat "Disk" embedded in an extended,

J.C. Barentine, *The Lost Constellations*, Springer Praxis Books,
DOI 10.1007/978-3-319-22795-5_1

more or less spherical "Halo".[1] A set of spiral arms, regions dense with stars and the materials from which they form, unwind from the center to the edge; the arms are not readily apparent in our night sky by virtue of the fact that the Solar System lies in their common plane. The central region of the Milky Way is characterized by a "bar," a transient kinematic feature kicked up by resonant interactions between stars in their orbits and the gravitational potential of the entire galaxy. From our perspective, we see the bar oriented at an angle of about 45° to the line of sight toward the Galactic center, one end of the bar pitched toward to us and the other directed away.

Since we live inside the Disk, our understanding of the geometry of the Galaxy comes from the structures we can infer from our vantage point, supplemented by studies of other galaxies seen from the outside that we think resemble the Milky Way. Combining information from both sources, we can make an educated guess as to both the shape of our galaxy and the location of our solar system within it. An artist's conception of the Milky Way is shown in Fig. 1.1, illustrating what a viewer situated high above the plane of the Disk might see looking down at it. The Sun, whose position is marked by a yellow circle, makes a leisurely orbit around the Galactic center once every 225–250 million years. Our motion through the Galaxy ensures that the Earth's night sky is never static, and that the stars of tonight's sky differ from those of the distant past and future.

Early astronomers presumed the Solar System was located at the center of this system given two pieces of evidence. First, when we look out into the night sky we see stars in every direction, but they tend to be concentrated into a fairly narrow band dividing the sky into two halves. This suggests a largely flattened system, and that we must be situated nearly in its middle. Second, this band of stars is fairly uniform in density around the sky in an azimuthal sense, so we must be near or at the center of the flat disk. Studies carried out in the twentieth century showed that we are not at the center of the Galaxy, but rather at some considerable radius. In fact, the Sun's place in the Milky Way is on a spur projecting from one of the spiral arms, a relatively quiet backwater compared to the swift streams of the main spiral arms and the crowded environment of the Galactic center.

Humanity's first view of the Milky Way as a single stellar system emerged in the eighteenth century. While the Galaxy has been known since time immemorial, an understanding of its composition awaited the invention of the telescope. In 1610 Galileo Galilei turned one to the Milky Way for the first time, resolving its luminous clouds into innumerable faint stars. A century later philosophers and scientists began to devise ideas about what the Galaxy physically represents. The English astronomer Thomas Wright published *An original theory or new hypothesis of the Universe* in 1750, in which he posited that the Milky Way was a large, rotating body composed of individual stars held together by gravity. He deduced it was a scaled-up version of the solar system, by then describable with relatively simple physics under Isaac Newton's universal theory of gravitation. In 1755, the German philosopher Immanuel Kant elaborated on Wright's

[1]By convention in astronomy, capitalized versions of "Disk", "Halo", and "Galaxy" refer specifically to the Milky Way and its constituent parts; lowercase versions of these same words refer generically to any external galaxy other than our own.

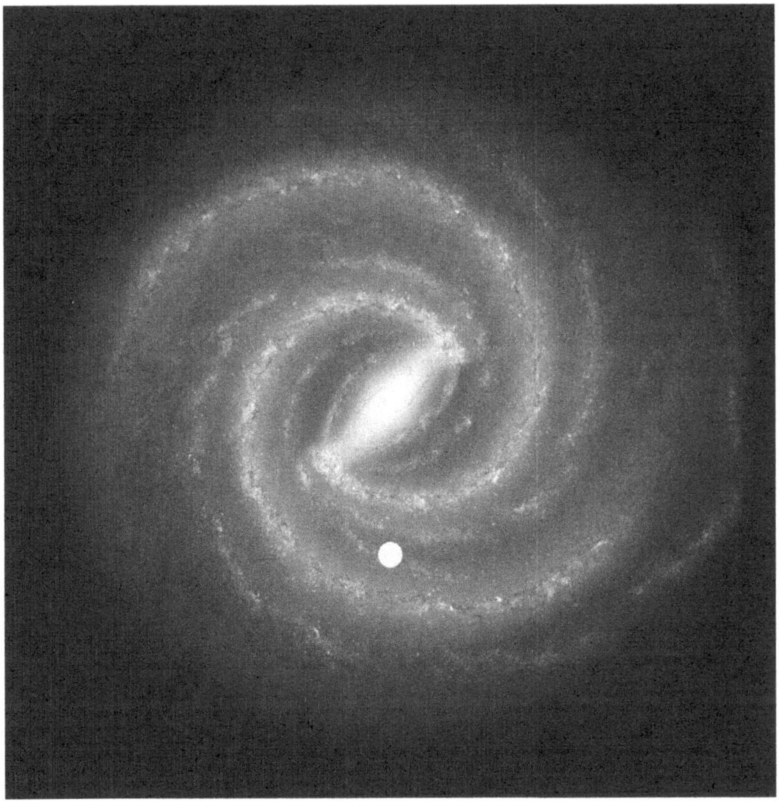

1.1 An artist's rendering of the Milky Way as it might appear viewed face-on from several hundred thousand light years above the midplane. The *solid white circle* indicates the location of the Sun (Credit: NASA/JPL-Caltech/R. Hurt)

hypothesis, suggesting that the Milky Way might have begun as a spinning cloud of gas that somehow condensed into stars. He further speculated that other similar systems might exist and that the faint and featureless nebulae seen in contemporary telescopes could be such "island universes" unto themselves.

Based on his careful telescopic counts of stars toward various directions in the night sky, the Anglo-German astronomer William Herschel made probably the first attempt at a structural model of the Milky Way (Fig. 1.2), correctly concluding that the solar system was embedded inside it. From this vantage point, Herschel wrote,

> the heavens will not only be richly scattered over with brilliant constellations, but a shining zone or milky way will be perceived to surround the whole sphere of the heavens, owing to the combined light of those stars which are too small, that is, too

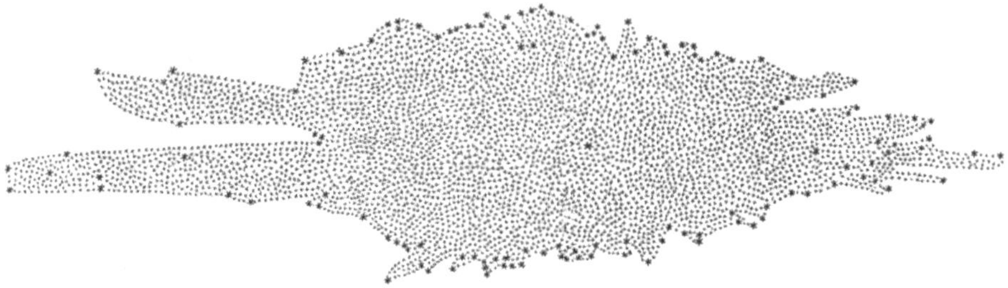

1.2 William Herschel's model of the Milky Way based on his star counts. The model is essentially a longitudinal slice through the Galaxy, reflecting the Solar System's position within its Disk. Figure 4 from "On the Construction of the Heavens," *Philosophical Transactions of the Royal Society of London*, Vol. 75, pp. 213–266 (1785)

> remote to be seen. Our observer's sight will be so confined, that he will imagine this single collection of stars, of which he does not even perceive the thousandth part, to be the whole contents of the heavens.

Historical interest in the makeup of the Galaxy recounted here is not about appreciating Galactic structure as a subject in its own right; rather, it is in understanding how the stars are scattered about our night skies. The bright stars that mark the familiar constellations tend to be remarkably close to Earth, and many are intrinsically faint given the rapid drop-off of observed brightness with distance, it is clear that most of the stars visible to the naked eye are within a relatively short distance of Earth. This is illustrated in Fig. 1.3, a representation of all known stars within 15 light years of the Sun in space that shows the random nature of the distribution of stars in the space near us. In order to see such low-luminosity stars, they cannot be very far away from us.

Considering the faint, integrated light of the entire Galaxy across our skies as a separate phenomenon, the *individual* stars we see sample a relatively small volume of the Milky Way. We do not live in a part of the Galaxy particularly close to where stars are actively forming; this point is important because such clustered environments are the only places where the distribution of stars in such small volumes of space is relatively uniform. Once stars leave their natal clusters, they drift away in random directions until their orbits around the Galactic center are no longer distinguishable from those of myriad other stars.

For all intents and purposes, the bright stars in Earth's night skies are distributed (semi-)randomly, and do not lend themselves to any *apparent* positional hierarchy imposed by the laws of nature. The constellations, therefore, are a distinctly human invention dictated by culture an imagination rather than being the result of any physical process. How and why we ended up with a sky full of mythical heroes and fantastical beasts says much more about the human condition than it tells us anything useful about how the heavens are constructed.

The tendency of the human brain to detect regular patterns has served our species very well in the realms of science and technology; the assembly of taxonomies is

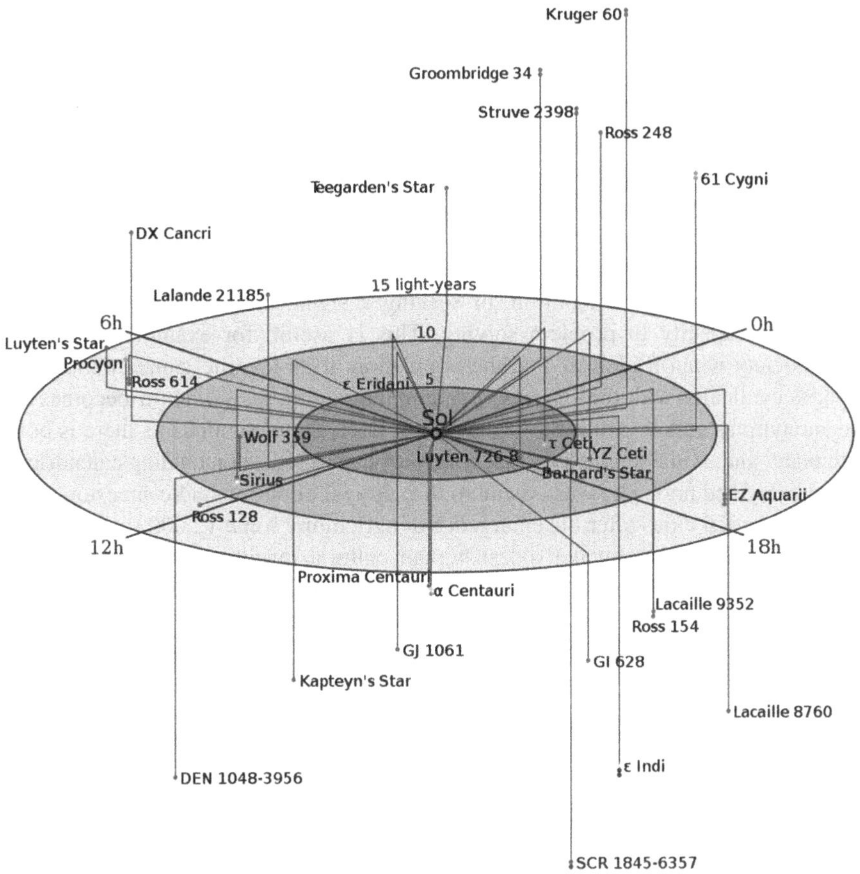

1.3 A schematic representation of all 32 known stars within 15 light years of the Sun ("Sol") in space. If a star is multiple, the components are shown stacked vertically with the actual position being that of the star closest to the center plane. The coordinate system is right ascension (RA) and vertical distance; hours of RA are marked, as well as distance in three concentric circles with radii of 5, 10 and 15 light years

often the first instinct of scientists when confronted with new and unfamiliar data. We look for similarities and bin them together in a form of intellectual stamp collecting (hopefully) before speculating on the underlying natural laws and principles that make for different categories of phenomena. This approach has been fantastically successful, yielding everything from the hierarchical, so-called "tree of life" arrangement of living organisms to the "tuning fork diagram" of galaxy classification and its early insights into the formation and evolution of galaxies. Under the assumption that, as in biology, "form follows function," a great deal can be learned about the nature of complex systems by searching for the order of repetitive structures or characteristics.

But not every regular pattern in nature indicates some sort of meaningful significance. Identifications of figures in the skies, composed by drawing imaginary lines between the bright stars, seems to be a natural result of the phenomenon of *pareidolia*, a predilection on the part of humans to sense significance in otherwise random stimuli. Pareidolia manifests itself most familiarly in the tendency to see shapes in clouds passing overhead on a summer day; it is involved in specific cases, among the most famous of which is the so-called "face" on Mars. What the constellations represented changed through history even as the figures remained the same and the civilizations that first named them disappeared.

Part of our overwhelming success as a species involves the superior capacity of the human brain for pattern recognition, of sensing a signal buried within the noise and applying that capacity to problem solving. This is useful, for example, to a hunter-gatherer society when it realizes that the population of its favorite game animals waxes and wanes cyclically with the seasons, or among agriculturalists when it becomes clear, as the anonymous author of the Biblical Book of Ecclesiastes reminds us there is both "a time to plant, and a time to pluck up that which is planted."[2] Understanding calendric cues in both the day and night sky was crucial to the survival of our nomadic ancestors and the transition to settled cities after the emergence of agriculture some 12,000 years ago.

Constellations are among the oldest human cultural inventions, certainly predating writing and, in all likelihood, civilization itself. The presumably oldest figures still in existence, such as the Hunter and the Bull, refer to a time in human history before the emergence of settled agricultural communities. It is probably no coincidence that Orion and Taurus reflect themes in the oldest extant works of art: the human form and game animals. Furthermore, it is likely that well-developed oral traditions about these figures long predate the emergence of written proto-language in the early Neolithic period, perhaps as long ago as the seventh millennium BC. Drawing such a conclusion about pre-literate humans is not a huge intellectual leap, considering what early modern humans had to do at night: long before the Internet, television, radio, or even books, before we gathered in insular communities behind the walls of shelters, we lived at night under the sky. Even though early lifeways may have been dictated by the rising and setting of the Sun, some fraction of life must have been lived at night. As the light of campfires died down, those still awake would have been confronted by night skies absolutely untouched by the modern scourge of light pollution. Those people had each other, their folklore and the stars; in such circumstances the constellations were born and later adapted to other uses. Perhaps 10,000 years ago, humanity began a journey in folklore associated with a religious tradition, developed it for practical purposes, and ultimately refined it in the empirical interest of science.

The oldest accounts of constellations and individual stars date to the earliest phases of Old Babylonian culture in the Middle Bronze Age, although the number and variety of Sumerian names found in extant catalogues suggests (but does not prove) the existence of an earlier, pre-literate tradition from which they were drawn. These earliest figures fall into either of two broad categories: gods and their symbols, and depictions of rustic activities

---

[2]Ecclesiastes 3:2 (King James Version).

associated with the practice of agriculture (Rogers, 1998). The motivation for identifying and naming them may have served a very practical purpose: already in the prehistoric period, humans discovered that the positions and movements of celestial bodies served a calendric function—rather important to societies dependent on correctly predicting the regular cycles of planting and harvesting. The creators of the constellations therefore may well have been the astronomer-priests of the Mesopotamian tradition who viewed the placement of mythological figures in the heavens as a fundamentally pragmatic act.

The earliest surviving written record referencing the Mesopotamian constellations is found in a text called "Prayer to the Gods of the Night" (Cooley, 2011), dating from about 1700 BC, which references the Arrow (the star Sirius), the Yoke Star (the star Arcturus), the "Stars" (the Pleiades star cluster), the True Shepherd of Anu (Orion), the Dragon (possibly the constellation Hydra), the Wagon (the "Dipper" stars of Ursa Major), the Goat Star (Vega) and the Bison (the composite figure depicted in the constellations Ophiuchus and Serpens).

A more thoroughly detailed account, consisting of lists and observations of nearly all the Mesopotamian constellations was carefully recorded in cuneiform script in the so-called "MUL.APIN" tablets (Watson and Horowitz, 2011) whose oldest dated version was written in the eighth century BC but is based on observations from before 1000 BC. The figures of the classical zodiac were established in the Old Babylonian period of the Near East, cast in their final form during the Neo-Babylonian era around the sixth century BC. The folklore that originated in the earliest societies of the Fertile Crescent was transmitted widely across the region, adapted to fit into local cultures; for instance, by around 1000 BC nearly every Near Eastern people adopted its own version of the Mesopotamian flood myth as though it were an original story.

Around the time of the rise of classical Greek culture, the author of the Biblical Book of Job referred[3] to several asterisms, including Ursa Major (the Great Bear), Orion (the Hunter) and the Pleiades. Expanding on earlier work by Hunger and Pingree (1999) on the MUL.APIN tablets, Schaefer (2006) recently concluded, somewhat controversially, that "most of the Mesopotamian constellations and observational data were made from near a latitude of 33–36° between 1300 and 1000 BC, by people we would call Assyrians." Adding additional constraints in the fourth century BC works of the Greek astronomer Eudoxus of Cnidus (408–355 BC), Schaefer further narrowed the time and place of the classical constellations' origin to 1130 BC and 36° north latitude.

This work suggests that the figures in the Western tradition that occupy the night skies of the northern hemisphere were essentially in place as we now know them around 3000 years ago and originated in or near the northern half of Mesopotamia. Textual evidence indicates that the Greeks came into possession of these constellations in the century or so preceding Eudoxus, and that their cultural influence spread the ideas into Europe by the Roman era. In the second century AD, Claudius Ptolemy (*c.* AD 90–*c.* 168),

[3]Job 9:9, 38:31–32.

a Greco-Egyptian writer in Alexandria, codified 48 constellations from the Middle Eastern canon in his *Almagest*, a work that set the Western standard of astronomy for nearly 1500 years.

We will likely never know for certain the ultimate origin of the oldest constellations in our night sky since they long predate written history. Humans showed an affinity for regular spatial structure in some of the earliest works of art dating to some half a million years ago; the recognition of constellations could well be as old, perhaps contemporaneous with the emergence of religious belief and the notion of mythical figures. However old the constellations, it is safe to conclude that they have long journeyed with us on our path to becoming fully human.

With belief comes dogma, and each world culture came to regard its vision of the night sky as authoritative. In the past century a version of the constellations regarded by professional astronomers as canonical emerged, guiding the nomenclature encountered in the scholarly literature. This version was informed distinctly by the cultural preferences of the West, and, whether fair or not, has become the recognized global standard. In the Western vision are preserved echoes of distant human history and extinct peoples, making the modern recension of ancient skies representative of the history of civilization itself. But the census of figures in the Western night sky through history is surprisingly dynamic, reflecting a variety of changing tastes and biases. The adoption of mythology as scientific convenience, and what was lost along the way, is what this book is about.

Even as Ptolemy's constellations held sway for some 1500 years, the figures believed to be represented by stars in the night sky were not entirely static. As Europe emerged from its intellectual dark ages and embarked upon audacious new voyages of exploration, it opened a new "discovery space" in the southern hemisphere. Upon encountering previously unknown stars, sailors drew their own patterns on the night sky as much for of the practical needs of navigation as the expression of imagination. They brought their culture and history to newly-conquered lands, spreading European sky lore across the globe. Their maps of Earth and sky present the world as a knowable place, with each continent and each constellation in its respective place. Finally, science emerged with its emphasis on structure and function. While human knowledge of the universe expanded, the familiar night sky remained the user interface to the cosmos. Inevitably it called for some standardization among the people trying to understand it.

With a sense of what a constellation is in hand, the question remains: who decides? From a literal world of interpretations of the patterns of the night sky emerged a set of 88 figures to be found on every star map today used by amateur and professional astronomers alike. That some constellations are considered "official" by astronomers implies that others are—or were—not, and that a history of night sky folklore existed that is now lost. Its story is fully told here for the first time.

2

The Contemporary Sky Emerges

The problem of what a constellation is—and on whose authority—proved to be a vexing question in historical astronomy for reasons that transcended mere aesthetics. The pursuit of certain topics in astronomy research, and in particular those whose studies were bolstered by the emerging technology of the photographic process, required a consistent nomenclature framed in part by being able to distinguish the membership of stars in one constellation or another. Before those memberships can be established the boundaries separating them must be demarcated, and that implies astronomers agree on a fixed canon of constellations to which none would be added or subtracted. But the problem at the end of the nineteenth century was that no such general agreement among astronomers existed. Its solution has much to do with the phenomenon of "lost" constellations in the first place.

FROM ANTIQUITY TO THE NINETEENTH CENTURY

The Ptolemaic Constellations

While they disagreed on other points, by 1800 essentially all European celestial cartographers agreed on one thing: there existed a set of 48 constellations charted by Claudius Ptolemaeus, or Ptolemy, a Hellenized Egyptian astronomer living at Alexandria in the second century AD. Ptolemy adhered to a model in place by at least the Classical Period in Greek history (*c.* 500–300 BC). Whether Ptolemy intended so or not, the list of constellations named in his *Almagest* (Table 2.1) represents the first recorded attempt to define something like a canon of constellations since the Bablyonian "Three Stars Each" texts dating approximately to the twelfth century BC and the MUL.APIN catalog compiled around 1000 BC. After Ptolemy, no new constellations were proposed for nearly 1400 years that are today considered official, and all of Ptolemy's constellations may be found on any modern star chart.

Ptolemy's list is divided into 21 constellations north of the ecliptic, 15 constellations south of the ecliptic, and 12 constellations along the ecliptic itself (the Zodiac). By the

J.C. Barentine, *The Lost Constellations*, Springer Praxis Books,
DOI 10.1007/978-3-319-22795-5_2

Table 2.1 The 48 Ptolemaic constellations.

North of the ecliptic	*South of the ecliptic*	*Zodiac*
Ursa Major	Cetus	Aries
Ursa Minor	Orion	Taurus
Draco	Eridanus	Gemini
Cepheus	Lepus	Cancer
Boötes	Canis Major	Leo/Coma Berenices
Corona Borealis	Canis Minor	Virgo
Hercules	Argo Navis	Libra
Lyra	Hydra	Scorpius
Cygnus	Crater	Sagittarius
Cassiopeia	Corvus	Capricornus
Perseus	Centaurus	Aquarius
Auriga	Lupus	Pisces
Ophiuchus	Ara	
Serpens	Corona Australis	
Sagitta	Piscis Austrinus	
Aquila/Antinoüs		
Delphinus		
Equuleus		
Pegasus		
Andromeda		
Triangulum		

turn of the nineteenth century, historians, philologists and archaeologists had developed a general sense that the figures representing many of the Ptolemaic constellations referred to much older stories derived from the mythology of the ancient Middle East, and so might the constellation patterns themselves have originated some three or four millennia in the past. But Ptolemy missed about one quarter of the sky, a region that never rose above the horizon of his observing position in Alexandria. The text of the *Almagest* indicates that Ptolemy understood the equinoxes moved relative to the "fixed stars," a phenomenon explained by the precession of the Earth's rotation axis; it is therefore somewhat surprising that he did not conclude that the effect would bring some constellations into view from Alexandria over time while carrying others permanently below his local horizon.

Prior to their voyages of discovery in the fifteenth and sixteenth centuries, the southernmost sky was unknown to Europeans, leaving a "hole" in early southern hemisphere maps. This effect is seen clearly in Albrecht Dürer's 1515 map of the Ptolemaic constellations in the southern celestial hemisphere (Fig. 2.1). A large empty space, roughly circular in shape, is seen to the left of the center of the figure where converging lines mark the position of the south celestial pole in Dürer's time. Given that Ptolemy's constellations more or less completely encircle this blank space, it is easy to see that the emptiness represents the declinations permanently below his midlatitude horizon during antiquity. Precession later carried the position of the pole away from the center of the space, leading to the off-center

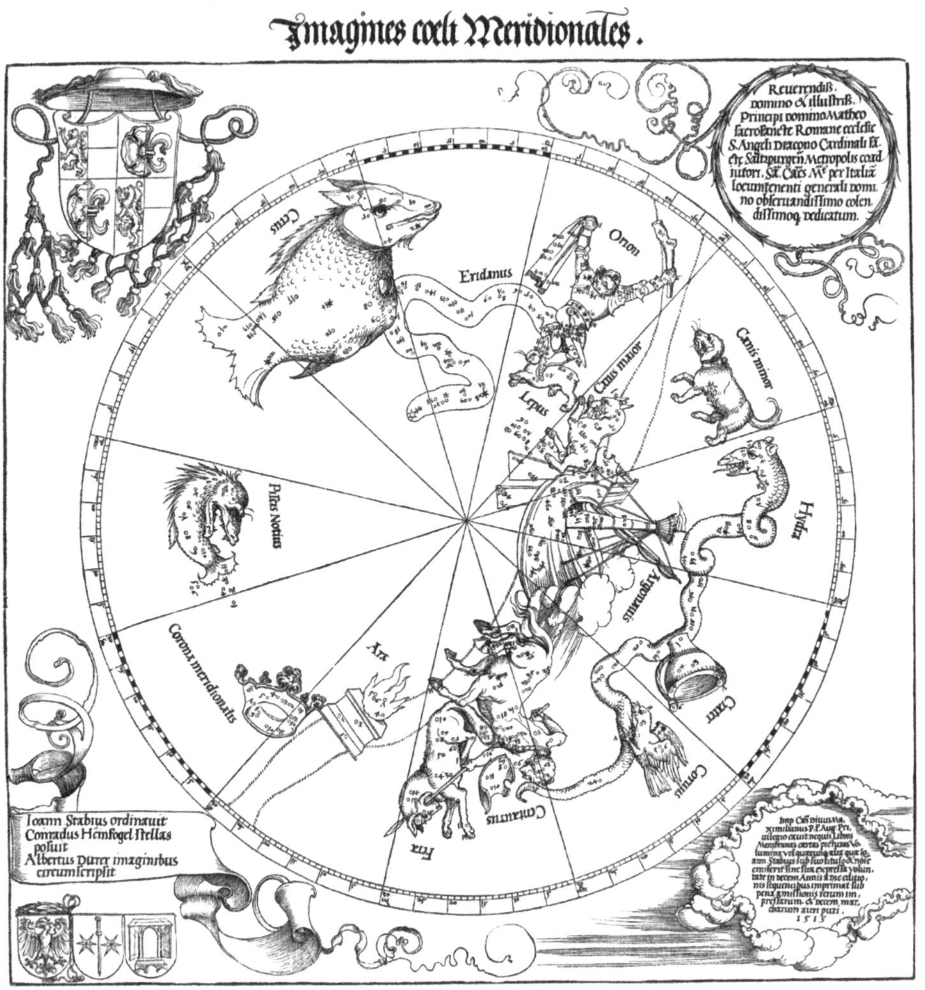

2.1 The constellations of the southern sky shown on Albrecht Dürer's *Imagines coeli Meridionales* (1515), the oldest extant, printed European sky map. The roughly circular empty space to the left of the south celestial pole (*center*) indicates the region of the sky permanently below the horizon at the latitude of Alexandria in Ptolemy's time

appearance of the figure. The southernmost constellations described by Ptolemy in the *Almagest* include, in rough order of increasing distance from the south celestial pole:

- **Crux** (the Cross), described by Hipparchus and Ptolemy as part of Centaurus
- **Ara** (the Altar), described by Aratus in the third century BC as a lighthouse (φάρος)
- **Corona Australis** (the Southern Crown), described by Aratus as one of two Στεφάνοι (*stefanoi*), or crowns, one each in the northern and southern sky

- **Centaurus** (the Centaur)
- **Lupus** (the Wolf)
- **Argo Navis** (the ship *Argo*), seen in its entirety above the horizon in Alexandria at the time of Ptolemy due to precession (Chap. 5)
- **Eridanus** (the River)

In this and the following lists in this chapter, constellations that remain part of the modern canon are indicated in **bold**, while those since discarded are shown in normal face.

Opening the Southern Sky

Cartographers began to fill in Ptolemy's hole during the sixteenth century on the basis of explorers' travels in the southern hemisphere. Pieter Dirkszoon Keyser (*c.* 1540–1596) and Frederick de Houtman (1571–1627), Dutch navigators in the employ of the Flemish cartographer Petrus Plancius (1552–1622), contributed 11 constellations mapped during their voyage to the East Indies via Madagascar in 1595–1597:

- **Apus** (the Bird of Paradise)
- **Chamaeleon** (the Chameleon)
- **Dorado** (the Swordfish)
- **Grus** (the Crane)
- **Hydrus** (the Water Snake)
- **Indus** (the Indian)
- **Musca** (the Fly)
- **Pavo** (the Peacock)
- **Phoenix** (the Phoenix Bird)
- **Triangulum Australe** (the Southern Triangle)
- **Tucana** (the Toucan)
- **Volans** (the Flying Fish)

All of the constellations introduced by Keyser and de Houtman are found on modern maps. Plancius himself is recognized as the first person to show Crux (the Cross) on a map as a constellation distinct from Centaurus in 1589. He also drew what purports to be Triangulum Australe on the same map, but Plancius' figure does not match the constellation by that name recognized on modern lists; see Chap. 21 for further discussion.

A complete set of southern figures did not emerge until the mid-eighteenth century. Nicolas Louis de Lacaille (1713–1762), a French abbé who measured the positions of nearly 10,000 southern stars between August 1751 and July 1752 from a simple observatory near Cape Town, South Africa. His stellar position measurements were presented to the Académie Royale des Sciences in *Table des ascensions droites et des déclinaisons apparentes des Etoiles australes…* (1752), although the new constellations he formed were only published the year after his death in *Coelum Australe Stelliferum*. These were:

- **Antlia** (the Air Pump)
- **Caelum** (the Chisel)
- **Circinus** (the Circles)
- **Fornax** (the Furnace)
- **Horologium** (the Water Clock)
- **Mensa** (Table Mountain)
- **Microscopium** (the Microscope)
- **Norma** (the Square)
- **Octans** (the Octant)
- **Pictor** (the Painter's Easel)
- **Reticulum** (the Reticle)
- **Sculptor** (the Sculptor)
- **Telecopium** (the Telescope)

In addition, Lacaille broke up the constituents of Ptolemy's Argo Navis (Chap. 5) into four distinct constellations: Carina (the Keel), Puppis (the Stern), Vela (the Sails), and Pyxis (the Compass). All 17 of Lacaille's inventions remain canonical. While Keyser and de Houtman placed in the heavens mostly images of animals and people of faraway lands, Lacaille chose figures to honor the arts and sciences of his eighteenth-century world.

Completing Maps of the Northern Sky

The classical constellations described in Ptolemy's *Almagest* tended toward the brighter stars in the sky, and the reasons why are not difficult to understand: bright stars tend to suggest importance and define more recognizable patterns. But his figures were not all-inclusive of the stars in the sky visible to the Greeks, leaving behind generally fainter stars that were considered either (1) non-definitive members of adjacent constellations, or (2) unaffiliated with any particular constellation. They became known as αμορφοτοι ("unformed" stars) that were left as cartographic orphans in the intervening centuries. As Europe emerged from the Medieval period with a newfound interest in all things classical and scientific, work was undertaken to more thoroughly characterize the northern sky by fashioning new figures from the "unformed" stars. Celestial mapmakers eager to establish (or capitalize upon) their reputations introduced new constellations to fill out the details of some of the most aesthetically beautiful maps of the night sky ever produced.

The German cartographer Caspar Vopel (1511–1561) led the way with the first depiction of Coma Berenices (Berenice's Hair) on a 1536 globe. While the group of faint stars in the region roughly bounded by Leo, Boötes, Virgo and Canes Venatici was recognized as an asterism by the Hellenistic Greeks, Ptolemy skipped over it and instead considered Coma's stars to represent a tuft of fur at the end of the Lion's tail. Vopel merely upgraded the ancient figure to constellation status, and the change stuck. Petrus Plancius took liberties with some of the more prominent unformed stars, creating from them the modern constellations Camelopardalis (the Giraffe), Columba (the Dove) and Monoceros (the Unicorn).

Gottfried Kirch (1639–1710), Astronomer Royal to Frederick III, Elector of Brandenburg (1657–1713; later Frederick I, Duke/King of Prussia) and first director of the Berlin Observatory, introduced three new constellations. Each figure was proposed in honor of a royal or imperial figure whose patronage he sought: Gladii Electorales Saxonici (the Crossed Swords of the Saxony Electorate), Pomum Imperiale (the Imperial Orb), and Sceptrum Brandenburgicum (the Sceptre of Brandenburg). However, none of these constellations survived the test of history and were abandoned by the end of the nineteenth century.

Johannes Hevelius (1611–1687), a Polish astronomer working at Danzig, a free city then in the Polish-Lithuanian Commonwealth, gave us seven constellations that remain part of the modern canon in his *Firmamentum Sobiescianum* (1687):

- **Canes Venatici** (the Hunting Dogs)
- **Lacerta** (the Lizard)
- **Leo Minor** (the Lesser Lion)
- **Lynx** (the Lynx)
- **Scutum Sobiescanum** (the shield of King John Sobeski; now referred to simply as **Scutum**)
- **Sextans** (the Sextant)
- **Vulpecula cum Anser** (the Fox with Goose; now referred to as **Vulpecula**)

and four that were discarded by the turn of the twentieth century:

- Cerberus (the Three-Headed Guardian of the Underworld, later merged with Ramus Pomifer; Chap. 7),
- Mons Maenalus (Mount Mainalo; Chap. 15),
- Musca Borealis (the Northern Fly; Chap. 16)
- Triangulum Minus (The Lesser Triangle; Chap. 28).

The last great, influential work of European celestial cartography before the advent of the photographic process that would revolutionize the work of mapping the heavens was *Uranographia* (1801b) by Johann Elert Bode (1747–1826), Kirch's successor at the Berlin Observatory in 1786. His beautiful charts, informed by the best astrometry of the day, filled in the few remaining gaps in the northern sky. Bode either invented or directly introduced though his published works seven new constellations, none of which remains canonical:

- Custos Messium (the Harvest Keeper; Chap. 8)
- Felis (the Cat; Chap. 9)
- Globus Aerostaticus (the Hot Air Balloon; Chap. 11)
- Honores Frederici (Frederick's Glory; Chap. 12)
- Lochium Funis (the Log and Line; see Volume 2)
- Machina Electrica (the Electrical Generator; Chap. 14)
- Officina Typographica (the Printshop; Chap. 17)

In addition, Bode included 14 constellations and asterisms invented by others that have also since disappeared from star charts:

- Anser (the Goose; Chap. 3)
- Antinoüs (the Favorite of Hadrian; Chap. 4)
- Caput Medusae (the Head of Medusa; see Volume 2)
- Musca Borealis (the Northern Fly; Chap. 16)
- Psalterium Georgianum (George's Harp; Chap. 18)
- Quadrans Muralis (the Mural Quadrant; Chap. 19)
- Ramus Pomifer (the Apple-Bearing Branch, merged with Cerberus; Chap. 7)
- Rangifer (the Reindeer; Chap. 20)
- Robur Carolinum (the Royal Oak; Chap. 22)
- Sceptrum Brandenburgicum (the Brandenburg Scepter; Chap. 24)
- Taurus Poniatovii (Poniatowski's Bull; Chap. 25)
- Triangulum Minus (the Lesser Triangle; Chap. 28)
- Turdus Solitarius (the Solitary Thrush; Chap. 29)

A handful of other northern constellations were proposed by various authors, which were also discarded before the modern constellations were defined in 1922:

- Cancer Minor (the Lesser Crab; Chap. 6)
- Gallus (the Rooster; Chap. 10)
- Jordanis (the River Jordan; Chap. 13)
- Malus (the Mast): An asterism defined as part of Argo Navis (Chap. 5) by Lacaille; Bode did not show it on his charts but formed Lochium Funis nearby.
- Phaeton (the Son of Apollo-Helios): an asterism within the constellation Eridanus that achieved independent mention on sixteenth century charts (see Volume 2)
- Polophylax (the Guardian of the Pole): a sixteenth century southern figure with an uncertain location (see Volume 2)
- Rhombus (the Bullroarer): also known as Quadratum, a southern figure that may have been primarily a navigational convenience (Chap. 21)
- Telescopium Herschelii Major (Herschel's 20-Foot Telescope) and Telescopium Herschelii Minor (Herschel's 7-Foot Telescope) (Chap. 26)
- Tigris (the River Tigris; Chap. 27)

Constellation Counts in Disagreement

The number of constellations listed in the previous section totals 107. Looking carefully at totals in various nineteenth century textbooks gives different results. For instance, James Ryan counted as many as 104 constellations in *The new American grammar of the elements of astronomy* (1827). He found a minimum of 94 constellations from his main lists considered in common circulation and a maximum 104 by adding ten figures from "modern astronomers" and others made "out of the unformed stars, and those stars in the southern hemisphere that were invisible to the ancient astronomers on account of never appearing above their horizon." George F. Chambers wrote a history of the constellations and presented a suggested canon of 109 constellations in his 1877 *A Handbook of Descriptive Astronomy*. He had all of Ryan's constellations plus the two Herschel telescopes and the Large and Small Magellanic Clouds, Chambers also counted the star Cor Caroli (α Canem Venaticorum) as a constellation unto itself. However, Ezra Otis Kendall counted only 89 constellations in *Uranography: or, a description of the heavens* (1845) in part because he doubled up: "Perseus et Caput Medusae" were one constellation, not two, as were "Vulpecula et Anser" and the "Triangula."

In each of these cases recall that only 48 were agreed upon by all writers, having descended from classical antiquity; all others suggested since the beginning of the early modern period were a matter of individual taste. Certain cartographers—Hevelius and Bode, in particular—were highly-influential tastemakers whose figures were copied by scores of other mapmakers of succeeding generations. But until well into the twentieth century, there was no list of "official" constellations recognized by the world's astronomers, and a lack of consensus presented a genuine obstacle to scientific research.

THE "MODERN 88"

The need for professional astronomers to settle on a final set of constellations recognized uniformly was born of an urgent need in early twentieth century observational astronomy. The field of variable star research was fashionable at the time, but a quirk of nomenclature created a problem in keeping proper track of any particular star. Newly-discovered variable stars are traditionally given designations including the name of the constellation in which they are found.[1] To assign names to objects according to the constellations to which they "belong" implies there are clear demarcations establishing the thresholds at which one constellation transitions to its neighbor. By about 1900 most professional astronomers agreed that such boundaries existed between constellations, but no clear consensus existed as to the exact location of these lines.

[1]This type of nomenclature extended to other phenomena. As Eugène Delporte wrote in (1930): "The precise demarcation of the constellations is of utmost importance for a great deal of astronomical work such as the systematic observation of meteors, the study and denomination of variables, the observation of novae, etc."

Although previous mapmakers tackled the problem, the solutions were largely unsatisfactory. Most notable among these efforts is the work of Benjamin Apthorp Gould (1824–1896), an American astronomer who founded both the *Astronomical Journal* and the Argentine National Observatory. In 1879 Gould published *Uranometria Argentina: Volume 1*, an atlas of 7756 stars within 100° of the south celestial pole whose positions he measured during his tenure at Córdoba, Argentina. Gould intended his catalog to be complete to a visual magnitude of +7, roughly the threshold of naked-eye visibility, although comparison of his magnitudes to modern photoelectric measurements shows that his estimates were roughly half a magnitude too faint. Like the eighteenth-century English astronomer John Flamsteed, Gould numbered the stars in his catalog separately by constellation, generally by increasing right ascension. His 66 constellations consisted of the aggregate of the more southerly of Ptolemy's figures (with Argo Navis broken up according to Lacaille) and the inventions of Keyser, de Houtman, and Lacaille.

Gould's place in the history of constellations has to do with his innovative method of defining their boundaries by lines of constant right ascension and declination. His obliques, or arcs of great circles, were defined by the coordinates of the extremities of the constellations as traditionally described, although the result was somewhat arbitrary as to which parallels Gould chose in each case to best circumscribe the constellation figures. The shortcoming of Gould's maps is that they did not cover the northern hemisphere. After the problem was solved for the entire sky and the results published, the prior situation was neatly summarized[2]:

> For the southern sky the constellations had already been delimited scientifically by B.A. Gould in 1877, but for the northern sky it has hitherto been impossible to make a precise statement of their boundaries, since their delimitation depended merely on the adoption of curves laid down on certain accepted charts; these curves were in many cases not susceptible of mathematical definition, and the boundaries were not authoritatively defined in any other way. Indeed, point is given to the unsatisfactory state of affairs by the fact that a certain number of individual stars have been allocated to different constellations by different authorities, and so appear under different names in different catalogues.

The authority of historic charts still stood in the 1920s, but astronomers realized their crucial fault: any attempt to fix definite boundary lines on a chart would inevitably pin the constellations onto a shifting night sky. The reason for this has to do with human attempts to superimpose a set of coordinates onto the sky. In order to map the sky, as to map the Earth, its limits must be defined. Given an essentially spherical Earth rotating about an imaginary axis, the points at which the axis pierces the Earth's surface—the poles—represent two extrema. In the frame of reference in which the center of the Earth is at rest, the poles are the only points on the Earth that do not move while the rest of the planet rotates; as a result, they make for a convenient set of origin points of a series of parallel lines known as longitude. A second sort of extremum is the line at the Earth's surface

[2] *Journal of the Royal Astronomical Society of Canada*, Vol. 25, p. 417 (1931).

everywhere perpendicular to the rotation axis at the radius of the sphere—the equator. In a spherical polar coordinate system with unit radius, the equator represents the direction perpendicular to the rotation axis at which the circumference of the sphere is a maximum. From this line, a set of parallel lines, latitude, are drawn at decreasing distance from the rotation axis, reaching a radius of zero at the poles.

The sky appears to us as a flat surface at constant, infinite distance from the surface of the Earth, reminiscent of the "sphere of fixed stars" in the Ptolemaic model of the universe. As the inner surface of such an imaginary sphere, the geometry of the sky lends itself to a similar coordinate system. In the frame of a stationary Earth, the sky seems to revolve around our planet with two singularities at which the rotation rate appears to be zero. These "celestial poles" are the projection of the Earth's rotation axis onto the sphere of the sky and lead to a set of connecting lines akin to longitude known as right ascension. Similarly, the Earth's equator projected onto the sphere of the sky is the "celestial equator" and lines corresponding to latitude are known as declination. The specification of the right ascension and declination of any object in the night sky precisely locates that object on maps of the night sky in the same sense that longitude and latitude locates a place on the surface of the Earth.

If that were the end of the story, then any set of arbitrary constellation boundaries would suffice indefinitely. To remain constant in time with respect to this geocentric coordinate system, the orientation of the poles must remain fixed with respect to the stars whose distance is taken to be infinite. But the Earth's rotation axis is not fixed; rather, like a slowly-spinning child's toy top, the direction of the axis executes a slow, circular precession with a period of about 25,700 years (Fig. 2.2). Again in the fixed coordinate frame of a stationary Earth, the poles appear to execute circles around the imaginary fixed point in the center of the figure of precession over this period of time. Lines of right ascension remain tied to the poles, and declination to the equator, but the position on the celestial sphere of any particular "fixed" star appears constantly in motion.

This creates an obvious problem if one astronomer wants to communicate the position of an interesting object to another: right ascensions and declinations are periodic functions of time. Astronomers therefore attach a date known as an "equinox" to a set of coordinates to indicate the time for which the quoted coordinates are valid. Equinoxes tend to be century and half-century years such as the most frequently used equinox at present, "2000.0" (12 h Universal Time on 1 January 2000). To compute the corresponding position on another date, allowing for precession, an astronomer specifies an "epoch." The equinox and epoch of a set of coordinates are inputs to transformation equations that yield an object's position corrected for precession. The standard equinox is traditionally updated every fifty years because inaccuracies in calculating the precessed position of objects after several decades can result in significant positional errors.

Any scheme to standardize the positions of constellation boundaries necessarily must account for this effect, for after some time the precessed positions of the stars will cause them to wander across any arbitrarily-drawn set of lines. Gould's solution to this problem was to draw the boundaries of southern constellations strictly along lines of constant right ascension and declination at the equinox 1875.0. The advantage of this approach is that the position of a star in any constellation is permanently fixed with respect to the

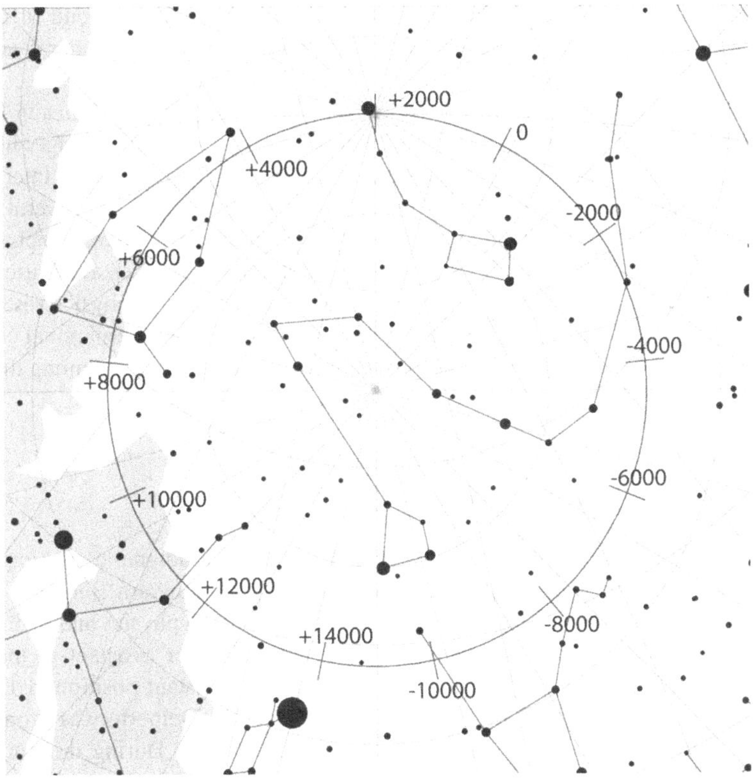

2.2 The apparent path of the north celestial pole through the sky during one complete period of precession. Numbers along the circle show dates in BC (negative numbers) and AD (positive numbers) including an artificial year "0." Two coordinate origins are shown: one for the position of the pole in AD 2000 near the star Polaris (*top*) and another at the "pole" of precession itself in the constellation Draco (*center*). Figure by Wikimedia Commons user Tau'olunga, licensed under CC BY-SA 2.5

boundaries of the constellation. However, precession causes both the boundaries and the stars contained within them to shift with respect to the poles and the equator. By the time of IAU Commission 3—nearly fifty years after Gould's chosen equinox—slight changes in the positions of stars and constellation boundaries were noticeable on ordinary star maps and atlases. Gould's approach would still result in a need for new maps in the future, but no longer would precession carry any given star from one constellation's space into another's and confuse its official designation.

All that remained to solve this problem was the establishment of an international governing body to decide on a permanent list of constellations and to define their boundaries; if all astronomers agreed to observe these definitions and boundaries, disputes over the association of particular stars and constellations would effectively vanish. That world body emerged from the ashes of western Europe following the end of the First World War as professional astronomers sought to establish a new global order in academic research.

In July 1919, European astronomers gathered at the Palais des Académies in Brussels for the Constitutive Assembly of the International Research Council during which they chartered a governing body for world professional astronomy known as the International Astronomical Union (IAU). The organization was originally composed of several international efforts in astronomy that came to be run under common governance, including the Bureau International de l'Heure (International Time Bureau), the extensive Carte du Ciel photographic sky-mapping project, and the "Solar Union" of solar physics researchers. The astronomers who met in Brussels set 1922 as the year for an international congress, to be held in Rome, Italy. The constellation boundary controversy was among the issues taken up at the first IAU General Assembly.

EUGÉNE DELPORTE, ARBITER OF THE HEAVENS

Eugéne Joseph Delporte was born on 10 January 1882 in Jodoigne, a municipality in the majority French-speaking Walloon Brabant province of southern Belgium. He was educated at the University of Brussels, receiving his doctorate in physics and mathematics (with honors) in 1903. The same year, he became a volunteer assistant at the Royal Observatory of Belgium in Uccle, obtaining an appointed assistant position in 1904. He married Yvonne Lambert in 1906; the couple had two sons. Delporte was promoted to assistant astronomer in 1909, holding that position until 1924. During this time came the first major success of his career with the publication, with Paul Stroobant,[3] of *Les Observatoires astronomiques et les Astronomes* (1907), a volume containing an exhaustive list of world observatories, astronomical societies and individual astronomers.

Also during these years he was employed in the use of the Askania meridian circle, assisting with improvements to the instrument. Part of his work involved geodetic measurements; in 1912 he accurately determined the angular distance along the surface of the Earth between the observatories at Paris and Uccle in order to obtain a precise longitude for Uccle by bootstrapping from the very precisely known longitude difference between Paris and Greenwich. This was no small feat in the era before the Global Positioning System, and determining the longitude of the observatory was crucial in order to properly interpret transit timing measurements. In 1914, the same year in which war broke out in Continental Europe, Delporte and Uccle astronomer Hector Philippot (1872–1925) published a catalog of 3553 stars between +21° and +22° in declination.

[3]Stroobant (1868–1936) was one of the foremost Belgian astronomers of his generation. He was a Professor of Astronomy at University of Brussels from 1896 to his death, directed the Royal Observatory in Brussels from 1925–1936, and presided over the Royal Belgian Academy's "Classe des Sciences" in 1931–1932. (*The Observatory*, Vol. 59, pp. 349–352, 1936.)

Yvonne died in 1918; 4 years later, Delporte married his second wife, Marie Otten. The couple had no children together. Delporte remained formally on the staff at Uccle until his retirement in 1947, but he continued working there as something of an emeritus, volunteering his mornings to examine plate photographs of the night sky taken as part of an Observatory program to monitor asteroids. He was at his desk studying plates on the morning of 19 October 1955 when he suddenly collapsed, went into cardiac arrest, and died. He was mourned by the worldwide astronomical community; in the half-century of his service to the Royal Observatory, he forged contacts and professional relationships with researchers at many institutions both foreign and domestic. He was eulogized by his friend and colleague August Kopff[4] in the May 1956 issue of *Astronomische Nachrichten*:

> His selfless devotion to science, his modest and reserved nature, the honest friendship with his colleagues, including those abroad, brought rich sympathy aroused everywhere. The writer of these lines was close to him for 40 years and considers to how many other colleagues his passing represents a painful personal loss.

Kopff estimated the importance of Delporte's work for the IAU in defining the constellation boundaries at the same level as his efforts in the publication of *Les Observatoires astronomiques et les Astronomes*. A year after his passing in October 1956, the French astronomer Gabrielle Renaudot Flammarion[5] wrote in *L'Astronomie*,

> To cite here all of his works – many of which relate specifically to Belgian astronomy – is impossible, as the work of E. Delporte is vast and varied ... His sudden death is mourned in the astronomical world, for in the variety of his scientific efforts he created relationships with observatories around the world.

IAU COMMISSION 3

At the 1922 Rome meeting, IAU members voted to constitute a number of "Commissions," effectively a series of working groups comprised of subject experts tasked with establishing a series of standards on which all working astronomers would agree. Commission 3 was on the topic of "Notations," formed in part to deal with the problem of constellation boundaries. The Commission's task was to decide on a final set of constellation boundaries given Gould's work in the southern hemisphere. The sensible approach seemed to be to adopt the boundaries from *Uranometria Argentina* and extend this model to the northern hemisphere.

---

[4]Kopff (1882–1960) was the Director of the Institute for Astronomical Calculation at Humboldt University in Berlin. He discovered a number of asteroids and comets, and is commemorated by the names of both the lunar crater Kopff and the asteroid (1631) Kopff.

[5]Flammarion (1877–1962) was the wife of Camille Flammarion (1842–1925) and served as the General Secretary of the Société Astronomique de France. She worked at a private observatory at Juvisy-sur-Orge, France, from which she published observations of planets, asteroids and variable stars. The crater Renaudot on Mars is named in her honor, as is the asteroid (355) Gabriella.

The First General Assembly laid the groundwork required for the standardization of constellation boundaries by adopting a list of what the professional astronomical community of the time considered widely-recognized constellations. This list comprised 88 figures summarized in the table in the Appendix. At the same time, each constellation was assigned a three-letter abbreviation to make uniform the short references to constellations in the names of variable stars. Some found the three-letter scheme sufficiently inadequate to tell the constellations apart that an extension to four letters was proposed. In the proceedings of the Fourth General Assembly (1932) then-President of Commission 3, Frank Schlesinger,[6] wrote:

> At the meeting held at Rome in 1922, the Union adopted a standard set of three-letter abbreviations for the constellations. It is proposed to introduce at the forthcoming meeting a resolution for the adoption of standard four-letter abbreviations for which there seems to be a considerable demand. The advantage of these is that they would suggest at once the name of the constellation, which is not always true of the three-letter abbreviations.

In the end this proposal proved problematic and abbreviations reverted to three letters, where they remain to this day.

Commission 3 was also charged with standardizing the nomenclature of the constellations; for example, prior to the existence of the IAU, astronomers tended to use the names "Scorpius" and "Scorpio" more or less interchangeably to refer to the constellation of "the Scorpion". The IAU decided that "Scorpius" was the more correct name, relegating "Scorpio" to a moniker now used exclusively by astrologers. Standard pronunciations for the constellation names were not suggested at the time; the American Astronomical Society (AAS), the professional society of U.S. astronomers, later took action on this.[7] The IAU evidently decided against standardizing pronunciations when it defined the modern constellations because the linguistic issue was not considered pressing as was the issue of variable star names.

Delporte's Pilot Study (1923–1925)

With a list of constellations and their abbreviations in hand from the First General Assembly, the Belgian National Astronomical Committee took up the issue of constellation boundaries in 1923, commissioning Eugène Delporte to study the problem and propose

[6]Schlesinger (1871–1943) was an American astronomer among the first to rely primarily on the use of photographic plates rather than traditional visual observations to collect data for research. He is best known for compiling and publishing the Yale Bright Star Catalogue.

[7]The AAS report "Pronouncing Astronomical Names" included suggested pronunciations of both constellations and individual stars. It was first published in the June 1943 issue of *Sky & Telescope* magazine. The pronunciations suggested by AAS are problematic for a variety of reasons summarized by *S&T*'s Tony Flanders in the article "Constellation Names and Abbreviations" (http://www.skyandtelescope.com/letsgo/helpdesk/Constellation_Names.html).

a solution. Delporte conducted the investigatory work into extending Gould's system of boundaries with L. Casteels (University of Ghent, Belgium) at Uccle. Delporte later recalled[8] their efforts in *Délimitation scientifique des constellations* (1930):

> [T]he starting points were as follows: to carry out a scientific demarcation of the constellations of the northern hemisphere, the boundaries being mathematically defined with respect to a determined equinox, and deviating as little as possible from the *non-defined* lines found in modern atlases, so as to avoid as much as possible moving stars from constellation to constellation, and with the *express condition* of preserving the existing names of catalogued variable stars.

The Belgian Committee was pleased with Delporte's suggestion and requested that the IAU put the issue of boundaries on the agenda for the Second General Assembly to be held in 1925 at Cambridge, England. Recounting events for the American Nature Association in 1948, Isabella Lewis wrote:

> At that meeting the recommendation of the committee that had been appointed to consider the matter was to adopt new boundaries for all of the constellations north of 12 degrees south of the equator. The new boundaries were to be hour circles and circles of declination, as were those adopted by Gould for his southern constellations.

The Belgian Committee tasked Delporte with first undertaking a proof-of-concept effort after the Rome IAU meeting that was to be completed and presented at Cambridge. The results of the pilot study were considered by a specially constituted subcommittee of Commission 3, consisting of Guillaume Bigourdan,[9] Casteels, Delporte, John Charles Duncan,[10] Edward Ball Knobel,[11] Mary Proctor,[12] and Reynold Kenneth Young.[13]

---

[8]"Les bases de d'epart, suivant les propositions de M. Delporte, furent les suivantes: Réaliser une deélimitation scientifique des constellations de l'hémisphére boréal, les limites étant mathématiquement définies par rapport à un équinoxe déterminé, ces limites s'écartant au minimum des tracés *non définis* figurant sur les atlas moderne, de facçon à éviter le plus possible les changements d'étoiles de constellations à constellations, et avec la *condition expresse* de conserver aux étoiles variables cataloguées le nom que ces étoiles avaient reçu." (p. 4).

[9]Bigourdan (1851–1932) was a member of the French Institute and is best known for measuring precise positions of 6380 nebulae and his participation in an effort to redetermine (with greater precision) the longitude difference between London and Paris at the turn of the twentieth century.

[10]Duncan (1882–1967) was director of the Whiting Observatory in Wellesley, Massachusetts. His chief contribution to astronomy was his photographic demonstration of expansion in the Crab Nebula in the 1930s.

[11]Knobel (1841–1930), FRAS, was a London businessman and amateur astronomer for whom Knobel Crater is named on Mars.

[12]Proctor (1862–1957), FRAS, was an Irish-American popularizer of astronomy for whom the lunar crater Proctor was named.

[13]Young (1886–1977) was Professor of Astronomy at the University of Toronto, and is remembered for his studies of stellar radial velocities and guidance in the design and construction of the 1.88-m telescope at the David Dunlap Observatory, Ontario, Canada.

The Commission 3 subcommittee looked favorably upon Delporte's initial efforts and decided as a matter of policy that establishment of a formal set of constellation boundaries would be of useful service to astronomy. A full study was endorsed, to be presented at the Second General Assembly. In the interest of consistency, the IAU essentially gave Delporte the exclusive right to conduct this work. It was thought that the best course of action was to consider a single, unified proposal, and either vote it up or down rather than pitting competing views against each other for the entire Union to sort out; in the aftermath of the First World War, nationalist factions within the Union could have attempted to derail the effort over perceived sleights meant to cause division. That the IAU granted the work exclusively to a Frenchman was already cause for some concern by the German-speaking IAU delegates; in his defense, Delporte sought to avoid promoting any sort of agenda that could have been considered nationalist in tone. Delporte wrote interim four reports to the subcommittee from October 1925 to September 1927 in which he communicated his progress and solicited committee members' advice on resolving uncertainties and internal disagreements to further reduce the effect of any possible bias on his part.

Delporte drew new boundaries for the 88 constellations recognized by the IAU in 1922 along hour circles defined for the equinox of 1875.0 to establish consistency with Gould's southern hemisphere charts in *Uranometria Argentina*. To meet the goal of retaining the classical associations of certain variable stars with particular constellations, he consulted the variable star lists of Prager (1928) and Guthnick and Prager (1928) such that none of the stars shifted out of their "traditional" constellations in the process; this approach explains the unusual shapes of some of the constellation boundaries, particularly in the southern sky. Delporte also decided against the possibility that two or more constellations should share any stars with overlapping designations, deciding "that Andromeda should retain her head (and hence the horse [Pegasus] would lose its navel), and that Taurus should retain the tip of its horn while Auriga sacrificed his foot." (Ridpath, 1989) Ophiuchus and Serpens traditionally shared several stars according to the classical definition of "Serpentarius;" Delporte assigned the shared stars to Ophiuchus and therefore caused Serpens to be split into two non-contiguous halves, Serpens Caput (the Serpent's Head) and Serpens Cauda (the Serpent's Tail). Serpens was, however, the only such instance in which a constellation was divided into non-contiguous pieces.

Charts showing the new constellation boundaries were drafted at Uccle using the same projection and scale as those of Gould. Delporte chose a limiting magnitude of +6.5 to be close to the limit that Gould actually achieved at Córdoba. Final drafting of the charts was done by Gaston Coutrez; Delporte wrote in 1930 that he was "pleased to have the opportunity to pay tribute to the dedication and perseverance of his collaborator, who made it possible to complete the work successfully and in due time." Fourteen charts were prepared for publication; thirteen charts covered individual segments of the northern sky, and the fourteenth chart showed the entire hemisphere with the constellation boundaries, but without stars. This final chart was devised such that "the indexed meridians and parallels facilitate the task of classifying special stars." Upon completion of the work, Delporte sent the new maps to the subcommittee, which in turn forwarded the charts with its recommendation of the proposed boundaries to Commisions 3, 22 (Meteors), and 27 (Variable Stars) for their examination. The matter was then officially placed on the agenda for the Third IAU General Assembly at Leiden in 1928.

Adoption of the Final Constellation Boundaries (1928)

At the Leiden meeting, the full Union approved the recommendation of Commission 3 to adopt Delporte's boundaries for constellations north of declination −12.5°. The following reproduces in its entirety the report of Commission 3 from *Transactions of the IAU*, Vol. 3 (1928):

> 3. COMMISSION OF RATINGS, UNITS, AND THE ECONOMICS OF PUBLICATIONS
>
> PRESIDENT: MR. STROOBANT, *Director of the Royal Observatory of Belgium, Uccle, Brussels*
> MEMBERS*: MM. Bigourdan, Castillo, Singing, Dyson, Gautier, Grabowski, Horn Arturo, Leuschner, Schlesinger, Strömgren*[14]
> *The following proposals were presented to the Commission No. 3.*
>
> 1. *Final adoption of rationale to set the boundaries of the constellations, and based on arcs of hour circles and parallels of declination. Published under the auspices of the International Astronomical Union, an atlas extending between declinations of +90 and −12.5 and containing the stars up to magnitude 6.5.*
> 2. *Periodic publication of a detailed list of observatories and astronomers, under the auspices of the International Astronomical Union.*
> 3. *Discussion about the adoption of a new unit of stellar distance and a new definition of the absolute magnitude.*
> 4. *Separate designations for the various types of variable stars and meteor showers.*
> 5. *Generalization of the use of the decimal division of nonagesimal degree.*
>
> P. STROOBANT
> *President of the Commission*

From the proceedings:

> The Commission adopts the proposal to fix the limits of the constellations while basing itself in time arcs and circles and celestial parallels and expresses the wish to see the work published, under the auspices of the U.A.I.[15] including an atlas

[14] Bengt Georg Daniel Strömgren (1908–1987), Dutch astronomer.

[15] "Union Astronomique Internationale," the French name for the IAU.

> extending from +90 to −12° and containing stars down to 6.5 magnitude. The adopted equinox is that of 1875. The limits of the constellations are selected so as to avoid the renaming of stars and variables, this condition will be also be examined by the Commission of variable stars of the A.G. *[General Assembly]*.

The Commission also requested funds from the Union to secure the publication of the atlas:

> The Commission expresses the wish to see publishing the precise indications concerning these limits and asks for a subsidy of 90 pounds for the impression of the atlas. If this were not currently possible, one could, in addition to the values of the co-ordinates of the hour circles and the limiting parallels, to give charts in a smaller format.

With this report the professional astronomy community in 1928 agreed on the number, name, and boundaries of 88 constellations found on maps to the present day.

At Schlesinger's suggestion, Delporte extended the scope of his charts down to the south pole, thereby taking into account Gould's boundaries and making adjustments as needed according to the needs of variable star observers. As part of this process, Delporte dispensed with the diagonal lines that Gould occasionally used, replacing them with regular arcs of right ascension and declination. Further adjustments were made to specifically retain TV Ophiuchi, UW Ophiuchi, DG Aquilae, RR Normae, T Circini and U Tucanae in their original constellations, but the changes were made without transposing any stars in Gould's catalog. Charts were then drafted for the Southern Hemisphere. June 1929 was selected as a sort of "epoch" of the boundaries, as described by astronomer Philip Childs Keenan[16] in his review[17] of Delporte's resulting *Délimitation scientifique des constellations* (1930):

> The boundaries are taken along arcs of hour circles and of parallels of declination in such a way that all variables discovered prior to the end of June, 1929, remain in the constellation to which they were originally assigned. This authoritative delimitation of the constellations satisfies a long-felt need in doing away with the confusion in nomenclature of stars hitherto classified as common to more than one constellation.

However, some small problems were left unsolved at Leiden and remained to be addressed informally between the members of Commisions 3, 22, and 27, and Delporte. According to the proceedings of the Fourth General Assembly at Cambridge, Massachusetts, in 1932, "Mr Delporte's boundaries were carefully examined at Yale Observatory in order to make sure that no variable stars would have their designations changed, and that a minimum of stars having Baeyer [sic] or Flamsteed designations would be moved to other constellations." There are two documented corrections to Delporte's

[16] Keenan (1908–2000) was an American spectroscopist who, with William Wilson Morgan (1906–1994) and Edith Kellman (1911–2007), developed the MKK stellar spectral classification system.

[17] *Astrophysical Journal* **75**, 68 (1932).

boundaries produced by the Yale Observatory review incorporated as errata into Delporte's charts; in one instance a boundary is marked out by hand and redrawn, probably by the printer as a last-minute change.

The constellation boundary project was not considered complete until spring 1930 when Commission 27 members returned their final comments on the proposed charts; Delporte made the suggested changes and relayed them to IAU General Secretary Frederick J.M. Stratton[18] for final approval. That approval was likely obtained in either February or March of 1930. The Union approved a requisition for the printing costs of what became Delporte's April 1930 *Atlas Céleste* and accompanying explanatory volume, *Délimitation scientifique des constellations*, both published by Cambridge University Press. Further corrections were made to the southern charts in an IAU Circular after they were printed including minor typesetting mistakes and changes to the line drawings. Later printings were issued with these errors corrected.

The results of Delporte's decade-long efforts were reviewed by Schlesinger in the proceedings of the Fourth General Assembly (1932):

> 3. COMMISSION DES NOTATIONS, DES UNITS ET DE L'ÉCONOMIE DES PUBLICATIONS
>
> PRÉSIDENT: M. F. SCHLESINGER, *Director of the Yale Observatory, New Haven, Conn., U.S.A.*
> MEMBRES*: MM. Bosler,*[19] *Chant,*[20] *De Vos van Steenwijk,*[21] *Fisher,*[22] *Grabowski,*[23] *Horn d'Arturo,*[24] *Ludendorff,*[25] *Russell,*[26] *Strömgren, Stroobant*
>
> *At the meeting held at Leiden on July 13, 1928, the Union approved the recommendation of Commission 3 to adopt new boundaries for the constellations north*

[18] Lieutenant-Colonel Frederick John Marrian Stratton (1881–1960), OBE, FRS, was a decorated officer of the British Army and Professor of Astrophysics at the University of Cambridge from 1928 to 1947.

[19] Jean Bosler (1878–1973) was a French astronomer and director of Marseille Observatory from 1923 to 1948.

[20] Clarence Augustus Chant (1865–1956) was a Canadian astronomer and physicist considered by many as the "father of Canadian astronomy."

[21] Baron Jacob Evert "Jaap" de Vos van Steenwijk (1889–1978) was a Dutch astronomer who hailed from a prominent political family and served as both mayor of Haarlem and his native Zwolle.

[22] Willard James Fisher (1867–1934), an American astronomer and Research Associate at Harvard College Observatory, was an authority on meteors.

[23] L. Grabowski was a Polish astronomer at the observatory at Lwów (now Lviv, Ukraine).

[24] Guido Horn d'Arturo (1879–1967) was an Italian observational astronomer who designed and built the first segmented-mirror telescope in 1935.

[25] Friedrich Wilhelm Hans Ludendorff (1873–1941) was a German astronomer and astrophysicist, the younger brother of General Erich Friedrich Wilhelm Ludendorff (1865–1937), chief manager of the German war effort in the First World War. From 1921 to his retirement in 1938 the younger Ludendorff was Director of the Astrophysical Observatory of Potsdam.

[26] Henry Norris Russell (1877–1957) was an American astronomer best known for developing the Hertzsprung-Russell diagram with the Danish chemist and astronomer Ejnar Hertzsprung (1873–1967).

of 12.5° south declination, these boundaries being, in each case, hour circles or parallels of declination. The Union also approved of a grant to cover the costs of publishing an atlas showing these new boundaries. The boundaries were defined by Mr Delporte of the Uccle Observatory from whom this proposal originally came. Mr Delporte's boundaries were carefully examined at Yale Observatory in order to make sure that no variable stars would have their designations changes, and that a minimum of stars having Baeyer or Flamsteed designations would be moved to other constellations.

After making a few minor alterations on these accounts, Mr Delporte sent his manuscript to the General Secretary, under whose supervision the Cambridge University Press published, in 1930, a volume entitled Délimitation Scientifique des Constellations. At the suggestion of the present writer this volume extends not merely to 12.5° south declination, but from pole to pole. Gould had, for the most part, defined the southern constellations by hour circles and parallels; in the few cases where this had not been done, Mr Delporte revised the boundaries accordingly.

With the approval of the Executive Committee, an Atlas Céleste was published in the same way by the Cambridge University Press, showing all stars down to the sixth magnitude in their new boundaries, and giving the positions in the text of bright stars together with the most important variables, double stars, nebulae, and clusters.

FRANK SCHLESINGER
President of the Commission

In his *Star Tales* (1989), Ian Ridpath rightly noted that *Délimitation Scientifique des Constellations* represents "an international treaty on the demarcation of the sky, to which astronomers throughout the world have conformed ever since."

After Commission 3

A handful of further changes were made after the publication of *Atlas Céleste*. At the Cambridge, Massachusetts meeting in 1932, the IAU officially changed the name of Corona Australis to Corona Austrina, but the new designation never gained wide acceptance. Despite the fact that the latter remains the official name, the IAU website implicitly endorses the former as an alternative designation. At the Twelfth General Assembly at Hamburg, Germany in 1964, the three-letter abbreviation of the constellation Hydrus was changed from "Hys" to the more standardized "Hyi" on the recommendation of the Czech astronomer Antonín Bečvář.[27] Furthermore at Hamburg the suggestion of the Large and Small Magellanic Clouds as separate constellations was approved, but the recommendation was never popular and few astronomers ever referred to them as such.

---

[27]Bečvář (1901–1965) is best known for his work compiling star charts, in particular the *Atlas Coeli Skalnate Pleso* (1951) better known simply as the "Skalnate Pleso" Atlas.

No further changes to Delporte's constellation boundaries have been made or adopted by the IAU, but in the nearly century and a half since epoch 1875.0 the relentless precession of the Earth's poles has continued. By the end of the twenty-first century, the drift of the constellations against Delporte's boundary lines will inevitably require changes to correct the distortions of those lines to bring the constellation figures back in line with the system of right ascension and declination. To make such changes would require constituting a new committee, for Commission 3 was formally dissolved at the Eleventh General Assembly at Berkeley, California, in 1961.

Part II
The Lost Constellations

3

Anser

The Goose

Genitive: Anseris
Abbreviation: Ans
Location: Immediately south of Vulpecula and north of Sagitta[1]

ORIGIN AND HISTORY

Johannes Hevelius created "Vulpecula cum Ansere" (the Little Fox with the Goose) out of unformed stars in the space between Cygnus and Aquila. The constellation first appeared in *Firmamentum Sobiescianum sive Uranographia* (1687) and the accompanying introductory text, *Prodromus astronomiae* (1690). To emphasize that it was his own creation, Hevelius dedicated Figure L to the pair, shown as a detail in Fig. 3.1. In atlases published before Hevelius, its stars are shown simply marking the flow of the Milky Way through the area. Ian Ridpath (1989) noted that "Hevelius placed the fox near two other hunting animals, the eagle (the constellation Aquila) and the vulture (which was an alternative identification for Lyra). He explained that the fox was taking the goose to neighbouring Cerberus, another of his inventions." Richard Hinckley Allen (1899) put the total number of constituent stars identified by Hevelius at 27, while the German astronomers Friedrich Argelander (1843) and Eduard Heis (1872) recorded 37 and 62 stars, respectively, in their later catalogs.

In the generations after Hevelius, there was a persistent ambiguity among cartographers as to Anser's independence from Vulpecula. John Hill (1754) did not consider Anser a standalone constellation but rather merely "part of a constellation." The other half, the

[1] *"Over the Eagle"* (Hill, 1754); *"Between the Arrow and Swan"* (Bode, 1801a); *"Between the Eagle and the Swan"* (Ryan, 1827); *"South of the Swan, and north of the Dolphin and the Eagle"* (Kendall, 1845).

J.C. Barentine, *The Lost Constellations*, Springer Praxis Books,
DOI 10.1007/978-3-319-22795-5_3

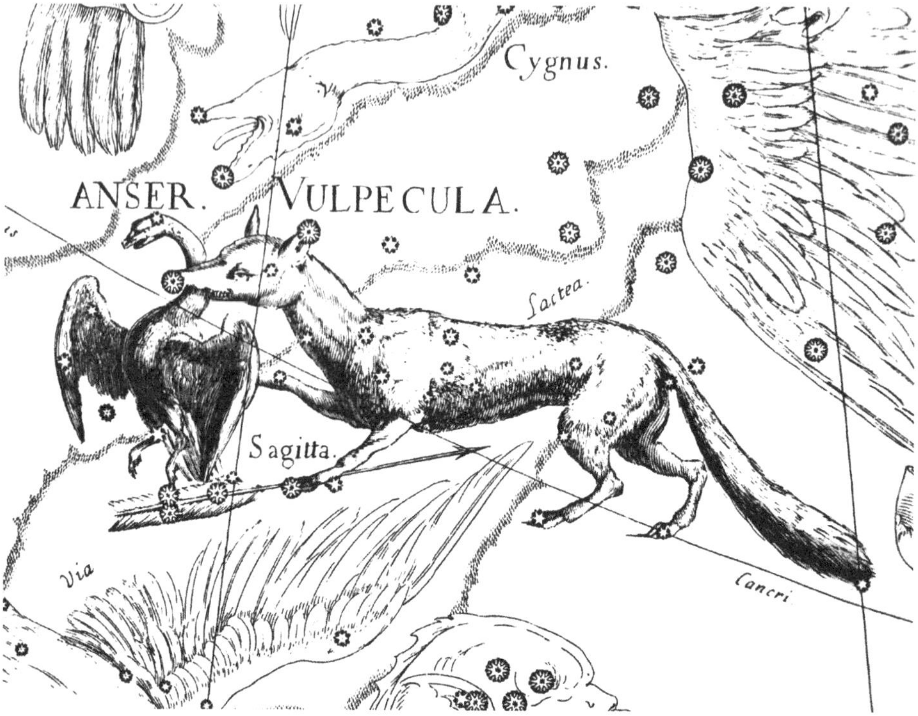

3.1 The debut of Vulpecula and Anser, creations of Johannes Hevelius, in Figure L of *Prodromus Astronomiae* (1690)

Fox alone, he identified by the name 'Vulpis.' "The whole is the Fox and Goose," he wrote, "Vulpecula et Anser." Johann Elert Bode included it in *Vorstellung der Gestirne* (1782), an elaboration on Flamsteed's earlier *Atlas Coelestis*, as "Gans". It appears to have equal billing with Vulpecula ("Fuchs"), as both names are given in all capital letters, even though a single boundary encompasses both figures. In *Allgemeine Beschreibung und Nachweisung der Gestirne*, the companion catalog to *Uranographia*, Bode described[2] Anser as

> located between the Arrow and Swan, in the middle of the split in the Milky Way,[3] and is one of the new constellations introduced by Hevelius. The stars in this figure are of the fourth and fifth magnitudes. Hevelius noted a new star in the head of the Fox in 1670, which remained visible for a few months.

---

[2]"Steht zwischen dem Pfeil und Schwan, mitten in der hier getheilten Milchstrasse, und gehört zu den neuen von Hevel eingeführten Sternbildern. Die kenntlichtsten Sterne in diesem Bilde sind nur von der vierten und fünften Grösse. Beym Kopf entdeckte Hevel im Jahr 1670 einen neuen Stern dritter Grösse, er war nur einige Monate sichtbar."

[3]The "Great Rift", a thin ribbon of obscuring dust that appears to divide the Galaxy into two branches in this region of the sky.

The "new star" was in fact Nova Vulpeculae 1670, now known as CK Vulpeculae (Fig. 3.3). It was discovered by a Carthusian monk in France named Père Dom Voiture Anthelme (ca. 1618–1683) on 20 June 1670 as a previously unseen star of the third magnitude. Hevelius independently noted it on the night of 25 July.[4] Jacob Green described Hevelius' observations in *Astronomical Recreations* (1824):

> In the head of the Fox there appeared, in 1670, a new star of the third magnitude which Hevelius observed during August, but which suddenly disappeared. In March, 1671, it became visible again, as a star of the fourth magnitude, and in March, 1672, it had diminished in its apparent light to a star of the sixth magnitude, since which time we have no account of its being visible.

Anthelme recovered it during the second maximum on 17 March 1671, and it peaked in apparent brightness at around visual magnitude +2.6 on 30 April. Both Hevelius and the Italian astronomer Giovanni Domenico Cassini (1625–1712) regularly observed the star until it again faded beyond the naked-eye limit in the late summer of 1671. The following March, Hevelius found it brightening yet again, observing it routinely until it disappeared again after 22 May; in the appearance of 1672, it was evidently at the threshold of naked-eye visibility near magnitude +6. It has not since achieved anything near the brightness seen during the outbursts of 1670–1672. The nova's nebulous remnant was found[5] in 1982. Given its erratic behavior and very high amplitude (near 18 magnitudes between its quiescent and outburst states), CK Vulpeculae may not a normal "classical" nova but rather some other type of cataclysmic variable star.

The Goose routinely appeared on charts of the eighteenth and nineteenth centuries, given the popularity of Hevelius' atlas and its wide publication reach. For the first hundred years after its introduction, it seems to have give equal billing to Vulpecula and Anser by most cartographers, such Johann Leonhard Rost in *Atlas Portatilis Coelestis* (1723, Fig. 3.2) and John Flamsteed in *Atlas Coelestis* (1729). From Bode's time forward, maps trended toward a combined depiction of the two figures in returning to something like the original notion of Hevelius'—"Vulpecula et Anser;" authors using this approach included Ryan (1827), Kendall (1845), Chambers (1877), Rosser (1879), and Cottam (1891). In describing the combined constellation, Green (1824) wrote

> Vulpecula Et Anser, or the Fox and Goose, is a modern constellation. It was arranged by Hevelius out of the unformed stars of the ancients. They are all too minute to be embraced in our plan ... This constellation crosses the Milky Way a little to the south of Cygnus, and to the north of the Arrow. On the breast of the Fox there is a small oval nebula.[6]

---

[4] *Philosophical Transactions of the Royal Society*, **5**, No. 65, 2087 (1670).

[5] Shara, M.M. and Moffat, A.F.J., "The recovery of CK Vulpeculae (Nova 1670)—The oldest 'old nova'" *Astrophysical Journal Letters* **258**, 41–44 (1982).

[6] This is the well-known planetary nebula Messier 27, aka the "Dumbbell Nebula."

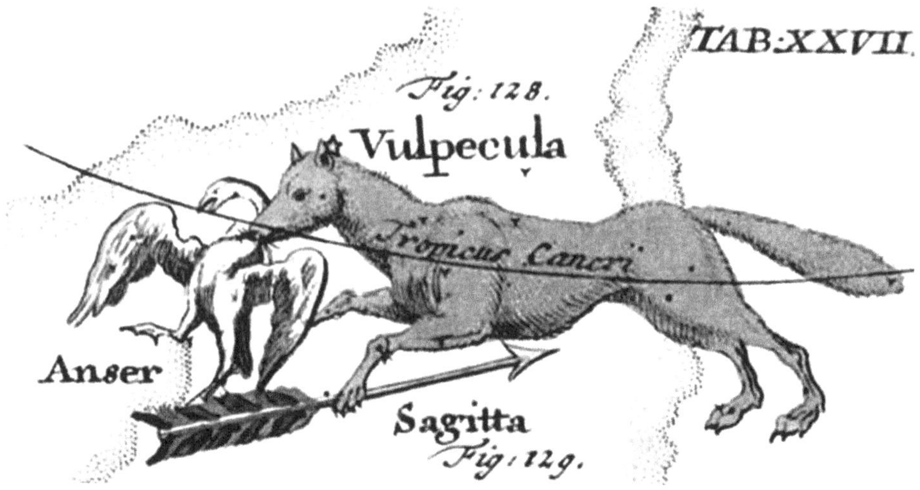

3.2 Vulpecula and Anser in Table XXVII, Figs. 128 and 129, of Johann Leonhardt Rost's *Atlas Portatilis Coelestis* (1723)

ICONOGRAPHY

As drawn in virtually every historical map, the figures of Fox and Goose are literally inextricable: The Fox carries what appears to be an already-dead Goose in its mouth; the Goose is always shown with wings slightly spread, its head flopping in the direction the Fox takes it. John Hill described the figure of the Fox "as having just seized upon his prey, and running away with it; the posture, in which he is drawn, is that of at once creeping and making off; and the Goose is represented as struggling as he holds her in his mouth, her wings are extended, and her head is bent to one side." The combined figure, he wrote, "is placed over the Eagle, with a little one called the Arrow between, and seems running toward Hercules." William Tyler Olcott, writing in 1911, indicated Hevelius' choice of animal for the unformed stars in the summer Milky Way was motivated by the context of the figures located nearby: "Hevelius is said to have selected this figure because of its appropriateness to its position, as the fox was a cunning and voracious animal, and was placed near the Eagle [Aquila] and vulture [Lyra] which are of the same rapacious and greedy nature."

In a well-developed folklore extending nearly back to antiquity, foxes have the distinct qualities of being marauding and ravenous. No friend of the farmer, they were known for preying relentlessly on domestic fowl and employing a hunting strategy of playing dead in order to lure birds, which they then seized (Fig. 3.4). In the seventh century, Isidore of Seville wrote[7]

[7] *Etymologies*, Book 12, 2:29.

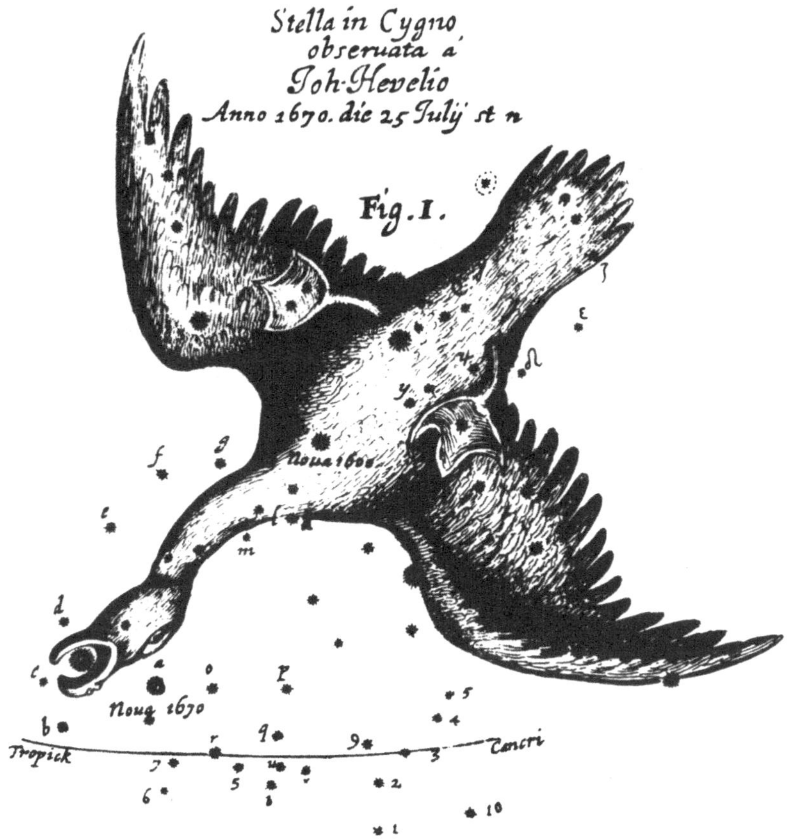

3.3 Hevelius' chart showing the position of Nova Vulpeculae 1670 (lower left, to the right of the head of Cygnus) published in *Philosophical Transactions of the Royal Society* in the same year. The stars below and to the right of the nova were included in Hevelius' later invention of "Vulpecula cum Ansere"

> Foxes (vulpes) are named as if pleasurable (volupes), because the fox flies with its feet (volat pedibus). They are deceptive animals that never run on a direct course, but only follow a winding path. To get food the fox pretends to be dead, then captures birds that come to feed on what they suppose to be its corpse.

These behaviors are commonly seen in medieval bestiaries, exemplified by the thirteenth century illuminated manuscript shown in Fig. 3.5. By depicting the Fox carrying off the Goose by the neck, Hevelius referenced a common motif in both medieval art and later naturalistic painting. Even at the close of the nineteenth century, a fox retreating into its "lair" with its ill-gotten bounty—a duck, in this case—was a subject for the English naturalist painter Heywood Hardy (Fig. 3.4). Foxes were also believed to be a scourge of

3.4 *Returning To The Fox's Lair* (1896) by the English naturalist painter Heywood Hardy (1842–1933). Oil on canvas, 69.8 × 81.8 cm

grape harvests, obliquely remembered in Aesop's fable of the Fox and the (sour) Grapes. According to the thirteenth century Franciscan scholar Bartholomaeus Anglicus,[8]

> Foxes lurk and hide themselves under vine leaves, and gnaw covetously and fret the grapes of the vineyard, and namely when the keepers and wards be negligent and reckless, and it profiteth not that some unwise men do, that close within the vineyard hounds, that are adversaries to foxes. For few hounds, so closed, waste and destroy more grapes than many foxes should destroy that come and eat thereof thievishly. Therefore wise wardens of vineyards be full busy to keep, that no swine nor tame hounds nor foxes come in to the vineyard.

[8]*De proprietatibus rerum*, book 17.

3.5 Two foxes depicted in a detail from an English illuminated manuscript from the second quarter of the thirteenth century, now in the Bodleian Library at Oxford University. The upper fox is shown carrying off a bird, which the lower fox plays dead to lure unsuspecting birds into its waiting jaws. MS. Bodl. 764, fol. 026r, tempera on parchment, original page size 29.8 cm × 19.5 cm

The art of the medieval period is replete with depictions of foxes in monastic garb, satirizing the begging monks of the era so disliked by the clergy. In commenting on thirteenth century carvings of foxes so dressed, Arthur H. Collins wrote in *Symbolism of Animals and Birds, Represented in English Architecture* (1913) (Fig. 3.6),

> The Bestiaries relate that the fox ensnares unwary fowls by pretending to be dead; in like manner the devil deceives unwary souls who love the corrupt things of the world. When geese are listening to a fox we suppose that they symbolize the silly souls who put their trust in the monk or friar, as the case may be. But, of course, the meaning is often simpler than that. Quite as frequently the fox is represented as preaching in a monk's or friar's habit to geese and other creatures, as on the stalls of Beverley Minster, S. Mary's Beverley, and Ely Cathedral (13th century). Generally such carvings are accompanied by others which represent Reynard[9] devouring his

[9]Reynard was the trickster hero of the ballad *The Romance of Reynard the Fox.*

3.6 Reynard the Fox as a bishop preaching to birds, in a detail from a late thirteenth/early fourteenth century French illuminated manuscript *Decretals of Gregory IX with glossa ordinaria* (also known as the "Smithfield Decretals"). British Library MS Royal 10 E IV fol. 49v, tempera on parchment, original page size 45 cm × 28 cm

> flock, or paying the penalty of his crimes on the scaffold: from which ordeal he sometimes emerges alive to try again! At Worcester Cathedral there are carved on a misericord foxes running in and out of holes. St. John the Evangelist stands near by with his Gospel in his hand, and his eagle at his feet. Here we can see an allusion to our Savor's words, 'Foxes have holes," etc., in S. Matt. viii. 20. It has been supposed that the object of this particular carving is to induce him who sees it to choose between good and evil."

The Fox is often shown with one paw resting on, or gripping, the constellation Sagitta (the Arrow). The most common explanation seems to be that Sagitta is, as Allen put it, "the weapon that slew the eagle of Jove, or the one shot by Hercules toward the adjacent Stymphalian birds."[10] But this is not always true; for instance, the atlas of Jamieson (1822) indicates a clear separation between the Arrow and the Fox, while the Fox's paw remains near it. This seems to imply that the Arrow was shot toward the Goose, but the Fox is deflecting it away from his prey. As noted previously, foxes are destructive to vineyards, but they are also known to steal hens and geese. The impression that the Fox is somehow evading the Arrow then becomes a subtle reference to the cunning Fox escaping a Hunter—

[10]The Stymphalian birds (Στυμφαλίδες ὄρνιθες; *Stymphalídes órnithes*) were fierce, man-eating birds that Eurystheus sent Heracles to defeat in his Sixth Labour. Kept as pets by Ares, the god of war, they were said to have beaks of bronze and sharp, metallic feathers they launched at their victims. As if to add insult to injury, their droppings were said to be toxic.

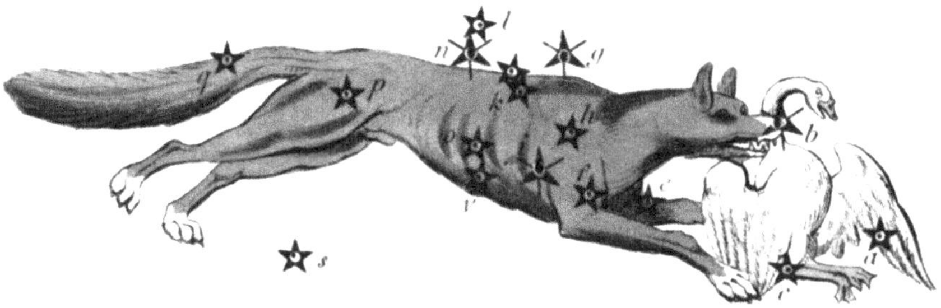

3.7 Vulpecula and Anser from Plate 14 in *Urania's Mirror* (1825)

perhaps the nearby Hercules—and bringing the stolen prize. In that context, the whole scene becomes an extended reference to the historic presumption that foxes were little more than evil, furry robbers (Fig. 3.7).[11]

DISAPPEARANCE

The process by which Anser fell into disuse unfolded slowly over the full span of the nineteenth century as authors variously depicted its figure as a constellation independent of (but associated with) Vulpecula, labeled it an asterism or subunit of Vulpecula, or simply omitted it altogether. The disappearance began as Thomas Young chose not to label Anser on his northern hemisphere map in Plate XXXVI, Fig. 517 of *A Course of Lectures on Natural Philosophy and the Mechanical Arts* (1807). Three years later, William Croswell left off both the Fox and the Goose in his *Mercator map of the starry heavens* . Argelander cataloged it as "Vulpecula cum Ansere" in *Neue Uranometrie* (1843), but showed only the label "Vulpecula" on Plates VI and XI of the accompanying atlas even though the figure of the Goose was drawn in. Alexander Keith Johnston, too, drew the figure of Vulpecula with Anser in its mouth on Plate 16, Map 4 of *Atlas of Astronomy* (1855), but labeled the combined figure "Vulpecula."

Richard Anthony Proctor proposed a compromise that foreshadowed the ultimate disposition of Anser in *A new star atlas for the library, the school, and the observatory* (1872): he combined Vulpecula and Anser into a new constellation "Vulpes". However, a mere 4 years later in *The Constellation-Seasons*, Proctor backed away from this name, dropped Anser entirely, and referred to the composite figure as simply "Vulpecula". Camille Flammarion followed Proctor's initial suggestion and used "Vulpes" for both figures in 1880s *Astronomie populaire*. In the last quarter of the century, most authors of popular-level texts gradually followed suit. Simon Newcomb made no mention of Anser in his hugely influential *Popular Astronomy* (1878) while discussing Vulpecula, and

[11] It may also be no coincidence that the character "Swiper" on the 1990s PBS children's show *Dora The Explorer* is a fox wearing a robber's face mask.

Sharpless and Philips drew the figure of the Goose in the Fox's mouth in charts included with *Astronomy for Schools and General Readers* (1882) but the entire figure is labeled "Vulpecula".

By end of the century the Goose was figuratively cooked and it vanished completely from maps. In 1899, Allen wrote *"The two members are sometimes given separately; indeed the Anser is often omitted... Astronomers now call the whole Vulpecula."* One of the last references to it in print was in William Tyler Olcott's *Star Lore of All Ages* (1911), in which it was referred to as "Vulpecula cum Ansere". At the first IAU General Assembly at Rome in 1922, the entire figure was designated "Vulpecula," and a single constellation boundary drawn around both Fox and Goose in Eugéne Delporte's *Atlas Céleste* (1930a). Yet it leaves behind a hint of its former place in the heavens: the *lucida* of Vulpecula (α Vulpeculae) is still known as "Anser" (Fig. 3.8).

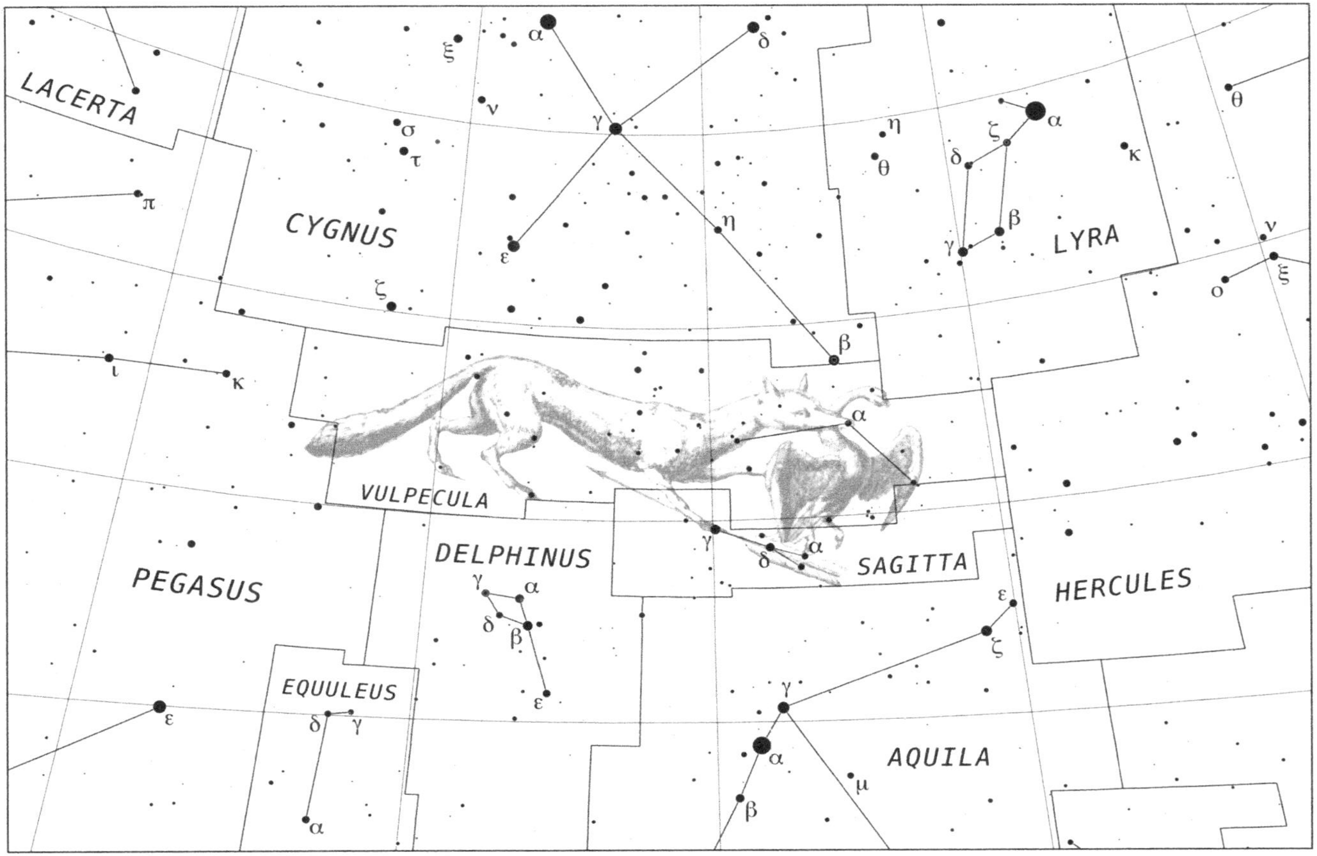

3.8 The figures of Vulpecula, Anser and Sagitta from Figure L of Johannes Hevelius' *Prodromus Astronomiae* (1690) overlaid on a modern chart

4

Antinoüs

The Favorite Of Hadrian

Genitive: Antinoi
Abbreviation: Atn
Alternate names: "Puer Adrianaeus," "Bithynicus," "Phrygius," "Troicus," "Novus Aegypti Deus," "Puer Aquilae," "Pincerna," "Pocillator" (the Cup-bearer) (Allen, 1899); "the Ithacan" (Lalande)
Location: Immediately southeast of the star Altair (α Aquilae)[1]

ORIGIN AND HISTORY

This constellation (Figs. 4.1 and 4.2) finds its beginnings in antiquity. It was first identified by the Greeks as Ganymede, brought to his lover, Zeus, by the eagle Aquila. The Romans repurposed the stars, borrowing heavily from the Greek legend in a brilliant piece of propaganda attributed to the emperor Hadrian. Publius Aelius Traianus Hadrianus Augustus (24 January AD 76–10 July AD 138) was born to an Italian family, probably in Italica, a city in the far-flung Roman province of Hispania. He was elevated to the purple

[1] *"Between the Eagle and Sagittary."* (Sherburne, 1675); *"Immediately south of Aquila ... bounded by Aquila, Scutum Sobiesci, Sagittarius, Capricornus, Aquarius, and Delphinus."* (Green, 1824); *"Under the Eagle"* (Kendall, 1845); *"East of Taurus Poniatowski and Scutum Sobieski, on the equinoctial"* (Bouvier, 1858); *"[L]ies in the Milky Way, directly south from the star Altair; the head of the figure at η and σ, the rest of the outline being marked by θ, ι, κ, λ, ν and δ, all now in Aquila."* (Allen, 1899); *"According to Ptolemy, Antinous consisted of six stars, which we now know as Eta, Theta, Delta, Iota, Kappa and Lambda Aquilae"* (Ridpath, 1989).

J.C. Barentine, *The Lost Constellations*, Springer Praxis Books,
DOI 10.1007/978-3-319-22795-5_4

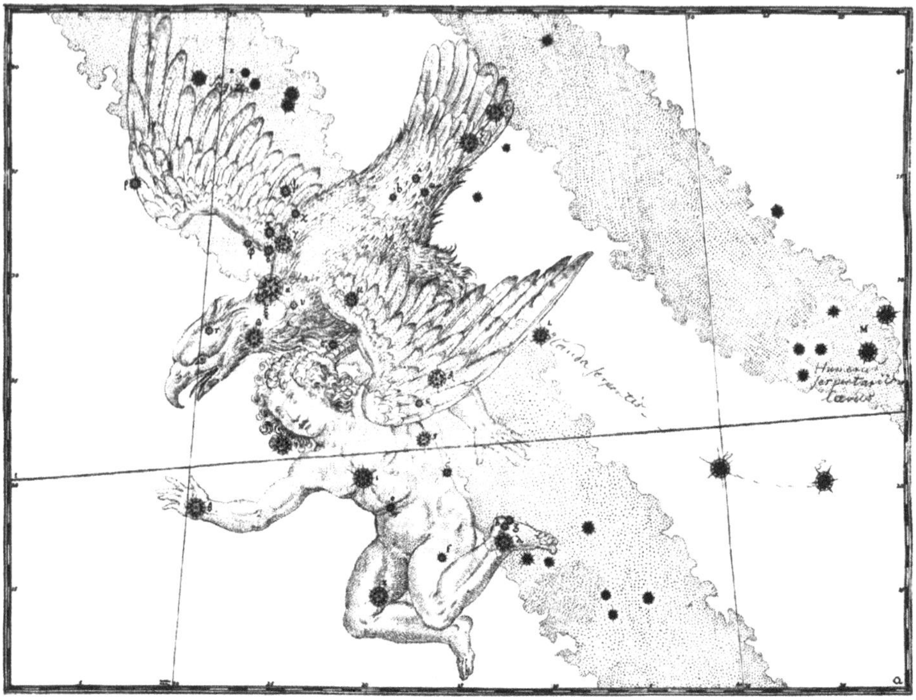

4.1 Aquila and Antinoüs shown on Leaf 16 (verso) of Bayer's *Uranometria* (1603)

on 10 August 117 at the death of Trajan, who according to Trajan's wife, Pompeia Plotina, named Hadrian his successor on his deathbed.

Little is known from second century sources of the life of the youth who so captivated Hadrian, and later authors embellished the details of his story far beyond believability. Antinoüs was born to a Greek family on 27 November, *c.* AD 111, in the small town of Mantineum, near the city of Bithynion-Claudiopolis, in the Roman province of Bithynia (now the city of Bolu, Turkey). The circumstances under which he was first introduced to Hadrian are not recorded, but he became associated with the emperor and joined his retinue at an early age. His image is still well known; he is depicted in many surviving works of Roman sculpture as a physically beautiful boy and then young man. His beauty is said to be the attribute that initially brought him to Hadrian's attention. It is widely thought by historians that their relationship was sexual, as Hadrian indulged himself in the Greek pederastic tradition as the ἐραστής (*erastês*, or older, dominant partner) with Antinoüs as the ἐρόμενος (*erômenos*, or younger, passive partner).

In 130, Antinoüs accompanied Hadrian on a journey up the Nile River from which he did not return. The lore has it an oracle predicted the emperor's life would be saved by the sacrifice of the object he most cherished, and Antinoüs evidently understood the prediction implied he was that object. The imperial fleet arrived in the ancient city of Hermopolis at

4.2 Aquila and Antinoüs from a plate in *Theatrum Mundi, et Temporis* by Giovanni Paolo Gallucci (1538–1621?), published at Venice in 1588. The page on which it appears (209) is headed with "Aquila seu vultur volans sydus 16" ("The Eagle, or flying vulture, 16 stars"), suggesting that Gallucci considered both Antinoüs and Aquila to form a single constellation

the time of the Nile's annual flood in late summer, celebrated locally in commemoration of the story of the death and resurrection of the Egyptian god Osiris. In the second century the Nile Valley was the breadbasket of the Empire, producing much of the grain it consumed, but by 130 the annual Nile flood had already failed for two years' running and threatened a famine. In such dire circumstances, beautiful youths were traditionally offered as a blood sacrifice to the gods by ritual drowning in the Nile, just as Osiris himself had drowned;

those who were sacrificed were often afterward deified, particularly if the following year's Nile flood was unusually voluminous.

After the Osiris-Nile festival, the Fleet progressed upriver until it reached a place called "Hir-wer", identified as the site of a minor temple to the nineteenth dynasty pharaoh Ramses II (*c.* 1303–1213 BC). On a date that year later celebrated annually on 28 October, Antinoüs fell into the Nile and drowned. It is unknown whether the fall was an accident or if Antinoüs voluntarily sacrificed himself to the river to fulfill the oracle's prophecy in hopes of saving Hadrian from danger. Hadrian grieved deeply (and very publicly) at the loss of his beloved, scoring a public relations coup in the process. Alluding to Alexander's deification of the dead Hephaistion, Hadrian deified Antinoüs after his death and established a religious cult devoted to him. The cult identified strongly with Egyptian culture and the deified Antinoüs was frequently depicted in Romano-Egyptian art in an aspect as the native god Osiris, associated with the 'rebirth' of the Nile through its annual flood. Hadrian also founded a city called Antinopolis (or Antinoë) on the site of Hir-wer, whose ruins are located near the present-day Egyptian town of Mallawi in the Minya governorate.

Something brought the stars in the region immediately south and east of the Greek constellation Aquila to the attention of Hadrian's court, or perhaps even the Emperor himself. According to the third-century, Greek-speaking Roman historian Cassius Dio,[2]

> [Hadrian] *declared that he had seen a star which he took to be that of Antinoüs, and gladly lent an ear to the fictitious tales woven by his associates to the effect that the star had really come into being from the spirit of Antinoüs and had then appeared for the first time.*

It is unclear whether this is intended to be understood figuratively, involving an otherwise known star, or if it refers to an historical event in the emergence of a "new" star (likely a nova or supernova) in around the year 130. There are no known supernovae for which reliable documentary evidence exists before AD 185.[3] Supernovae tend to leave behind remnant shells of gas visible for up to several tens of thousands of years after the stellar explosion, yet no remnants of about the right age are found among the stars that once comprised either Aquila or Antinoüs. Another possibility is that a smaller-scale explosion called a classical nova, which would not necessarily leave behind an obvious remnant, occurred around the time of Antinoüs' death. Until and unless further evidence

[2] *Epitome*, Book LXIX, published on page 447 of Vol. VIII of the Loeb Classical Library edition, translated by Earnest Cary (1925).

[3] This object's remnant is catalogued as RCW 86; see D.H. Clark and F.R. Stephenson, "The remnants of the supernovae of AD 185 and AD 393," *The Observatory* Vol. 95, pp. 190–195 (1975) and R. Stothers, "Is the supernova of AD 185 recorded in ancient Roman literature?" *Isis* Vol. 68, pp. 443–447 (1977). Y.N. Chin and Y.L. Huang (*Nature*, Vol. 371, Issue 6496, pp. 398–399, 1994) argue that the supernova identification for this event is erroneous and support a translation from Chinese sources indicating the "guest star" was rather a comet. B. Schaefer (*Astronomical Journal*, Vol. 110, p. 1793, 1995) suggests that neither explanation is sufficient, advocating that the Chinese observations of 185 are best explained as "a concatenation of two events in the same region of the sky"—both a nova and a comet.

is produced supporting the (super)nova hypothesis, the story recounted by Cassius Dio should be regarded as more myth than fact.

Exactly when Hadrian named the stars is also unclear; Allen ventures a guess (AD 132), but he cites no source for the date. *Poole Bros' Celestial Handbook* (1892) identifies 132 as the year the constellation was created (*"During the reign of Emperor Adrian, 132 years A.D."*), but infers that date on the belief that it was the same year as Antinoüs' death[4]. In 1624 Jacob Bartsch speculated in *Usus Astronomicus* that the constellation was formed by Ptolemy on orders directly handed down from Hadrian himself[5]:

> ANTINOUS, the boy placed under the Eagle, whose unformed stars were once numbered to Aquila by Ptolemy on orders of the Emperor Hadrian, of whom he was beloved. To others he is said to be Ganymede, suspended under the claws of the eagle, by which Jupiter stole him away into the sky.

Less than 20 years after the death of Antinoüs, Hadrian's constellation was still evidently in circulation, at least in Egypt: Ptolemy mentioned it as an asterism within Aquila ("ἐφ ὢν ὁ ἀντίνοος") in the *Almagest*, but did not formally list it among his 48 canonical constellations. Ptolemy deferred to the Greek tradition, which already saw the youth Ganymede clutched in the Eagle's talons, but perhaps in recognition of the political realities of the era included Antinoüs by way of a mention. It is not obvious why Ptolemy handled the naming of its stars as he did; Hadrian died in 138, but the religious cult dedicated to Antinoüs the god was still quite active when Ptolemy was writing.

The identification of certain stars near Aquila with Antinoüs survived the Medieval period despite its scandalous association with a heathen emperor and his male lover, but the figure did not appear on Western maps and globes until the sixteenth century. The first known, surviving depiction of Antinoüs from this period is on a celestial globe made by Caspar Vopel (1511–1561) in 1536; 15 years later, the Dutch cartographer Gerardus Mercator (1512–1594) included it on his globe. Its big break was the appearance of Antinoüs in Johann Bayer's *Uranometria* (1603)—but for the entirely wrong reason. Bayer labeled the figure "Ganymedes", referring to the earlier (pre-Roman) mythology surrounding the time.

Around the same time, Tycho Brahe's protégé Johannes Kepler posthumously published Tycho's comprehensive star list, the first since Ptolemy's *Almagest*, in *Astronomiae Instauratæ Progymnasmata* (1602), in which he elevated Antinoüs to full-constellation status by giving it a listing in his tables separate from Aquila. It is unclear whether this was Tycho's decision or one taken by Kepler; a few years later, in 1606, Kepler first labeled the stars in Antinoüs as distinctly separate from those in Aquila, giving it arguably more legitimacy as a constellation in its own right. In any case, several later authors, including Johann Elert Bode (1801a), Ezra Otis Kendall (1845), George

[4]There is some chance that the ancient authors confused the story of the Antinoüs star with a "temporary star," probably a nova, that appeared in nearby. Ophiuchus in AD 123 according to Allen (1899) and Olcott (1911).

[5]"ANTINOVS puer Aquilae subiicitur, cuius stellae olim informes ad Aquilam numeratae a Ptolemaeus post iussu Adriani imperatoris formatae, cuius is fuit amasius. Aliis Ganymedes dicitur, de unguibus aquilae suspensus, quem Iupiter in coelum rapuit."

Chambers (1877), Arthur Cottam (1891), and William Tyler Olcott (1911) attributed the introduction of Antinoüs as a constellation to Tycho, probably as a result of the diffusion of Kepler's 1602 work. Through the seventeenth century, the figure continued to appear in many popular atlases and books, including Christen Longomontanus' *Astronomica Danica* (1622) and Kepler's *Rudolphine Tables* (1627). Edward Sherburne, in his 1675 commentary on Marcus Manilius' *Astronomica*, wrote that

> ANTINOUS and GANYMED are one and the same Constellation for the Asterism which by the *Greeks* is feigned to represent *Ganymed* rap'd by the Eagle and carried up to Heaven to serve *Jupiter* as a Cup-bearer; the *Romans* in Honour of *Antinous* (the beloved Favourite of *Hadrian* the Emperour) will have to be the Representation of that beautiful *Bithynian*, who dying a voluntary Death for the Welfare of the Emperor, was by him honoured with Statues, Temples, Priests, and a Place among the Celestial Constellations; between the *Eagle* and *Sagittary*. It consists according to *Kepler* of seven Stars, according to *Baierus*[6] of eleven, and comes to the Meridian at Midnight about the Middle of *July*.

John Hill (1754), writing nearly a century after Sherburne, thought that the separation of Aquila and Antinoüs as fully-fledged constellations was a modern invention: [W]hile some of the moderns count them separate, others follow the antients, and making the Eagle and Antinous one constellation, count them together.

Through the eighteenth century opinions varied considerably on the point and authors wavered. John Flamsteed referred to the pairing three different ways in his *Historiae coelestis britannicae* (1725): *"Aquila Antinous"*, *"Aquila vel Antinous"* (Aquila, or Antinoüs), and *"Aquila cum Antinoo"* (Aquila with Antinoüs). Bode added to the confusion in *Vorstellung der Gestirne* (1782), in which he attempted to draw boundaries around the constellations based on Flamsteed's star identifications (Fig. 4.3). His boundary for Aquila and Antinoüs, however, extends as far as to encompass the constellation Scutum (*"Die Sobieskische Schild"*). Antinoüs *seems* to have top-billing, its name rendered in all capital letters like other familiar constellations featured on the same chart page (Delphinus, Vulpecula, Anser, and Sagitta).

Bode wrote[7] in *Allgemeine Beschreibung und Nachweisung der Gestirne* (1801a):

[6]Johann Bayer (1572–1625), author of *Uranometria* (1603).

[7]"Nach verschiedenen Dichtern der Vorzeit war dies der Adler, welcher den schönen Knaben Ganymedes, einem Sohn des Phrygischen Königs Tros, am Berge Ida für den Jupiter raubte. Antinous war gleichfalls ein schöner Knabe aus Bithynien, den der Kayser Hadrian an seinem Hofe hatte; nach andern ist hier gleichfalls Ganymedes verstirnt. Uebrigens hat erst Tycho den Antinous unter die Gestirne gebracht. Der Adler fliegt mitten in der Milchstrasse unterhalb den Schwan, nach Often. Er hat einen Stern erster Grösse, Altair gennant, am Halse, mitten zwischen zwey andern dritter und vierter Gröse. Ostwärts stehen zwey der dritten Gröse am Schwanz. Im Antinous sind Sterne dritter und vierter Gröse ostwärts bey der Milchstrasse sehr kenntlich. Sie bilden zum Theil ein verschobenes Viereck."

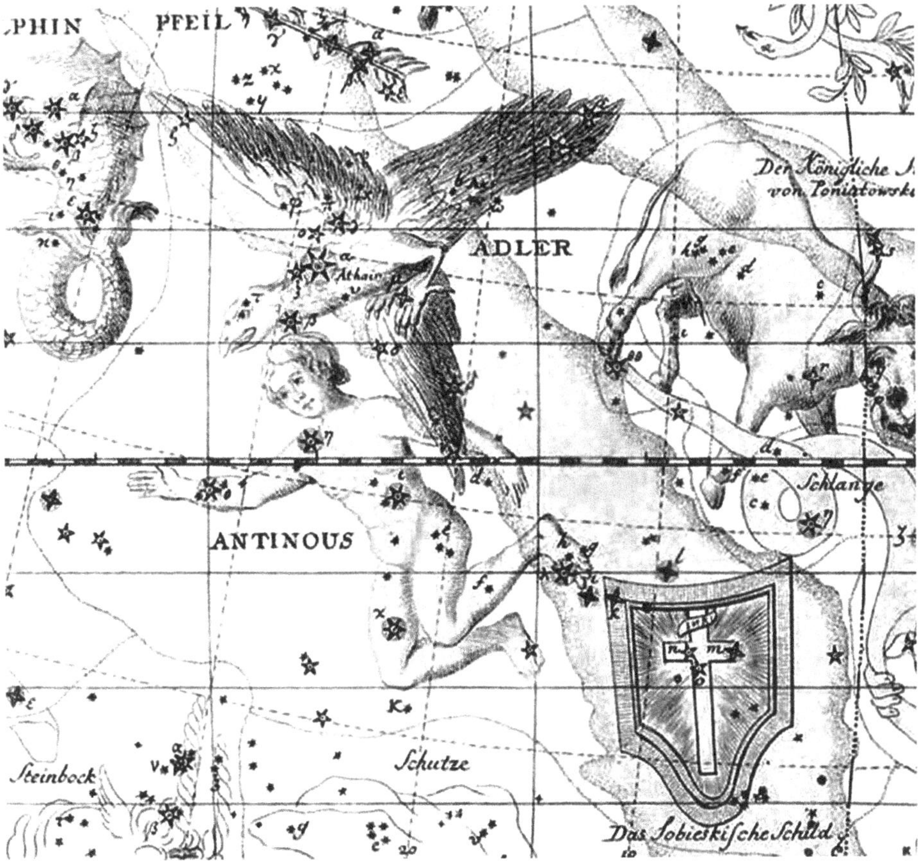

4.3 Johann Elert Bode's depiction of Aquila and Antinoüs in *Vorstellung der Gestirne* (1782) includes boundaries based on Flamsteed's star designations, extending so far as include Scutum Sobieski (*lower right*)

According to various poets of antiquity this was the eagle that stole the beautiful boy Ganymede, a son of the Phrygian king Tros, at Mount Ida for Jupiter. Antinous was also a beautiful boy from Bithynia that Emperor Hadrian had at his court; according to others, it must be seen in the figure of the young man Ganymede and not Antinous. Incidentally, it was Tycho who placed Antinous among the stars. The eagle flies in the middle of the Milky Way below the Swan, toward the east. It has a first-magnitude star, called Altair, on its neck, midway between two others of the third and fourth magnitudes. To the east are two stars of the third magnitude at the tail. Eastward in the Milky Way, in Antinous, are very recognizable stars of the third and fourth magnitude. They form part of a displaced square.

Jacob Green showed Antinoüs on Plate 7 of *Astronomical Recreations* (1824) and in the accompanying text asserted the constellation was introduced by Tycho, who

> placed this constellation in the heaves to perpetuate the memory of a youth much esteemed by the emperor Adrian. Antinous was a native of Bithynia in Asia Minor. So greatly was his death lamented by Adrian, that he erected a temple to his memory, and built in honour of him a splendid city on the banks of the Nile, the runs of which are still visited by travellers with much interest.

However, he expressed some doubt that the figure stood on its own as a constellation independent of Aquila, ascribing the combined figure to classical mythology: "As this constellation is often units to Aquila, and the whole considered but one group, some writers assert that the figure which accompanies the Eagle is not Antinous, but Ganymede; thus referring the whole to one of the exploits of Jupiter."

ICONOGRAPHY

The Greeks saw the figure of Ganymede (Γανυμεδες) among the stars later identified with Antinoüs. In their mythology, Ganymede was a divine hero with origins in Troy. He was described by Homer (*c.* eighth century BC) as among the most beautiful of mortals, a son of Tros by Calirrhoë, and a brother of Ilus and Assaracus. In one version of the myth,[8] he is abducted by Zeus, who takes the form of an eagle, to serve him as a cup-bearer in Olympus (Fig. 4.4):

> Tros, who was lord of the Trojans, and to Tros in turn there were born three sons unfaulted, Ilos [Ilus] and Assarakos [Assaracus] and godlike Ganymedes who was the loveliest born of the race of mortals, and therefore the gods caught him away to themselves, to be Zeus' wine-pourer, for the sake of his beauty, so he might be among the immortals.

Homer notes that Zeus paid appropriate compensation to Tros in taking his son: "These [the horses of Aeneias] are of that strain which Zeus of the wide brows granted once to Tros, recompense for his son Ganymedes, and therefore are the finest of all horses beneath the sun and the daybreak."[9] The story was similarly rendered in an Homeric hymn to Aphrodite[10] (*c.* seventh to fourth centuries BC):

> Verily wise Zeus carried off golden-haired Ganymedes because of his beauty, to be amongst the Deathless Ones and pour drink for the gods in the house of – a wonder to see–, honoured by all the immortals as he draws the red nectar from the golden bowl … deathless and unageing, even as the gods.

[8] *Iliad*, 20. 232ff (trans. Lattimore).

[9] *Ibid.*, 5. 265ff.

[10] Homeric Hymn 5 to Aphrodite, 203 ff (trans. Evelyn-White).

4.4 Antinoüs depicted in Corbinianus Thomas' *Mercurii philosophici firmamentum firmianum* (1730)

In his first Olympian ode (fifth century BC), the lyric poet Pindar wrote[11]:

> He [Poseidon] seized upon you [Pelops], his heart mad with desire, and brought you mounted in his glorious chariot to the high hall of Zeus whom all men honour, where later came Ganymede, too, for a like love, to Zeus.

With the near-wholesale appropriation of Greek myth by the Romans, the Ganymede story became a tale popular in the Republican era; for example, Ovid referenced Ganymede in the *Metamorphoses*[12]:

[11]Olympian Ode 1. 40ff (trans. Conway).
[12]*Metamorphoses* 10. 152ff (trans. Melville).

4.5 Third-century AD Roman mosaic depicting the abduction of Ganymede by Zeus in the form of an eagle. Sousse Archaeological Museum, Sousse, Tunisia; Image ©Ad Meskens, used with permission

> But now I need a lighter strain, to sing of boys beloved of gods and girls bewitched by lawless fires who paid the price of lust. The King of Heaven once was fired with love of Ganymedes Phrygius, and something was devised that Jupiter would rather be than what he was. Yet no bird would he deign to be but one that had the power to bear his thunderbolts. At once his spurious pinions beat the breeze and off he swept Iliades [Ganymedes of Ilion] *who now, mixing the nectar, waits in heaven above, though Juno frowns, and hands the cup to Jove.*

In Hadrian's own time, the novelist Apuleius wrote,[13]

> Highest Jupiter's royal bird appeared with both wings outstretched: this is the eagle, the bird of prey who recalled his service of long ago, when following Cupidos' guidance he had borne the Phrygian cupbearer [Ganymede] *to Jupiter.*

The abduction of Ganymede by the aquiline Zeus was a common motif in Roman art of the Empire period, as exemplified by the third-century Roman mosaic shown in Fig. 4.5. The myth was a model for the Greek pederastic tradition that proved irresistible to Hadrian; it was a ready-made story, complete with a figural representation in the night

[13] *The Golden Ass* 6. 15ff (trans. Walsh).

sky, that would cast him as a living deity if he could only urge popular interpretation of Antinoüs as Ganymede. The gamble worked, and Ganymede as the Eagle's quarry was soon forgotten. He may have been assisted by certain Roman authors who evidently considered the constellation Aquarius (the Water-Bearer) to be the figure of Ganymede on Olympus. This is suggested by, e.g., Pseudo-Apollodorus, second century AD ("Zeus kidnapped Ganymedes by means of an eagle, and set him as cupbearer in the sky"),[14] Gaius Julius Hyginus, c. 64 BC—AD 17 ("Mortals who were made immortal ... Ganymede, son of Assaracus, into Aquarius of the twelve signs"),[15] and the fifth century epic poet Nonnus of Panopolis. ("The mixing-bowl from the sky [the constellation Crater], from which Ganymedes mixes the liquor and ladles out a cup for Zeus and the immortals.")[16] In fact, Hyginus seems to ignore the classical identification of the stars below Aquila with Ganymede, referring to him exclusively in the figure of Aquarius. Of Aquila, he wrote in *Astronomica*,

> This is the eagle which is said to have snatched Ganymede up and given him to his lover, Jove ... And so it seems to fly above Aquarius, who, as many imagine, is Ganymede.[17]

and of Aquarius:

> Many have said he is Ganymede, whom Jupiter is said to have made cupbearer of the gods, snatching him up from his parents because of his beauty. So he is shown as if pouring water from an urn.[18]

As a result Hadrian perhaps thought he was not disturbing ancient traditions by restoring a human figure below Aquila, particularly as his invention fit well within the Antinoüs narrative. The notion of Ganymede as Aquarius persisted well past antiquity; on his map *Hemisphaerium meridionale et septentrionale planisphaerii coelestis* (ca. 1706), Carel Allard notes next to the label for Aquarius: "Ganymedes quem Jupiter coelestem fecit Pocillatorem" ("*Ganymede, whom Jupiter made the [cup] bearer of heaven.*")

The constellation of Antinoüs was drawn of stars immediately south of Altair (α Aquilae). Allen has "the head of the figure at η and σ, the rest of the outline being marked by $\theta, \iota, \kappa, \lambda, \nu$ and δ, all now in Aquila. Flamsteed omitted σ and ν from his catalogue, but added ι." The youth is shown grasped in the talons of Aquila, sometimes separately labeled, but most cartographers depict the pair as "Aquila et Antinoüs". The pose of the figure indicates a person being carried by the Eagle while in flight, as described by Hill (1754):

> It is represented in the schemes of the heavens in figure of a naked youth, of very good proportion, and in a posture that is neither standing, sitting, kneeling, nor lying,

[14] *Bibliotheca*, 3.141 (trans. Aldrich).
[15] *Fabulae*, 224 (trans. Grant).
[16] *Dionysiaca*, 47.98 ff (trans. Rouse).
[17] *Astronomica*, 2.16.
[18] *Astronomica*, 2.29.

4.6 Aquila and Antinoüs depicted on Mercator's 1551 celestial globe

> but seems as if he were falling through the air. The whole figure is represented naked, the head is covered with hair, and the body bulky rather than thin, the legs are bent backwards, and the arms expanded.

Jacob Bartsch suggested an interpretation[19] filtered through Christian theology in which the youth was perhaps Christ himself: "To us the Eagle can either be of the Roman Empire or the sign of John the Evangelist: the boy is a subject of or the new-born Christ himself, Luke chapter Two,[20] or some other sign of the Evangelist Matthew."

[19]"Nobis Aquila esse potest vel Romani imperij, vel Iohannis Evangelistae signum: Puer subiectus vel recens natus puer Christus, Luc. 2 vel alterius Evangelistae Matthaei signum."

[20]Luke 2 is concerned with the birth of Jesus, his presentation in the Temple, and his early life up to the age of twelve.

4.7 Aquila and Antinoüs shown in Figure R of Johannes Hevelius' *Prodromus Astronomiae* (1690)

Gerardus Mercator's 1551 celestial globe (Fig. 4.6) shows the youth with his legs bent, kneeling on a plinth as though the Eagle is depositing him there, but he doesn't seem to indicate that the plinth is defined by any particular stars. Hevelius (1690) transformed Antinoüs from a mere flailing youth, helpless in the claws of Aquila, into an archer drawing an arrow in a bow, aiming in the general direction of Aquarius and Capricornus (Fig. 4.7). Why Hevelius added this innovation is unclear, as there is no known, specific association between either the mythical Ganymede or the historic Antinoüs and archery (notwithstanding Apuleius' mention of Cupid in relating the story of Zeus and Ganymede), nor does it make sense that he would target the Zodiac for any particular reason. It is possible that Hevelius intended to allude to Ganymede's description as the "Eros" of homosexual love and desire; Plato, for instance, refers to Ganymede as "ημεϱος " (sexual desire):

> And when his feeling continues and he is nearer to him and embraces him, in gymnastic exercises and at other times of meeting, then the fountain of that stream, which Zeus when he was in love with Ganymede named Himeros (Desire), overflows upon the lover, and some enters into his soul, and some when he is filled flows out again.[21]

The same bow-and-arrow motif is repeated in *Urania's Mirror* (1825), where Antinoüs clutches a bow and several arrows in his right hand and a single arrow in his left, giving the impression he was about to launch an arrow from his bow when the Eagle swooped down upon him (Fig. 4.8). The action, a moment later, is frozen in time on the card as Antinoüs looks back over his left shoulder in shock. William Tyler Olcott thought that the bow and arrow had something to do with the nearby constellation Sagitta (the Little Arrow):

> On Burritt's map, Antinous is represented as grasping a bow and arrows as he is borne aloft in the talons of the Eagle. In this connection there may be a significance in the position of the asterism Sagitta, the Arrow, just north of Aquila."

Allen gives a variety of alternate names for the stars representing the figure of Antinoüs, quoting Jérôme Lalande, all of which appear to be epithets of the historical Antinoüs: *Puer Adrianaeus* ("Hadrian's Boy"), *Bithynicus* ("The Bithynian"), *Phrygius* ("The Phrygian"), *Troicus*, *Novus Aegypti Deus* ("The New Egyptian God"), *Puer Aquilae* ("The Aquiline Boy"), *Pincerna* or *Pocillator* ("The Cup-bearer"). He further wrote,

> Caesius[22] saw in it the Son of the Shunammite raised to life by the prophet Elisha[23]; and La Lande said that some had identified it with the bold Ithacan, one of Penelope's suitors slain by Ulixes.[24]

DISAPPEARANCE

Bode included Antinoüs in *Uranographia* (1801b), one of the last major works of "artistic" celestial cartography. During the nineteenth century, progressively fewer authors accorded it independent status as a constellation (Fig. 4.9), but most mentioned it as an asterism associated with Aquila. James Ryan stated this succinctly in 1827's *The new American grammar of the elements of astronomy*: "Antinous is generally reckoned a part of the

[21] *Phaedrus* 255 (trans. Fowler).

[22] Allen probably means Philipp von Zesen (aka Philippus Caesius; 1619–1689), among those in the seventeenth century who attempted to introduce new, Biblically-themed constellations in place of the ancient pagan originals.

[23] 2 Kings 4:8–37.

[24] The Latinized form of Odysseus of Ithaca, son of Laertes. Antinous, son of Eupeithes, was killed by Odysseus in Book XXII of the *Odyssey*. Allen's source for Lalande's reference is unclear.

4.8 Delphinus, Sagitta, Aquila and Antinoüs depicted on Plate 13 of *Urania's Mirror (1825)*

constellation Aquila." Sixteen years later, Argelander omitted it by title from *Uranometria Nova*, but illustrated it as a part of Aquila.

Toward midcentury, references to Antinoüs began to disappear from some popular astronomy texts (Proctor, 1876; Newcomb, 1878), while others kept it (Sharpless and Philips, 1882; Cottam, 1891). The second edition of Joel Dorman Steele's *Popular Astronomy*, published in 1899, referred to Antinoüs and Aquila as "a double constellation," each component receiving a clear label on his Map No. 6. Allen's 1899 synopsis of

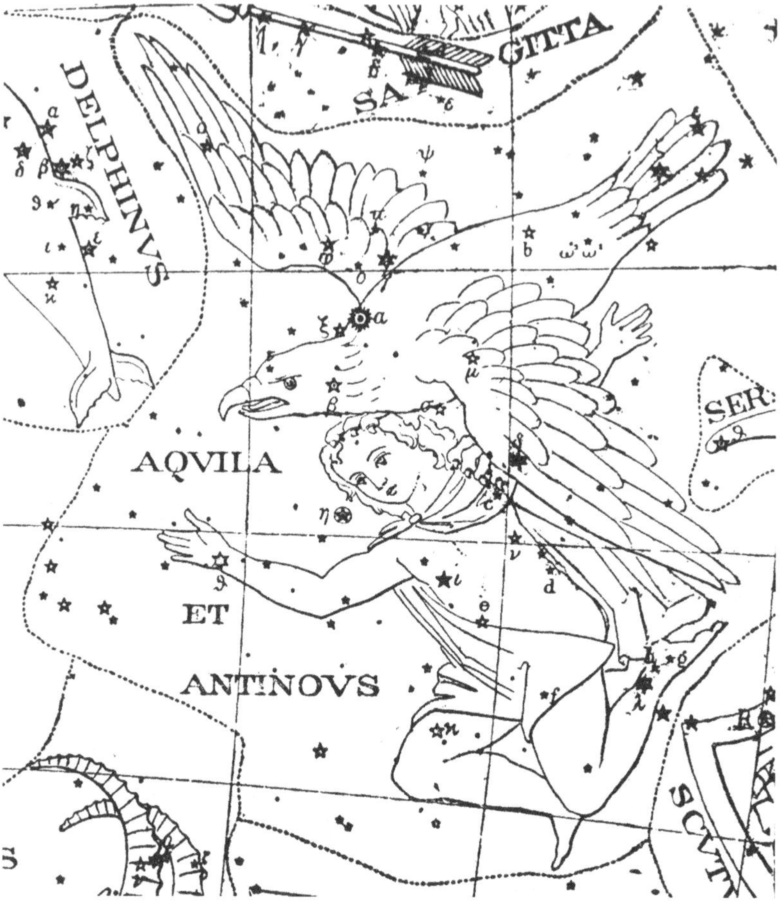

4.9 Aquila and Antinoüs depicted on Map 12 in Sharpless and Philips' *Astronomy for Schools and General Readers* (1882)

the constellation's history pronounces it dead at the end of the nineteenth century ("It is now hardly recognized, its stars being included with those of [Aquila]"), but a few holdouts carried it past 1900. Among the latest authors to mention it is Olcott (1911), who give it only asterism status ("Aquila is generally joined with Antinoüs."). At Rome in 1922, the IAU chose to cement that status and did not include it in the modern list of 88 constellations; its fate was sealed as Eugène Delporte's 1930 atlas drew Aquila's boundaries around its stars, eliminating Antinoüs even as an asterism (Fig. 4.10).

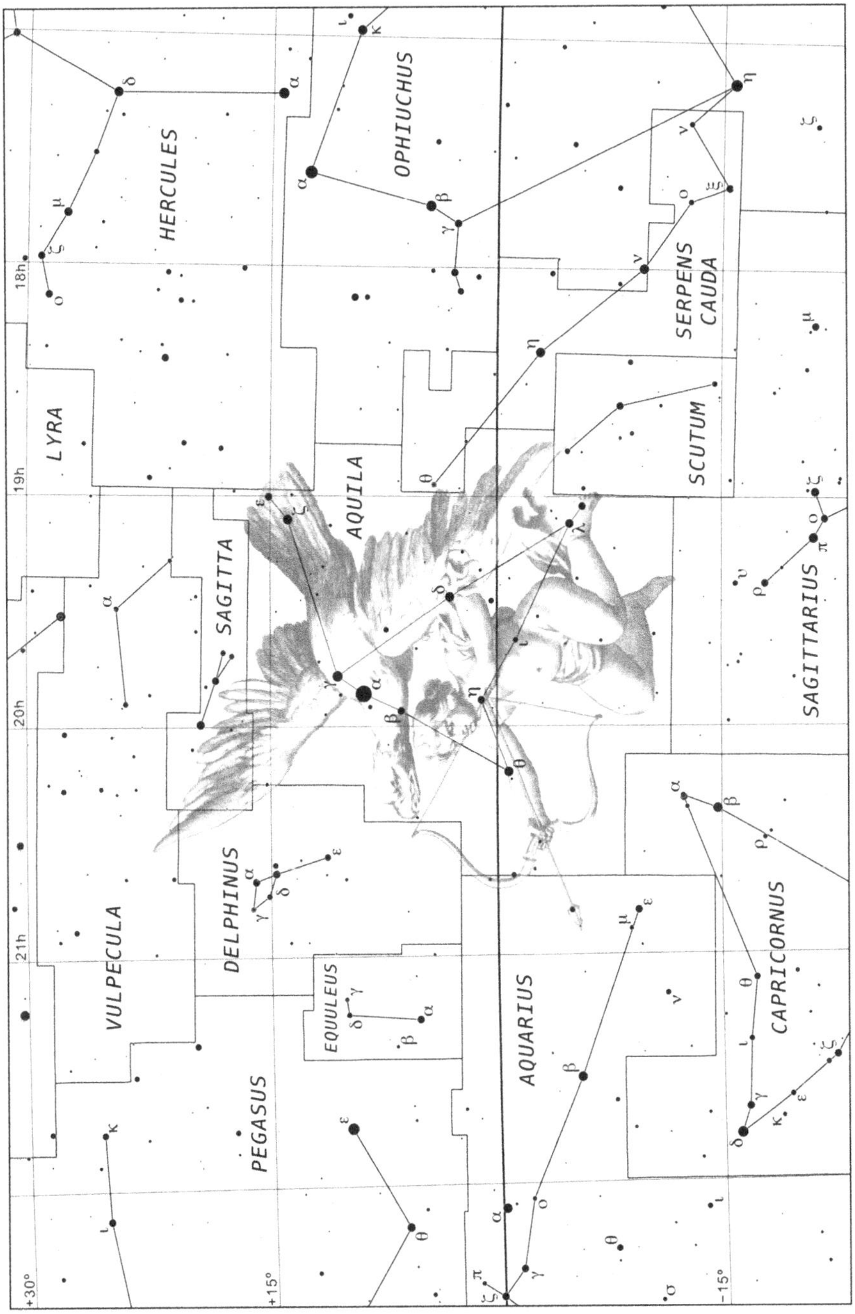

4.10 The figures of Aquila (*above center*) and Antinoüs (*below center*) from Figure R of Johannes Hevelius' *Prodromus Astronomiae* (1690) overlaid on a modern chart

5

Argo Navis

The Ship *Argo*

Genitive: Argūs Navis
Abbreviation: Arg
Location: The modern constellations Carina, Pyxis, Vela, and Puppis[1]

ORIGIN AND HISTORY

The figure of Argo Navis occupies an entirely unique position among all of the constellations in this book; as Ian Ridpath (1989) put it, "Argo is a constellation that is not so much disused as dismantled." Of an ancient origin, it persists to the current day albeit in a reduced form. In the mid-eighteenth century it was broken up into a series of adjacent figures still recognized as "official" by the International Astronomical Union[2]: Carina (the Keel), Puppis (the Poop Deck), and Vela (the Sails). It is also the only of the 48 constellations listed in Ptolemy's *Almagest* (second century AD) that is no longer officially recognized as a constellation by the IAU (Fig. 5.1).

[1] *"A large space in the southern hemisphere, to the south-east of Canis Major"* (Green, 1824); *"Situated south-east of Monoceros and Officina, and east of Columba"* (Bouvier, 1858) *"Eastward and S.E.-ward of Canis Major, extending from Dec. 10 S. to 60 S., between R.A. 7h. and 10 1/2 h"* (Rosser, 1879); *"It lies entirely in the southern hemisphere, east of Canis Major, south of Monoceros and Hydra, largely in the Milky Way"* (Allen, 1899); *"East of Canis Major and south of the Unicorn and Hydra"* (Olcott, 1911).

[2] The modern constellation Pyxis (the Compass) comprises a group of stars that were identified with Argo's mast in antiquity. One occasionally finds a modern astronomy book that asserts the Compass was placed in the stars by the Greeks but this is impossible because magnetic compasses were not used for navigation in the West prior to the Medieval period.

J.C. Barentine, *The Lost Constellations*, Springer Praxis Books,
DOI 10.1007/978-3-319-22795-5_5

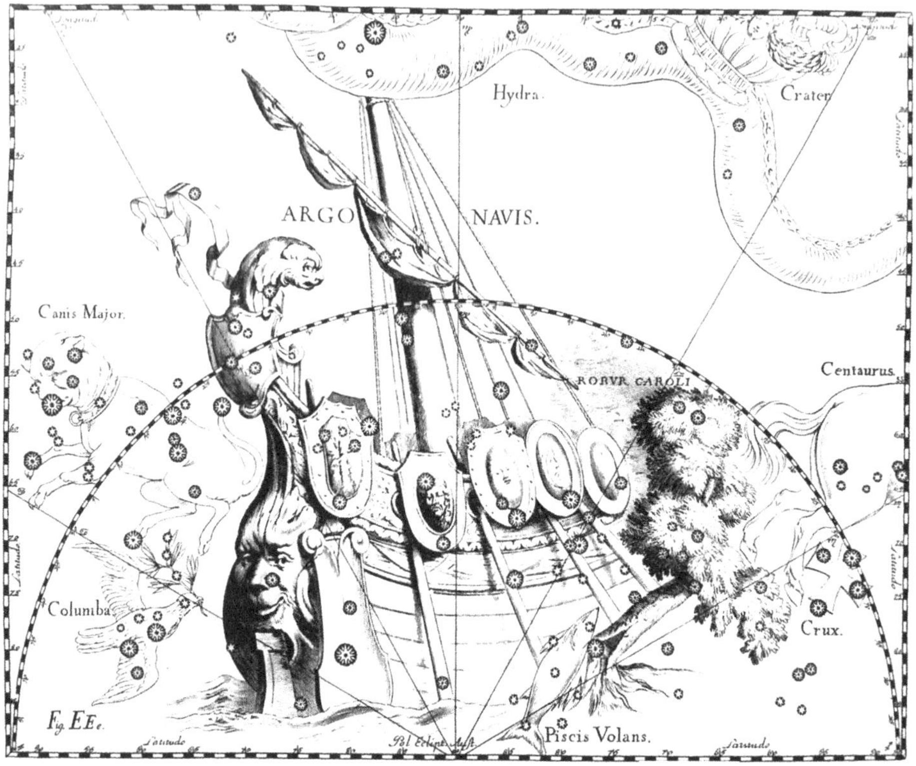

5.1 Argo Navis shown on Figure Eee of Johannes Hevelius' *Prodromus Astronomiae* (1690)

Jacob Green recounted the basics of the figure in *Astronomical Recreations* (1824):

> *Position, &c.* – This ship occupies a large space in the southern hemisphere, to the south-east of Canis Major. It contains a considerable number of fine stars, the most brilliant of which is called Conopus *[sic]*. This is of the first magnitude, but is never seen above our horizon; a very small part of this constellation being visible for us. Near the tail of Canis Major may be seen three stars of the third magnitude, which locate the prow of the vessel. The Milky Way passes directly through the middle of this group.

Johann Elert Bode described[3] the ship and its associated legend in *Allgemeine Beschreibung und Nachweisung der Gestirne* (1801a):

[3]"Dies Gestirn soll das Andeken des im Alterthum berühmten Schiffs verewigen, welches nach den Fabeln der Dichter auf Befehl der Minerva und des Neptuns in Thessalien vom Argo erbauet wurde, und dessen sich jene von Jason angeführten griechischen Helden, die Argonauten, zu ihrer damals unerhörten Seefahrt bedienten, um aus der am östlichen Ufer des schwarzen Meers gelegenen Landschaft Colchis das sogenannte goldene Vliess abzuholen."

> This figure commemorates the famous ship of antiquity, which was built according to legend at the command of Minerva and Neptune in Thessaly from Argo, and it is that which the Greek hero Jason and the Argonauts used to collect the Golden Fleece from the place on the eastern shore of the Black Sea known as Colchis.

He also noted that, much as in antiquity, the figure of the *Argo* disappears into the southern horizon as seen from the latitudes around the Mediterranean Sea, and that more or less of it was visible throughout history[4]:

> This constellation takes up a large part of the southern sky, from the Greater Dog in a southeasterly direction, and is replete with many bright stars, of which one of the first magnitude, Canopus, placed thoughtfully at the helm, but no longer rises here.[5] Only the northernmost part of the Ship, to the left [east] of the Greater Dog, appears on our horizon, and there are marked some stars of the third and fourth magnitude in the Milky Way, which passes through the middle of this constellation.

This is consistent with the description given[6] by Jacob Bartsch in *Usus Astronomicus* (1624) (Fig. 5.2):

> ARGO, SHIP, αϱγώ, Jason's ship, which had brought the Argonaut heroes to Colchis by the sea, he seized the Golden Fleece and carried it back to his homeland with Medea. It is ironically called in Latin "Celox[7]," a little slow and sluggish, and is therefore fashioned such that its keel never emerges in its entirety above the horizon (and not on account of any turbulent waves). In modern times it is depicted as a Dutch sailing ship. To us it is Noah's Ark (Genesis 6 and 7)[8] especially, with the Raven *[Corvus]* and the olive branch-bearing Dove *[Columba]* not far away. It has a multitude of stars, but very few rise *[at our latitude]*. Among those stars is the entirety of the ship's stern and, just grazing the horizon at Rhodes, its helmsman is said to be Canobus or Canopus [α Carinae].

---

[4]"Dieses Gestirn nimmt einen grosse Raum am südlichen Himmel, vom grossen Hund südostwärts ein, und ist mit vielen hellen Sternen besetzt, worunter einer der ersten Gröse Canopus am Steuerruder besindlich, bey uns aber nicht mehr aufgeht. Es kömmt nur der nordlichste Theil vom Schiff, zunächst links beym grossen Hund über unsern Horizont, und macht sich daselbst an einigen Sternen dritter und vierter Grösse in der Milchstrasse, die mitten durch dies Gestirn geht, kenntlich."

[5]Indicating precession in the ~1500 years since the end of antiquity had taken Canopus below the horizon as seen in Germany.

[6]"ARGO, NAVIS, αργώ, Navis Iasonis, quâ vecti Argonautæ heroës in Colchidem per mare, raptum vellus aureum cum Medeâ in patriam reportarunt. Dicitur ita per antiphrasin, quasi minimè segnis & pigra, Latinis Celox: vel ideò efformata, quod eius carina, ut non ex undarum fluctibus, ita nunquam tota supra horizontem emergat. Recentioribus navis Batavica pingitur. Nobis sit Arca Nohæ, Gen. 6. & 7. præsertim cùm Corvus & Columba olivifera non procul absint. Stellas habet plures, sed paucissimæ nobis oriuntur. Inter illas toto ferè cœlo illustrissima fertur ea, quæ in clavo, seu temone navis, radens horizontem Rhodiensium, & ab eius gubernatore quodam Canobus vel Canopus dicitur." (pp. 63–64)

[7]Denotes a type of cutter or otherwise swift-sailing ship.

[8]These chapters recount the traditional Biblical telling of the Flood Myth.

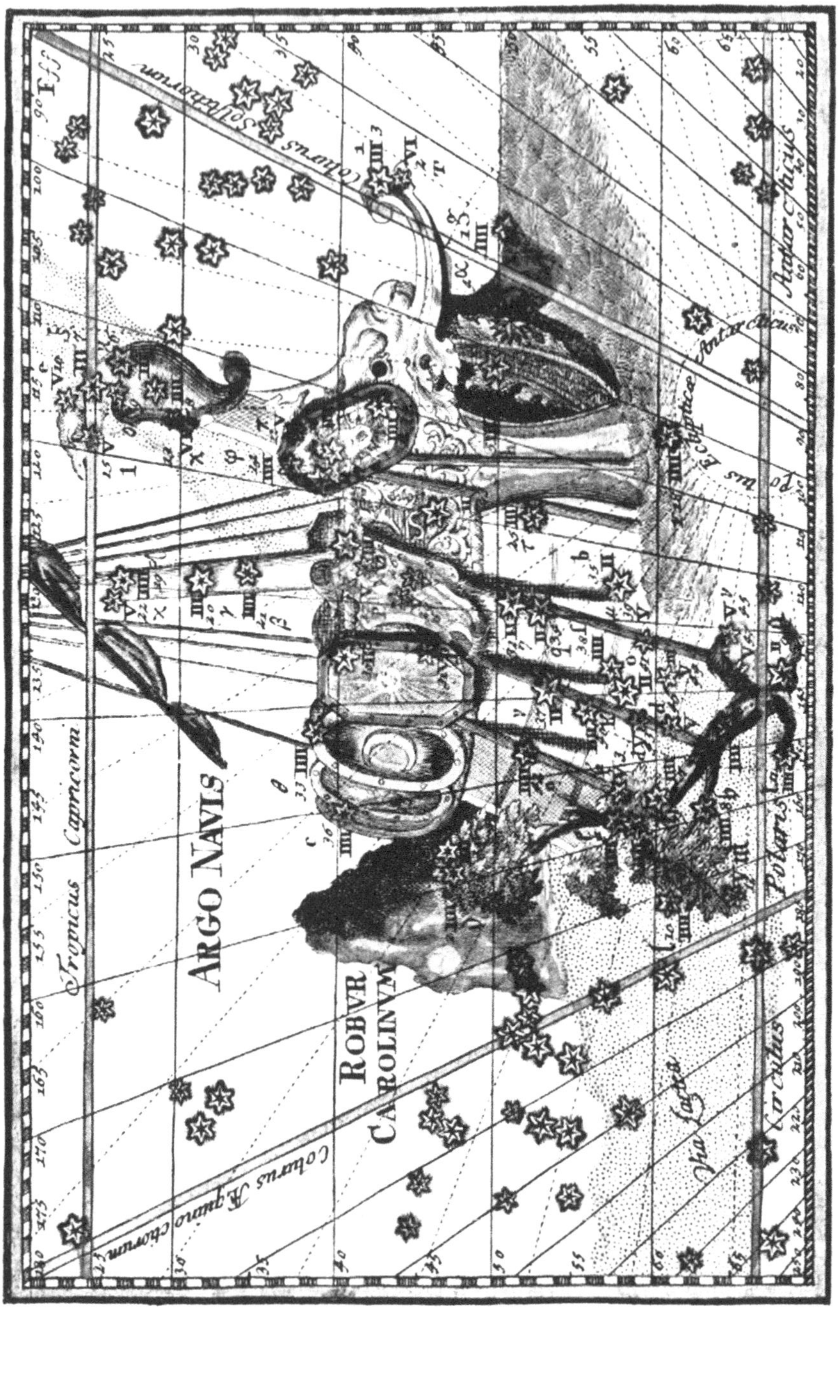

5.2 Argo Navis shown in Corbinianus Thomas' *Mercurii philosophici firmamentum firmianum* (1730)

Corbinianus Thomas further explained the mythological origin of its *lucida* in *Mercurii philosophici firmamentum firmianum* (1730). "Argo" was, he wrote[9]

> Either the name of the shipbuilder, or of the Greek ἀργός, referring to the ship's swiftness, or in fact, according to Cicero, because of the more than 50 Argive Heroes (consequently known as the Argonauts) who with Jason set out to plunder the Golden Fleece,
>
> > That ship built with the protection of warlike Minerva,
> > and first to run these unknown waters.[10]
>
> Now, the huge bulk, carried higher, landed between stars and there reached a peaceful shore. The brightest star at the helm, which the Egyptians call "Canopum", or after their hallowed divinity, or from "Canopo" the helmsman of Menelaus' ship, who was felled by the snake's bite,[11] or, finally, because from the town of Canopus the star is seen just grazing the horizon, of which Manilius [wrote] (Fig. 5.3):
>
> > One will nowhere find the glittering of Canopus
> > until one comes to the Nile by the waves of the sea.[12]

Argo Navis appears in a list of constellations in the *Phaenomena* of Eudoxus of Cnidus (*c.* 390–340 BC), the earliest extant reference on the ancient Greek night sky. Its mythology is certainly older than Eudoxus and was in place by the Homeric era, as the *Argo* was known[13] to the author of the *Odyssey*. Another clue to its history may be found in Richard Hinckley Allen's alternate "first ship" history given in *Star Names* (1899):

> Another Greek tradition, according to Eratosthenes, asserted that our constellation represented the first ship to sail the ocean, which long before Jason's time carried Danaos with his fifty daughters from Egypt to Rhodes and Argos and, as Dante wrote,[14] "Startled Neptune with the aid of Argo."

---

[9]"Aut ab Architecto hujus nominis, aut à Graeco ἀργός, quod *celeritatem* significat, aut denique, teste Cicerone, sic dicta, quia ultra 50. Argivorum Heroës (qui & inde Argonatuæ) duce Jasone in Colchidem ad diripiendum aureum velus profecti, *Nam rate, quæ, curâ pugnacis facta Minervæ, Per non tentatas prima cucurrit aquas.* Iam, qua & mole maxima, altius sublata inter stellas appulit, placidum ibi littus tenens. Lucidissimam in gubernaculo stellam tenet, quam Ægyptii *Canopum* dixere, aut horum Numini sacratum, aut derivata ab oppido Ægypti, aut à *Canopo* navis Menelai gubernatore, aspidis ictu extincto, aut tandem, quia circa Canopum oppidum primò conspici incipiat horizontem radens, de quo Manilius: – – – *nusquam invenies fulgere Canopum, Donec Niliacus per pontum veneris undas.*

[10]Ovid, *Tristas* III.9.

[11]Conon, *Narrations* 8; Strabo, *Geography* 17.1.17.

[12]Manilius, *Astronomica* I.216–I.217.

[13]"Only one ocean-going vessel has passed between them [Scylla and Charybdis], the celebrated *Argo* fleeing from Aeetes, and the waves would have quickly broken her on the massive crags, if Hera had not seen her through, because of her care for Jason." Book XII, line 69 (trans. A.S. Kline).

[14]*Paradiso*, Canto XXXIII, line 96.

5.3 Argo Navis shown in a late Medieval woodcut from Julius Firmicus Maternus' *Matheseos Liber* 1499; the figure itself was copied from a version of Hyginus' *Poeticon* published by Erhard Ratdolt at Venice in 1482. That the traditional interpretation of the Ship omitted the stern is evident

Fixing a period of time for the factual basis of Argo's existence has busied historians for centuries. The events of the story take place a generation before the Trojan War (*c.* 1300 BC in literature), and persisted in oral tradition until the time of Homer some five centuries later. But did the Greeks invent a ship constellation, or was it received from elsewhere? Jacob Green (1824) suggested that the *Argo* myth might predate the Greeks:

> For ourselves we are persuaded that this group of stars originally referred to the Ark of Noah and the Deluge. Dr. Bryant supposes the Argonautic expedition to be a Grecian fable, founded on some Egyptian traditions, which referred to the preservation of Noah and his family during the flood.

Here Green referred to Jacob Bryant (1715–1804), an eminent British mythographer who built a scholarly career trying to determine the historical truth behind various pagan myths via Judaeo-Christian tradition. In particular, Bryant went to lengths to establish the Bible as a primary source by demonstrating that ancient Greek and Egyptian myths were substantially drawn from narratives in the Book of Genesis. In *A New System* (1774–1776), Bryant advanced a theory that the Biblical account of the flood myth in the role of Noah's Ark was misinterpreted by the Greeks as the *Argo*:

> The Argo, however, that sacred ship, which was said to have been framed by divine wisdom, is to be found there; and was certainly no other than the ark. The Grecians supposed it to have been built at Pagasse in Thessaly, and thence navigated to Colchis. I shall hereafter shew the improbability of this story. ... In respect to the Argo, it was the same as the ship of Noah, of which the Baris of Egypt was a representation.

Allen (1899) suggested that the identification with the Ark predated Bryant by a considerable length of time:

> The biblical school of course called it Noah's Ark, the Arca Noachi, or Archa Noae as Bayer wrote it; Jacob Bryant, the English mythologist of the last century, making its story another form of that of Noah. Indeed in the 17th century the Ark seems to have been its popular title.

The historical understanding of *Argo* as Noah's Ark is strengthened by the appearance nearby of Columba, a dove that matches the Biblical narrative of the flood myth. While modern authors such as Ridpath (1989), Bakich (1995), and Kanas (2007) attributed the formation of Columba to Petrus Plancius in 1592, Allen suggested that the roots of the figure extend back to antiquity itself:

> The following from Caesius[15] may indicate knowledge of its stars, and certainly of the present title, seventeen centuries ago. Translating from the Paedagogus[16] *of Saint Clement of Alexandria,*[17] *he wrote:* Signa sive insignia vestra sint Columba, sive Navis coelestis cursu in coelum tendens sive Lyra Musica, in recordationem Apostoli Piscatoris.[18]

In this we may have a glimpse of how early Christians interpreted the classical Greek constellations, anticipating by nearly 1500 years the remaking of the heavens in Biblical imagery by Julius Schiller in *Coelum Stellatum Christianum* (1627).

Dating mythical events continued to be important work for understanding the context of Biblical stories. In his *Short Chronicle*[19] (1728), Isaac Newton calculated a specific date for the ship's famous expedition, placing it in the Homeric era:

> The ship Argo *is built after the pattern of the long ship in which Danaus came into Greece. And this was the first long ship built by the Greeks. Chiron who was born in the golden age, forms the Constellations for the use of the Argonauts & places the solstitial & equinoctial points in the fifteenth degrees or middles of the constellations*

[15]Presumably refers to Philipp von Zesen, also known as Filip Cösius or Caesius (1619–1689), a German poet and writer.

[16]Written c. AD 198.

[17]c. 150–c. 215.

[18]'The signs or emblems of you are the Dove, or the course of the Heavenly Ship stretching across in the sky or the musical Lyre, in remembrance of the Fisherman Apostle [Peter].'

[19]*A short Chronicle from the first memory of things in Europe to the conquest of Persia by Alexander the great.*, MS Add. 3988, Cambridge University Library, Cambridge, UK.

of Cancer Chelæ[20] *Capricorn & Aries. Meton*[21] *in the year of Nabonassar*[22] *316 observed the summer solstice in the eighth degree of Cancer, & therefore the solstice had then gone back seven degrees. It goes back one degree in about seventy & two years & seven degrees in about 504 years. Count these years back from the year of Nabonassar 316, & they will place the Argonautic expedition about 936 years before Christ.*

Newton's estimate is not unreasonable. Like many Greek tales, the *Argo* myth probably dates to the Heroic era of the Bronze Age, from just before the beginning of the written tradition back to around 2000 BC. It appears at first reasonable to assume that the Greeks received the figure of the ship from Mesopotamia, and it is tempting to connect up ship imagery with the (pre-)Sumerian Flood Myth.[23] One would therefore expect a ship constellation to appear in ancient star lists. However, there is no evidence that the Sumerians or any later Mesopotamian culture considered the same stars identified with the *Argo* by the Greeks to be a ship. Instead, among the stars of Argo Navis the Mesopotamians saw several constellations, including the Goddess of Motherhood (NIN.MAH), the Harrow (GAN.UR), and the Bow (BAN) and Arrow (KAK.SI.KI). Furthermore, no ship figure appears in any known Mesopotamian star list, which in any case date to no later than the second millennium BC. It is therefore likely that Bronze Age Greeks had already recognized a ship in the night sky, coincident with the later identification of Argo Navis, before receiving the constellation lore of Mesopotamia.

If not from Mesopotamia, did the Greeks borrow their ship constellation from elsewhere? It is much more believable to conclude that Argo Navis arrived from Egypt sometime around 1000 BC, given that trade ties between Egypt and mainland Greece were established by at least the Geometric Period (*c.* 900–700 BC). The Greeks had by then adopted a ship design called a "galley" that originated in the reed boats of Egypt and were adapted to timber construction by the Phoenicians (see Fig. 5.7 for reference). This provides a plausible route for a Greek ship constellation. Evidence among later authors support the idea; as an example, Plutarch identified[24] Argo Navis with the Egyptian constellation called the "Boat of Osiris:"

Moreover, they give to Osiris the title of general, and the title of pilot to Canopus, from whom they say that the star derives its name; also that the vessel which the Greeks call Argo, *in form like the ship of Osiris, has been set among the*

[20]An alternate name for the constellation Libra, indicating the "claws" of Scorpius.

[21]Meton of Athens (*c.* fifth century BC) was a Greek mathematician and astronomer for whom the 19-year lunar Metonic cycle is named.

[22]Nabonassar (Nabû-nāṣir) was the king of Babylon from 747 to 732 BC.

[23]See, e.g., John J. McHugh, *The Deluge: A Mythical Story that was Projected onto the Constellations*, master's thesis, Department of Anthropology, Brigham Young University, 1999 (Harold B. Lee Library call number GN 2.02 .M33 1999).

[24]*Isis and Osiris* 22, trans. F.C. Babbitt (from *Moralia*, Loeb Classical Library edition, Vol. V, 1936).

constellations in his honour, and its course lies not far from that of Orion and the Dog-star; of these the Egyptians believe that one is sacred to Horus and the other to Isis.

Similarly, Allen (1899) wrote “Egyptian story said that it was the ark that bore Isis and Osiris over the Deluge.”

Associated Constellations

Robur Carolinum

In the eighteenth and nineteenth centuries, a few ancillary parts of the Ship were added by various cartographers to deal with faint, unformed stars that were otherwise not associated with particular neighboring constellations. One central, unsolved problem left over from the ancient conception of the constellation was what to do about the fact that the Ship appeared to be missing its prow; no bright stars marked what should have been the front end of the Ship, so mapmakers generally dealt with the problem by awkwardly showing its front end disappearing into a cloud of mist. The English astronomer Edmond Halley (of eponymous comet fame) took it upon himself to solve the problem by pandering to his patron, the restored King Charles II; in 1678, Halley resolved the mist ahead of Argo into a tree. He named the new constellation “Robur Carolinum” (Charles’ Oak) after the Royal Oak in which Charles was hidden during his flight from Oliver Cromwell’s army in the English Civil War; its history is treated more fully in Chap. 22. It gained some initial popularity on account of its inclusion in Hevelius’ *Firmamentum Sobiescanum* (1690; Fig. 5.1), but was rejected by many for its manifestly nationalist overtones, and disappeared off maps before the end of the nineteenth century (Fig. 5.4).

Malus

Another set of stars considered part of the Ship was largely problematic from antiquity until the International Astronomical Union recognition of the “modern” constellations. These formed what the ancients saw as the Ship’s mast, commonly referred to as “Malus.” This name appeared on some charts until the beginning of the twentieth century (e.g., Fig. 5.5). The available evidence suggests that Malus was treated more as an asterism within Argo Navis by most cartographers, and never enjoyed an existence as an independently recognized constellation in its own right.

When Nicolas Louis de Lacaille broke up Argo into its constituent components in the mid-eighteenth century, Malus was the only part that lost its classical identification. Rather than the Ship’s mast, Lacaille identified a handful of stars suspended above Puppis not as a mast, but rather as *la Boussole* (the “Marine Compass”), giving Bayer-style designations to ten stars now known as α to λ Pyxidis (notably skipping ι). He labeled the new form “Pyxis Nautica” on his 1763 chart (Fig. 5.6).

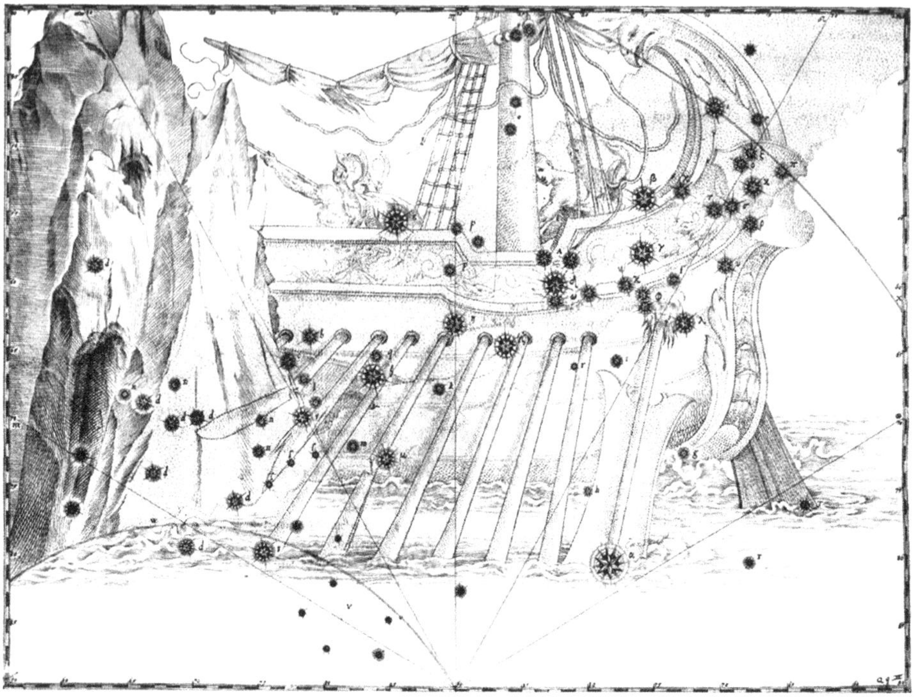

5.4 Argo Navis depicted on Figure Qq of Johannes Bayer's *Uranometria* (1603). Before the invention of Robur Caroli in the late seventeenth century, most cartographers showed the prow of the ship disappearing into a cloud of mist or, as in Bayer, behind a conveniently placed rock

Some eight decades after Lacaille's death, John Herschel suggested retaining the name "Malus" for these stars rather than otherwise following Lacaille's convention for the components of the Ship. Writing in the *Monthly Notices of the Royal Astronomical Society of London* (1843), Herschel suggested "that Argo be divided into four separate constellations, as partly contemplated by Lacaille; retaining his designations of Carina, Puppis, and Vela; and substituting the term Malus for Pixis Nautica, since it contains four of Ptolemy's stars that are placed by him in the mast of the ship." But then he recommends retaining Argo for some stars:

> That the original constellation Argo, on account of its great magnitude and the subdivisions here proposed, be carefully revised in respect of lettering, in the following manner:– First, In order to preserve the present nomenclature of the principal stars, all the stars in Argo (that is, in the general constellation, regarded as including the subdivisions above-mentioned) indicated by Greek letters, by Lacaille, to be retained, with their present lettering, under the general name Argo. Secondly, All the remaining stars, to be designated by that portion of the ship in which they occur, such as Carina, Puppis, Vela, and Malus, and to be indicated by the Roman

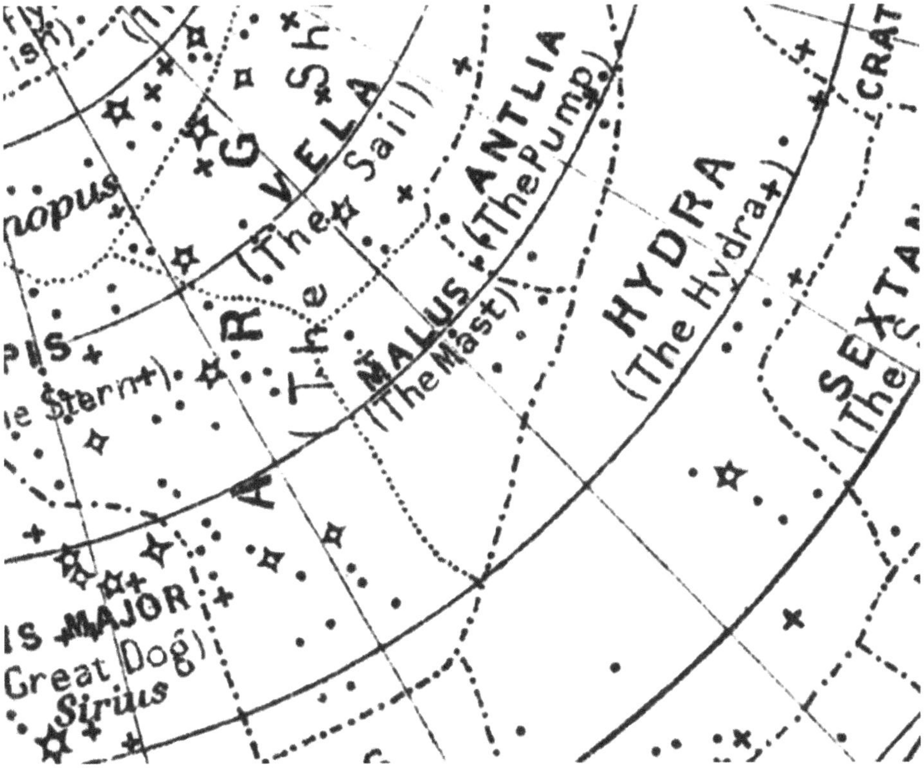

5.5 Malus (*center*) is shown as part of a dismembered Argo Navis in an 1894 edition of Camile Flammarion's *Astronomie Populaire*

> letters adopted by Lacaille, as far as the fifth magnitude inclusive. And no two stars, far distance from each other in the same subdivision, to be indicated by the same letter; but, in cases of conflict, the greater magnitude is to be preferred; and, when they are equal, the preceding star to be fixed upon.

Despite Herschel's influence in the astronomical community, his proposal failed to gain traction. By the time Richard Hinckley Allen was writing (1899), Malus seems to have been popularly discarded in favor of Pyxis. Lacaille, he wrote,

> formed from stars in the early subordinate division Malus, the Mast, Pyxis Nautica, the Nautical Box or Mariner's Compass, the German *See Compass*, the French *Boussole* or *Compas de Mer*, and the Italian *Bussola*; and this is still recognized by some good astronomers as Pyxis.

Upon the adoption of the modern canon of constellations by the IAU in 1930, Pyxis was given official recognition while Malus was finally laid to rest.

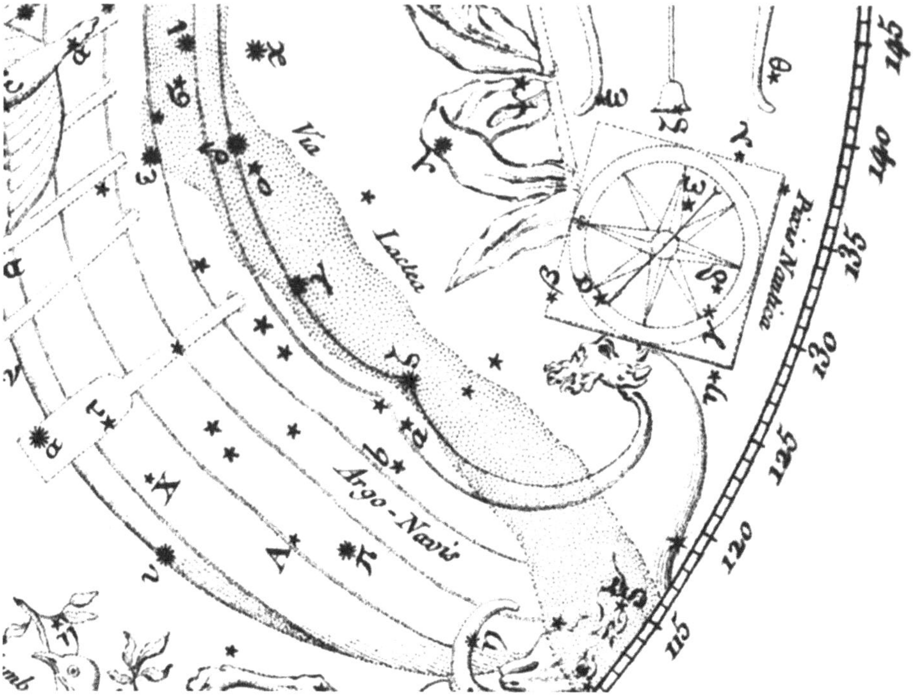

5.6 "Pixis Nautica" shown on the southern planisphere published in the posthumous second edition (1763) of Lacaille's *Coelum Australe Stelliferum*

Lochium Funis

Bode made a finer distinction regarding this figure in his *Uranographia* (1801b), dealing with fainter stars yet in the space between the upper end of Malus/Pyxis and nearby Felis (Chap. 9). For these stars, he introduced the figure of Lochium Funis, or the Log and Line (see Volume 2). A log and line is a nautical measurement device consisting of the "log", a heavy (yet buoyant) piece of flat wood, attached to the "line", a long piece of rope into which knots were tied at regular intervals. It was used to measure the speed of a ship at sea; after throwing the log overboard, sailors would count the number of knots laid out in the half-minute span timed by a purpose-built sand glass, allowing for a straightforward calculation of speed after multiplying the number of knots by the interval of length between them. Bode's designation was clever: the constellation is depicted on his chart such that the stars marking the line are the knots in the rope. As Ian Ridpath points out,

> Bode treated Pyxis and Lochium Funis (constellation) as a combined figure; on his atlas he enclosed them both within the same constellation boundary and listed

their stars together in his accompanying catalogue, *Allgemeine Beschreibung und Nachweisung der Gestirne*. As Bode put it, the log measured the speed of the ship while the compass gave the direction.

ICONOGRAPHY

The Origin of the *Argo*

There is one point on which essentially all authors agree: In Greek mythology *Argo* was the ship on which Jason and the Argonauts sailed to retrieve the Golden Fleece. However, there has long been disagreement about the proper origin of the ship's name, whether an adjective describing its speed, the eponym of its builder, a proper place name, or a demonym derived from that place. In *Cyclopædia, or an Universal Dictionary of Arts and Sciences* (1728), one of the first English-language general encyclopedias, Ephraim Chambers wrote

> The Criticks are divided about the Origin of the Name: Some will have it thus called from the Person who built it, *Argus* ; others, from the Greek Word *Argos*, swift, as being a light Sailer ; others, from the City *Argos*, where they suppose it built: Others, from the *Argives*, who went on board it, according to the Distich quotes from an antient Latin Poet by *Cicero*, in this first Tusculan ; *Argo, quia Argivi in ea Delecti Viri/Vecti, petebant pellem inauratam Arietis.*[25]

The most common rendition of the myth is that *Argo* was named for its builder, Argus (Ἄργος), who is said to have constructed the ship under the supervision of Athena.[26]

Much of the *Argo* mythology comes to us by way of the *Argonautica*, an epic poem by the third century BC poet Apollonius Rhodius. The only surviving epic poem of the Hellenistic era, it incorporates Rhodius' original research into geography, Homeric literature, and ethnography and served as Virgil's model in his composition of the Roman epic, the *Aeneid*. While scholarly, making use of the best source materials of the age, Apollonius' most enduring contribution to the epic tradition in the *Argonautica* is the exposition of the love between hero and heroine.[27] *Argonautica* exerted an important influence on later Latin poetry; in addition to Virgil, its traces are evident in the works of Ovid and Catullus.

---

[25] "From Argos she did chosen men convey, bound to fetch back the Golden Fleece, their prey." (*Tusculan Disputations* Book 1, Sect. XX, trans. C.D. Yonge, 1877.)

[26] "She herself too fashioned the swift ship; and with her Argus, son of Arestor, wrought it by her counsels. Wherefore it proved the most excellent of all ships that have made trial of the sea with oars." (Apollonius Rhodius, *Argonautica* I, 110, trans. R.C. Seaton. *Loeb Classical Library Volume*, London: William Heinemann Ltd., 1912.)

[27] See e.g., A.W. Bulloch, "Hellenistic poetry" in *The Cambridge History of Classical Literature Volume 1: Greek Literature* eds. P.E. Easterling and Bernard M.W. Knox, Cambridge University Press (1985).

According to various sources, the *Argo* was constructed with the help of Athena's wisdom, and contained in its timbers wood from a tree felled in the forest of Dodona[28] imbued with magical powers.[29,30] But what kind of ship was the *Argo*? While the origin of Argo Navis antedates surviving depictions of ancient Greek ships, some informed guesses can be made. Early Greek ships descended from models that had already plied the waters of the Mediterranean Sea for nearly two millennia. The *Argo* would have been some variety of galley, an oceangoing craft with a shallow draft, low profile, and long, narrow hull. They could be propelled by means of sails or oars, but it was human strength that primarily moved them through the waters of the ancient world. While transportation by galley was much more intensive on their crews than the case for other types of ships, it gave these vessels considerable freedom to move under conditions when sailing by winds was unfavorable.

The simplest of the early Greek galleys had only a single bank of oarsmen on each side of the ship, a model called a monoreme. In particular, the number of oars involved suggests a particular kind of monoreme. Crew lists compiled from various ancient sources[31] yield a total of 49 men—and one woman, Atalanta—for a total of 50 oarsmen. This number suggests a specific kind of galley known from the Archaic period on: the penteconter (πεντηκόντορος, 'fifty-oared'). An example of this type of ship is shown at the top of Fig. 5.7. Given reasonable assumptions as to how closely-packed such a ship's crew might be situated, penteconters should have had lengths on order of 30 m, widths of 4 or 5 m, and may have been capable of top speeds in excess of 9 knots (18 km/h).

There remains the uncomfortable fact that *Argo* is shown on maps without a bow, since no convenient set of stars happens to exist in that part of the sky to complete the pattern of a ship. Various cartographers dealt with this problem by showing the figure of the Ship disappearing behind rocks, vanishing into a cloud of mist, or covered up by a tree (see Chap. 22). In the *Phaenomena* (*c.* 250 BC), Aratos solved the problem by simply having *Argo* sail stern-first:

> Sternforward Argō by the Great Dog's tail
> Is drawn: for hers is not a usual course.
> But backward turned she comes, as vessels do
> When sailors have transposed the crooked stern

[28] An oracle at Epirus in northwestern Greece devoted to a Mother Goddess identified with Dione.

[29] "*Ovid* calls *Argo* a sacred Ship, *sacram conscendis in Argum* ; by reason, say some, that *Minerva* contrived the Plan, and even assisted int he building thereof : Or rather, on account of a piece of Timber in its Prow, which spoke, and render'd Oracles—Several Authors make mention of the Piece of Timber, which is said to have been hewn in the sacred Forest of *Dodona*." (Chambers, 1728)

[30] "[T]he oracles were delivered by the priests, who, by artfully concealing themselves behind the oaks, gave occasion to the superstitious multitude to believe that the trees were endowed with the power of prophecy. As the ship Argo was built with some of the oaks of the forest of Dodona, there some beams in the vessel which gave oracles to the Argonauts, and warned them against the approach of calamity." (John Lemprière, *A Classical Dictionary*, London: T. Cadell and W. Davies, 1820.)

[31] For example, Apollonius Rhodius, *Argonautica*, 1, 23–228; Pseudo-Apollodorus, *Bibliotheca* 1. 9. 16; Hyginus, *Fabulae*, 14.

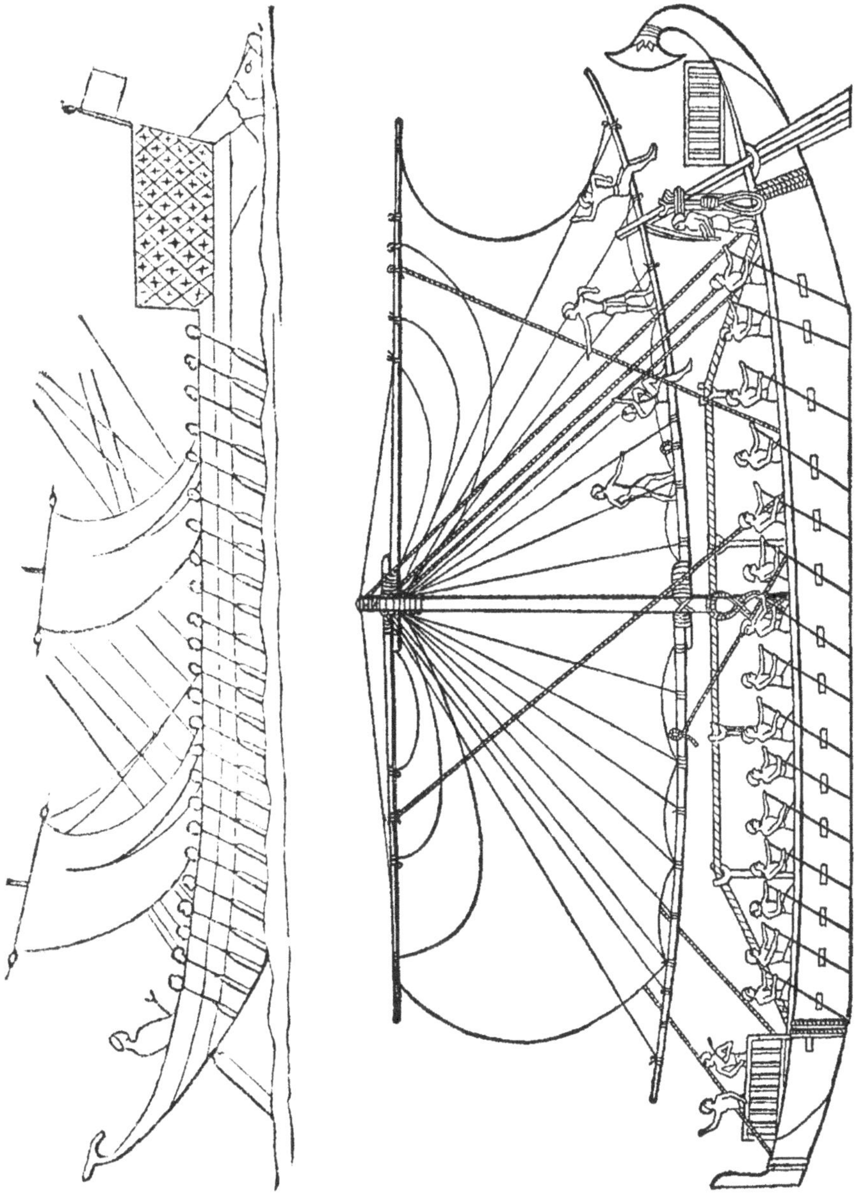

5.7 *Top*: Sketch of a Greek penteconter from Ernst Wallis' *Illustrerad verldshistoria* (Stockholm: Centraltryckeriets förlag, 1877). *Bottom*: "Egyptian ship on the Red Sea, about 1250 B. C." from Cecil Torr's *Ancient Ships* (Cambridge, UK: Cambridge University Press, 1895) reproduced in H.G. Wells, *The Outline of History* (Garden City, New York: Garden City Publishing Co., Inc., 1920)

On entering the harbor; all the ship reverse,
And gliding backward on the beach it grounds.
Sternforward thus is Jason's Argō drawn.
And part moves dim and starless from the prow
Up to the mast, but all the rest is bright.
The slackened rudder has been placed beneath
The hind-feet of the Dog, who goes in front.[32]

Jason, the Argonauts, and the Golden Fleece

The *Argo* features most prominently in a story whose origins are certainly in the preliterate era. Jason was the son of Aeson, king of Iolcus in Thessaly, and his wife, Alcimede. Theirs was a happy existence in prosperous times until the king's brother, Pelias, forcibly removed Aeson from the throne. In fear of their lives, Aeson and Alcimede fled, taking the infant Jason with them. Taking refuge outside of Iolcus bus fearing their discovery, they entrusted Jason to the wise old centaur Chiron, whom they asked to raise up as an able warrior who would retake his rightful place on the throne.

Jason's Early Life

Chiron raised Jason to be quick-witted, strong and skillful, but Chiron kept the secret of his high birth from him until he was old enough to understand the story of his family and the wrongs inflicted on his parents by the usurper Pelias. Jason became filled with rage, upon which he vowed to either defeat his uncle or die trying. Judging him ready to undertake the task and fulfill his destiny, Chiron sent Jason on his way with particular instructions: he was to recall that while his injury was the fault of Pelias alone, he retained a responsibility to use his powers equally to help humanity in any way he could. Jason strapped on his sandals and girded himself with his sword, setting out for Iolcus.

It was early springtime when Jason's journey began, and streams across the countryside were running high with snowmelt from mountain elevations. He came to a particular stream which was virtually impassible and was about to try fording it when he saw an old woman on the same bank, looking helplessly across the raging torrent she was unable to cross. Remembering Chiron's commandment, he offered to help the woman across the stream by carrying her on his back provided that she would give him her staff on which to lean for support. Gladly accepting the offer, they were soon underway but Jason struggled mightily against the current. While he made it across safely, depositing the woman on the opposite bank he discovered that during the crossing he had lost one of his sandals. As he

[32]Translated by Brown (1885).

stood on the bank ruing the loss, the woman was suddenly transformed into the figure of Hera; he bowed before her and asked her aid and protection, which she graciously promised before disappearing.

Jason continued on, arriving at Iolcus in time for a festival headed up by Pelias. He climbed a hill leading to the main temple, pushing his way through the crowd until he was confronted by the king himself. Pelias gave him brief consideration before continuing his sacrifice; at the conclusion of the ceremony, the king noticed Jason's lone bare foot at which he grew pale in horror. At that instant he remembered the warning he once received from an oracle to beware of a man who appeared before him wearing only one sandal. Jason, revealing his true identity, confronted Pelias and demanded the return of his rightful throne. While Pelias had no such intention, he hedged and invited Jason to discuss the matter at a banquet prepared for the festival. "Quickly the king saw him and pondered," wrote[33] Apollonius Rhodius, "and devised for him the toil of a troublous voyage, in order that on the sea or among strangers he might lose his home-return."

During the sumptuous meal, bards sang of great heroes of the past and recited the story of Phryxus and Helle, the son and daughter of Athamas and Nephele. To escape the cruelty of their stepmother Ino, they mounted a winged ram with a golden fleece sent to rescue them by Poseidon. The ram bore them over land and sea; Helle grew frightened at the sight of the waves and lost her hold on the fleece, falling to her demise over the eponymous Hellespont. Phryxus held on and arrived safely in the kingdom of Colchis on the far side of the Euxine Sea and in gratitude to the gods scarified the ram and hung the golden fleece on a tree. He positioned a dragon to guard the fleece, which the bards said remained draped over the tree to that very day.

Jason listened intently, growing more enthusiastic for details. His reaction did not go unnoticed by Pelias, who saw his opportunity to dispatch the threat posed by his nephew. Lamenting the fact that none of the young men of the current generation were brave enough to risk their lives in the glorious cause of retrieving the fleece, Pelias was cut off as Jason leapt up and vowed to return with it. Certain that Jason would lose his life in the attempt, Pelias gladly sent him on his way.

The Quest for the Fleece

The next morning, refreshed by sobriety and a good night's sleep, Jason immediately regretted making his vow but felt bound by his word and prepared to depart for Colchis. He knew he would require Hera's help to be successful, so he visited her shrine at Dodona and consulted her oracle who was locally known as the "Speaking Oak." The Oak assured him of Hera's protection and instructed him to cut off some of its own mighty limbs and carve from them a figurehead for the ship which Athena, at Hera's request, would arrange to have built from pine trees grown on the slopes of Mount Pelion. Jason finished the figurehead, finding it had the gift of speech to provide periodic words of wisdom to

[33]*Argonautica* 1, 16–17, trans. R.C. Seaton.

help guide his journey. Upon the completion of the *Argo*, Jason outfitted it with a crew, dubbed the Argonauts, consisting of as many heroes as he could find: Hercules, Castor, Pollux, Peleus, Admetus, Theseus, Orpheus, and many more. Hera bargained with Aeolus to ensure favorable winds to speed the *Argo* on to Colchis.

En route to the far east they made periodic landings to renew their stock of provisions, but nearly every delay brought misfortune; included among these episodes are stories of Hylas and the Nymphs and Phineas and the Harpies. Jason and his crew fought an attack by the Stymphalian birds, having been previously banished to an island in the after their defeat by Hercules during the Sixth Labour. The Argonauts encountered the Symplegades, a group of floating rocks that continually crashed together and destroyed all objects caught between them; Jason ordered his men to row at a certain speed after observing the safe passage of a dove through the rocks and knowing that the *Argo* sailed as fast as a dove on the wing. Having failed to destroy the *Argo*, their capacity for evil was drained from them and they were held fast to the bottom of the sea near the mouth of the Bosporus, from then on posing a threat no different than other ordinary rocks.

At last the Argonauts reached Colchis and presented themselves (and their demand) before King Aetes. Understandably loathe to part with his most prized possession, Aetes declared that before he would be allowed to claim the Golden Fleece, Jason would have to catch and subdue two fire-breathing bulls dedicated to Hephaestus and, after harnessing them, use the bulls to plow a stony field sacred to Ares. Alluding to the story of Cadmus, Jason was then to sow the field with dragon's teeth, which would raise up an army of giants. He had to slay all the giants and the guardian dragon of the Fleece itself, or the Fleece would never be his.

Jason, undaunted, decided to ask the figurehead for advice. On the way to the shore, he met the king's daughter, Medea, a beautiful young sorceress who was willing to use her magic to help him provided that he agreed to marry her. Jason was enchanted by her beauty and agreed to the request. He harnessed Hephaestus' bulls and plowed the requested field, but soon grew uneasy as the glittering spears and helmets of the giants appeared to push their way from the ground. When he saw their ranks in full armor he considered backing down but knew that fleeing the scene would ensure his downfall. He stood his ground and as the phalanx of giants approached, he threw a handful of sand in their faces. Temporarily blinded, the giants attacked each other, "And the Earthborn, like fleet-footed hounds, leaped upon one another and slew with loud yells; and on earth their mother they fell beneath their own spears, likes pines or oaks, which storms of wind beat down."[34] In short order the whole lot was exterminated.

The Flight from Colchis

Jason and Medea then made their way to the Golden Fleece and its watchful dragon. Medea prepared a potion using her magic that, when administered to the dragon, made it fall into

[34] *Argonautica* 3, 1340 et seq., trans. R.C. Seaton.

a profound sleep. Jason approached closely enough to sever the creature's head, rendering it harmless. He then pulled the Fleece down and bore it triumphantly back to the *Argo*. Awaiting his imminent return the Argonauts were already seated at their oars, ready to depart on a moment's notice. With Jason and Medea aboard, and in possession of the king's only son, Absyrtus, they shot at once out of the Colchian harbor.

As the following morning dawned, Aetes awoke to find his dragon dead, his daughter gone, the Fleece missing and the *Argo* conspicuously absent. He ordered a vessel readied at once and personally set off across the in pursuit of the Argonauts. While the Colchian crew were good, strong sailors, they didn't track down the *Argo* until both were near the mouth of the Danube River. Aetes begged Medea to return home with him, but she had no wish to be torn from Jason.

Instead of responding to her father's demands, she encouraged the Argonauts to row harder in flight. But the Colchian rowers were more powerful and quickly drew alongside the *Argo*. Sensing the end was near, she killed Absyrtus and cut his body into pieces, dropping one piece at a time overboard. Aetes was overcome with grief and busied himself collecting the fragments of Absyrtus' body from the sea; in the process, he lost sight of the *Argo* and all hope of recovering Medea. He returned in grief to Colchis where he buried his son.

In an episode echoing the *Odyssey;* the *Argo* was led off course. As punishment for killing Absyrtus, Zeus sent a series of storms that wracked the ship. As usual, Jason consulted the figurehead, which recommended he seek ritual purification with Circe, a nymph living on the island of Aeaea. After performing the task, the Argonauts continued their journey home. Before reaching Iolcus, they were further beset by the same Sirens that plagued Odysseus, but Orpheus drew his lyre and played so loudly that he drowned out their alluring song. When they reached Crete, they found it guarded by Talos, a man made of bronze who hurled huge stones at the ship to keep it away from the shore. But Talos had a fatal flaw: a single blood vessel spanning his neck to his ankle that was closed by one bronze nail. Medea used her magic to tranquilize Talos to the point where she could pull out the nail, causing the man to bleed to death. The *Argo* sailed on.

The Return to Thessaly

Meanwhile, Pelias reigned ably over Thessaly in Jason's absence, confident he would surely never return. But soon he was told that the *Argo* had arrived, bearing Jason and the Fleece. Before he could devise a plan to thwart the inevitable, Jason appeared before him and compelled him to relinquish the throne to the rightful king, Aeson. However, by that time Aeson was so old and infirm that he had no interest in taking the reins of power; Jason begged Medea to use her magic to restore his youth and vitality. He was quickly returned to his former strength and grace.

As soon as Pelias' daughters learned of this magic, they begged Medea to divulge her formula so that they might similarly rejuvenate their father. Sensing an opportunity to dispatch Jason's rival once and for all, she instructed them to cut Pelias' body into pieces and boil them with certain herbs, promising it would restore their father accordingly. But being all too credulous, the daughters ultimately killed the father they all dearly loved. Acastus, Pelias' son, cast Jason and Medea out of the kingdom for the murder.

Jason and Medea resettled and lived a happy life in Corinth, but after some time his fire for her faded and he became enamored of Creusa, daughter of the king of Corinth, and became engaged to marry her. When Medea discovered the engagement she confronted Jason, who blamed Aphrodite for causing him to fall in love with Creusa. Furious that Jason broke his vow to be forever hers, Medea plotted her revenge; she gifted to Creusa a magic robe for her wedding, which burned her body as soon as she put it on. Both Creusa and her father, Creon, died as he tried to save her by removing the robe. Unsatisfied with the outcome, Medea killed the two children she had with Jason out of fear they would be punished for her actions. By the time Jason discovered their deaths, Medea was already gone, borne away to Athens in a chariot led by dragons sent by her grandfather, Helios.

Various versions of the story diverge here. In some, Jason joins Peleus, father of Achilles, to defeat Acastus and reclaim the throne of Iolcus; Jason is then later succeeded as king by his son, Thessalus. In another, darker rendition, Jason becomes a victim of his own despair and remorse for the loss of Medea, Creusa and his children. Each day he wanders down to the shore to sit in the shade of the *Argo*'s rotting hulk. One day, while sitting there and musing over his youthful adventures, he falls asleep and is killed instantly when a beam from the *Argo*'s stern detaches and falls on him. Yet another telling ends up both with a happy ending and an explanation for how *Argo* became a constellation. At the end of the adventure, Ephraim Chambers' *Cyclopædia,* relates, "*Jason* having happily accomplished his Enterprize, consecrated the Ship *Argo* to *Neptune*, in the *Isthmus* of *Corinth* ; where it did not remain long before it was translated into Heaven, and made a Constellation."

In *Myths of Greece and Rome*,[35] H.A. Guerber framed the interpretive context of the myth of Jason and the Argonauts:

> The Argonautic expedition is emblematic of the first long maritime voyage undertaken by the Greeks for commercial purposes ; while the golden fleece which Jason brought back from Colchis is but a symbol of the untold riches they found in the East, and brought back to their own native land.

DISAPPEARANCE

Argo Navis is the only of Ptolemy's original 48 constellations that did not survive into the twentieth century in its originally form. Its fate is unique among the constellations included here: it went extinct only in name, while most of its constituent pieces survived unscathed and are considered canonical by the International Astronomical Union. Its breakup was proposed by Nicolas Louis de Lacaille in the mid-eighteenth century after his expedition

[35]New York: American Book Company (1893), p. 274.

to the southern hemisphere. He wrote the following in *Mémoires of the Académie Royal des Sciences* (1752; published 1756), quoted here in the translation by Evans (1992):

> The Constellation of Argo was composed of more than 160 easily visible stars, and I first gave Greek letters to its brightest stars. In the end I divided it into three parts – the poop [Puppis], the body [?Carina?] and the sails [Vela]. The poop is separated from the body of the vessel by the rudder, and I have called the sails everything outside the vessel between the edges and the horizontal mast, or the spar on which the sail is reefed.

Almost a century later, John Herschel (1843) proposed that Lacaille's plan be formalized by the community of astronomers, suggesting

> 5°. That *Argo* be divided into four separate constellations, as partly contemplated by Lacaille; retaining his designations of *Carina*, *Puppis*, and *Vela*; and substituting the term *Malus* for *Pixis Nautica*, since it contains four of Ptolemy's stars that are placed by him in the *mast* of the ship.
>
> 6°. That the original constellation *Argo*, on account of its great magnitude and the subdivisions here proposed, be carefully revised in respect of lettering, in the following manner:– First, In order to preserve the present nomenclature of the principal stars, all the stars in Argo (that is, in the general constellation, regarded as including the subdivisions above-mentioned) indicated by Greek letters, by Lacaille, to be retained, with their present lettering, under the general name Argo. Secondly, All the remaining stars, to be designated by that portion of the ship in which they occur, such as *Carina*, *Puppis*, *Vela*, and *Malus*, and to be indicated by the Roman letters adopted by Lacaille, as far as the fifth magnitude inclusive. And no two stars, far distance from each other in the same subdivision, to be indicated by the same letter; but, in cases of conflict, the greater magnitude is to be preferred; and, when they are equal, the preceding star to be fixed upon.

After Herschel's proposal, many authors stubbornly retained the Ship in its entirety, perhaps out of deference to Ptolemy and tradition. Ezra Otis Kendall listed it intact among a list of constellations described in *Uranography* (1845) as "no other than the Ark of Noah;" however, he also describes Pyxis as though it were a separate, independent constellation rather than an asterism within Argo. Richard Anthony Proctor, in *The Constellation-Seasons* (1876), kept Argo together on his maps within a single set of boundary lines; it is the only now-lost constellation he depicts, and in fact his maps look very modern in terms of names. Two years later Argo was shown intact on charts in Simon Newcomb's classic text *Popular Astronomy*. This is reinforced by his reference to Eta Carinae as "η Argus". Other authors who kept Argo together toward the end of the nineteenth century include Rosser (1879), Sharpless and Philips (1882), Colas (1892), and Flammarion and Gore (1894).

Harbord (1883) described Argo in his *Glossary of Navigation* as "a very extensive constellation of the southern hemisphere, of which the several parts are named Carina, 'the Keel;' Puppis, 'the Poop;' Malus, 'the Mast;' and Vela, 'the Sails'," and labeled

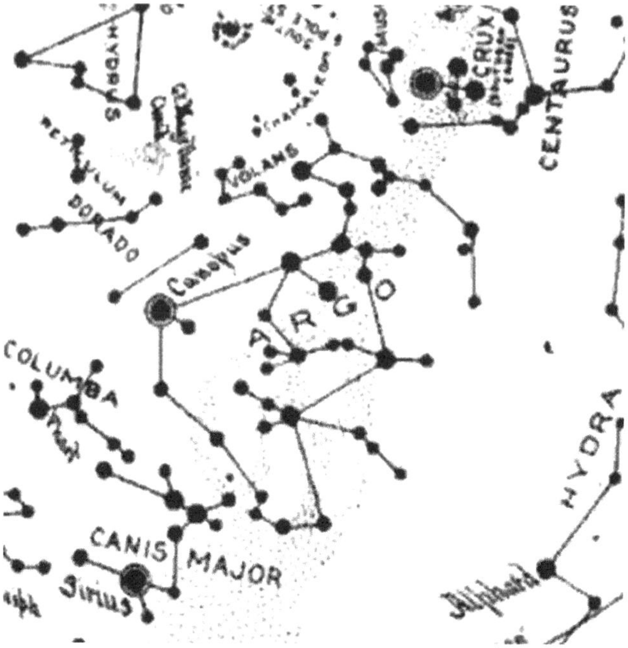

5.8 One of the last popular depictions of Argo Navis in *The Monthly Evening Sky Map* (1916)

the bright star Canopus, now α Carinae, "α Argûs". The second edition of Joel Dorman Steele's *Popular Astronomy* (1899) has "Ship Argo" intact on is charts, and describes the still-brilliant η Carinae: "Here is the magnificent constellation Argo, in which we find Canopus, looked upon anciently as next to Sirius in brilliancy: η, a variable star, now surpasses it in brightness." The same year, Allen discussed Argo as a unit in the context of describing its past history, although he mentioned Lacaille's efforts to separate the Ship into its constituent components.

Argo stubbornly persisted on charts and in popular astronomy works into the first two decades of the twentieth century, proving the staying power behind its tradition. William Tyler Olcott considered it a "minor constellation" in *Star Lore of All Ages* (1911), "and owes its place under such a heading to the fact that in these latitudes but a very small part of it is visible, so that only a brief reference to it is necessary… The Ship is figured without a prow, one of the best evidence that chance had no part in the invention of the constellation." Two years later, George F. Chambers wrote in *Astronomy* that Argo Navis

> is not an easy constellation to describe because of its great extent, and, by way of facilitating work in its confines, it has been by common consent cut up into 4

> divisions, respectively called *Carina* (The Keel), *Malus* (The Mast), *Puppis* (The Poop), and *Vela* (The Sails), to which some add a 5th, *Pyxis Nautica*, a part of Malus.

One of its last appearances on popular charts was in Leon Barritt's periodical *The Monthly Evening Sky Map* (1916; Fig. 5.8), where it is shown both labeled "Argo", "Argo Navis", and by the names of components Puppis and Vela.

In the end, Argo's ship sailed by the time Eugéne Delporte began deciding on constellation boundaries for the IAU in the late 1920s. Upon the publication of *Atlas Céleste* in 1930, Argo Navis was discarded for good with the recognition of its components Carina, Puppis, Pyxis, and Vela (Fig. 5.9).

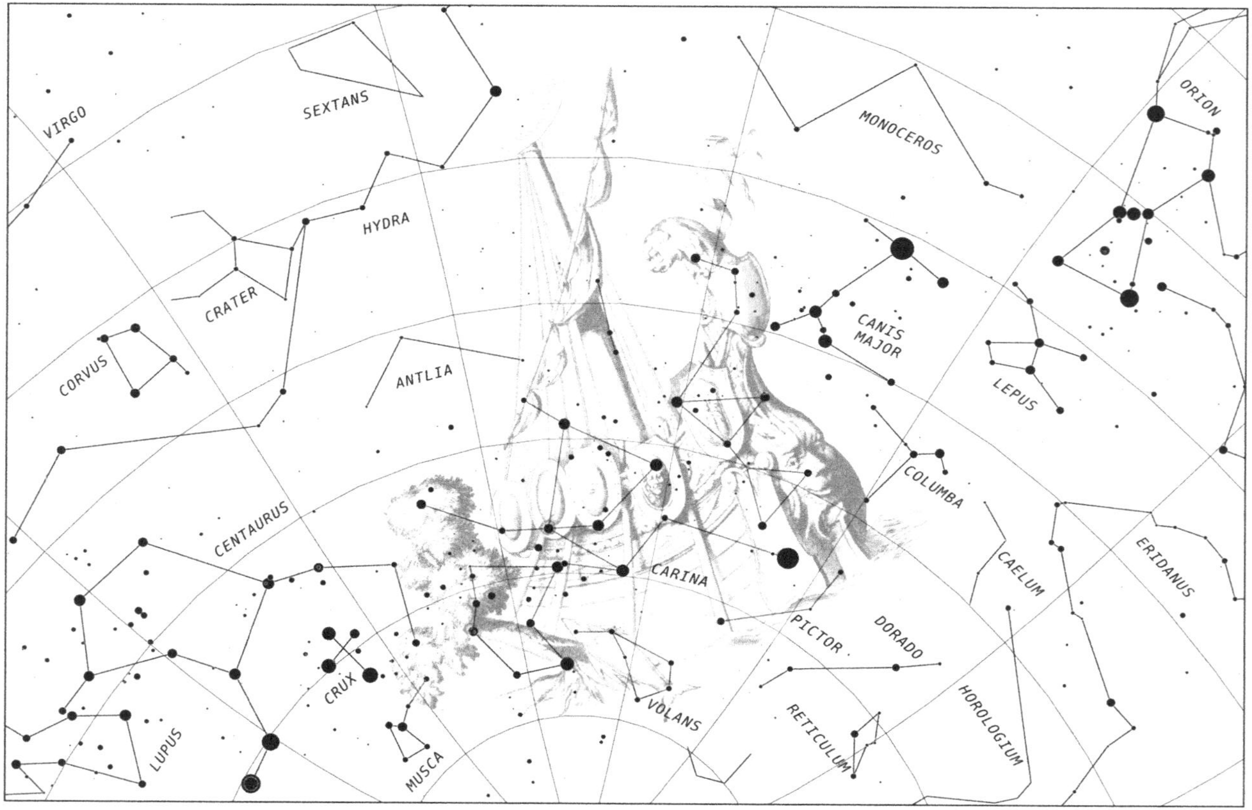

5.9 The figures of Argo Navis (*center*), Robur Carolinum (below and left of Argo; Chap. 22), and Piscis Volans (below Argo) from Figure Eee of Johannes Hevelius' *Prodromus Astronomiae* (1690) overlaid on a modern chart. Constellation boundaries and Bayer designations of stars are omitted for clarity

6

Cancer Minor

The Lesser Crab

Genitive: Cancri Minoris
Abbreviation: CnM
Location: An arrow-shaped figure in the gap between Cancer and Gemini, consisting of the 5th magnitude stars HIP 36616, and 68, 74, 81 and 85 Geminorum

ORIGIN AND HISTORY

The constellation was introduced by Petrus Plancius on maps published in 1612–1614 to fill a perceived empty space between Cancer and Gemini (Fig. 6.1). Some authors attribute its formation to the Polish nobleman and astronomer Stanislaus Lubieniecki (1623–1675); Richard Hinckley Allen (1899) wrote: "[Jacob] Bartschius and Lubienitzki, in the 17th century, made it into a Lobster, and the latter added toward Gemini a small shrimp-like object which he called Cancer minor." However, Bartsch did not mention the constellation in 1624s *Usus Astronomicus*, nor did he show it in the charts published in *Planisphaerium stellatum* (1661; Fig. 6.2). Similarly, Lubieniecki did not show it in his magnum opus *Theatrum cometicum* (1681), although he included its stars (Fig. 6.3).

Here is a case where the few faint stars in this area could have been easily subsumed into either constellation, but Plancius evidently believed referring to the classical Cancer would lend the weight of history to his invention. Most prominently, it appears in Andreas Cellarius' *Harmonia Macrocosmica* (1661; Fig. 6.1); were it not for this instance, the constellation would be considered single-sourced. The influence of Cellarius' work meant that many people saw the suggested figure, but it appears that no other mapmakers adopted it after the mid-seventeenth century. Many maps continued to show its stars, but not Plancius' figure.

J.C. Barentine, *The Lost Constellations*, Springer Praxis Books,
DOI 10.1007/978-3-319-22795-5_6

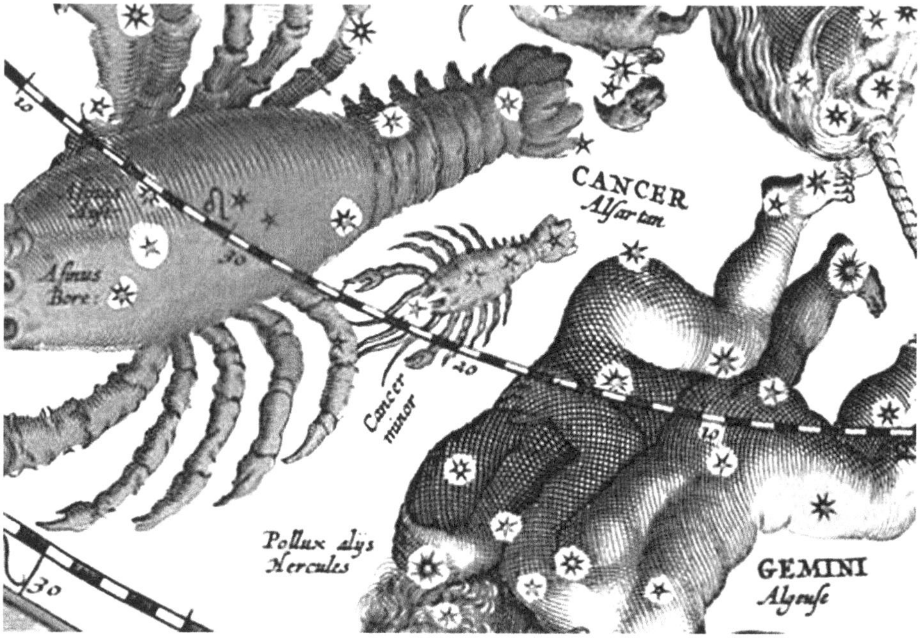

6.1 Cancer, Cancer Minor, and Gemini as shown on Plate 27 of Andreas Cellarius' *Harmonia Macrocosmica* (1661)

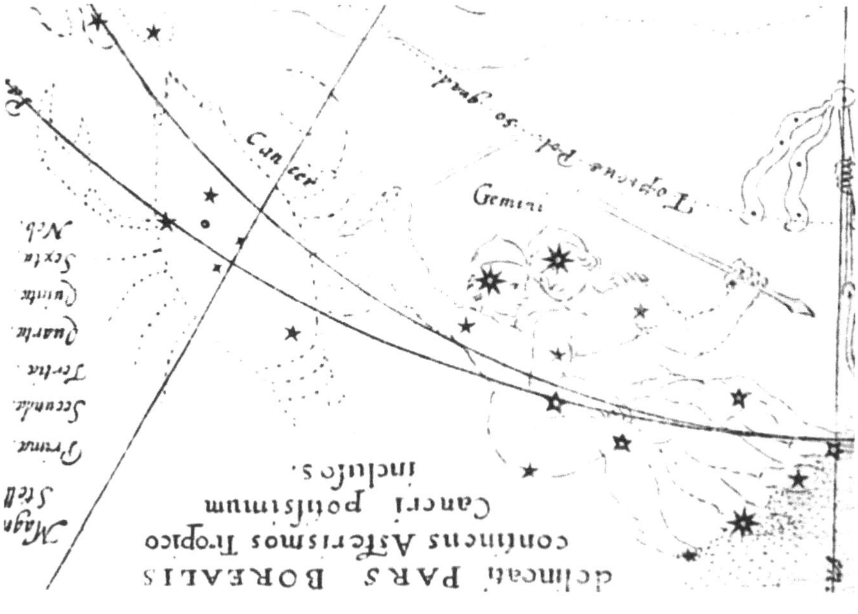

6.2 Cancer and Gemini shown on the northern celestial hemisphere in Jacob Bartsch's *Planisphaerium stellatum* (1661)

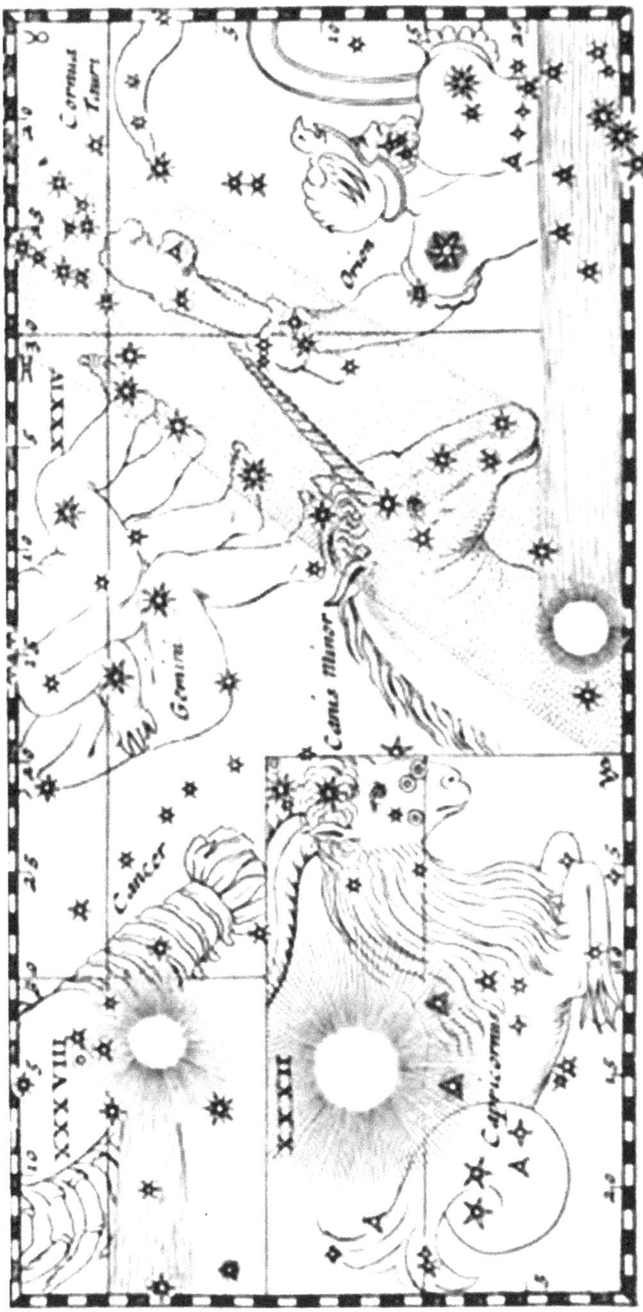

6.3 Figure 63 from Volume 2 of Stanislaus Lubieniecki's *Theatrum cometicum* (1681), showing the locations of some bright historical comets in antiquity. The stars comprising Cancer Minor are shown at *upper left* near the label "Cancer"

ICONOGRAPHY

Major/Minor Constellation Pairs

Cancer Minor follows something of a tradition from antiquity that named pairs of animals with large/small components (Canis Major/Canis Minor and Ursa Major/Ursa Minor). In each pair, both animals are proximate to each other in the sky and share either a common origin story, mythology, or both.

Of these, the oldest is certainly the Bears. Ursa Major's origins probably trace back to the shadowy world of the Proto-Indo-Europeans; Mallory and Adams (2006) call Ursa Major "The most solidly 'reconstructed' Indo-European constellation." Berezkin (2005) wrote of "the mythological motif of the Cosmic Hunt ... [that] forms the core of the tales typical for northern and central Eurasia and for the Americas but is rarely, if at all, known on other continents." Ursa Major figures prominently in one variant of the Hunt. Schaefer (2006) argued that Ursa Major was known to the first hunter nomads who reached the Americas during the Pleistocene. Ursa Minor, on the other hand, is likely a comparatively recent invention of the Greeks; according to Allen, it

> was not mentioned by Homer or Hesiod, for, according to Strabo,[1] it was not admitted among the constellations of the Greeks until about 600 B.C., when Thales, inspired by its use in Phoenicia, his probable birthplace, suggested it to the Greek mariners in place of its greater neighbor, which till then had been their sailing guide

While it is not historically established, the "dipper" shape formed by the brightest stars of both Bears seems to suggest why they were grouped together as a pair, and their endless circling of the pole provided the context for their Greek mythology. However, the analogy breaks down for Canis Major and Minor, the latter of which as usually drawn looks nothing like an animal at all. Allen contended that "It was not known to the Greeks by any comparative title, but was always προκύον, as rising before his companion Dog, which Latin classic writers transliterated Procyon." Its *lucida* aside, he wrote, its ultimate origin as a constellation is lost: "Who traced out the original outlines of Canis Minor, and what these outlines were, is uncertain, for the constellation with Ptolemy contained but two recorded stars, and no ἀμόρφωτοι."[2] However constructed, its associated mythology (see Allen and Ridpath 1989) makes clear that its stars were first identified as something having to do with a dog, to which existing stories were fit in an inversion of the usual means of founding classical constellations. Perhaps so inspired, when Johannes Hevelius formed Canes Venatici in *Firmamentum Sobiescianum* (1687) he followed the convention of Canis Minor and manage to make the form of two dogs out of a short, stubby structure of two bright stars.

---

[1] "[I]t is likely that in the time of Homer the other Bear had not yet been marked out as a constellation, and that the star-group did not become known as such to the Greeks until the Phoenicians so designated it and used it for purposes of navigation." *Geography*, I.1.6 trans. H.L. Jones, published in Vol. 1 of the Loeb Classical Library edition (1917).

[2] Ptolemy's "unformed stars" not belonging to any particular constellation.

Later cartographers introduced figures representing either member—or both members—of major/minor pairs. In the only such instance in which a pairing was created after Ptolemy's time that later became canonical, Hevelius added a small figure north of Leo called Leo Minor in *Firmamentum Sobiescianum* to complement the classical Lion. "It was formed by Hevelius from eighteen stars between the greater Lion and Bear," Allen wrote, "in a long triangle with a fainter line to the south, and thus named because he said it was 'of the same nature' as these adjoining constellations." The 'nature' so referred had to do strictly with an imitation of the figure of classical Leo, which at least gives some insight as to why Hevelius suggested a constellation with essentially no backstory. In any case, the name stuck, and the figure remains canonical to this day. In 1789, Maximilian Hell, director of the Vienna Observatory, introduced a major/minor pair of telescopes as constellations (Chap. 26) honoring his contemporary William Herschel. The "minor" telescope went extinct before the turn of the nineteenth century, but its "major" counterpart persisted on charts until nearly the time that the modern constellations were defined in the 1920s.

Cancer

Cancer Minor's paired animal is the zodiacal constellation Cancer, a catasterism that predates the Greeks even though it played a small role in the mythology of the Herculean Labours (see Chap. 7). After Hercules successfully slew the Nemean lion, Eurystheus sent him to dispatch the Hydra, a multi-headed water monster raised by Hera for express purpose of killing Hercules. Hercules traveled to Lake Lerna and covered his mouth and nose to protect himself from the poisonous vapors of the swamp where the Hydra lived. To drive the beast out of its lair, Hercules fired a series of flaming arrows into the deep cave associated with the spring of Amymone. As Hercules attacked and removed each of the Hydra's heads, he found that two grew back in its place; its weakness was that it remained invulnerable provided it still had at least one head. Pseudo-Apollodorus (*c.* first to second century AD) picks up the story[3] in the *Bibliotheca*:

> … *[Hercules]* seized and held it fast. But the Hydra wound itself about one of his feet and clung to him. Nor could he effect anything by smashing its heads with his club, for as fast as one head was smashed there grew up two. A huge crab also came to the help of the Hydra by biting his foot. So he killed it, and in his turn called for help on Iolaus who, by setting fire to a piece of the neighboring wood and burning the roots of the heads with the brands, prevented them from sprouting. Having thus got the better of the sprouting heads, he chopped off the immortal head, and buried it, and put a heavy rock on it, beside the road that leads through Lerna to Elaeus. But the body of the Hydra he slit up and dipped his arrows,in the gall. However,

[3] *Bibliotheca* 2.5.2, trans. J.G. Frazer.

> Eurystheus said that this labour should not be reckoned among the ten because he had not got the better of the Hydra by himself, but with the help of Iolaus.

Hera was furious at the outcome of the struggle, and immortalized both the Hydra and the crab by placing them into the firmament. Thus while the crab itself was only a minor player in the story, it is one of four constellations associated with the Herculean Labours, the others being the adjacent (Nemean) Lion, Leo, Hydra and the extinct Cerberus et Ramus Pomifer (Chap. 7).

Long before this, the Akkadians referred to its stars as the home of the "Sun of the South," probably because in their time the northernmost point of the ecliptic was found among its stars whereas precession has since carried it westward into Taurus. "According to Chaldaean and Platonist philosophy," Allen wrote, "it was the supposed Gate of Men through which souls descended from heaven into human bodies." The Babylonians called it MUL.AL.LUL, the Tortoise, a figure that was probably in place by around 4000 BC. Egyptian records from around 2000 BC describe it as "Scarabaeus," the sacred scarab beetle, viewed as an emblem of immortality (Fig. 6.4).

An open cluster known as the Praesepe (the "Beehive Cluster," Messier 44) figures prominently at the center of the constellation. Aratus calls the cluster Achlus ("little mist") in *Phainomena*. Hipparchus mentioned the cluster as a *nephelion* ("little cloud") in his second century BC star catalog, while Ptolemy referred to it as the "nebulous mass in the breast" of Cancer in the *Almagest*. The Greeks and Romans saw the cluster as a manger where two donkeys (the adjacent stars γ Cancri, Asellus Borealis, and δ Cancri, Asellus Australis) are eating; the donkeys are interpreted as those that Dionysos and Silenus rode as they battled the Titans.

Allen wrote that Cancer "is the most inconspicuous figure in the zodiac... Showing but few stars, and its *lucida* being less than a 4th-magnitude, it was the Dark Sign, quaintly described as black and without eyes." Dante referred[4] to this tradition in *The Divine Comedy*, in which he wrote

> Then a light among them brightened,
> So that, if Cancer one such crystal had,
> Winter would have a month of one sole day.

DISAPPEARANCE

Cancer Minor enjoyed a brief heyday in the mid-seventeenth century, but seems to have gone extinct shortly thereafter. It does not appear on some of the important charts of the early eighteenth century, such as Carel Allard's map *Hemisphaerium meridionale et septentrionale planisphaerii coelestis* (1706) or Johannes de Broen's *Hemelskaart voor de noordelijke* ... (1709). Johann Gabriel Doppelmayr showed its constituent stars in *Atlas Coelestis* (1742), carefully drawn between the figural lines of Cancer and Gemini, but

[4] *Paradiso*, Canto XXV, lines 100–102; trans. H.W. Longfellow.

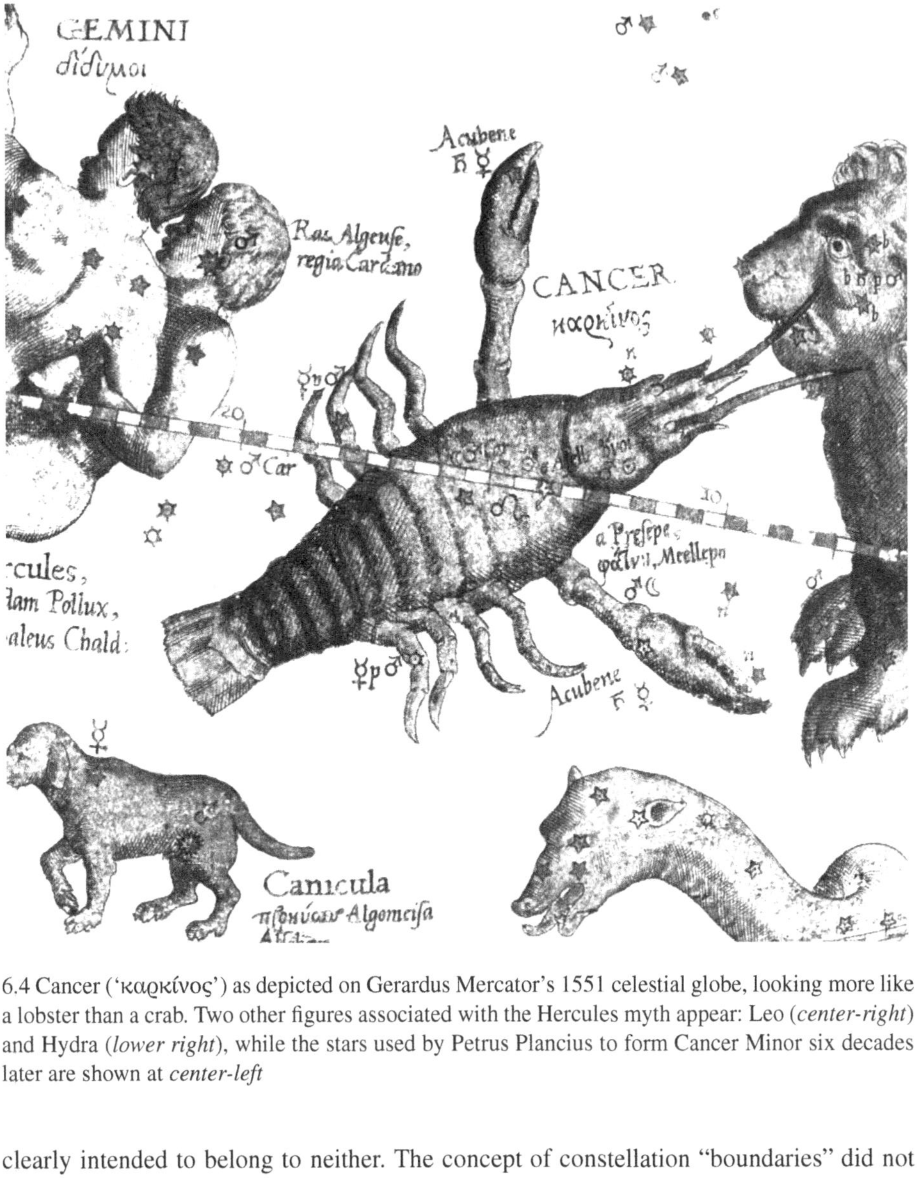

6.4 Cancer ('καρκίνος') as depicted on Gerardus Mercator's 1551 celestial globe, looking more like a lobster than a crab. Two other figures associated with the Hercules myth appear: Leo (*center-right*) and Hydra (*lower right*), while the stars used by Petrus Plancius to form Cancer Minor six decades later are shown at *center-left*

clearly intended to belong to neither. The concept of constellation "boundaries" did not yet practically exist, so Doppelmayr simply treated such faint stars almost like a decorative background onto which the proper figural constellations were pasted. Boundary lines came into being around the time of Bode's *Uranographia* (1801b); in Fig. 6.5, Bode intentionally drew the boundary of Cancer to include the stars that once were marked as Cancer Minor. It is evident that by the start of the nineteenth century, Cancer Minor was merely an oddity on old charts drawn a hundred years before, and disregarded entirely when the modern constellation boundaries were established (Fig. 6.6).

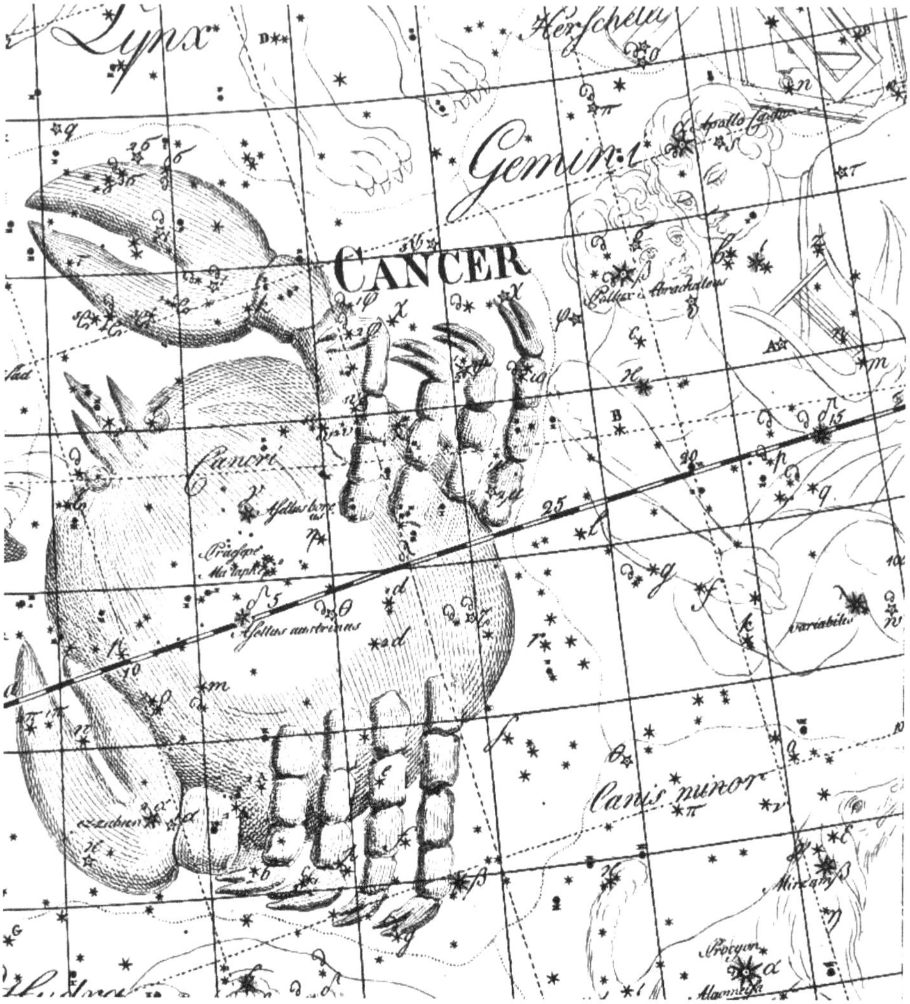

6.5 Cancer as shown on Plate 13 of Johann Elert Bode's *Uranographia* (1801b). The stars formerly attributed to Cancer Minor are shown just below and slightly right of center, under the number "25" marked on the *ecliptic line*. Bode carefully drew the suggested boundary of Cancer to include these stars

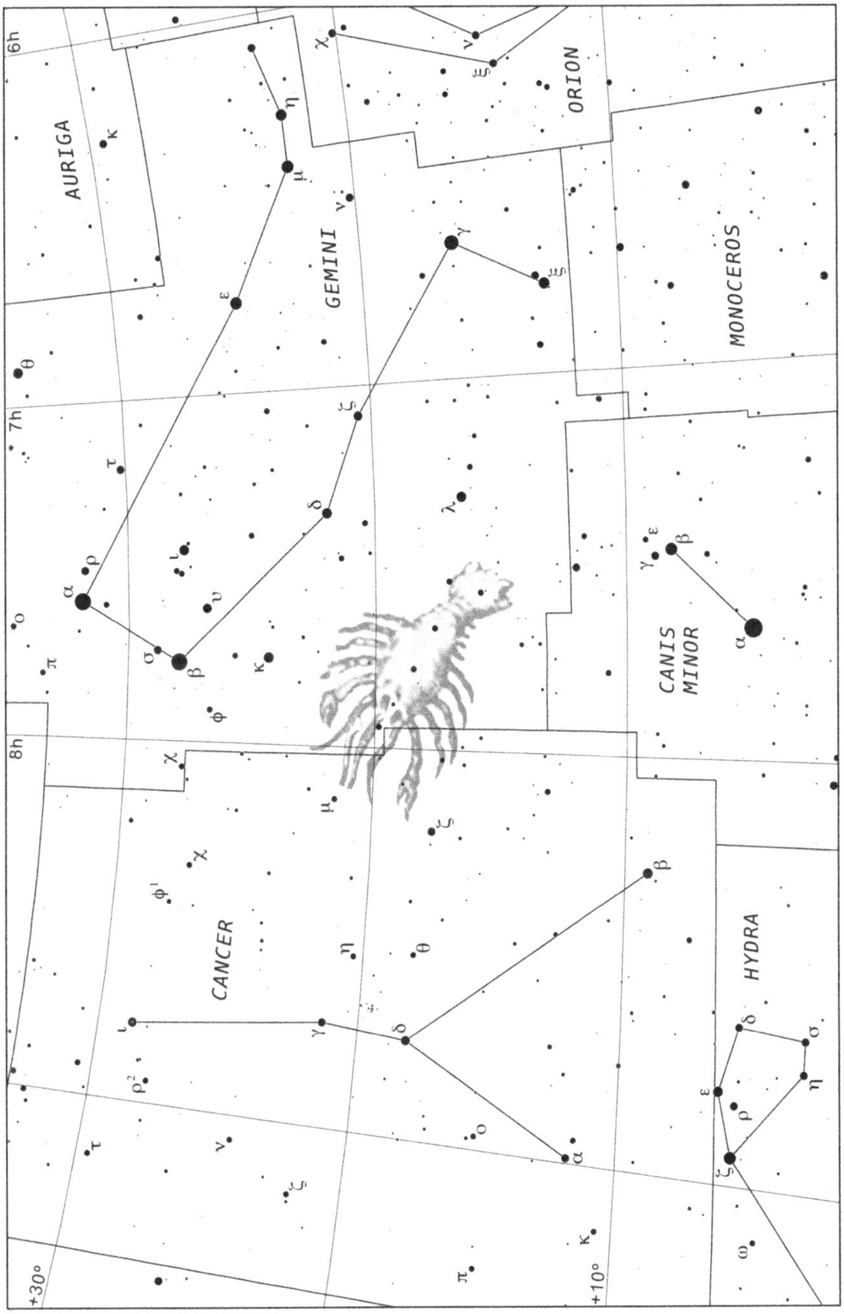

6.6 The figure of Cancer Minor from Plate 27 of Andreas Cellarius' *Harmonia Macrocosmica* (1661) overlaid on a modern chart

7

Cerberus et Ramus Pomifer

The Guardian of the Underworld and the Apple-Bearing Branch

Genitive: Cerberi et Rami Pomifer
Abbreviation: CeR
Alternate names: "Cerberus et Ramus Pomosus" (Jamieson, 1822)
Location: Unformed stars in the space between Cygnus and Hercules.[1]

ORIGIN AND HISTORY

Among the constellations of the Ptolemaic canon, certain figures clearly predate the Greeks and were received by them from some significantly older source thought to be the civilization of ancient Mesopotamia. Some Mesopotamian constellations received by the Greeks were simply repurposed to fit Greek legends, such as the Perseus-Pegasus-Andromeda-Cepheus-Cassiopeia-Cetus group, while others like Orion were interpreted generically without becoming attached to any particular mythological figures. Hercules is among the former, "one of the oldest sky figures, although not known to the first Greek astronomers under that name," according to Richard Hinckley Allen,

> for Eudoxos had 'Ενγούνασι; Hipparchos, 'Ενγόνασι, i.e. ὁ ἐν γόνασι καθήμενος, Bending on his Knees; and Ptolemy, ἐν γόνασιν. Aratos added to these designations 'Οκλάζων, the Kneeling One, and "Ειδωλον, the Phantom, while his description in the *Phainomena* well showed the ideas of that early time as to its character:

[1] *"East of Hercules"* (Bouvier, 1858); *"The 4th- to 5th-magnitude stars that Hevelius assigned to it are Flamsteed's 93, 95, 96, and 109, lying half-way between the head of Hercules and the head of the Swan"* (Allen, 1899); *"Flamsteed 93, 95, 96, and 109, lying halfway between the head of Hercules and the head of Cygnus"* (Bakich, 1995).

J.C. Barentine, *The Lost Constellations*, Springer Praxis Books,
DOI 10.1007/978-3-319-22795-5_7

> . . . like a toiling man, revolves
> A form. Of it can no one clearly speak, Nor to what toil he is attached; but, simply,
> *Kneeler* they call him. Laboring on his knees,
> Like one who sinks he seems; . . .
> . . . And his right foot
> Is planted on the twisting *Serpent's* head.
>
> But all tradition even as to
>
> Whoe'er this stranger of the heavenly forms may be,
>
> seems to have been lost to the Greeks, for none of them, save Eratosthenes, attempted to explain its origin, which in early classical days remained involved in mystery.

White (2008) suggests that the constellation Hercules is a version of a Babylonian constellation known as MUL.DINGIR.GUB.BA.MESH ("Standing Gods"). According to his interpretation, Standing Gods was represented as a chimeric figure consisting of a man's torso, arms, and head (the Greek constellation Hercules) with a serpent's lower body (the Greek constellation Draco). White claims that the 'Kneeler' label is the result of a conflation of that and a similar "Sitting Gods" figure by the Greeks. Allen made a similar inference with respect to the association between Hercules and Draco, but drew a connection between the Kneeler and the quasi-historical figure of Gilgamesh:

> Some modern students of Euphratean mythology, associating the stars of Hercules and Draco with the sun-god Izhdubar and the dragon Tiāmat, slain by him, think this Chaldaean myth the foundation of that of the classical Hercules and the Lernaean Hydra. . . . Izhdubar was identified with Nimrud, and known, too, as Gizdhubar, Gilgamesh, or Gi-il-ga-mes, the Γίλγαμος of Aelian. He was aided in his exploits by his servant-companion, the first Centaur, Ea-bani, or Hea-bani, the Creation of Ea.

To explain the kneeling posture of their Hercules constellation, the Greeks appealed to the Hercules narrative following the Tenth Labour. As recounted by Dionysus of Halicarnassus[2] among others, en route home to Mycenae from Iberia, Hercules passed through Liguria in northern Italy. There he fought a battle with the giants Albion and Bergion (alternately Dercynus), both sons of Poseidon. Since both giants were exceptionally strong, Hercules prayed to his father Zeus for help, assuming a kneeling position. Zeus intervened and, with his help, Hercules won the battle. The Greeks adopted the Kneeler as Hercules in his moment of prayer. The Hercules myth is more fully discussed below.

Later representations of Hercules as a constellation would gradually add various accoutrements to the figure referring to specific elements of his mythology as a way of identifying him on globes and charts. In the few graphical sources surviving from antiquity,

[2]Dionysius (c. 60 BC—aft. 7 BC) was a Greek historian and rhetorician.

such as the celestial globe of the Farnese Atlas, Hercules is depicted as a naked man, in a kneeling posture, holding nothing in his hands. The faint stars to the east between Hercules and Lyra were left unformed by Ptolemy at the time of the *Almagest*. When celestial atlases were next commonly drawn in the Renaissance, familiar items associated with the stories of Hercules were added to figural representations. Two early examples from the sixteenth century are shown in Fig. 7.1. In both panels of the figure, Hercules is drawn with two elements of the First Labour, in which he killed the Nemean Lion. The top panel shows Petrus Apianus' depiction of Hercules from his map *Imagines Syderum Coelestium* (1536) in which the hero wields a club in his raised right hand as he grasps at the lion with his left. Two decades earlier, on the map *Imagines coeli Septentrionales*... (1515), Albrecht Dürer showed the same scene but evidently after the completion of the First Labour, for Hercules has the lion's skin draped over his left forearm as he reaches with an open hand toward Lyra. It appears that Apianus was the first cartographer to assign an image to the stars in the otherwise empty space immediately east of Hercules comprising his extended left arm. However, many mapmakers of the sixteenth and seventeenth centuries, such as Bartsch and Cellarius, simply showed Hercules empty-handed. In rare instances, such as Petrus Plancius' 1594 world map (with star chart insets), Hercules is left off altogether.

In Bayer's depiction of Hercules on Plate G of his landmark atlas *Uranometria* (1603), he put the club in Hercules' raised left hand and showed the lion's skin worn as an article of clothing as shown in Fig. 7.2. However, on this plate Bayer introduced a novel Herculean symbol: in the hero's right hand, he placed the branch of an apple tree, referencing the Golden Apples of the Hesperides from the Eleventh Labour. While unlabeled on Bayer's atlas, later cartographers would also show the same greenery, which came to be known as Ramus Pomifer (the "apple-bearing branch"). Allen wrote:

> Bayer shows the strong man kneeling, clothed in the lion's skin, with his "all brazen" club and the Apple Branch. This last he called Ramus pomifer, the German *Zweig*, placing it in the right hand of Hercules, on the edge of the Milky Way; but this even then was an old idea, for the Venetian illustrator of Hyginus in 1488 showed, in the constellation figure, an Apple Tree with a serpent twisted around its trunk. Argelander followed Bayer's drawing, but Heis transfers the Branch to the left hand, with two vipers as a reminder of the now almost forgotten stellar Cerberus with serpents' tongues, which Bayer did not know.

Given the influence of Bayer on other mapmakers of the seventeenth century, Ramus Pomifer became the standard way of depicting these stars with a figure, although it seems to have been thought of only as an asterism within Hercules and not a constellation unto itself. In *Usus Astronomicus* (1624), Jacob Bartsch described[3] Hercules holding the Branch (Fig. 7.3):

> The figure is laboring at, threatening, or attempting a blow, leaning upon one knee while one foot tramples the head of the Dragon [Draco], and raising the club with his right hand, he holds with his left hand the branch bearing apples or quinces.

[3]"Imago est laboranti, minanti, ictum ve tentanti similis, dum innitens alteri genu, pede altero calcat caput Draconis, manuque dextra clavam sinistrâ, ramos pomorum aut cotoneorum tenet."

7.1 Before and after: Two depictions of Hercules involving the Nemean lion of his First Labour. *Top*: Petrus Apianus showed Hercules about to club the lion on *Imagines Syderum Coelestium* (1536). *Bottom*: Evidently having completed the job, Hercules is shown on Albrecht Dürer's *Imagines coeli Septentrionales* … (1515) with the lion's skin draped over his left forearm

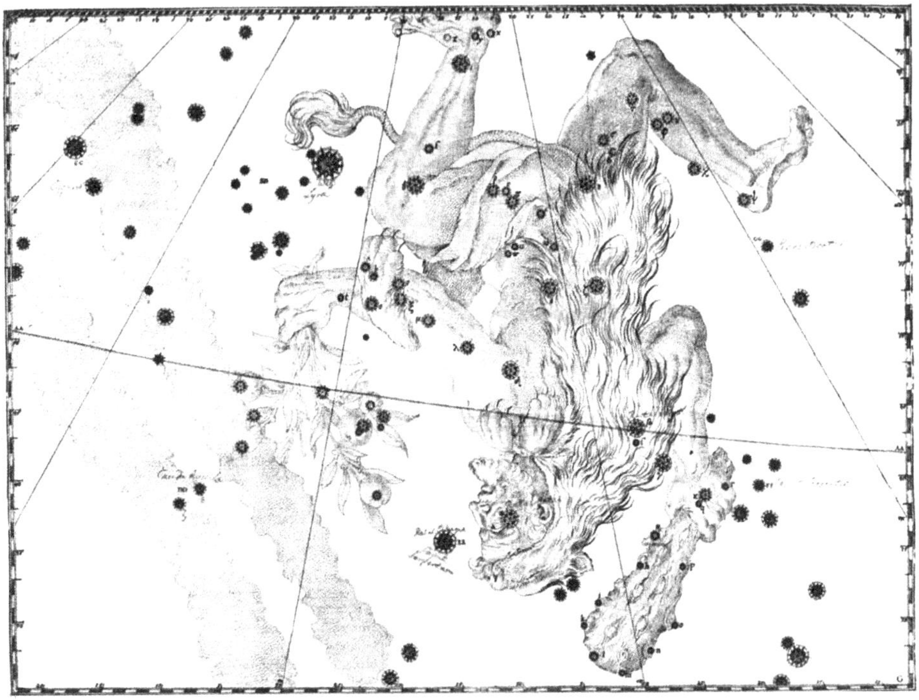

7.2 Johannes Bayer's depiction of Hercules holding Ramus Pomifer on Plate G of *Uranometria* (1603)

These same stars appeared in the century's other highly influential work, Johannes Hevelius' *Prodromus Astronomiae* (1687). Hevelius was not shy about appropriating existing figures and rebranding him as his own, and Bayer's apple branch was a slow-moving target. He replaced with another memorable Herculean symbol: the three snakes' heads of Cerberus, the mythological guardian of the underworld that Hercules famously captured using only his bare hands in the Twelfth Labour. In his depiction of Cerberus on Figure H of *Prodromus*, Hevelius solved the problem of how to show this "hellhound" with a serpent's tail and a lion's claws by simply omitting the body and showing only its three heads (Fig. 7.4).

In introducing Cerberus in place of Ramus Pomifer, Hevelius essentially threw down the gauntlet, beginning a dispute among cartographers that would persist until the eventual extinction of the figure in the late nineteenth century. On Plate 5 of Ignace-Gaston Pardies' *Globi coelestis* (2nd edition, 1693), Philippe de La Hire's *Planisphère céléste septentrionale* (1702), and Carel Allard's *Hemisphaerium meridionale et septentrionale planisphaerii coelestis* (1706), the Branch is shown with no reference to Cerberus; other

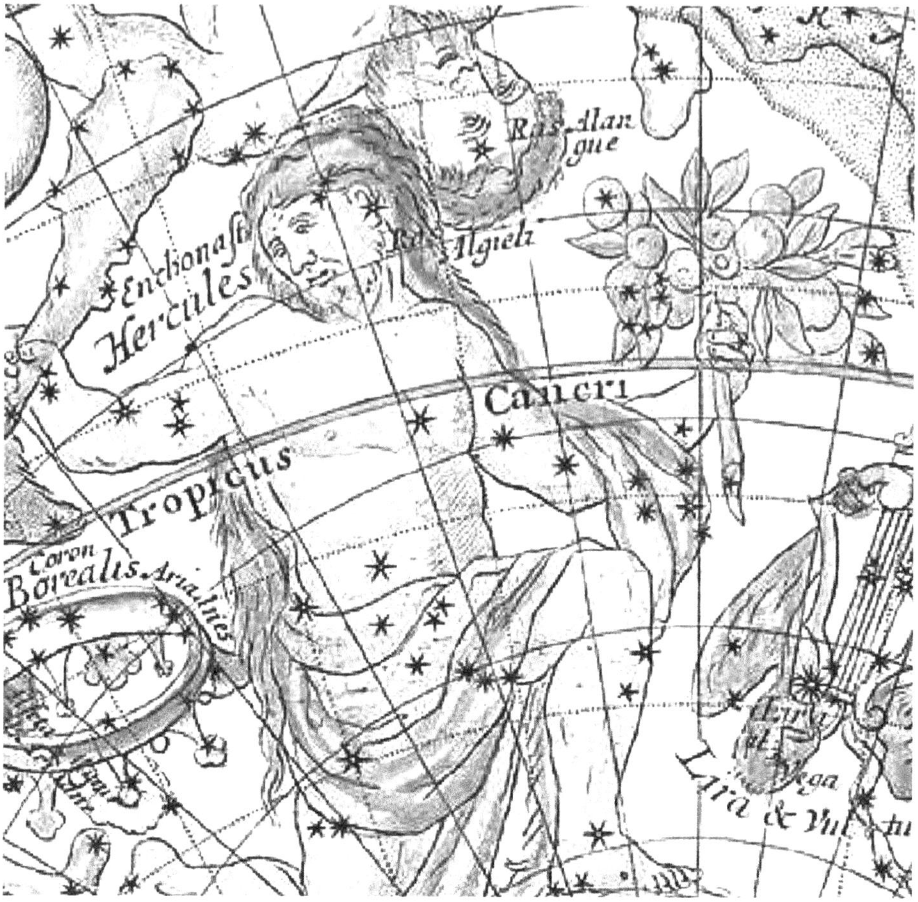

7.3 Ramus Pomifer (unlabeled, without Cerberus) in Carel Allard's *Hemisphaerium meridionale et septentrionale planisphaerii coelestis* (1706)

mapmakers, such as Johann Gabriel Doppelmayr, showed Cerberus with no mention of Ramus (*Atlas Coelestis*, 1742; Fig. 7.5).

According to Ridpath (1989), the English engraver John Senex[4] attempted a compromise that mixed the references into a combined figure he called "Cerberus et Ramus," in which the heads of Cerberus were intertwined with the tendrils of the Branch. John Flamsteed adopted Cerberus et Ramus on his *Atlas Coelestis* (1729), reprints of which influenced later mapmakers such as Johann Elert Bode, who included it as "Cerberus u Zweig" in *Vorstellung der Gestirne* (1782) and *Uranographia* (1801b). As a result,

[4]Senex (1678–1740), a London engraver and geographer to Queen Anne, was one of the principal cartographers of the eighteenth century.

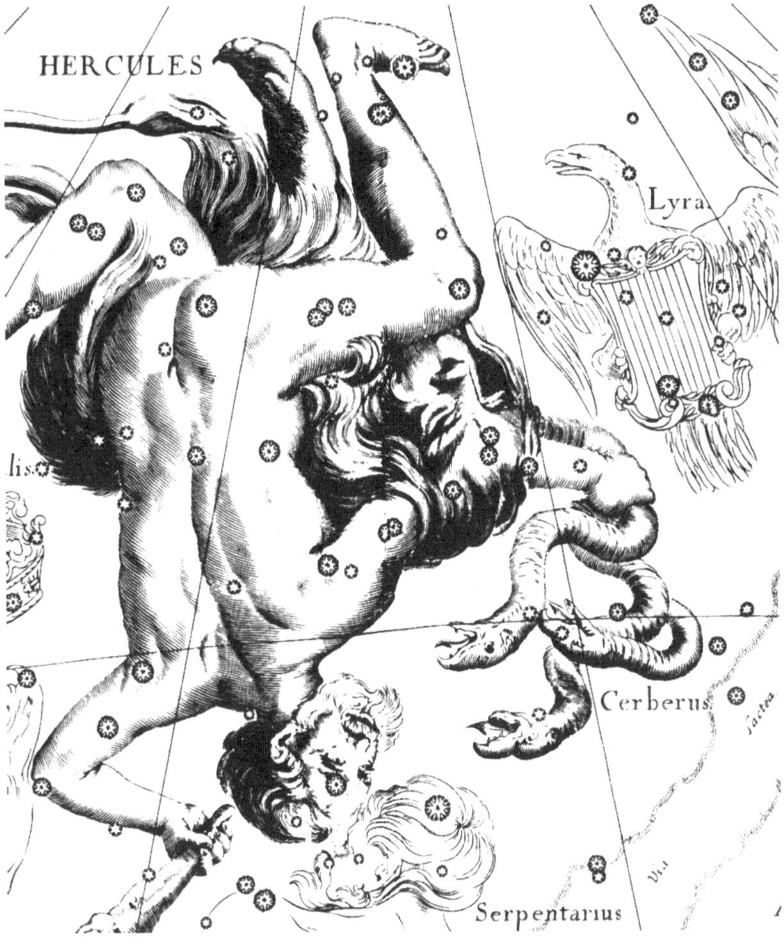

7.4 Detail of Hercules and Cerberus depicted in Figure H of Johannes Hevelius' *Prodromus Astronomiae* (1690)

nineteenth century maps tended to favor the combined figure, such as in Jacob Green's *Astronomical Sketches* 1824; Fig. 7.6).

Allen suggested that the combined figure, rather than implying some literary awkwardness in juxtaposing elements of different Herculean Labours, was actually intended to stand for "the story of the Golden Fruits of the Hesperides with their guardian dragon," and that "the serpent and apples" may have referred to the Biblical account of the betrayal in the Garden of Eden, with the Kneeler representing Adam.

7.5 Hercules and Cerberus shown in Johann Doppelmayr's *Atlas Coelestis* (1742)

Bode asserted[5] that the Apple-Bearing Branch preceded the introduction of Cerberus, which he attributed to Hevelius:

> This constellation is intended to perpetuate the memory of the Theban Hercules, son of Alcmene and Amphitryon, and famous in antiquity for his heroic deeds and extraordinary strength. Hevelius is the first to have placed in his hands the Cerberus or three-headed serpent; the branch of the Hesperides appears here on older star charts. The figure of Hercules is upside-down on the sky for us Europeans, where it

[5]"Dieses Gestirn soll den such Klugheit, Heldenthaten und ausserordentliche Stärke im Alterthum berühmt gewesenen Thebanischen Herkules, ein Sohn des Amphitryo und der Alkmene, verewigen. Hevel hat ihm erst den Cerberus oder die dreyköpfige Schlange in die Hand gegeben, der Apfelzweig kömmt schon in Bayers Himmels-Charten vor. Dieser uralte Held steht für uns in einer verkehrten Stellung am Himmel, und nimmt einen grossen Raum südwärts vom Drachen zwischen der Krone und Leyer ein. Mit dem öftlichen Fuss tritt er auf den Kopf des Drachen. Viele kenntliche Sterne dritter und vierter Grösse zeigen sich in dieser Gegend an der Brust, den Schultern und Lenden des Herkules, und der südlichste Stern dritter Grösse Ras Algethi steht an dessen Kopf westwärts bey dem der zweyten Grösse am Kopf des Ophiuchus."

7.6 The combined figure of "Cerberus et Ramus" shown on Plate 5 ("Constellations near Cygnus") in Jacob Green's *Astronomical Sketches* (1824)

occupies a vast space in addition to the Serpent, between the Crown and the Lyre, and his eastern foot rests on the head of the Serpent. In this region of the sky, upon the chest, shoulders and thighs of Hercules are many large number of remarkable stars between the third and fourth magnitude, the southernmost star of the third magnitude (Ras Algethi) is placed at the head of Hercules, not far from the second-magnitude star in the head of Ophiuchus toward the east.

Admiral Smyth wrote in the *Bedford Catalog* (1844a):

The early Venetian editions of Hyginus figure Hercules as going to attack a snake coiled round the trunk of an apple tree; and Bayer depicted a mystic apple-branch in the Theban's hand. Hevelius transformed it into a bunch of snakes, under the name of Cerberus, from the watch-dog of the infernal portals ; with the fox carrying a goose for his breakfast, as shown in the *Prodromus Astronomicae*. Some have considered the emblem as typifying the serpent which infested the vicinity of Cape Taenarus, whence a sub-genus of Ophidians still derives its name. ... This symbol of the 'tricapitem canem infernalem voracem'[6] figures among the new constellations which follow Hevelius, in his homage to Urania and the great astronomers, in the elaborate frontispiece to his Uranographia. Bode has adopted both the apple-branch and the snakes, in his Atlas, under the style and title of Cerberus et Ramus.

Allen suggested that the stars identified as Cerberus by Hevelius might have been known as such to the ancients, noting that while it "is supposed to have originated with Hevelius in his Firmamentum Sobiescianum, ... Flammarion asserts that it was on the sphere of Eudoxos with the Branch." However, if Eudoxus was the source of the constellations shown on the Farnese Atlas, this cannot possibly be true.

ICONOGRAPHY

Both Cerberus and the Hesperidean Apples figure in a series of stories collectively known as the Labours of Hercules, first attested in Ἡράκλεια (*Heracleia*), a now-lost epic poem composed by Peisander (fl. *c.* 640 BC) around the end of the seventh century BC. The stories were certainly older, and Clement of Alexandria (*c.* 150–*c.* 215) claimed[7] that Peisander merely plagiarized the poem from an otherwise unknown "Pisinus of Lindus." Peisander is thought to have fixed the number of Labours at twelve and introduced the notion of Hercules clothed in the lion's skin, replacing the tradition of heroes ensconced in armor.

---

[6] 'Gluttonous, infernal, three-headed dog'.

[7] *Stromata*, Book VI, Chap. 2.

7.7 Detail a third century AD Roman sarcophagus in the Palazzo Altemps, Rome, showing a sequence of the Herculean Labours. From *left to right*, they are: the Nemean Lion, the Lernaean Hydra, the Erymanthian Boar, the Ceryneian Hind, the Stymphalian birds, the Girdle of Hippolyta, the Augean stables, the Cretan Bull and the Mares of Diomedes

The Herculean Labors

Hercules, son of Zeus by the mortal woman Alcmene, was long despised by Zeus' wife, Hera. She harried him relentlessly, driving Hercules to madness; while in this state, he murdered his wife and six sons. Later, when he recovered his sanity, he deeply regretted the killings and sought wisdom from the oracle at Delphi to redeem himself. The oracle told him to travel to Tiryns and enter the service of King Eurystheus—another royal descendant of Perseus—for 12 years, obeying any command given; the reward would be his own immortality. While Hercules hated the notion of serving a man to whom he felt far superior, he feared the displeasure of his father. Eurystheus ordered Hercules to perform ten specific tasks, called the Labours (Fig. 7.7).

Upon completion of the first ten Labours, Eurystheus was dissatisfied with Hercules' performance in two of them: In the Second Labour, Hercules slew the Lernaean Hydra (with the help of Iolaus), while in cleaning the Augean stables in the Fifth Labour—meant to be both a humiliating and ultimately impossible job—he received payment for the work, which was really done by the rivers. Having decided these two didn't count, Eurystheus gave him two more tasks: First, to steal the Golden Apples from the garden of the Hesperides, nymphs who were daughters of Hesperus, and second, to capture alive (and with his bear hands) Cerberus. These final two Labours became the basis for the figure of stars shown held in the outstretched hand of the constellation Hercules.

The Eleventh Labour: The Golden Apples of the Hesperides

The first of these two additional Labours was to steal the apples from the Garden of the Hesperides. The Hesperides tended an Eden-like garden in a remote location somewhere in the western part of the world, said to be anywhere from Libya to the Atlas Mountains

7.8 "The Garden of the Hesperides" (*c.* 1892) by Frederic Leighton, 1st Baron Leighton (1830–1896). Oil on canvas, diam. 169 cm (66.5 in.). The Bridgeman Art Library, Object 5326

of Morocco.[8] Hercules first had to learn where the Garden was located, so he found and caught the Old Man of the Sea, a shape-shifting figure variously identified as one of a number of water gods whose mythic origins probably predates the Homeric era (Fig. 7.8).

As usual, in the course of his travels Hercules was beset by various figures trying to impede his progress. Pseudo-Apollodorus described[9] his harassment in Book 2 of the *Bibliotheca*:

[8]"These apples were not, as some have said, in Libya, but on Atlas among the Hyperboreans." (Pseudo-Apollodorus, *Bibliotheca* 2.5.11, trans. J.G. Frazer.)

[9]*Ibid.*

> So journeying he came to the river Echedorus. And Cycnus, son of Ares and Pyrene, challenged him to single combat. Ares championed the cause of Cycnus and marshalled the combat, but a thunderbolt was hurled between the two and parted the combatants.

After learning the location of the Garden,

> he traversed Libya. That country was then ruled by Antaeus, son of Poseidon, who used to kill strangers by forcing them to wrestle. Being forced to wrestle with him, Hercules hugged him, lifted him aloft, broke and killed him; for when he touched earth so it was that he waxed stronger, wherefore some said that he was a son of Earth.

Along the way, Hercules stopped in Egypt where he was detailed by a king called Busiris, a son of Poseidon and Lysianassa, herself a daughter of Epaphus. Hercules' timing was unfortunate. Long prior, the country was stricken with famine for 9 years. A soothsayer from Cyprus, called Phrasius, foretold that the famine would end if the Egyptians made an annual sacrifice of a foreigner to Zeus. Unfortunately for Phrasius, he was the first such target; Busiris thereafter proceeded to sacrifice all foreigners arriving in Egypt. Hercules was seized and led to the sacrificial altar but broke free from his chains and killed Busirus, his son and members of his retinue.

In one version of the story, on arriving at the Garden Hercules tricked Atlas into fetching some of the apples for him under a ruse in which he would hold up the heavens in the meantime; this ran the risk for Hercules of voiding the Labour because he received help in performing it. When Atlas returned with the apples, he did not want to resume his former duties and offered to deliver the apples to Eurystheus himself. Hercules tricked him again, asking that Atlas temporarily replace him in holding up the heavens while Hercules made himself more comfortable by adjusting his cloak. Hercules simply walked away with his prize.

In another version, Hercules entered the Garden and was confronted by Ladon, the serpentine guardian of the apples; he slew Ladon (Fig. 7.9) and departed with the objects of his quest. Apollonius Rhodius (fl. first half of third century BC) related[10] described the aftermath in *Argonautica* (Fig. 7.10):

> Then, like raging hounds, *[the Argonauts]* rushed to search for a spring; for besides their suffering and anguish, a parching thirst lay upon them, and not in vain did they wander; but they came to the sacred plain where Ladon, the serpent of the land, till yesterday kept watch over the golden apples in the garden of Atlas; and all around the nymphs, the Hesperides, were busied, chanting their lovely song. But at that time, stricken by Heracles, he lay fallen by the trunk of the apple-tree; only the tip of his tail was still writhing; but from his head down his dark spine he lay lifeless; and where the arrows had left in his blood the bitter gall of the Lernaean hydra, flies withered and died over the festering wounds. And close at hand the Hesperides, their

[10]*Argonautica*, Book 4, lines 1393–1386, trans. R.C. Seaton.

7.9 Fragment of a late Roman terra cotta plate showing, in relief, Hercules fighting the serpent Ladon for the tree bearing the Hesperidean Apples. Staatliche Antikensammlungen, Munich

> white arms flung over their golden heads, lamented shrilly; and the heroes drew near suddenly; but the maidens, at their quick approach, at once became dust and earth where they stood.

The Twelfth Labour: The Capture of Cerberus

Eurystheus was understandably furious at Hercules' accomplishment, one thought impossible to achieve. On returning with the apples, Eurystheus sent him off to his final Labour: the capture of Cerberus, the hound standing guard at the gates of the Underworld. He was to return Cerberus alive, a prospect making this Labour the most difficult and dangerous of them all (Fig. 7.11).

Cerberus was a frightful creature; according[11] to Pseudo-Apollodorus it "had three heads of dogs, the tail of a dragon, and on his back the heads of all sorts of snakes." It was described[12] in works as early as Homer's *Iliad* as Ἀιδης στυγερός, 'the hound of

[11] *Bibliotheca* 2.5.12, trans. J.G. Frazer.

[12] *Iliad* 8. 366 ff., trans. Lattimore.

7.10 A branch bearing the Golden Apples of the Hesperides is shown in the grip of Hercules in Corbinianus Thomas' *Mercurii philosophici firmamentum firmianum* (1730)

the grisly death god'. Hesiod (*c.* 700 BC) was the first to give it a name[13]: Κερβερος (*Kerberos*). "Typhaon … was joined in love to her [Ekhidna] … And next again she bore the unspeakable, unmanageable Kerberos, the savage, the bronze-barking dog of Haides, fifty-headed, and powerful, and without pity." Naturally, Dante encounters[14] Cerberus on his guided tour of hell in the *Divine Comedy*:

[13] *Theogony* 310 ff., trans. Evelyn-White.

[14] *Inferno*, canto VI, lines 12–32; trans. H.F. Cary.

7.11 Detail of Hercules and the combined figure Cerberus et Ramus Pomifer depicted in Alexander Jamieson's *Celestial Atlas* (1822)

Cerberus, cruel monster, fierce and strange,
Through his wide threefold throat barks as a dog
Over the multitude immers'd beneath.
His eyes glare crimson, black his unctuous beard,
His belly large, and claw'd the hands, with which
He tears the spirits, flays them, and their limbs
Piecemeal disparts. Howling there spread, as curs,
Under the rainy deluge, with one side
The other screening, oft they roll them round,
A wretched, godless crew. When that great worm
Descried us, savage Cerberus, he op'd
His jaws, and the fangs show'd us; not a limb

Of him but trembled. Then my guide, his palms
Expanding on the ground, thence filled with earth
Rais'd them, and cast it in his ravenous maw.
E'en as a dog, that yelling bays for food
His keeper, when the morsel comes, lets fall
His fury, bent alone with eager haste
To swallow it; so dropp'd the loathsome cheeks
Of demon Cerberus, who thund'ring stuns
The spirits, that they for deafness wish in vain.

In preparation for his meeting with the fearsome beast, Hercules needed to learn how to return alive from the underworld; he traveled to Eleusis and was initiated into the Eleusinian Mysteries, by way of which he was also absolved of his past crime of killing centaurs during the Fourth Labour.[15] Finding the entrance to the underworld at Tanaerum in Laconia, he was assisted across the entrance by Hermes and Athena. Hestia helped negotiate his passage across the Styx from Charon. As he approached the gates of Hades he met Theseus and Pirithous, imprisoned in the underworld for attempting to rescue Persephone. According to Pseudo-Apollodorus,[16] upon sight of Hercules "they stretched out their hands as if they should be raised from the dead by his might. And Theseus, indeed, he took by the hand and raised up, but when he would have brought up Pirithous, the earth quaked and he let go."

Hercules located Hades and asked his permission to bring Cerberus back to the surface. Hades agreed to the proposal provided that Hercules could manage to overpower the creature using only his own might and without the help of weapons. Again, Pseudo-Apollodorus relates[17] the tale: "Hercules found him at the gates of Acheron, and, cased in his cuirass and covered by the lion's skin, he flung his arms round the head of the brute, and though the dragon in its tail bit him, he never relaxed his grip and pressure till it yielded." He dragged Cerberus to the surface through a cavern entrance on the Peloponnese and laid it before Eurystheus' feet. His final Labour completed, Hercules was released from Eurystheus' service and left to join the Argonauts in their search for the Golden Fleece.

In describing the figure of Cerberus in relation to the constellation of Hercules, Ezra Otis Kendall recounted the story in *Uranography* (1845):

CERBERUS, according to Hesiod, was a dog with a hundred heads, though our mythologists give him only three; he was reputed to belong to Pluto, and to be stationed at the gates of the infernal regions as a guard. Cerberus is represented in this plate as a three-headed serpent. From this situation, Hercules, as his concluding labor, dragged him up to the realms of day, when he went to redeem Alceste. Some suppose this to be an astronomical fable, relating to the sun on his arrival at the autumnal equinox.

---

[15] Pseudo-Apollodorus, *Bibliotheca* 2.5.4, trans. trans. J.G. Frazer.

[16] Ibid.

[17] Ibid.

Richard Hinckley Allen (1899) related a slightly different version of the myth:

> Others have said that the figure typified the serpent destroyed by the Hero while it was infesting the country around Taenarum, the Μέτοπον of Greece, the modern Cape Matapan.

Allen further wrote of the place in the heavens occupied by Cerberus that

> the royal poet James I designated the infernal Cerberus as "the thrie headed porter of hell" and the heavenly one has been so figured, although with serpents' darting tongues; but the abode and task of the creature would seem to render very inappropriate his transfer to the sky, so that it probably was only made for the purpose of mythological completeness, as the death of this watch-dog of Hades fitly rounded out the circle of Hercules' twelve labours.

DISAPPEARANCE

An identity for this handful of stars separate from Hercules began to fall apart in the last quarter of the nineteenth century. It does not appear in the constellation list in William Henry Rosser's *The Stars and Constellations* (1879), but was mentioned as "Cerberus and Ramus" in the table "Old And New Constellations In Chronological Order" in *Poole Bros. Celestial Handbook* (1892), attributed to "Eudoxus; Aratus" with the caveat: "These constellations, not being recognized by the B.A. Catalogue, are not described in this Handbook; the stars by which they were formed are inserted in the constellations from which they were taken." In the second edition of *Popular Astronomy* (1899), Joel Dorman Steele hedged as to whether Cerberus was a proper constellation in its own right or merely an asterism within Hercules. The hero "is represented as a warrior clad in the skin of the Nemean lion, holding a club in his right hand and the dog Cerberus in his left," Steele wrote. Later, when describing Antinoüs (Chap. 4) and Aquila, Steele refers to Cerberus with a sense of independence: "A similar row [of stars] denotes the tail of the eagle ; the first star of which is named ζ, and the last star lies in Cerberus."

By the end of the century, Allen called Cerberus a "sub-constellation" and "former adjunct of Hercules, now entirely disregarded by astronomers." Among its last mentions, it was discussed in the context of Aquila in William Tyler Olcott's *Star Lore of All Ages* (1911) (Fig. 7.12):

> In the modern representations of the figure, Hercules swings a club in one hand, and holds fast a branch, or the three-headed dog Cerberus, in the other, so that there does not seem much reason or opportunity for him to pick up stones... Bayer represents Hercules as holding in addition to his club an apple branch, possibly to indicate his connection with the myth of the Golden Apples of the Garden of Hesperides. For his eleventh labour he was ordered to procure them. Those who claim that Hercules represents Adam certainly have much to substantiate their theory, for associated with the figure we find a serpent, a garden, and the apple."

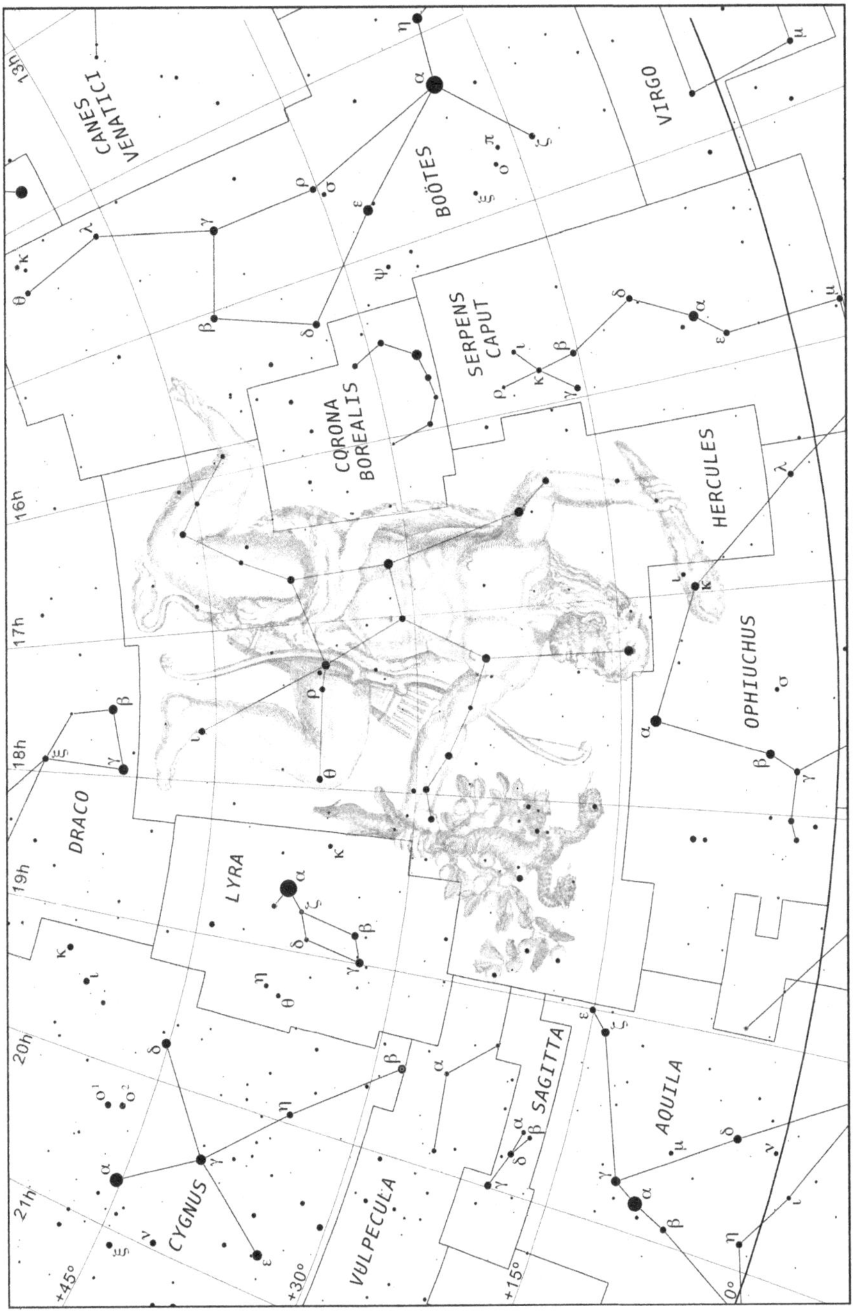

7.12 The figures of Hercules (*center*) and Cerberus et Ramus Pomifer (center-left) from Plate 8 of Alexander Jamieson's *Celestial Atlas* (1822) overlaid on a modern chart

8

Custos Messium

The Harvest Keeper

Genitive: Custodis Messium
Abbreviation: CsM
Alternate names: "Vineyard Keeper" (Young, 1807); "Messium" (Colas, 1892)
Location: Near the north celestial pole, encompassing stars now within the boundaries of Cassiopeia, Cepheus, and Camelopardalis.[1] 40 Cassiopeiae was one of its principal stars (Fig. 8.1).

ORIGIN AND HISTORY

Custos Messium was introduced on his 1775 celestial globe by the French astronomer Joseph Jérôme Lefrançois de Lalande (1732–1807) "on the occasion of the comet which appeared near the North Pole."[2] This comet was C/1774 P1, also known as Comet Montaigne; Messier described it in his handwritten document *Notice de mes comètes*[3] ("Notes on my comets"):

> Comet, found with the refractor, observed during the months August, September and October; it began to appear between the Reindeer [stars between Camelopardalis and Cassiopeia], Cepheus and Cassiopeia, passed through Lacerta and Andromeda, traversed Pegasus and ceased to appear at the knee of Aquarius. 41 days of

[1] *"Between Cassiopeia, Cepheus and Camelopardalis"* (Lalande 1776); *"It is bounded by Cassiopeia, Cepheus, Tarandus, and Camelopardalis"* (Green, 1824); *"It lies between the constellations of Cassiopoeia and the Reindeer"* (Kendall, 1845).
[2] *Catalogue of Scientific Papers (1800–1900): Ser. 1, 1800–1863*, ed. Henry White, Herbert McLeod, and Henry Forster Morley, London: George Edward Eyre and William Spottiswoode, Vol IV., p. 353 (1870).
[3] Paris Observatory manuscript No. C2-19, *c.* 1805, translated by Hartmut Frommert.

J.C. Barentine, *The Lost Constellations*, Springer Praxis Books,
DOI 10.1007/978-3-319-22795-5_8

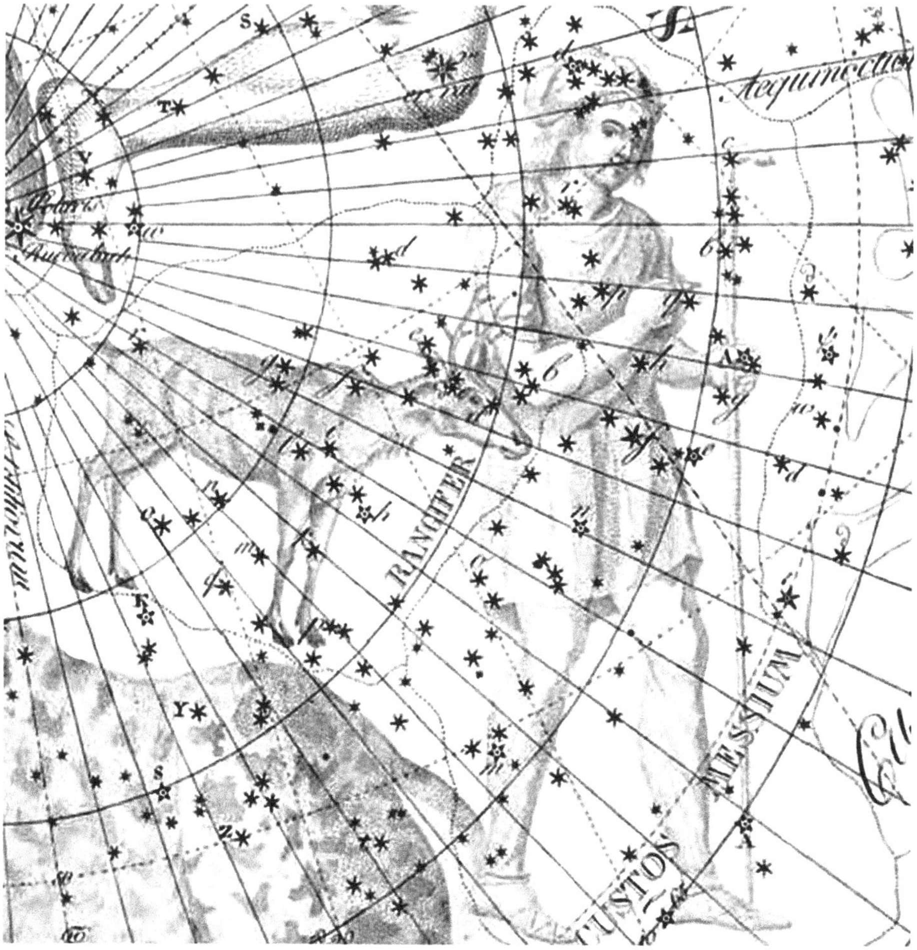

8.1 Custos Messium and Rangifer (Chap. 20) depicted on Plate 3 of in Bode's *Uranographia* (1801b)

> observations. The memoir of the observations, and the chart of its track [are] in the vol. of the *ac*. for the year 1775. M. Montaigne discovered this Comet on the eleventh of April at Limoges.

Lalande was, as Richard Hinckley Allen (1899) put it (Figs. 8.2 and 8.3),

> [T]he enthusiastic astronomer who would spend nights on the Pont Neuf over the Seine, explaining the wonders of the variable Algol to all whom he could interest in

8.2 Nineteenth century engravings of Joseph Jérôme Lefrançois de Lalande (*left*) and Charles Messier (*right*)

8.3 A late-nineteenth century postcard showing the exterior of the Hôtel de Cluny, now known as the Musée national du Moyen Âge, Paris. Messier worked in a wooden enclosure atop the octagonal tower at far right, removed by the time of this photograph

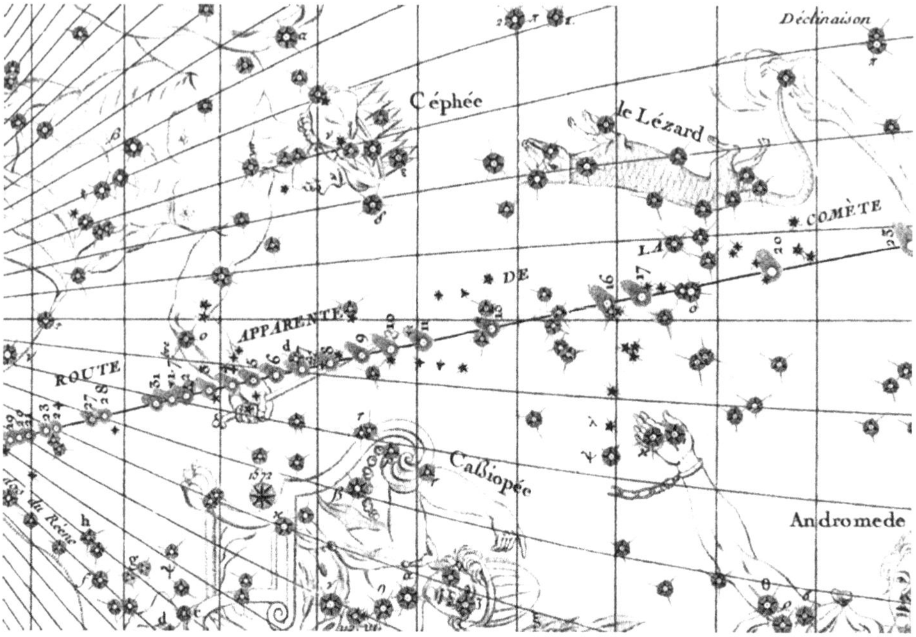

8.4 The path of Comet C/1774 P1 (Montaigne) in a detail of Plate XI from *Mémoires de l'Académie royale des sciences*, p. 476 (1775). This chart shows part of the northern sky that was later commemorated by Lalande in his constellation Custos Messium

> the subject, and whose seclusion in his observatory, amid the turmoil of the French Revolution, enabled him to "thank his stars" that he had escaped the fate of so many of his friends.

"Though he was being sincere about his friend," Philip Pugh (2012) wrote, "Lalande quite often proposed new constellations (often when inebriated)."

Lalande's globe was accompanied by an explanatory note titled *Explication des nouveaux globes céleste et terrestre* ("Explanation of new celestial and terrestrial globes"). This work was reviewed in the French scientific *Journal des Sçavans* in November 1776, summarizing Lalande's rationale (Fig. 8.4):

> Ce petit Ouvrage, que l'on peut avoir séparément des Globes, contient un detail sur l'usage des Sphères et des Globes avec divers problêmes, et une Dissertation sur la forme de quelques constellations.
>
> A l'occasion de la Comète de 1774, découverte dans une partie du ciel, où il y a plusieurs étoiles qui n'avoient aucun nom sur les Cartes, M. de la Lande a cru devoir placer, dans son globe céleste, une nouvelle constellation sous le nom de Messier, Custos Messium.

> On appele Messier, en François, celui qui est proposé à la garde des moissons ou des trésors de la terre. Ce nom semble naturellement se lier avec celui de M. Messier, notre plus infatigable Observateur qui, depuis vingt ans, est comme préposé à la garde du ciel et a la découverte des Comètes. On a cru pouvoir rassembler, sous le nom de Messier, les étoiles sparsiles ou informes, situées entre Cassiopée, Cephée et la Giraffe, c'est-à dire, entre les Princes d'un peuple agriculteur et un animal destructeur des moissons; et cette nouvelle constellation rappelera un même temps au souvenir et à la reconnoissance des Astronomes à venir, le courage et la zèle de celui dont elle porte le nom, et qui s'occupe actuellement à en mieux déterminer les étoiles.

Rendered in English:

> This minor work, which can be separate from the Globes, contains a detail on the use of Spheres and globes with various problems, and a dissertation on the form of a few constellations.
> On the occasion of the Comet of 1774, discovered in this part of the sky on which there are several unnamed stars on celestial maps, M. de la Lande felt obliged to put on his celestial globe a new constellation known as Le Messier, (or) Custos Messium.
> Called "Messier" in French, it refers to the guardian of the harvest or 'treasures of the earth'. This name seems to naturally mesh with that of Mr. Messier, our most tireless observer for 20 years, as an astronomer and discoverer of comets. It was believed to bring together, under the name Messier, the unformed stars located between Cassiopeia, Cepheus and the Giraffe, that is to say, between the rulers of a farming people and an animal that destroys the harvest, and this new constellation will, at the same time, bring to mind the memory and gratitude of astronomers, the courage and zeal of the man whose name it bears, and those who are currently working to better determine its stars.

By way of a subtle play on words, the suggested constellation was a thinly-veiled attempt by Lalande to place a living figure among the stars who was neither a ruling head of state nor a patron of astronomical research. The French alternate name, *Le Messier*, was considerably more blunt; Allen quotes Admiral Smyth, who wrote that the title was given "... in poorish punning compliment to his friend, the 'Comet ferret' ", invoking the nickname King Louis XV once bestowed upon Messier for his comet-hunting prowess. By fortunate coincidence, Messier's surname derived from the occupation name of someone who kept watch over harvested crops, discouraging both animal and human thieves (from the Late Latin *messicarius*, an agent derivative of *messis*, 'harvest').

Jacob Green (1824) described the circumpolar situation of the figure in *Astronomical Recreations*:

> LA LANDE introduced this asterism in honor of the celebrated astronomer *Messier*, and in allusion to his name it has been called the Guardian of the Harvests. The stars of this asterism are almost all invisible to the naked eye. It is bounded by Cassiopeia, Cepheus, Tarandus, and Camelopardalis.

Custos Messium served as a sort of northern hemisphere counterpart to the southern Polophylax (Guardian of the Pole; see Volume 2). Whether Lalande intended this parallel is unknown, but it is remarkable that both guardian figures were circumpolar constellations and fell into disuse before the end of the nineteenth century. Lalande seems to have been quite sincere in the attempt to honor his friend Messier. Whatever the case, as "gatherer and keeper of the harvest of comets" (in Allen's words), the proposed constellation seemed at first destined for posterity. Ridpath (1989) notes that there was no connection between Messier and the Comet of 1774; while Messier made many observations of that particular comet, its discovery is attributed to another French astronomer, Jacques Montaigne.

Prior to Lalande's time, the area of the sky in which he formed Custos Messium consisted of a number of faint, unformed stars at the junction of Camelopardalis, Cassiopeia, and Cepheus. Allen wrote that "the Phoenicians are said to have imagined a large Wheat Field in this part of the sky," and speculated that Lalande's invention may have been "induced by the fact that the two neighboring royal personages [Cepheus and Cassiopiea] were rulers of an agricultural people, and the Giraffe [was] an animal destructive to grainfields." Most mapmakers in the century or so before Lalande considered its stars formally part of Camelopardalis, as seen in Corbinianus Thomas' map of that constellation in Fig. 8.5.

The popularization of Custos Messium and its inclusion on later maps is almost certainly due to its early adoption by the German astronomer Johann Elert Bode (1747–1826), who initially incorporated it into his *Vorstellung der Gestirne* (1782), an expanded German version of the 1776 French edition of John Flamsteed's *Atlas Coelestis* (1729) (Fig. 8.6). But it was the appearance in Bode's *Uranographia* (1801b), one of the last artistic masterpieces of the age of celestial cartography, that secured at least a limited lease on life for Lalande's fanciful creation. In *Allgemeine Beschreibung und Nachweisung der Gestirne*, Bode wrote[4] of Lalande's creation (Fig. 8.7)

> This was consecrated by Mr. de la Lande in memory of Mr. Messier, *[an]* astronomer distinguished by the application with which he surveyed the heavens, and discovered nearly all comets that appeared in the sky over a period of thirty years. This asterism is located between Cassiopeia and the Reindeer.

Allen (1899) listed its names in other European cultures, giving some indication as to how it was interpreted by non-French speakers. He wrote that the Germans referred to it as *Erndtehüter*, a word that does not actually exist in German. Allen probably meant *Erntebehüter*, meaning "harvest watcher". To the Italians, it was *Mietitore* ("reaper"). Kendall (1845) claimed that "the Shepherd" was an alternate name for Custos Messium, although this is not attested in other sources. It is, however, easy to see how the harvest-keeper figure could be conflated with a shepherd, given that he was typically depicted in star atlases as holding a shepherd's crook. This suggests a "watcher" figure as a shepherd

[4]"Ist von Herrn de la Lande zu Ehren des Herrn Messier der bey seinen unermüdeten Durchmusterungen des Firmaments, in 30 Jahren sast alle erschienene Cometen entdekte, zwischen der Cassiopeja und dem Rennthier ans Firmament gebracht."

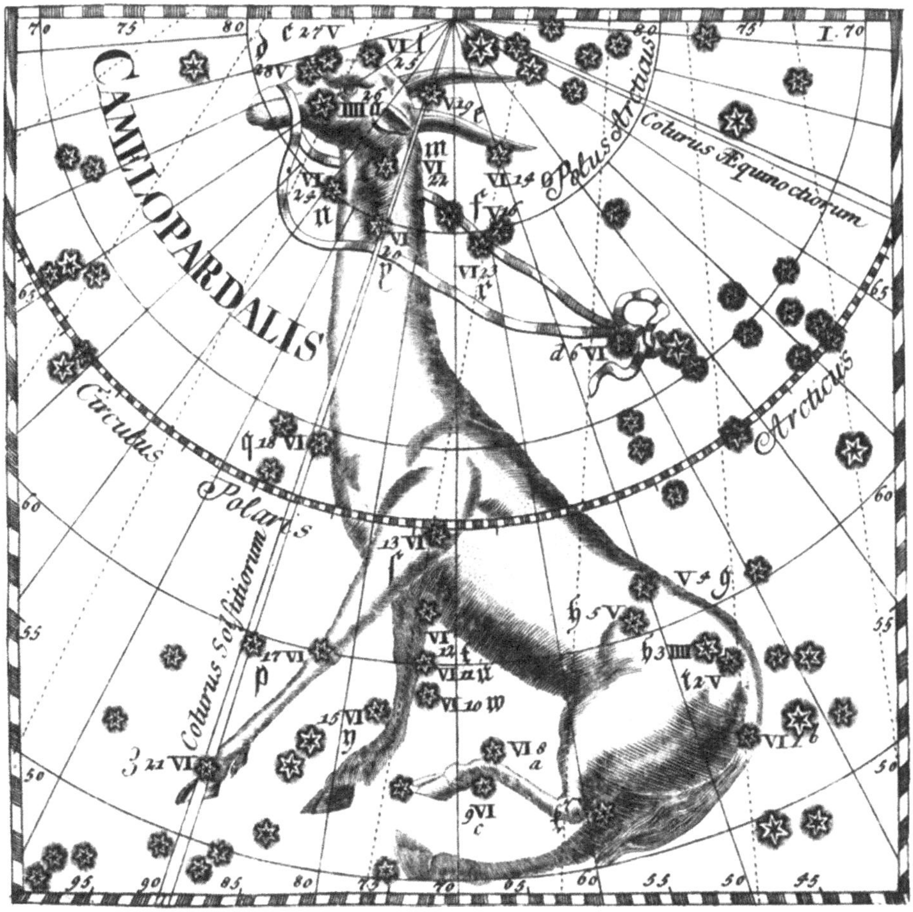

8.5 Prior to the introduction of Custos Messium by Lalande, its stars were considered unformed and identified with Camelopardalis (the Giraffe), as shown in Corbinianus Thomas' *Mercurii philosophici ...* (1730). The stars that Lalande made into Custos Messium are just to the right of the hind end of Camelopardalis; the stars above its back became Rangifer (the Reindeer; Chap. 20)

watches over and manages his flock. Perhaps to clarify his role, and to suggest the "reaper" attribute in the figure's Italian moniker, he appears in Jamieson's 1822 atlas and *Urania's Mirror* (1825; Fig. 8.8) perched on the back of Camelopardalis holding a crook in one hand and a sickle in the other, while nearly being bowled over by Rangifer.

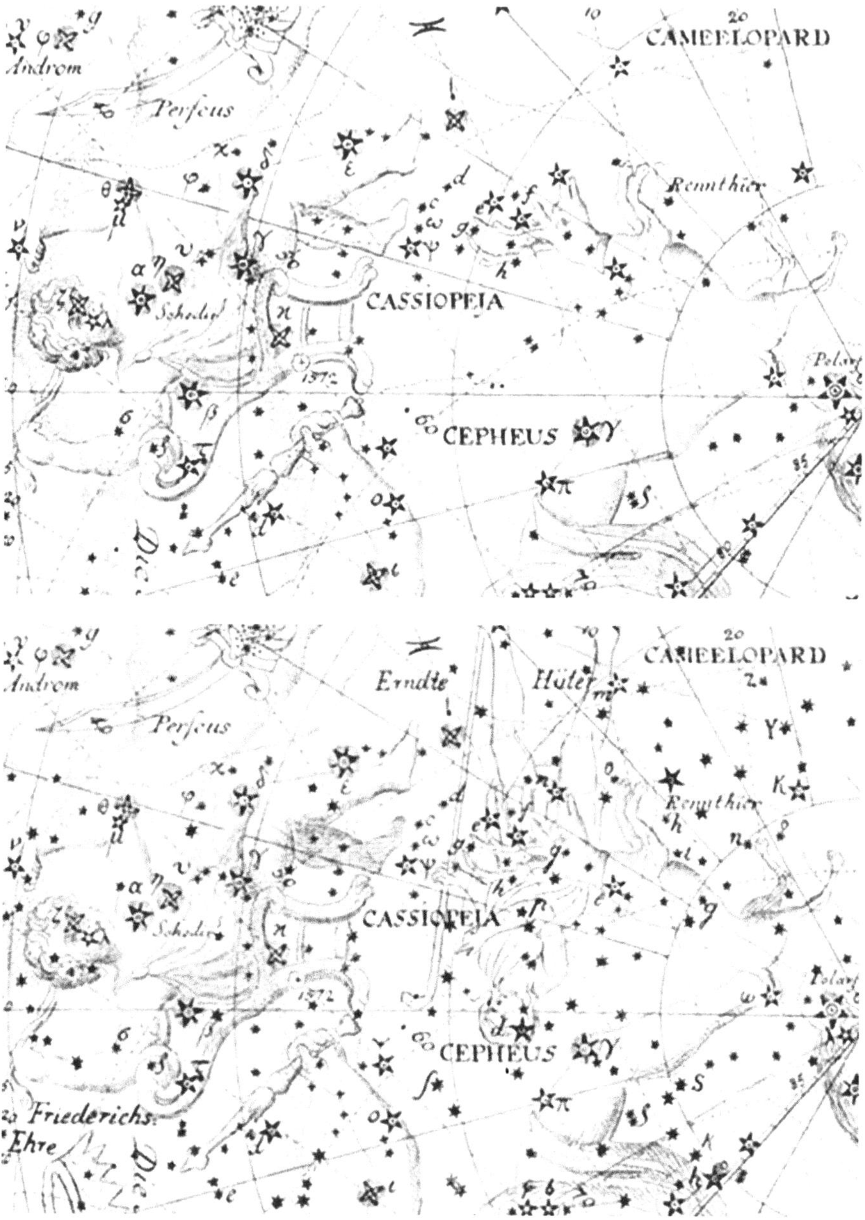

8.6 Before and after: Detail of Plate 2 from two versions of Bode's *Vorstellung der Gestirne* showing the addition of Custos Messium ("Erndte Hüter"): 1782 (*top*) and 1805 (*bottom*). Bode had already adopted the adjacent figure of Rangifer ("Rennthier"; Chap. 20) by the time of the first edition. Note also part of the now-extinct constellation Honores Fredrici ("Friederichs Ehre"; Chap. 12) at extreme lower left of the *bottom panel*

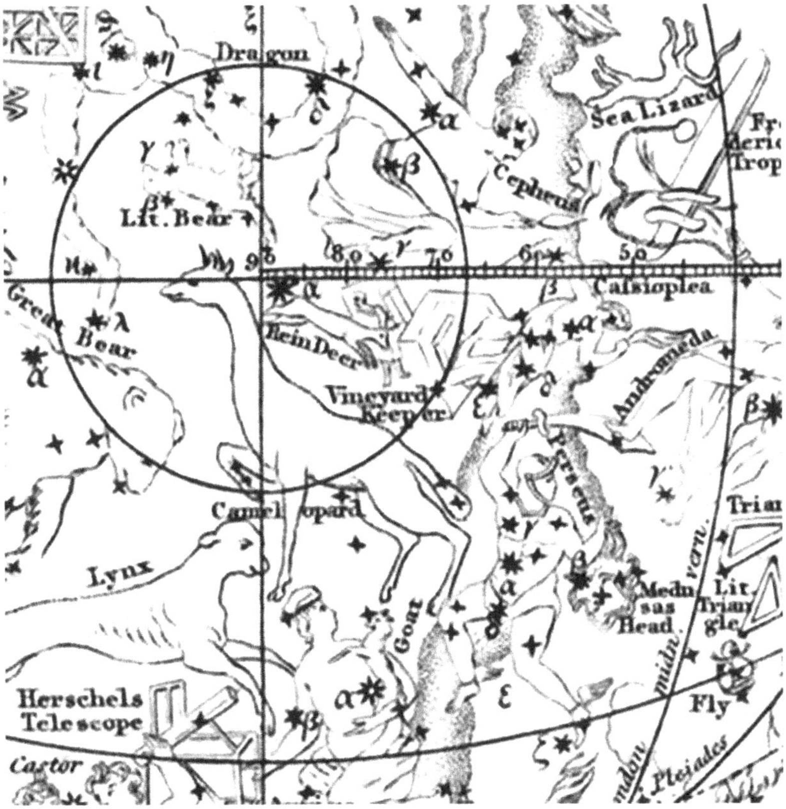

8.7 Custos Messium (as "Vineyard Keeper") shown on Plate XXXVI, Fig. 517 of Thomas Young's *Course of Lectures on Natural Philosophy and the Mechanical Arts* (1807). Many other now-lost constellations discussed in this book are also depicted on Young's chart: part of Quadrans Muralis (Chap. 19) at *far upper left*; Telescopium Herschelii Major ("Herschels Telescope," Chap. 26), Rangifer ("ReinDeer," Chap. 20), Honores Fredrici ("Frederic's Trophy," Chap. 12), Triangulum Minus ("Little Triangle," Chap. 28), and Musca Borealis ("Fly," Chap. 16)

ICONOGRAPHY

Custos Messium was situated in the northern sky among faint, anonymous stars between Cassiopeia, Cepheus, and Camelopardalis; the double star 40 Cassiopeiae was once one of its principal stars. The extinct constellation Rangifer, the Reindeer (Chap. 20), was originally situated roughly between Custos Messium and the north celestial pole, as shown in Fig. 8.1. The modern constellation boundaries are drawn such that its stars are now entirely subsumed within Cassiopeia.

The placement of this constellation between the King and Queen of Ethiopia (Cepheus and Cassiopeia) and the Giraffe itself has symbolic meaning, given that the royalty

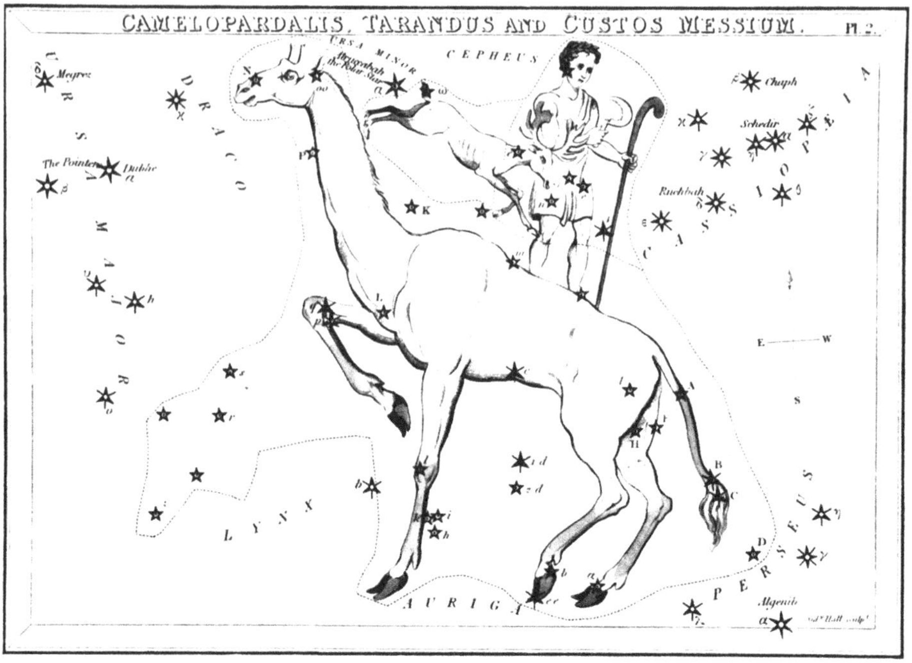

8.8 Camelopardalis, Custos Messium, and Rangifer depicted on Plate 2 of *Urania's Mirror* (1825)

rules over an agricultural society and the animal is known as a pest to African farmers that destroys crops. Allen (1899) further cements this understanding as he notes the Phoenicians associated the same part of the night sky with a vast wheat field. This fact, along with the role of "guardian" or "watcher" figures in mythology and the history of the constellations, leads to a plausible theory explaining the origin of Custos Messium.

Guardian/Watcher Figures

In Astronomy

The notion of certain stars, asterisms or constellations as "guardians" or "watchers" of other stars or groups of stars is certainly as old as the idea of identifying constellations in the first place. In ancient Persia the sky was thought to be divided into four districts, each guarded by a so-called "royal" star. "The Persians," wrote[5] Joel Stebbins,[6] "are said to have considered 3000 years ago that the whole heavens were divided up into four great

[5]"The Constant Stars," *Publications of the Astronomical Society of the Pacific*, Vol. 55, No. 325, p. 177 (1943).

[6]Stebbins (1878–1966) was a pioneer of photoelectric photometry during his tenure as director of Washburn Observatory at the University of Wisconsin-Madison from 1922–1948.

districts, each watched over by one of the 'Royal Stars'. These districts and stars are still there." Each star was identified with its particular region of the heavens. Of Regulus, for example, Allen wrote "[I]t was the leader of the Four Royal Stars of the ancient Persian monarchy, the Four Guardians of Heaven." The other Royal Stars were Aldebaran (α Tauri; "Watcher of the East," associated with the vernal equinox), Antares (α Scorpii, "Watcher of the West," autumnal equinox), and Fomalhaut (α Piscis Austrinus, "Watcher of the South," winter solstice); Regulus itself was the "Watcher of the North," associated with the summer solstice. These stars were assigned to the four "colures," or points on the celestial sphere marking the intersection of meridians drawn between the poles and the equinoxes and solstices; thus each star lags the Sun in right ascension by roughly the same amount on the indicated dates and is seen in the west after sunset on the corresponding equinox or solstice. But at a much earlier epoch, precession would have positioned each star considerably closer to its respective colure; in *Star Lore of All Ages* (1911), William Tyler Olcott wrote that these alignments were "only possible in antediluvian times," testifying as to the ancient origin of the Persian stories. However, the notion of "royal" Persian stars has been criticized[7] as a relatively recent invention that may have resulted from a misinterpretation of the original sources.

In certain instances, constellations are referred to as "watchers" based on their proximity to other figures; this propensity is highest toward the poles, where it seems that a constant cat-and-mouse game takes place as the constellations appear to forever chase each other. The *lucida* of the Ptolemaic figure Boötes, the Herdsman, is Arcturus (α Boötis), whose name derives from ancient Greek αρκτουρος (*arktouros*) meaning "Guardian of the Bear." Other names for the star, according to Allen,

> were Ἀρκτοφύλαξ and Ἀρκτοῦρος, the Bear-watcher and the Bear-guard, the latter first found in the Ἔργα καὶ Ἡμέραι, the Works and Days, 'a Boeotian shepherd's calendar,' by Hesiod, eight centuries before our era. But, although these words were often interchanged, the former generally was used for the constellation and the latter for its lucida, as in the *Phainomena* and by Geminos and Ptolemy. Still the poets did not always discriminate in this, the versifiers of Aratos confounding the titles notwithstanding the exactness of the original; although Cicero in one place[8] definitely wrote:

[7] See, e.g., George A. Davis, Jr., "The So-Called Royal Stars of Persia", *Popular Astronomy*, Vol. 53, No. 4, April 1945.

[8] "Arctophylax, commonly said to be Boötes." Aratus, *Phainomena*, 96.

> Arctophylax, vulgo qui dicitur esse Boötes.

> Transliterated thus—or Artophilaxe—and as Arcturus, both names are seen for the constellation with writers and astronomers even to the eighteenth century.

Arcturus picked up its name because, according to Isidorus of Seville,[9]

> it follows Arctos, that is, the Great Bear. People have also called this constellation Boötes, because it is attached to the Wain. It is a very noticeable sign with its many stars, one of which is Arcturus. Arcturus is a star located in the sign of Boötes beyond the tail of the Great Bear. For this reason it is called 'Arcturus,' as if it were the Greek ἄρκτου οὐρα, because it is located next to the heart of Boötes.

The breakdown of the Ἀρκτοφύλαξ gives the stem -φύλαξ (phylax), the third declension of φύλαξ, φύλακος, meaning a guard or sentry, from the verb φυλάσσω, "to keep watch and ward, keep guard."[10] The word construction was used similarly in the name of the obsolete constellation Polophylax, or Guardian of the (South) Celestial Pole (see Volume 2).

Sometimes even particular stars stand as guardian figures. In his eponymous *Celestial Handbook* (1978), Robert Burnham, Jr., relates a story about two stars in the bowl of the "dipper" asterism of Ursa Minor. First he quotes a few lines of William Shakespeare[11]:

> The wind-shak'd surge, with high and monstrous mane
> Seems to cast whiter on the burning Bear,
> And quench the guards of th' ever fixed pole...

The 'guards' to which Shakespeare referred were

> the stars Beta and Gamma Ursae Minoris, long known as the traditional 'Guards' or 'Guardians of the Pole'. About 3000 years ago the true Pole was closer to Beta than to the present Polaris, so it is likely that some of the very early references to the 'Star of the North' actually refer to Beta Ursae Minoris instead.

Burnham refers to the Watchman's monologue in the opening of Aeschylus' *Agamemnon* (*c.* 458 BC) and argues that "If this memorable scene took place in the traditional year of the conquest of Troy, 1184 BC, the closest star of any prominence to the true Pole was Beta Ursae Minoris, then about 6° distant."

---

[9] *Etymologies* III.71.8–9, trans. S.A. Barney, W.J. Lewis, J.A. Beach and O. Berghof (2006).

[10] From Henry George Liddell and Robert Scott's *A Greek-English Lexicon*, edited by Sir Henry Stuart Jones and Roderick McKenzie, Oxford: Clarendon Press (1940). Liddell and Scott gave two definitions of φύλαξ, φύλακος: (I) a "watcher, guard, sentinel" in the context of a military, or (II) a "guardian, keeper, protector" or an "observer".

[11] *Othello, the Moor of Venice* Act II, Scene I, lines 15–17.

In Mythology

Guardians are invoked in classical mythology as protectors of valuable goods and/or figures intended to thwart the intention of heroes. These guardians are often envisioned as horrific monsters such as Cerberus, the Lernaean Hydra and Scylla. Others, such as the Gorgons, were possessed of apotropaic magic, their images used to convey protection to people and places. Some figures had mortal associations, such as Aeacus, the former mortal king of Aegina, who became the guardian of the keys of Hades and one of three judges of humanity. In certain stories, guardians were invoked to protect mythological figures, human or divine; for example, the Curetes and Dactyls were a group of spirits appointed by Rhea to guard the infant Zeus in a cave on Mount Ida in Crete. To protect him from the cannibalistic predilections of his father, Cronus, they masked the boy's cries with the sounds of clashing shields and spears in a frenzied dance.

An entirely different type of guardian is the benevolent guide whose protection is extended to mortal or god, usually with a special purpose in mind. Biblical sources,[12] whose origins are probably in the ancient folklore of the greater Near East, refer to "watchers" (ἐγρήγοροι). A similar figure, Vanth, appears in Etruscan mythology with no evident Greek counterpart. Vanth is a psychopomp, from ψυχοπομπός (literally, "guide of souls"), who shepherds newly deceased human souls through the Underworld.

Some guardians bear direct relation to the constellations, as described above; further, some apply particularly to obsolete constellations. Cerberus (Chap. 7), for example, is the vicious hound guarding the gates of the Underworld in Hercules' final Labour. A dragon stands watch over the Golden Fleece in the myth of Jason and the Argonauts in their Colchian adventure to which they sailed in the ship *Argo* (Chap. 5).

Relation of Custos Messium to Roman Guardian Deities

Does Custos Messium bear a direct relation to the concept of guardian figures in classical mythology? A plausible construct exists in which it does, implying Lalande at his most clever. It maybe that he borrowed from the tradition of the Lares, guardian deities in the ancient Roman religion whose origins are uncertain. Their antecedents were likely hero-ancestors worshipped by the Etruscans[13] and assimilated into the Roman world along with many other elements of Etruscan belief and culture. The Lares were popular figures from the Republican period through late Imperial times (Fig. 8.9).

While most often thought of as household gods to whom ordinary Romans kept shrines in their homes, there were a great many Lares, each associated with a particular physical

---

[12]Examples are found in the Book of Daniel (second century BC) and the apocryphal Books of Enoch (second to first centuries BC), the latter of which refers to both good and bad Watchers.

[13]See, e.g., Inez Scott-Ryberg, "Rites of the State Religion in Roman Art," *Memoirs of the American Academy in Rome*, Vol. 22, pp. 10–13 (1955). Scott-Ryberg cited as examples (1) a wall painting in the Tomb of the Leopards at Tarquinia in which offerings are made to Lar-like figures or deified ancestors prior to the celebration of funeral games, and (2) a black-figure Etruscan vase featuring the iconography associated with the later Roman cult.

8.9 First century AD Roman bronze sculpture of a Lar Familiaris holding a cornucopia from Lora del Rio, Spain. Museo Arqueológico Nacional, Madrid; 8 × 5 × 14 cm. Photo ©Luis García, licensed under Creative Commons CC BY-SA 3.0

domain to which their protection was attributed. In that sense they were highly "local" gods whose function very much represented the needs of particular places. Examples include the Lares Compitalicii (the Lares of local neighborhoods celebrated at the Compitalia festival) and the Lares Praestites (the Lares of the city—and later the state and society—of Rome).

Of relevance to the Custos Messium question are the Lares Rurales, the Lares of the fields. They were identified with the protection of agricultural fields from plunder by man and beast alike, and were invoked for the benefit of bountiful harvests. The power of the Roman state in part hinged on its ability to provide for the many millions of people under its control, so any disruption to the supply of food had the potential to bring about social and political instability. The Latin elegiac poet Albius Tibullus (*c.* 55–19 BC) mentions the Lares of the fields in his first book of poetry in recounting[14] the loss of his landed property: "Vos quoque, felicis quondam, nunc pauperis agri custodes, fertis munera vestra, Lares." ("And you, Lares, guardians of once prosperous fields, accept your gifts.") Similarly, Cicero suggests[15] an association between the Lares and rural areas: "In urbibus delubra habento; Lucos in agris habento et Larum sedes." ("Let there be temples in the cities; let there be sacred groves and abodes of the Lares in the countryside.")

The combination of Tibullus' characterization of certain Lares as *custodes agri*, the presence of "a large Wheat Field in this part of the sky," and the derivation of Messier's surname from the Latin word for "harvest" leads to the reasonable conclusion that Lalande alluded to a Roman tradition and employed a play on words to honor the work of Messier without directly placing a living figure among the stars. We have no direct evidence that this is specifically what Lalande intended, but it is not unrealistic to imagine that a man as well-read and cosmopolitan as Lalande could have drawn such a connection, lending the weight of the classical tradition of forming constellations from mythological subjects.

Charles Messier

The object of Lalande's proposed honor, Messier was born on 26 June 1730 in Badonviller, a village in Lorraine region of northeastern France. At the time of Messier's birth Badonviller was part of the Principality of Salm, a small independent state in the Vosges mountains between the Duchy of Lorraine and the Kingdom of France. Charles was the tenth of twelve children born to Nicolas (1682–1741) and Françoise (Grandblaise) Messier (d. 1765). Nicolas was a court usher, a position now roughly equivalent in most western countries to a bailiff, who died when Charles was 11. Six of his siblings also died young.

Messier became interested in astronomer as a teenager when he saw the six-tailed Great Comet of 1744 (C/1743 X1) and an annular solar eclipse visible over Badonviller on 25 July 1748. He pursued an astronomical career in the French Navy, at first in the employ of Joseph-Nicolas Delisle (1688–1768), the Navy's astronomer, who trained him to keep careful records of his work. Messier's first recorded observation was of a Mercury transit on 6 May 1753. By 1759, Messier was appointed the chief astronomer of the Marine Observatory, and later followed Delisle as Astronomer to the Navy in 1771. During this time he searched for (but missed) the first predicted return of Comet 1P/Halley; however,

[14] Tibullus, 1.1.19–24.

[15] *De Legibus* 2.8.

a mistake in the Naval calculations led Messier to search the wrong part of the night sky and the recovery credit went to Johann Georg Palitzsch (1723–1788) on the evening of Christmas Day, 1758.

Messier began earning scientific accolades in his thirties. He was made a fellow of the Royal Society in 1764 and a foreign member of the Royal Swedish Academy of Sciences in 1769, and he was elected to the Académie des Sciences in 1770. In the same year, at age 40, Messier married Marie-Françoise de Vermauchampt (b. 1730), but suffered a double loss 2 years later when Marie-Françoise died after giving birth to their only child, son Antoine-Charles, who died 3 days later. Messier never remarried.

Messier's great love in life was searching for comets, with which he experienced more success than any of his contemporaries. He discovered 13 comets outright between 1750 and 1785, and independently co-discovered a further seven, accounting for nearly half of all comets discovered and co-discovered in Europe during the eighteenth century by a single individual. The process was slow and repetitive, and involved sweeping the sky every clear night with a 100 mm (4-in.) refracting telescope on the roof of the Hôtel de Cluny in Paris (Fig. 8.3).

It also required an intimate familiarity with the contents of the night sky developed over many years of repeatedly observing the same patches of sky devoid of comets. When first discovered visually, comets are generally faint and indistinct 'blobs' of light against a starry background that manifest themselves remarkably like the same bright galaxies, nebulae, and clusters of stars observable by Messier's telescope in the Paris of his times. But comets betrayed their nature by virtue of their slow relative motion against the distant "fixed stars," and any candidate comet could be verified within a couple of hours by coming back to the discovery field and looking for evidence of apparent motion. Any object that did not move relative to the stars was a permanent feature, and an annoyance to the potential comet discoverer.

On the night of 28 August 1758, Messier stumbled across one of these false alarms while sweeping the morning sky in the constellation Taurus.[16] While it had the appearance of a new comet, observations over several nights showed no apparent motion; Messier noted it for future reference and moved on. In 1760 he found a similar object in Aquarius[17] previously noted by the French astronomer Jean-Dominique Maraldi (1709–1788) in 1746 while observing a comet with Jacques Cassini (1677–1756).

The third time was the proverbial charm for Messier and his future fame in the history of astronomy. On the night of 3 May 1764, he found another false alarm in the space between Boötes and Coma Berenices.[18] Duly noted, Messier decided it would be best for his own work as well as that of other astronomers to systematically catalog these stationary (but otherwise comet-like) objects so that no one would even briefly mistake them for comets.

[16]Messier recorded this as object one in his later catalog; it is now familiar as the Crab Nebula supernova remnant (Messier 1).

[17]Messier 2, a globular star cluster.

[18]Messier 3, a globular star cluster.

He later wrote[19] that he undertook the resulting work "so that astronomers would no more confuse these same nebulae with comets just beginning to appear." He continued:

> I observed further with suitable refractors for the discovery of comets, and this is the purpose I had in mind in compiling the catalog. After me, the celebrated *[William]* Herschel published a catalog of 2000 which he has observed. This unveiling the heavens, made with instruments of great aperture, does not help in the perusal of the sky for faint comets. Thus my object is different from his, and I need only nebulae visible in a telescope of two feet *[focal length]*. Since the publication of my catalog, I have observed still others: I will publish them in the future in the order of right ascension for the purpose of making them more easy to recognize and for those searching for comets to have less uncertainty.

Messier's list was compiled from bright objects observable from Paris, down to a declination of about −35°; true to his stated intent, the objects are not grouped into any kind of scientific taxonomy nor are they arranged according to their places in the sky. The first version of the list, consisting of 45 objects, was published in 1774 in the journal of the Académie des Sciences. Of the 45 objects, only 17 were original discoveries by Messier. The list continued to grow through the 1770s and by 1780 contained 80 objects. The final version of the list as published in Messier's lifetime included 103 objects,[20] forty of which were his discoveries. Between 1921 and 1967, later historians of astronomy noted seven additional objects that were recorded by either Messier or his assistant, Pierre Méchain (1744–1804), in the years following publication of the 103-object list. These objects are widely considered valid additions to Messier's published lists.

Charles Messier died in Paris on 12 April 1817 at the age of 86. He was buried in Sect. 11 of Père Lachaise Cemetery, not far from the grave of Frédéric Chopin. Later, an eponymous crater on the Moon and the asteroid (7539) Messier were named in his honor.

DISAPPEARANCE

Custos Messium remained in circulation among European and American star atlases for roughly a century after its introduction. Alexander Jamieson included it in his atlas 1822 and it appeared in the contemporaneous *Urania's Mirror* 1825, but by mid-century its use was in decline. Kendall (1845) included it in his list of constellations as mentioned previously, referring to it as a "new constellation", while Chambers (1877) referred to Custos Messium without naming it: "Lalande placed Messier's name in the heavens, by forming a constellation in his honour, near Tarandus."

[19]In *Connaissance de Temps*, a French nautical almanac (1801).

[20]"Catalogue des Nébuleuses & des amas d'Étoiles" in *Connaissance des Temps* for 1784, pp. 227–267 (1781).

It soon fell off of lists of constellations in contemporary astronomy texts (e.g., Rosser 1879). By the end of the nineteenth century, the Harvest Keeper had moved on from the skies over Paris and "... passed out of the recognition of astronomers." (Allen, 1899) Among its last mentions in a reference work is an entry[21] in Volume 2 of *The Century dictionary and cyclopedia* (1913): "Custos Messium, a constellation proposed by Lalande in 1775. It embraced parts of Cepheus, Cassiopeia, and Camelopardalis, and had a star of the fourth magnitude stolen from each of the last two constellations," but by that time it no longer appeared on maps. In the constellation boundaries adopted by the IAU in 1928, the lines demarcating Cassiopeia were carefully drawn to incorporate almost all the stars that once belonged to Custos Messium (Fig. 8.10).

[21] *The Century dictionary and cyclopedia: with a new atlas of the world : a work of general reference in all departments of knowledge*, ed. William Dwight Whitney and Benjamin Eli Smith, The Century Co., New York, Volume 2, p. 1414 (1913).

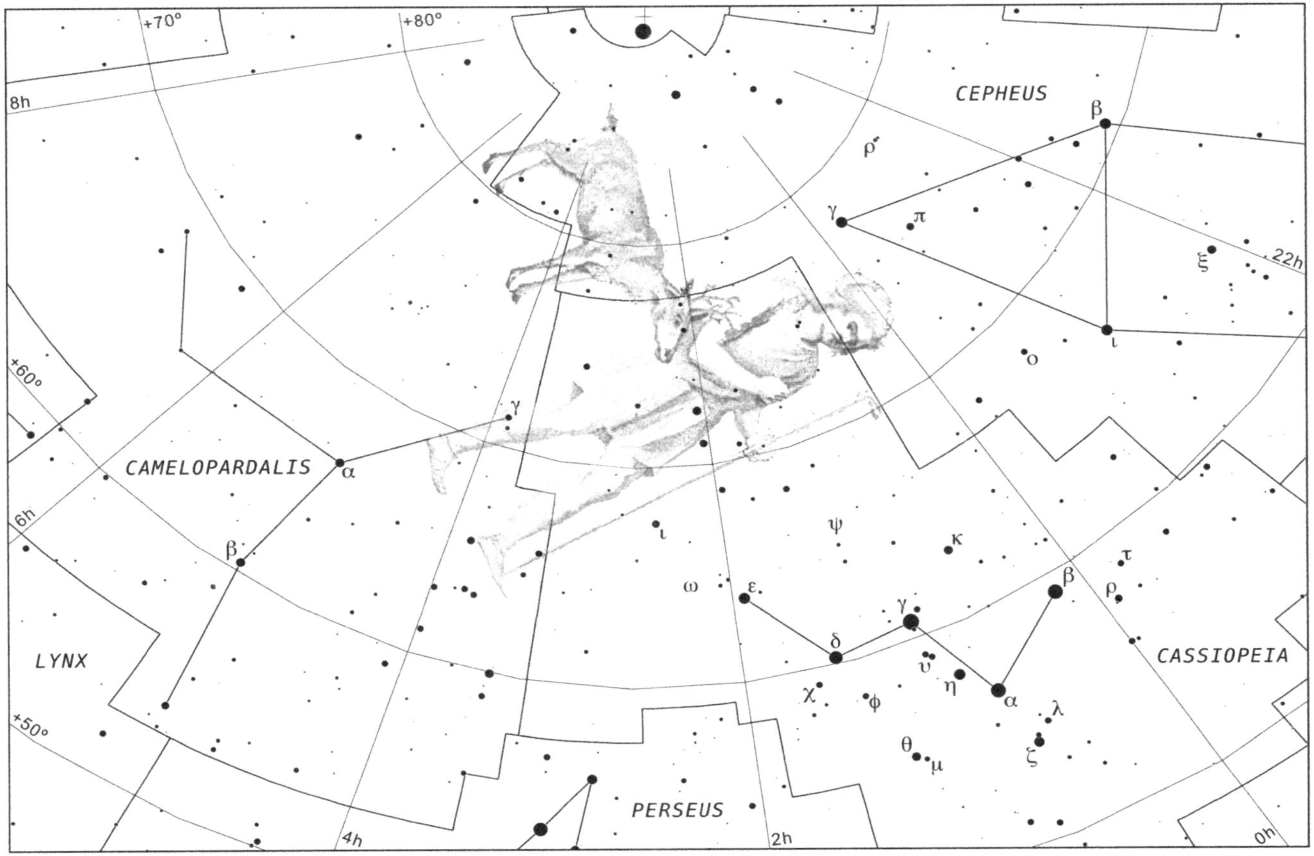

8.10 The figures of Rangifer (*top*; Chap. 20) and Custos Messium (*above center*) from Plate 3 of Bode's *Uranographia* (1801b) overlaid on a modern chart

9

Felis

The Cat

Genitive: Felis
Abbreviation: Fel
Location: Between Antlia and Hydra[1]

ORIGIN AND HISTORY

In around 1799, Joseph Jérôme Lefrançois de Lalande formed a new constellation out of a few faint stars below the main body of Hydra, but above the principal stars of Antlia. He named the figure Felis, the Cat, which he formally published in *Bibliographie Astronomique* (1805). Lalande often introduced new constellations, frequently under the influence of intoxicating beverages (see Chap. 8), although Felis was certainly one of his most whimsical suggestions. As he later wrote in *Histoire abrégée de l'astronomie* (1803),

> The large number of stars I supplied to M. Bode gave me some right to shape new constellations. There were already thirty-three animals in the sky; I put in a thirty-fourth one, the cat.

[1] *"Between Hydra and the Compass"* (Kendall, 1845); *"On the meridian with Regulus, the principal star in Leo, and ... south of Hydra"* (Bouvier, 1858); *"Formed from stars between Antlia and Hydra"* (Bakich, 1995); *"Between Antlia and Hydra"* (Ridpath, 1989).

J.C. Barentine, *The Lost Constellations*, Springer Praxis Books,
DOI 10.1007/978-3-319-22795-5_9

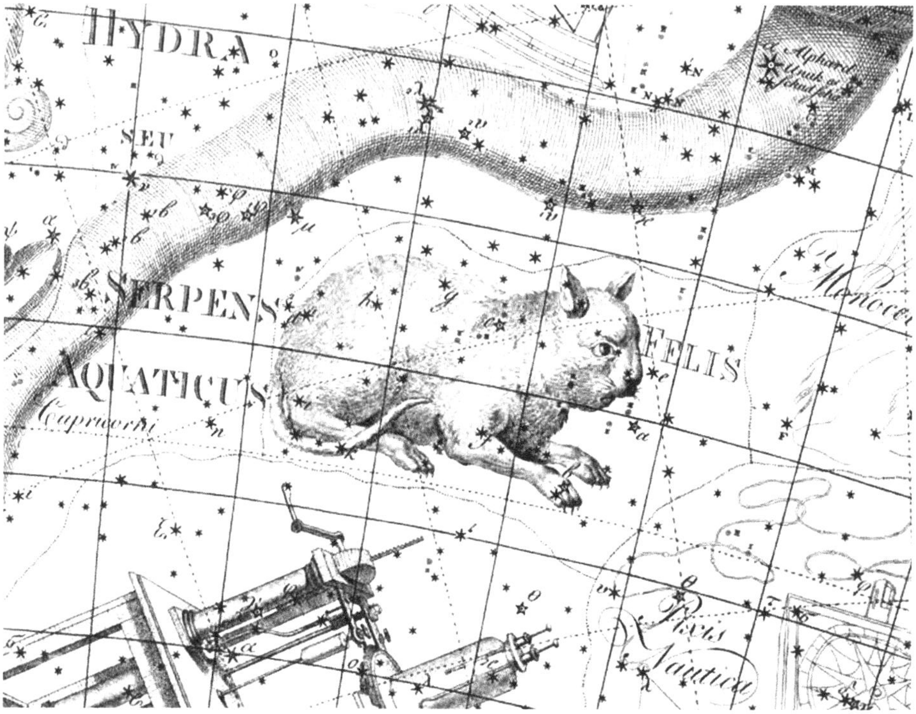

9.1 Felis as shown on Plate 19 of Bode's *Uranographia* (1801b)

"M. Bode" was Johann Elert Bode, Lalande's friend and frequent recipient of Lalande's sometimes drunken missives. However, Bode appreciated Lalande's enthusiasm and placed several of his suggestions onto charts beginning with 1801's *Uranographia* (Fig. 9.1). Bode described[2] the origins of Felis in *Allgemeine Beschreibung und Nachweisung der Gestirne* (1801a) (Fig. 9.2):

> This constellation has been only recently proposed by de la Lande to fill the previously unoccupied area south of the neck [throat] of Hydra. It consists only of stars of the fifth magnitude and below.

According to Karl Shuker,[3] Lalande was a well-known lover of cats "who had lamented the domestic cat's absence in a sky populated by no less than three different domestic dog constellations (Canis Major, Canis Minor, and Canes Venatici), as well as three wild cats

[2]"Dies Gestirn ist erst ganz neulich von de la Lande zur Ausfüllung des bis dahin noch unbesetzten Raums südlich unterm Hals der Hydra, eingeführt. Es besteht nur aus Sternen der fünften und geringern Grösse."

[3]From Shuker's forthcoming *Last Night I Saw The Strangest Cat: A Cat-alogue of Feline Magic, Mythology, and Mystery.*

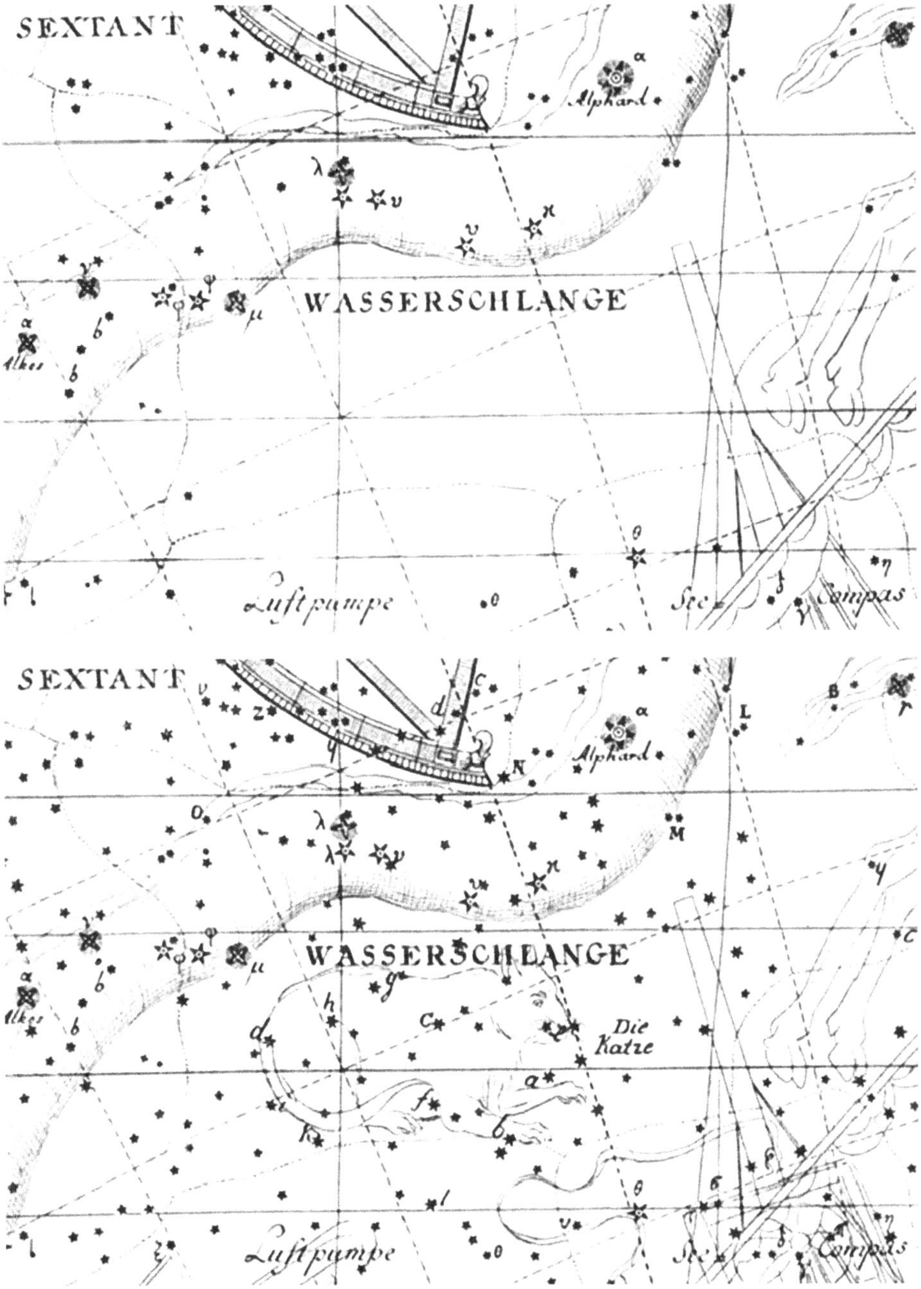

9.2 Before and after: Detail of Plate 26 in two editions of Bode's *Vorstellung der Gestirne*: 1782 (*top*) and 1805 (*bottom*), showing the introduction of Felis ("Die Katze") in the intervening years

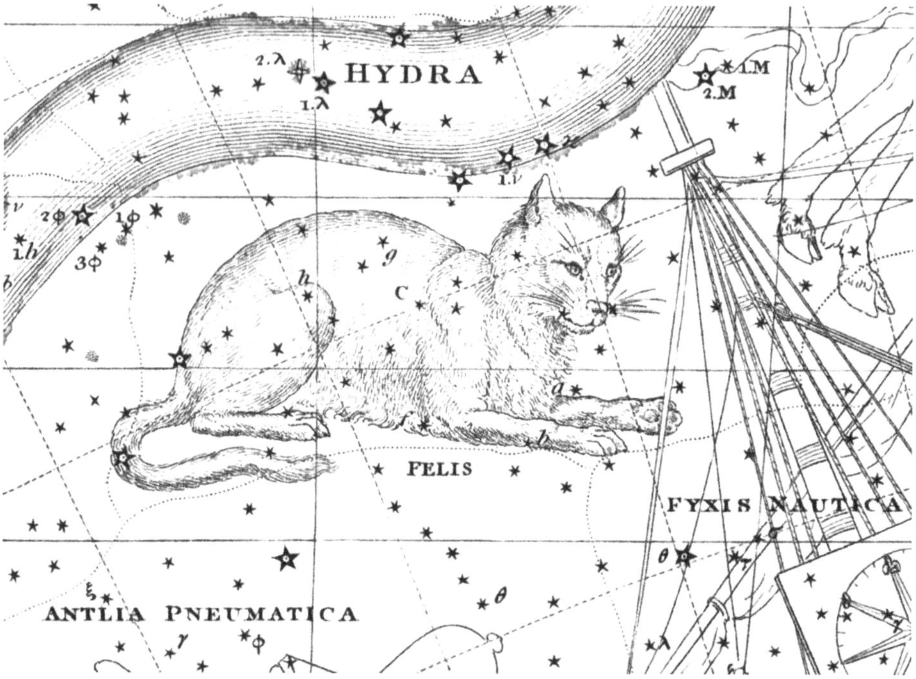

9.3 Felis depicted on Plate XXVI of Alexander Jamieson's *Celestial Atlas* (1822)

(Leo, Leo Minor, and Lynx)." Ridpath (1989) noted that Lalande said he was inspired to add his own cat to the firmament upon being inspired by a recent poem about cats by Claude-Antoine Guyot-Desherbiers.[4]

As a result of its publication in Bode's atlases, Felis gained a following among cartographers through the nineteenth century. In the same year that Lalande published it in *Bibliographie Astronomique*, August Gottleib Meissner showed it as "Katze" in Figure XXXV, Table 33 of his *Astronomischer Hand-Atlas*. Alexander Jamieson drew a rather detailed rendition in his *Celestial Atlas* (1822; Fig. 9.3), and it featured prominently with its only two bright stars on Plate 32 of *Urania's Mirror* (1825). James Middleton included Felis on Plate 3 ("Spring") of his *Celestial Atlas* (1842), and in 1845, Ezra Otis Kendall mentioned it in *Uranography* as lying "between Hydra and the Compass." On Plate 2 of *Himmels-Atlas* (1849), Christian Gottlieb Riedig curiously labeled it "Haase," an obsolete spelling of the German *Hase* (hare), but correctly labeled the figure "Katze" on Plate 19.

The figure was fairly well established by midcentury and appeared on most popular charts. Hannah Bouvier believed it to have been "introduced by Bode" according to her *Familiar Astronomy* (1858) and described it as "represented by a cat crouching." In 1862,

[4]French writer and politician (1745–1828).

Ludwig Preyssinger showed the figure as "Chat" on Plate II of *Astronomie Populaire ou Description des Corps Célestes*. George Chambers attributed it to "Bode's maps" without citing a date in *A Handbook of Descriptive Astronomy* (1877), and the following year it made an appearance in Simon Newcomb's widely-read *Popular Astronomy*. Richard Hinckley Allen suggested that Father Secchi showed Felis on a celestial globe around this time ("Except with Secchi, who included it as Gatto in his planisphere of 1878, it has long been discontinued in the catalogues and charts."), but no trace of this globe is now found. Among its last appearances on a popular chart is on the southern hemisphere all-sky map in Eliza A. Bowen's *Astronomy By Observation* (1888).

ICONOGRAPHY

A Brief History of Cats

The common cat (*Felis catus* or sometimes *F. silvestris catus*) is the domesticated species within the family *Felidae* familiar to humans around the world. Cats are kept by people for both their companionship as well as useful features like the ability to hunt small vermin such as mice. Their predatory and crepuscular activity habits position them in an ecological niche which is both at once dependent on, and independent of, humans. They exploit characteristics such as sharp, retractable claws, flexible bodies, sharp reflexes, excellent hearing and highly sensitive night vision make them well adapted to hunting small prey both in the wild and indoors. As obligate carnivores, cats have evolved to efficiently digest meat and they experience difficulty with digesting plant matter. Rather, their skulls and specialized jaw structure are finely tuned to both killing prey and tearing meat. The typical adult cat measures 20–25 cm in height and 45 cm in body length, with an additional ~30 cm of tail length, and weigh between about 4–5 kg.

Domestic cats are part of the genus *Felis*, containing about seven species depending on classification scheme.[5] Other species in the same genus include the Chinese mountain cat (*F. bieti*), the Arabian sand cat (*F. margarita*) and the southeast Asian "jungle cat" (*F. chaus*). Phylogenetic analysis of domestic cats show they are a subspecies of the wildcat, *F. silvestris*. Cats have about 20,000 genes distributed among 38 chromosomes,[6] and are subject to a wide variety of inherited diseases.

Felidae have evolved rapidly in the recent past, and share a common ancestor with much larger cars such as tigers, lions and cougars roughly between 10 and 15 million years ago; all cats in *Felis* share a common Asian ancestor that lived about 6–7 million years

[5] W. Johnson and S. J. O'Brien, "Phylogenetic Reconstruction of the Felidae Using 16S rRNA and NADH-5 Mitochondrial Genes" *Journal of Molecular Evolution* 44 (0): S98–S116 (1997).

[6] J.U. Pontius, J.C. Mullikin, and D.R. Smith, "Initial Sequence and Comparative Analysis of the Cat Genome," *Genome Research* 17 (11): 1675–1689 (2007); W. Nie, J. Wang, and P.C. O'Brien, "The Genome Phylogeny of Domestic Cat, Red Panda and Five Mustelid Species Revealed by Comparative Chromosome Painting and G-banding," *Chromosome Research* 10, 3, pp. 209–222 (2002).

ago.[7] Because of cats' appearance in ancient Egyptian art, it was long assumed that their domestication took place in the early history of Egyptian civilization, but recent research[8] suggests that domestication may have occurred around the Mediterranean as early as about 7500 BC. Genetic evidence[9] indicates a divergence date of about 8000 BC at which early *F. catus* separated into a distinct line from African wildcats (*F. silvestris lybica*). At least five *F. s. lybica* females were domesticated around this date, being the ancestors of all modern domestic cats.

It is thought that cats initially formed attachments to humans as early as 10,000 BC by hunting rodents that fed off stored grain after the advent of settled agricultural communities; however, observations of modern domestic cats interbreeding with their feral cousins suggest that domestication was a very slow process. As they have changed little in the time since domestication, they retain the ability to hybridize with other species within *Felis* .[10] That cats have not become entirely dependent on humans, and thus subject exclusively to their artificial selection, is borne out by the much slower rate of evolutionary change among cats relative to dogs and other domestic species[11]; in a sense, cats don't really need us, and retain enough wild behavior to sustain themselves in the absence of humans. Some have even suggested[12] that certain cat characteristics and behaviors evolved in the wild, and their social nature, small size and comparatively high intelligence may have 'preadapted' them to become more easily domesticated.

Cats have a long history in human culture. Their first appearance in art was in ancient Egypt, where cats were considered sacred (Fig. 9.4). They were identified with the goddess Bastet, originally a fierce lioness figure worshipped as early as the Second Dynasty (*c.* 2900 BC) whose image slowly evolved into that of a cat as domestic cats became increasingly important in Egyptian life. Befitting their sacred position in society, cats came to be mummified after death, and the respect with which cats were treated in death mirrored the regard in which humans held them in life. Bastet was seen in Lower Egypt as Bast, the goddess of warfare. By the time of the New Kingdom, a religious cat cult had developed in the Nile Delta region, and Shoshenq I (*c.* 943–922 BC) built up the city of Bubastis east

---

[7]W.E. Johnson, et al., "The Late Miocene Radiation of Modern Felidae: A Genetic Assessment," *Science*, 311, 5757, pp. 73–77 (2006).

[8]J.D. Vigne, et al., "Early taming of the cat in Cyprus," *Science* 304, 5668, p. 259 (2004). The work involved discovery of a Neolithic human burial on Cyprus including the remains of a young adult cat. Because the island was settled by agrarian colonists from Turkey, the first cat domestication may have taken place earlier on the mainland.

[9]C. A. Driscoll, et al., "The Near Eastern Origin of Cat Domestication," *Science* 317, 5837, pp. 519–523 (2007).

[10]R. Oliveira, et al., "Hybridization Versus Conservation: Are Domestic Cats Threatening the Genetic Integrity of Wildcats (*Felis silvestris silvestris*) in Iberian Peninsula?" *Philosophical Transactions of the Royal Society of London, B: Biological Science* 363, 1505 (2008).

[11]M.J. Lipinski, et al., "The Ascent of Cat Breeds: Genetic Evaluations of Breeds and Worldwide Random-bred Populations," *Genomics* 91, 1, pp. 12–21 (2008); C. Cameron-Beaumont, S.E. Lowe, and J.W.S. Bradshaw, "Evidence Suggesting Pre-adaptation to Domestication throughout the Small Felidae," *Biological Journal of the Linnean Society*, 75, 3, pp. 361–366 (2002).

[12]Cameron-Beaumont et al. (2002).

9.4 Relief carvings in a detail of the sarcophagus of Ta-miu, a cat belonging to Crown Prince Thutmose, eldest son of the Egyptian pharaoh Amenhotep III (r. *c.* 1386–1350 BC) and his consort, Queen Tiye (*c.*1398–1338 BC). Photo by Rafaèle Larazoni, licensed under CC BY 2.0

of the Delta into an important city for the worship of Bast. Her cult persisted until it was officially banned by the Roman government in the fourth century AD.

A folklore of cats developed in the classical world based on their characteristics, both real and perceived. After Egypt, Greece was the next most important center of cat culture in the ancient world. The classical Greek word for cat from the fourth century BC, αἴλουρος, was derived from the words αιολος ("waving") and ουρος ("tail"). The fabulist Aesop celebrated cats' hunting ability, portraying them as calculating and cunning in their interactions with birds, notably domestic fowl; Engels (2001) describes cats as 'typecast' in the fables: "It is a crafty predator, an especial terror to mice, an opportunist who is willing and capable of taking advantage of another animal's mistakes or distress to obtain a meal."

In the Roman world, cats were regarded for both their aesthetic qualities as exotic pets as well as for their vermin-hunting abilities. Romans associated cats' aloofness with liberty, and the goddess Libertas was often depicted with a cat at her feet. They were given free reign to roam around temples, the only animal so permitted. Cats are also sometimes

9.5 Roman mosaic showing a cat attacking a pheasant, from the House of the Faun (IV, 12, 2) at Pompeii (*c.* first century AD). Naples National Archaeological Museum, Inv. 9933. Photo by Marie-Lan Nguyen (2011)

seen in the iconography of Diana, goddess of the hunt. The Romans inherited a great deal of the Greeks' cultural legacy, and cats featured in their art and literature as well (Fig. 9.5). Plutarch wrote[13] that "the cat is excited to frenzy by the odour of perfumes," implying that their penchant for personal cleanliness rendered them highly reactive to offensive scents. Pliny the Elder noted[14] their reproductive habits in *Natural History* implying in them a symbolism for the lustfulness of humans.

In later European history, cats retained a special place in society. Norse mythology associated the goddess Freyja with two companion cats who pulled her chariot; farmers seeking protection of their crops sometimes left pans of milk in the fields for Freyja's cats. Vikings kept cats for many of the same reasons as their European predecessors, both for companionship as well as their rat-catching prowess. During the Medieval period, a folklore developed linking cats to witches, and cats were sometimes tortured or killed in public festivities as a means of warding off bad luck. Their reputation led to an unintended acceleration in the incidence of plague during the Black Death in the mid-fourteenth century; fearing their evil influence Europeans killed cats en masse, allowing the rapid proliferation of plague-carrying fleas on overabundant vermin. Later, black cats were specifically identified as unlucky in Europe and America, and it was commonly held that the deaths of infants were attributable to cats suffocating the children by "stealing" their breath. However, popular folk tales arose depicting cats as sympathetic and even heroic

[13]*Coniugalia Praecepta* 44, trans. F. C. Babbitt.
[14]*Naturalis Historia* 10.83.

9.6 Cats being instructed in the art of mouse-catching by an owl. Lombard School, *c.* 1700; oil on canvas within a painted lunette, 83.5 cm × 110.5 cm

characters, such as Puss In Boots and Dick Whittington's cat. By Victorian times, the domestic cat's image was fully rehabilitated, and cats took their place as one of the most popular pets in the world.

A Census of the Cosmic Zoo

Cats are only one member of a menagerie of animals represented among the constellations through human history. Including obsolete figures, a total of 52 constellations proposed since antiquity represent some kind of animal; of these, 41 appear on the IAU's list of "official" constellations while 11 have fallen into disuse. The extant animal figures constitute a full 46.5 % of all constellations, making them by far the most prevalent general type.

The 52 current and former animal constellations are gathered into a taxonomy of five broad categories shown in Table 9.1. Some figures fall into more than one category, while in the specific instance of Turdus Solitarius/Noctua,[15] both versions of the figure have been counted independently.

[15] A similar example of multiple identifications of the same group of stars is Apis/Musca Borealis, but there are fewer compelling reasons to differentiate these similar insects.

Table 9.1 A taxonomy of animal constellations. Notes for column two: (a) Ganymede/Antinoüs and Zeus; (b) Orion; (c) Biblical Noah; (d) Hercules; (e) Centaurus; (f) Perseus; (g) Asclepius; (h) Zeus; (i) Callisto; (j) Arcas. Notes for column six: (1) Hevelius, 1690; (2) Keyser & de Houtman, before 1600; (3) Plancius, 1613; (4) Plancius, 1592; (5) Lalande, 1799; (6) Plancius, 1598; (7) Jamieson, 1822; (8) Le Monnier, 1743; (9) Croswell, 1810; (10) Poczobut, 1777; (11) Le Monnier, 1776

Monsters	*Mythical*	*Exotic*	*Familiar*	*Classical*	*Late*
Capricornus	Aquila[a]	Apus	Anser	Aquila	Anser[1]
Cerberus	Canis Major[b]	Camelopardalis	Aries	Aries	Apus[2]
Cetus	Canis Minor[b]	Chamaeleon	Canes Venatici	Canis Major	Camelopardalis[3]
Draco	Columba[c]	Dorado	Canis Major	Canis Minor	Canes Venatici[1]
Hydra	Leo[d]	Grus	Canis Minor	Capricornus	Cerberus[1]
Monoceros	Lepus[b]	Hydrus	Columba	Cetus	Chamaeleon[1]
Pegasus	Lupus[e]	Leo	Corvus	Corvus	Columba[4]
Phoenix	Pegasus[f]	Leo Minor	Cygnus	Cygnus	Dorado[2]
	Scorpius[b]	Pavo	Delphinus	Delphinus	Felis[5]
	Serpens[g]	Rangifer	Equuleus	Draco	Gallus
	Taurus[h]	Sciurus Volans	Felis	Equuleus	Grus[2]
	Ursa Major[i]	Scorpius	Gallus	Hydra	Hydrus[2]
	Ursa Minor[j]	Tucana	Lacerta	Leo	Lacerta[2]
		Turdus Solitarius	Lepus	Lepus	Leo Minor[1]
		Volans	Lupus	Lupus	Lynx[1]
			Lynx	Pegasus	Monoceros[3]
			Musca	Pisces	Musca[2]
			Musca Borealis	Piscis Austrinus	Musca Borealis[6]
			Noctua	Scorpius	Noctua[7]
			Pisces	Serpens	Pavo[2]
			Piscis Austrinus	Taurus	Phoenix[2]
			Serpens	Ursa Major	Rangifer[8]
			Taurus	Ursa Minor	Sciurus Volans[9]
			Taurus Poniatovii		Tucana[2]
			Ursa Major		Taurus Poniatovii[10]
			Ursa Minor		Turdus Solitarius[11]
			Vulpecula		Volans[2]
					Vulpecula[1]

- **Mythical or "impossible" monsters**. Fantastic or chimerical animals of classical mythology (8 total; 7 extant, 1 obsolete).
- **Mythical animals**. Otherwise ordinary animals featured in specific stories from classical mythology. Their associated (human) mythological figures are indicated in the table caption with lettered notes (13 total; 13 extant, 0 obsolete).
- **Exotic animals of faraway places**. Typically unfamiliar creatures from places other than Europe, often first encountered during voyages of discovery (15 total; 12 extant, 3 obsolete).
- **Familiar animals**. Common animals often encountered by Europeans (27 total; 21 extant, 6 obsolete).
- **Classical figures**. Creatures introduced before Ptolemy (23 total, 23 extant, 0 obsolete).
- **Late inventions**. Creatures introduced after Ptolemy. Their inventors are indicated in the table caption with Arabic-numbered notes (27 total; 18 extant, 9 obsolete).

The animals may be further categorized by a variety of criteria. For example:

- **Real vs. impossible**. Real animals total 40, of which 32 are extant and 8 obsolete. Of chimerical or otherwise "impossible" creatures from mythology, there are 8 total, of which 7 are extant and 1 obsolete.
- **Domesticated vs. wild**. Of real animals, 9 are domesticated (6 extant, 3 obsolete) and 31 are wild (26 extant, 5 obsolete).
- **Primary place of habitation**. Of both real and impossible animals, there are three primary places they inhabit: Land or trees (30 total, 22 extant, 8 obsolete), water (9 total, 9 extant, 0 obsolete), and the air (11 total, 9 extant, 2 obsolete).
- **Type of animal**. The real animals represented are broadly characterized as fishes (5 total, 5 extant, 0 obsolete), reptiles (6 total, 6 extant, 0 obsolete), birds (13 total, 9 extant, 4 obsolete), insects (3 total, 2 extant, 1 obsolete) and mammals (19 total, 15 extant, 4 obsolete). Of the mammals, three are dogs (all extant).

Was Lalande justified in his complaint that the heavens didn't adequately represent cats? Of the 52 animals ever proposed as constellations, four were members of *Felidae*: Felis, Leo, Leo Minor, and Lynx. Only Leo was an ancient figure, whereas Hevelius introduced two (Leo Minor and Lynx, in 1690), and Lalande one (Felis). Of the four cats, all but Lalande's remain in the IAU's list of "official" constellation.

DISAPPEARANCE

While Felis had its (mostly European) fans, interest was not sufficiently sustained to see this constellation through the turn of the twentieth century. Most early American cartographers seemed loathe to adopt Lalande's creation. Ever the contrarian, William Croswell did not show Felis on his *Mercator Map of the Starry Heavens* (1810); in fact, he did not show any stars at all in the position of Lalande's cat. Neither did Jacob Green make any mention of Felis in his *Astronomical Recreations* (1824), and it is not shown on his Plate 17 even though he noted the presence of nearby, then-recently introduced constellations such as Robur Carolinum and Antlia. And in 1835, Elijah Hinsdale Burritt left the figure of Felis off of Plate 4 in *Atlas Designed to Illustrate the Geography of the Heavens* even though he plotted its stars.

The Cat began to disappear from the pages of many popular atlases in the last two decades of the nineteenth century. It does not appear in the constellation lists in William Henry Rosser's *The Stars and Constellations* (1879), and Flammarion (1882) made a strong case for removal of the last traces of "Lalande's cat" from modern charts[16]:

> Lalande's cat – the latest constellation created – is more than superfluous and it has disappeared... from modern charts. The small stars which were used to form it have returned to Hydra and to the Air Pump [Antlia Pneumatica].

Around the turn of the twentieth century, no one regarded Felis as a proper constellation anymore; Allen (1899) recounts its history but says "it has long been discontinued in the catalogues and charts." When the IAU settled on boundaries for the canonical list of modern constellations in 1928, the stars formerly belonging to Felis were subsumed into the neighboring constellations of Hydra, Antlia, and Pyxis (Fig. 9.7).

[16]Translated in Muriel Beadle's *The Cat: History, Biology, and Behavior*, New York: Simon & Schuster, p. 85 (1977).

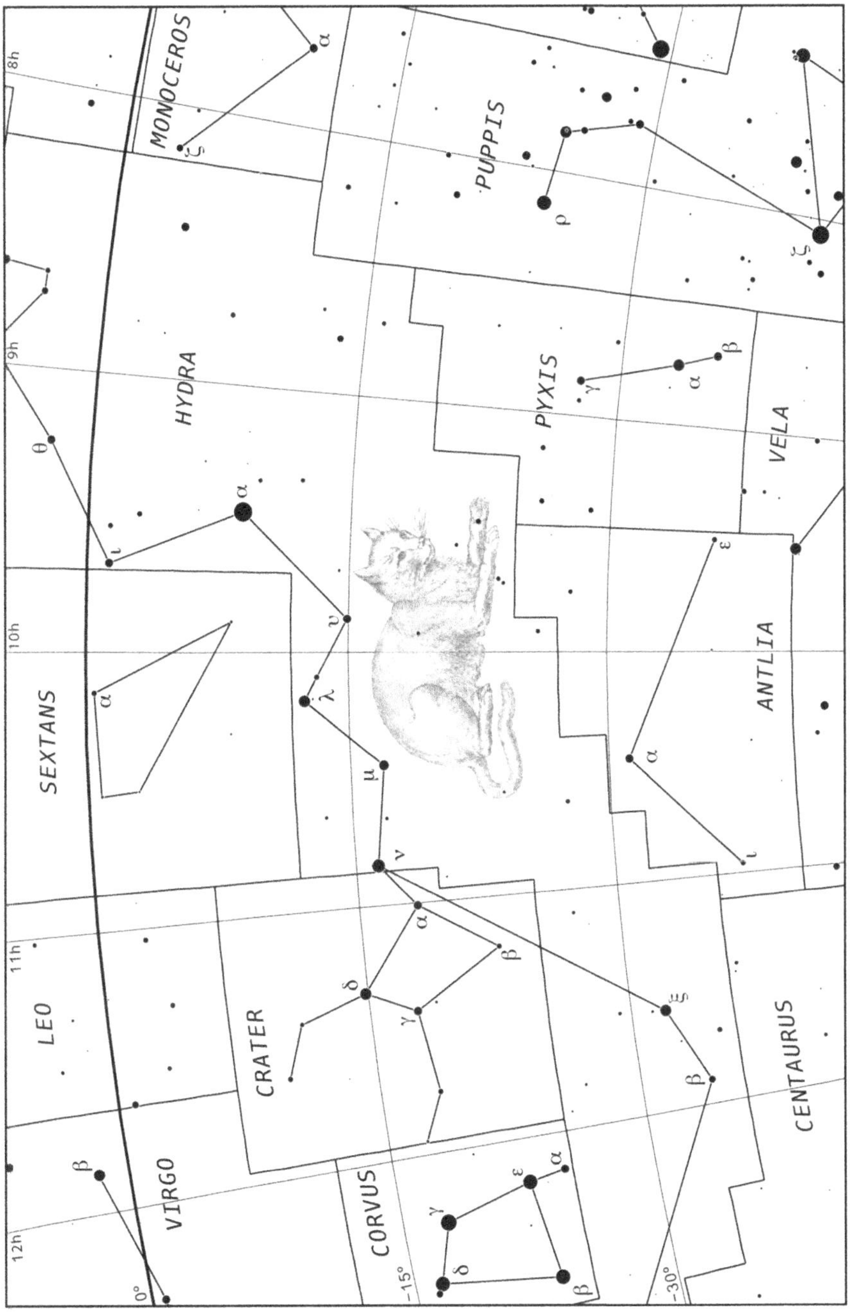

9.7 The figure of Felis from Plate 26 of Alexander Jamieson's *Celestial Atlas* (1822) overlaid on a modern chart

10

Gallus

The Rooster

Genitive: Galli
Abbreviation: Gal
Location: In the northern part of what is now Puppis, south of the celestial equator in the Milky Way.[1]

ORIGIN AND HISTORY

Gallus was introduced by Petrus Plancius on a globe he produced in 1612, although he left no written record as to why he chose the Rooster as the figural interpretation of these stars nor his intended meaning in doing so (Fig. 10.1). The constellation might have been lost to history altogether if not for the efforts of Jacob Bartsch, who described Gallus in *Usus astronomicus planisphaerii stellati* (1624), depicting it on the accompanying charts (Fig. 10.3). Bartsch[2] interpreted Gallus to represent the rooster heard crowing as Peter denied Christ for the third time in the canonical Gospels:

[1] *"Formed… out of stars between Argo Navis and Canis Major"* (Bakich, 1995); *"[I]n the Milky Way, south of the celestial equator in the northern part of what is now Puppis"* (Ridpath, 1989).

[2] "GALLUS ἀλεκτρυὼν *vel* ἀλεκτωρ prope tergum Canis majoris: Noviter adjectus: *Olim ad Arg. navem pertinuerunt ejus stellae, inprimis 2. lucidiores. Mihi sit Gallus Petri, Matth, 26. v. penult. Marc. 14. v. ult."* (p. 97)

J.C. Barentine, *The Lost Constellations*, Springer Praxis Books,
DOI 10.1007/978-3-319-22795-5_10

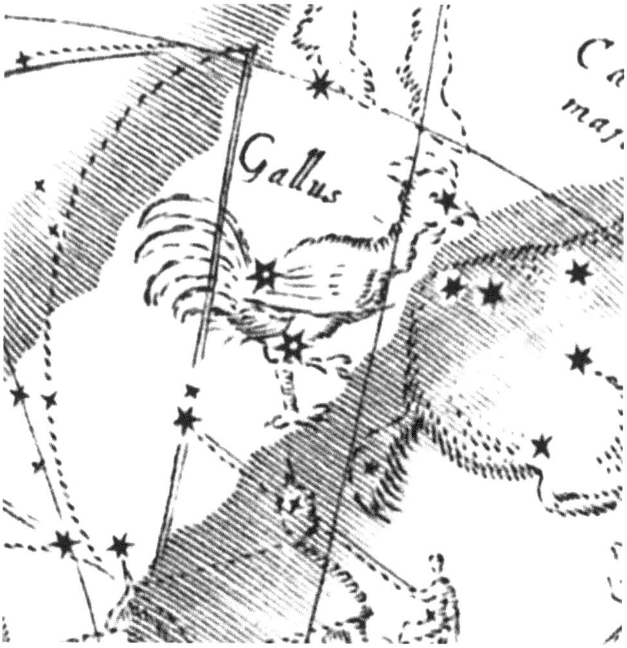

10.1 Gallus as shown on Isaac Habrecht's *Globus Coelestis* (1621)

> THE ROOSTER *alektruōn* or *alektor near the back of Canis Major: Newly added:* Once its stars belonged to Argo Navis, in particular the brightest two. It is, to me, the Petrine Rooster, Matthew 26:74.[3] Mark 14:72.[4]

He was, however, evidently unaware that Plancius originally devised the constellation; his source for Gallus was a 1621 globe made by Isaac Habrecht and, according to Ridpath (1989), "mistakenly attributed to him." Bartsch was motivated to devise Biblical interpretations of the constellations wherever possible, and decided that Gallus was the rooster that crowed after the third time Peter denied knowing Jesus as related in the canonical gospels (Figs. 10.4 and 10.5).

Gallus was not accepted by most cartographers, but there remained the problem of what to do with its relatively bright stars that were otherwise not formally incorporated into neighboring Argo Navis (Fig. 10.2). While not naming it as a separate figure, Johannes

[3]"Then he began to call down curses on himself and he swore to them, 'I don't know the man!' Immediately a rooster crowed." (NIV)

[4]"Immediately the rooster crowed the second time. Then Peter remembered the word Jesus had spoken to him: 'Before the rooster crows twice you will disown me three times.' And he broke down and wept." (NIV)

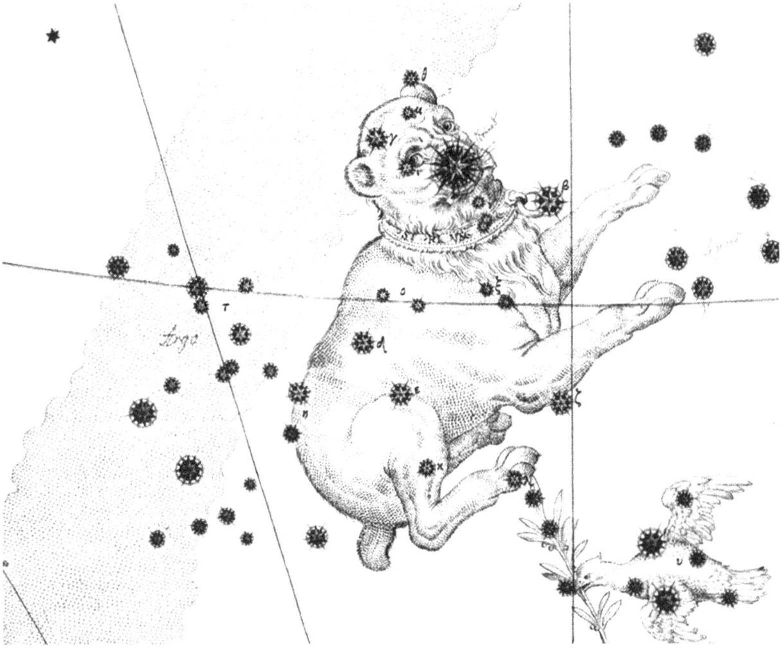

10.2 Prior to (and again, long after) Plancius' introduction of Gallus, cartographers typically left its contents as unformed stars belonging to Argo Navis, as seen here on Leaf 38 (verso) of Bayer's *Uranometria* (1603). The stars from which Plancius created Gallus are seen east (*left*) of Canis Major (*center*) in the shaded band of the Milky Way, labeled simply "Argo"

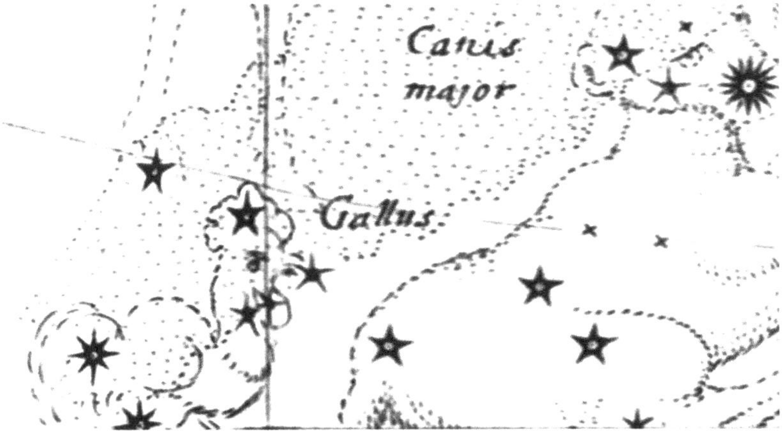

10.3 Gallus and Canis Major depicted in Jacob Bartsch's *Planisphaerium stellatum*... (1661)

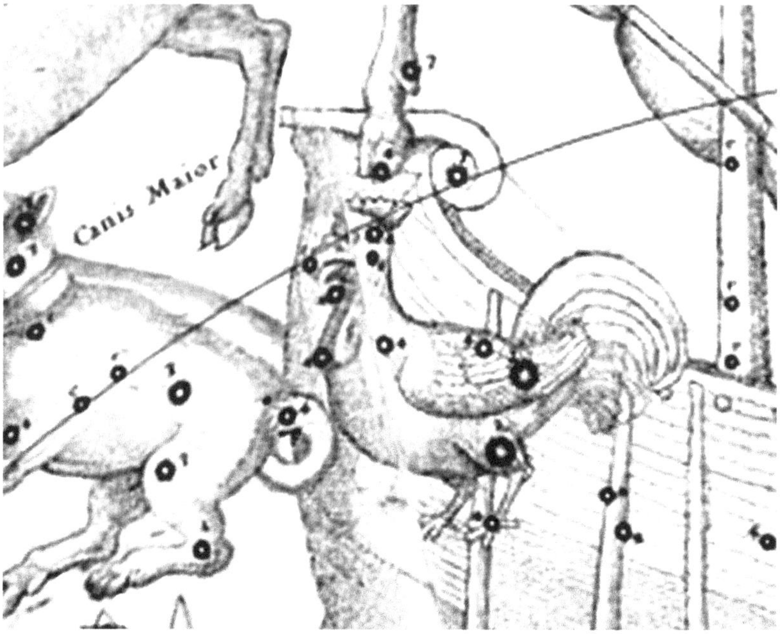

10.4 Gallus depicted on globe gores printed in 1695 by the Italian mapmaker Domenico di Rossi from plates engraved in 1636. The figure is reconstructed from gores labeled "1[a]" (*left side*) and "2[a]" (*right side*) by overlaying images of each digitally and distorting the overlays to lie along the common seam

de Broen invented a shield represented by the stars that he attached to one of the masts of the Ship in his *Hemelskaart voor de noordelijke en zuidelijke sterrenhemel uitgevoerd in Mercatorprojectie* (1709) (Fig. 10.7).

ICONOGRAPHY

Without written evidence of Plancius' intent in forming Gallus we are left to speculate on his choice of a common barnyard animal to fill what he perceived to be a gap in the space above and behind the Ship (Fig. 10.6).

Bartsch's interpretation of Gallus as the Petrine Rooster may, in fact, be reasonably connected back to the original intent of Plancius on account of nearby Columba (the Dove), seen prominently in de Broen's map along with the characterization of Argo Navis as "Arca noehi" (Noah's Ark). Columba was created by Plancius in 1592 from unformed stars below Lepus; he depicted the Dove on both the small planispheres of his large 1592 map as well as on a smaller world map of 1594 and some of his early globes. Plancius originally named the constellation Columba Noachi ("Noah's Dove"), referring to the

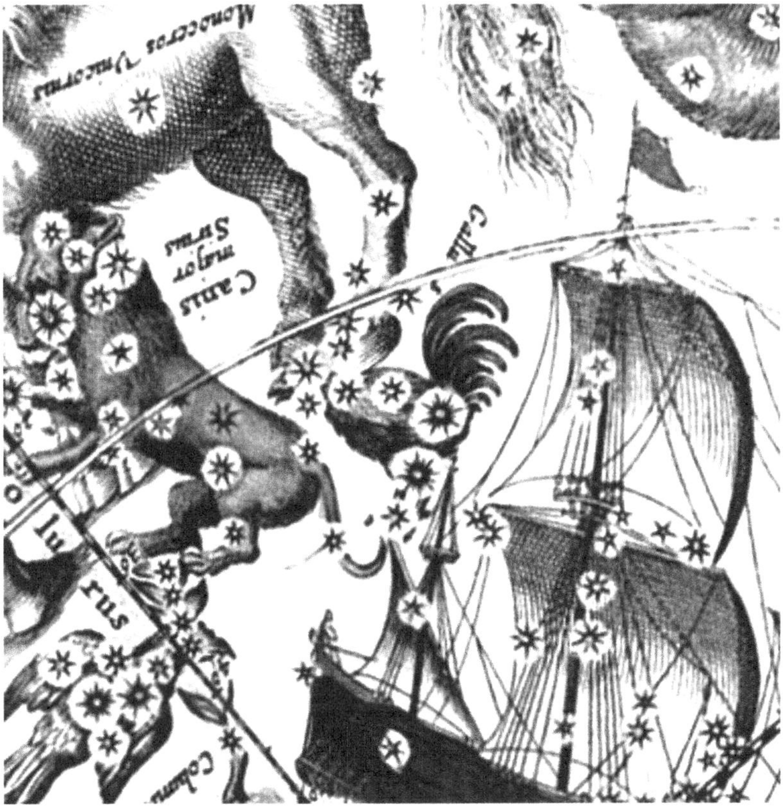

10.5 Gallus shown on Plate 27 of Andreas Cellarius' *Harmonia Macrocosmica* (1661). The hind end of Canis Major is at left and Argo Navis (Chap. 5) sails away toward the lower right corner. The Rooster is being trampled by the hooves of Monoceros (the Unicorn)

olive-branch-bearing dove that informed Noah of the receding waters of the Great Flood.[5] The placement of Columba near Argo Navis leaves little doubt that Plancius thought the Ship to be Noah's Ark, a popular interpretation in the early sixteenth century when some cartographers sought to remake the constellations of the firmament in the image of various Biblical figures. With a nearby, iconic bird referenced in a well-known Old Testament story, it is not a stretch to believe that Plancius intended Gallus to represent the most famous bird in the New Testament, although this conclusion remains unproven.

[5]"When the dove returned to him in the evening, there in its beak was a freshly plucked olive leaf! Then Noah knew that the water had receded from the earth." (Genesis 8:11; New International Version)

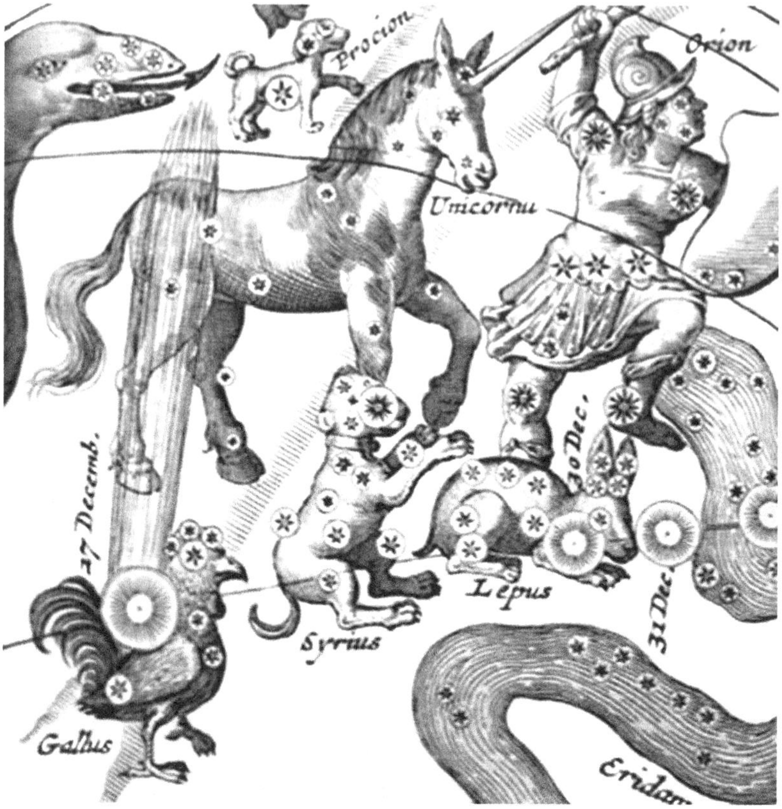

10.6 The path of the Great Comet of 1680 (C/1680 V1) shown in Volume 2 of Stanislaus Lubieniecki's *Theatrum Cometicum* (1681). The comet passed through the area of Gallus near its greatest brilliance on 27 December 1680

DISAPPEARANCE

By the mid-eighteenth century Gallus had all but vanished from popular astronomy texts and maps. John Hill, an Englishman with a penchant for obscure and fanciful constellations, argued that the stars comprising Gallus were by then rightly considered essential elements of Argo in *Urania: or, A Compleat View of the Heavens* (1754):

> GALLUS. A constellation formed by some authors out of the stars about the stern of the Ship, which they have thrown together under the figure of a cock: but the generality of writers continue to reckon them among the stars of the Ship.

That its stars were similarly absorbed into Argo Navis by Bode sealed the fate of Gallus as a separate figure. It appears in neither his atlas, *Vorstellung der Gestirne* (1782) nor his masterwork *Uranographia* (1801b; Fig. 10.8).

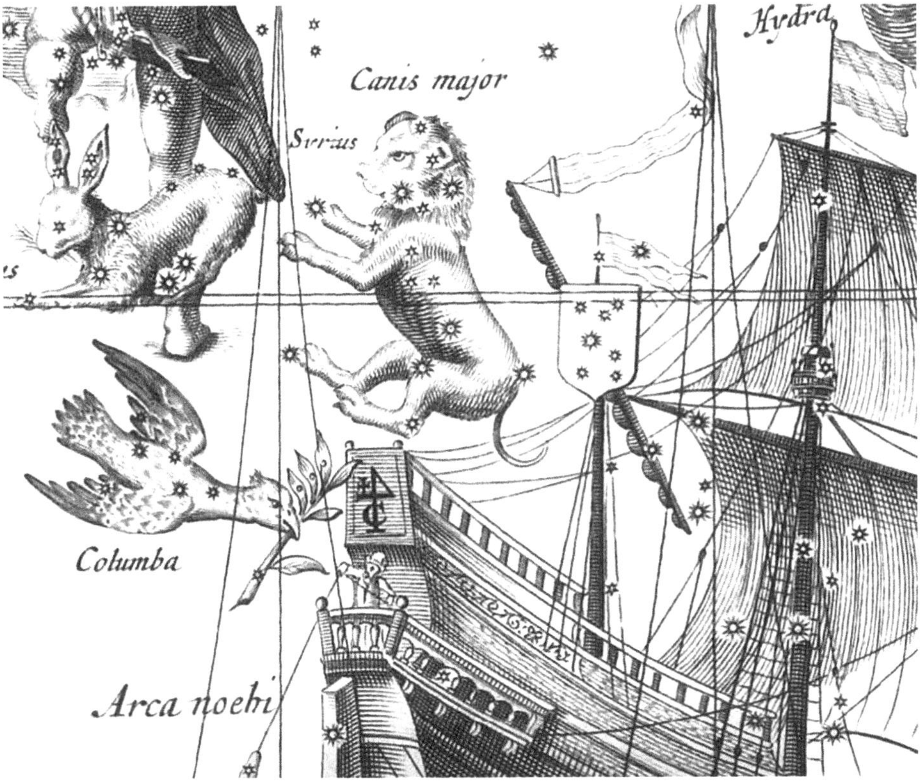

10.7 The stars formerly assigned to Gallus are shown here in Johannes de Broen's *Hemelskaart* ... (1709) as a shield-like device affixed to the masts of Argo Navis (*right of center*)

Gallus' chicken-run in the sky clearly came to an end by 1800. Heinrich Olbers mentioned it in "On a Reformation of the Constellations, and a Revision of the Nomenclature of the Stars," appearing in *Monthly Notices of the Astronomical Society of London* (1841), in the past tense: "The Cock, which was formed from a portion of the ship Argo, has likewise disappeared." At the end of the nineteenth century, Richard Hinckley Allen barely mentioned the figure in his discussion of Argo Navis in *Star Names*: "From stars in Argo, behind the back of the Greater Dog, was formed by Bartsch the small asterism Gallus, the Cock, but it has long since been forgotten." (Fig. 10.9)

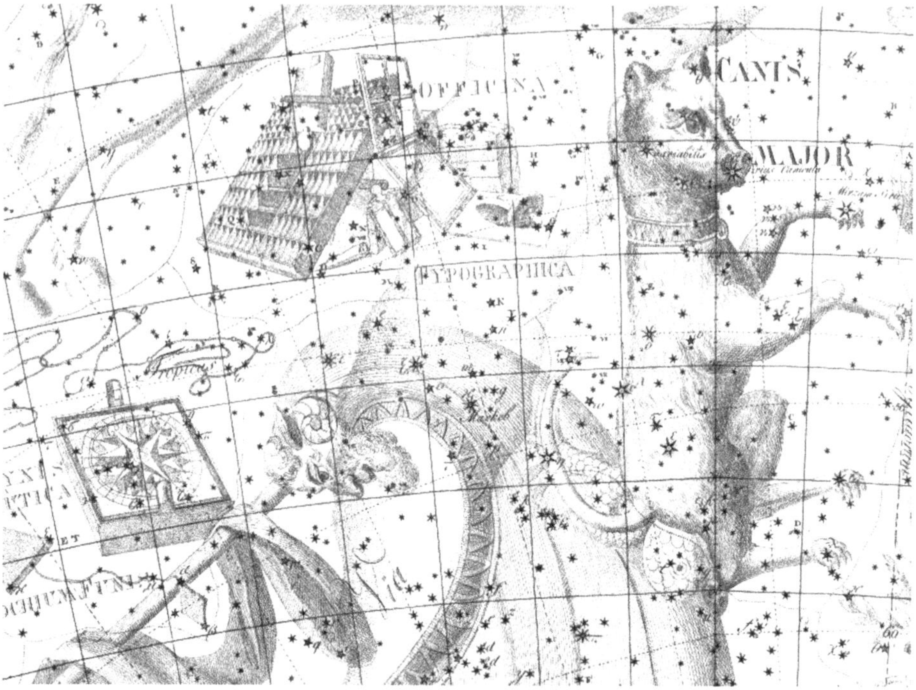

10.8 Bode repurposed the stars formerly in Gallus as the ornate, curving stern of Argo Navis (*center, above*) in *Uranographia* (1801b). The similarly extinct constellation Officina Typographica (Chap. 17) appears just above these stars, while Lochium Funis (see Volume 2) is shown at *lower left*

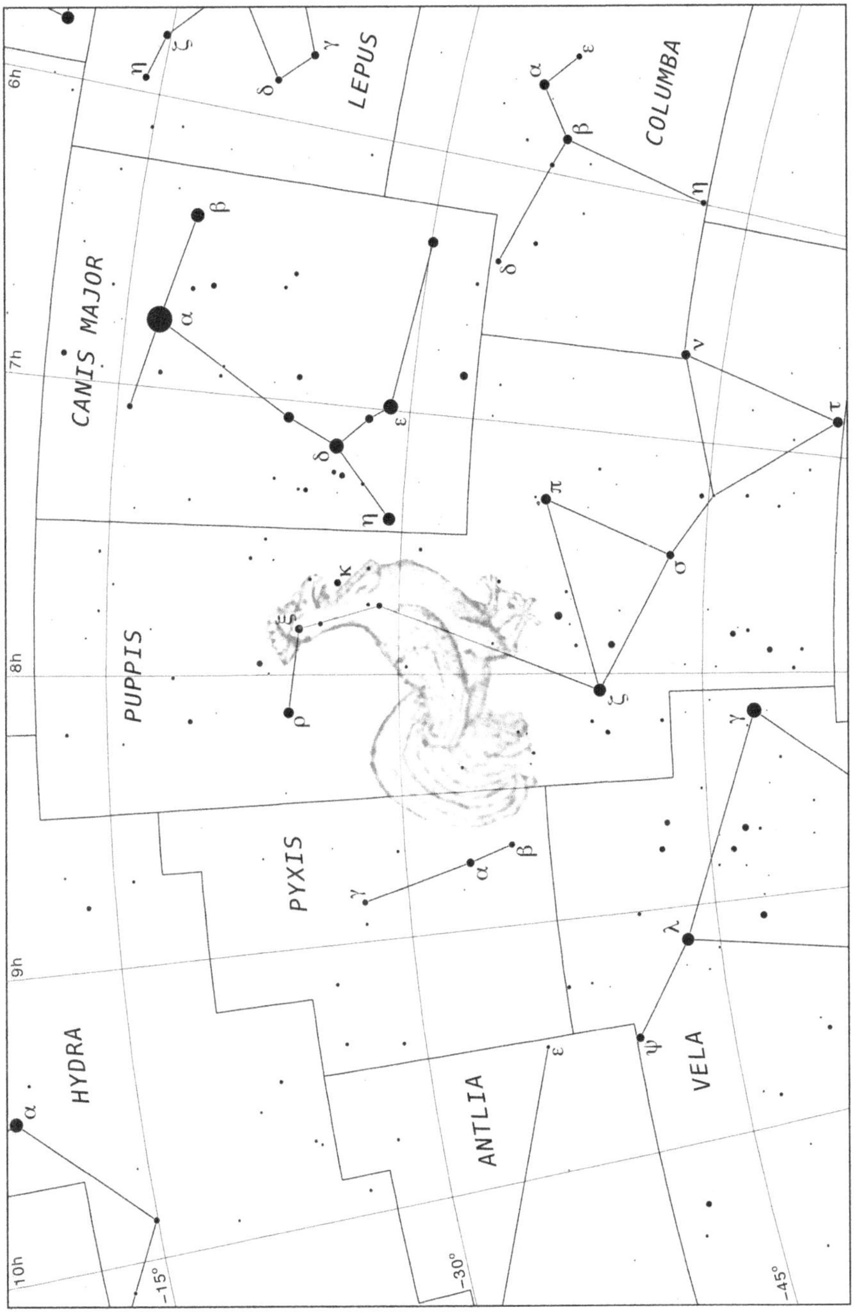

10.9 The figure of Gallus from Domenico di Rossi's 1695 celestial globe overlaid on a modern chart

11

Globus Aerostaticus

The Montgolfière Balloon

Genitive: Globi Aerostatici
Abbreviation: GlA
Alternate names: "Balloon" (Young, 1807); "Ballon Aerostatique" (Hall, 1825); "Aetherius" (Allen, 1899)
Location: Northwest of the body of Piscis Austrinus, and below Capricornus[1]

ORIGIN AND HISTORY

Globus Aerostaticus was devised by Joseph Jérôme Lefrançois de Lalande in about 1798 from a set of unformed stars tucked into the space immediately below Capricornus and northwest of Piscis Austrinus. He took as inspiration the southern constellations introduced by Lacaille nearly 50 years before, referring to the modern apparatus of science (Norma, Circinus, Telescopium, Microscopium, Fornax, Horologium, Octans, Reticulum, and Antlia). In keeping with the theme Lalande felt moved to honor what he called "he greatest discovery of the French," namely, the hot air balloon invented by the Montgolfier brothers; he suggested the Balloon to Johann Elert Bode, who included it in *Uranographia* (1801b; Fig. 11.1). At the same time, Lalande also suggested an equivalent honor for Germany recognizing the historic importance of Gutenberg's printing press (see Chap. 17).

[1] *"A little to the west of Fomalhaut and south of Capricornus"* (Green, 1824); *"This constellation is situated under Capricornus and west of the Southern Fish"* (Kendall, 1845); *"East of Microscopium and south of Capricornus"* (Bouvier, 1858); *"It lay east of the Microscope, between the tail of the Southern Fish and the body of Capricorn"* (Allen, 1899).

J.C. Barentine, *The Lost Constellations*, Springer Praxis Books,
DOI 10.1007/978-3-319-22795-5_11

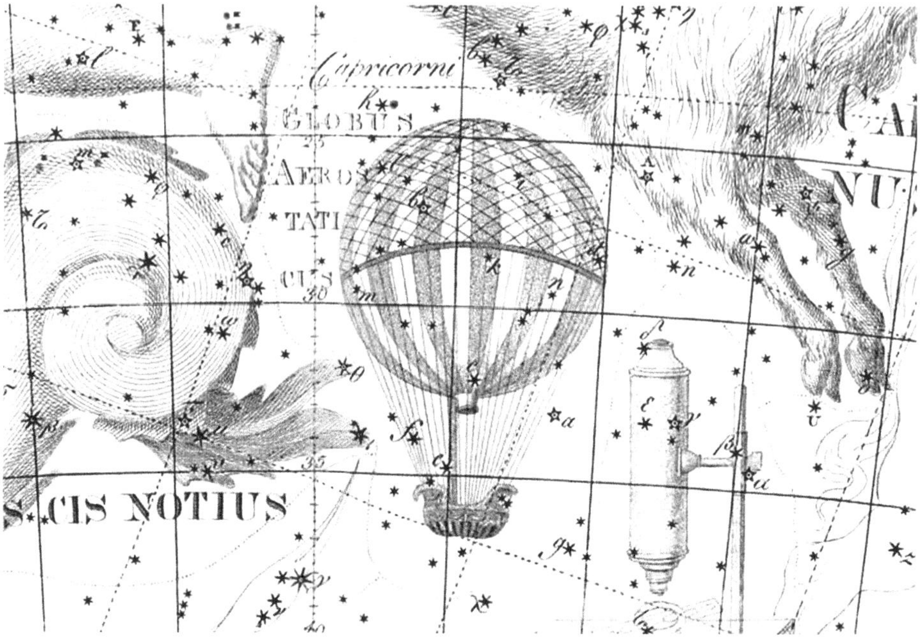

11.1 Globus Aerostaticus shown on Plate 16 of Bode's *Uranographia* (1801b)

Richard Hinckley Allen seems to be the only author to label it with the alternate name "Aetherius" in print (Fig. 11.2):

> Globus Aerostaticus, vel Aetherius, the Balloon, was formed by La Lande in 1798. ... Bode published it in his Die Gestirne as the Luft Ballon, Ideler's Luft Ball, with twenty-two stars; and Father Secchi still had it in his maps as the Italian Aerostáto. *With the French it was the* Ballon Aérostatique.

Bode described[2] the Balloon's origins in *Allgemeine Beschreibung und Nachweisung der Gestirne* (1801a):

> It was proposed as a new constellation by de la Lande in 1798 to commemorate the Montgolfier balloons. It is situated beneath Capricorn to the west of the Southern Fish, and is composed from several faint stars formerly part of those constellations.

The Balloon gained a small amount of popularity during the nineteenth century. Alexander Jamieson included it on Plate 21 of his *Celestial Atlas* (1822; Fig. 11.3) as "Le Ballon Aerostatique". It is depicted on Plate 26 in the influential *Urania's Mirror* set

[2]"Wurde von de la Lande im Jahr 1798 zum Andenken des von Montgolfier ersundenen Aerostaten als ein neues Sternbild vorgeschlagen. Er steht unterhalb dem Steinbock westwärts beym südlichen Fisch, und ist aus einigen ehemals zum Theil zu diesen Bildern gehšrigen kleinen Sternen formirt."

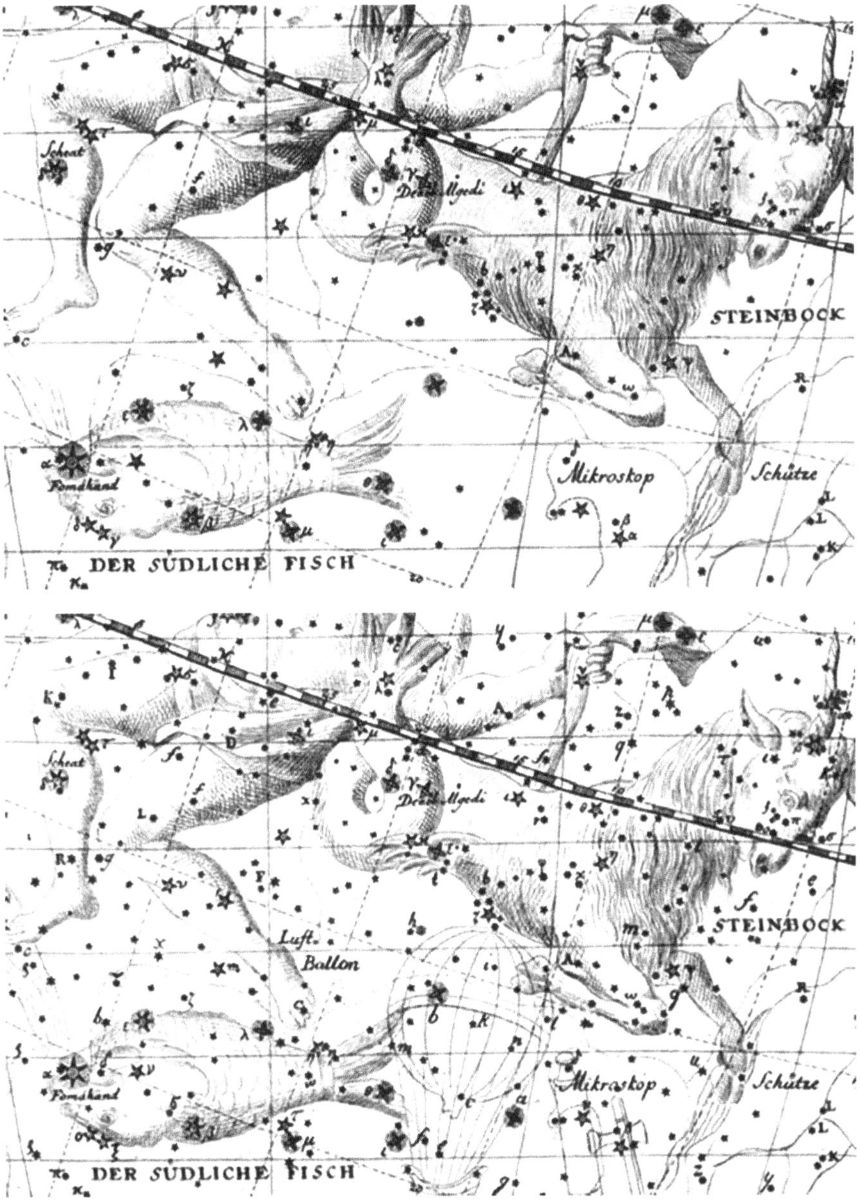

11.2 Before and after. Detail of Plate 21 in two editions of Bode's *Vorstellung der Gestirne*: 1782 (*top*) and 1805 (*bottom*), showing the introduction of Globus Aerostaticus (*bottom-center*) in the intervening years

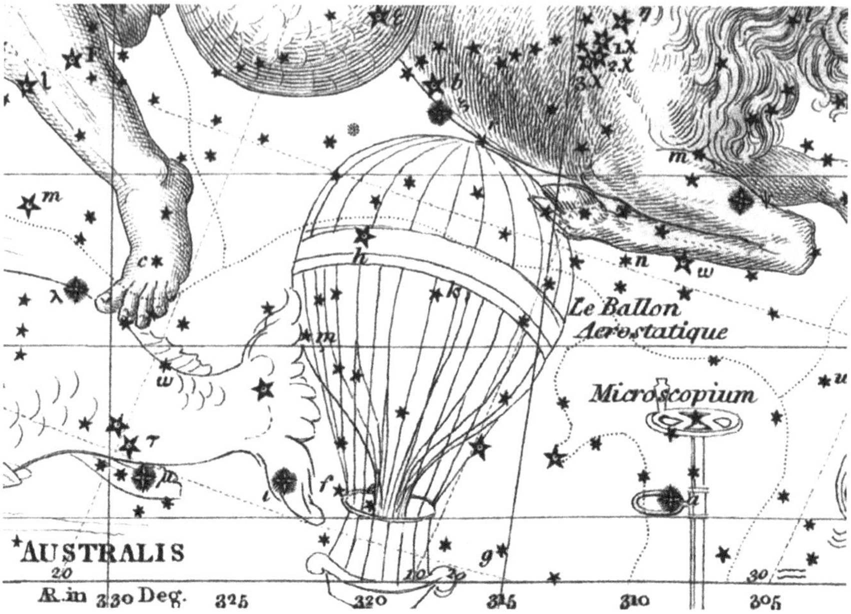

11.3 Globus Aerostaticus as shown on Plate 21 of Alexander Jamieson's *Celestial Atlas* (1822)

of atlas cards in 1825 as "Ballon Aerostatique". In each case, the cartographers referred to its French origin by name; in the latter, the boundary between Globus Aerostaticus and Capricornus was drawn deliberately to exclude ζ Capricorni from Globus. Jamieson, on the other hand, drew the boundary haphazardly, placing roughly one third of the stars of Globus within Capricornus.

Ezra Otis Kendall included "Globus Ærostaticus" in a list of constellations in *Uranography* (1845), describing it as "introduced by Lalande." Later, George Chambers mentioned "Globus Aërostaticus" in *A Handbook of Descriptive Astronomy* (1877) and attributed it to "Bode's maps" without citing a date.

ICONOGRAPHY

Since distant antiquity humanity dreamed of harnessing the power of flight and moving as freely through the air as birds. That a hot air balloon was proposed among the stars is a very telling indication of the times during which it was proposed, a period that saw the first practical passenger balloons in human history.

The capacity for unpowered flight hinges on the physical principle of buoyancy, a force counteracting that of gravity. It is exerted on an object by a fluid that opposes the weight of an immersed object whose volume displaces a certain amount of the fluid. Discovery of the buoyant force is traditionally assigned to Archimedes of Syracuse (*c.* 287 BC–*c.* 212 BC) in about the year 250 BC. In *On Floating Bodies*, he wrote[3] that "Any solid lighter than a fluid will, if placed in the fluid, be so far immersed that the weight of the solid will be equal to the weight of the fluid displaced."

This observation has come to be known as Archimedes' Principle and is familiar to anyone with some nautical knowledge: if the weight of a ship is less than the weight of the water its bulk displaces, then the buoyant force of the water on the ship is greater than the gravitational force and the ship floats. It also explains why ships sink. Opening a hole in the hull of a ship allows in water, decreasing the volume (and hence weight) of water displaced by the hull. When a critical equilibrium is surpassed, the weight of the boat exceeds the weight of water still displaced, and the ship is doomed.

To create lift due to the buoyant force an aircraft must weigh less than the weight of the surrounding air it displaces. This creates two engineering challenges: First, the material from which the craft is constructed must be both lightweight and durable; and second, the craft must contain or enclose some substance having a bulk density less than that of the surrounding air. Solving these problems in any meaningful sense would wait until the eighteenth century.

Early Ballooning

Most early dreams of flight given any serious consideration involved imitations of birds with flapping, human-powered wings. Ideas along these lines are attested by Roman sources, but planning involving anything remotely resembling modern engineering design is not known before the Medieval period. In *Wonderful Balloon Ascents; or, The Conquest of the Skies* by Fulgence Marion[4] (1870), a typical flying contraption was described:

> Roger Bacon,[5] in the thirteenth century, inaugurated a more scientific era. In his "Treaty of the Admirable Power of Art and Nature," he puts forth the idea that it is possible "to make flying-machines in which the man, being seated or suspended in the middle, might turn some winch or crank, which would put in motion a suit of wings made to strike the air like those of a bird."

None of these devices is known to have worked, in the sense that no craft achieved sustained flight and those that experienced any lift were merely gliders. During the period of technological innovations in Western Europe of the seventeenth century, it was realized that merely lifting an object into the air through buoyancy was an easier accomplishment than powered flight.

The technical capacity for balloon flight probably existed for centuries. Floating paper lanterns powered by small lamps were invented in ancient China, but there appears to be

[3] *On Floating Bodies*, Book I, Proposition 5. Translated in T.L. Heath, *The Works of Archimedes*, Cambridge University Press, Cambridge (1897).

[4] A pseudonym of the French astronomer and author Nicolas Camille Flammarion (1842–1925).

[5] Roger Bacon, OFM (*c.* 1214–June 1292?), English philosopher.

no evidence that they were used practically to lift payloads or that their creators understood why they floated. The earliest European ideas for buoyant flight appear to have been suggested by a Jesuit priest named Francesco Lana de Terzi (1631–1687) of Brescia, Lombardy, in about 1670. Lana de Terzi was a mathematician and naturalist who thought that buoyancy could be achieved by the absence of air. His "flying boat" concept (Fig. 11.4) involved fitting a sailing ship with a set of lightweight metal spheres that would be evacuated, making the weight of the sphere less than that of the air it displaced. His design, though published, never made it off the drawing board. While the theory was sound, Lana del Terzi's concept would have faced a significant practical implementation problem: no material known at the time for constructing the spheres had sufficient strength to resist the force of atmospheric pressure. Had the spheres been constructed and evacuated, they would have immediately imploded.

A disputed attribution for the first lighter-than-air balloon flight is made to a Brazilian-born Portuguese priest named Bartolomeu de Gusmão (1685–1724). In his youth, Gusmão began his Jesuit novitiate but left in 1701 without taking orders. He emigrated to Portugal shortly thereafter, completing his studies for a Doctor of Canon Law degree at the University of Coimbra. He devised an airship design around this time that was presented with a petition seeking the royal favor of King John V in 1709. Building on Lana de Terzi's "flying boat" concept, Gusmão's design (Fig. 11.5) involved a sail spread over the hull of a boat containing tubes through which air would be blown against the sail via a set of bellows. The vessel was to be propelled through the force of magnets encased in two hollow metal balls within the hull. Records indicate a public test was proposed on 24 June 1709, but for unclear reasons did not happen as planned.

Some kind of demonstration *did* take place before the royal Court at Casa da Índia in Ribeira Palace, Lisbon, on 8 August 1709, during which Gusmão managed to lift a small paper balloon about four meters in the air through the use of heated air. John V rewarded the accomplishment by appointing Gusmão professor at Coimbra and making him a canon. Gusmão never made the airship work, but continued to dream up other designs such as a triangular pyramid enclosing a lighter-than-air gas. While his 1709 balloon demonstration is regarded in the Portuguese-speaking world as history's first proper balloon flight, the claim is recognized generally neither by the aviation history community nor ballooning's international governing body, the Fédération Aéronautique Internationale.

Prospects for buoyant flight improved by mid-century. In 1766, Henry Cavendish (1731–1810) discovered hydrogen after mixing together iron, tin, zinc shavings and sulphuric acid, finding its weight only one-tenth that of air. On the basis of Cavendish's discovery and Robert Boyle's (1627–1691) work on the properties of gases, the Scottish chemist Joseph Black (1728–1799) proposed that a balloon filled with hydrogen should float and provide lift for a payload. Jacques Charles (1746–1823) studied the work of Cavendish and Black, and along with brothers Anne-Jean (1758–1820) and Nicolas-Louis (1760–1820) Robert proceeded to build a hydrogen balloon in the Roberts' workshop at the Place des Victoires in Paris. The fabrication process, dissolving rubber in a turpentine solution and painting the material onto sheets of silk, resulted in an airtight gas bag when the sheets were stitched together to make their balloon's envelope. The distinctive

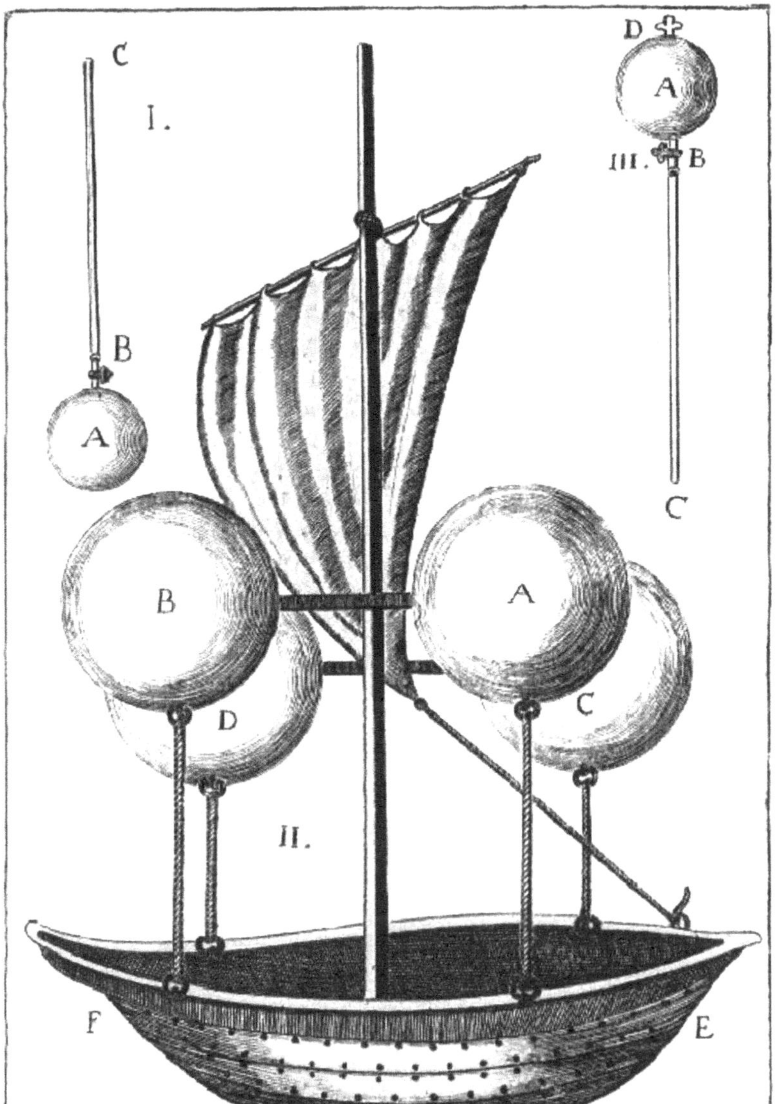

11.4 Francesco Lana de Terzi's design for a "flying boat" involving evacuated metal globes. Illustration from his *Del modo di fabbricare una nave che cammini sostentata sopra l'aria a remi ed a vele, quale si dimostra poter riuscire nella pratica* ("On the means of manufacturing a vessel that flies through the air through rowing and sails, which proves to be successful in practice"), published in 1784 by Giuseppe Galeazzi in Milan

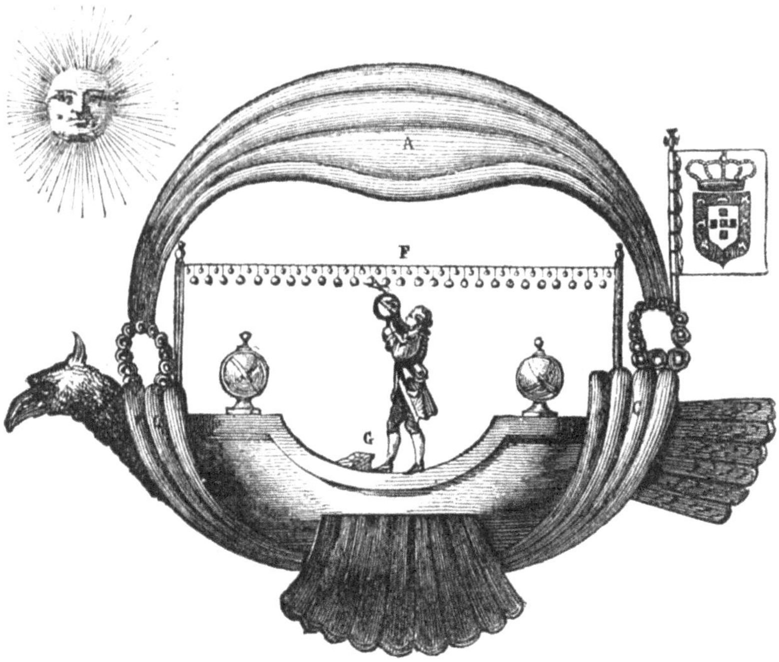

11.5 Bartolomeu de Gusmão's balloon design of 1709. Illustration from *Wonderful Balloon Ascents; or, The Conquest of the Skies* by Fulgence Marion (pseudonym of Camille Flammarion), Cassel Petter & Galpin, London, 1870

alternating red-and-yellow stripes of these earliest balloons were accidental; the sheets of silk used for the envelope were red and white, but the rubberizing process caused the white silk to yellow.

The first practical demonstration of a hydrogen balloon was conducted by Charles and *Les Frères Robert* at the Champ du Mars in Paris on 27 August 1783. The balloon used in the test was small—a mere 35 m^3—with a lifting capacity of only about 9 kg. Hydrogen to fill the envelope was produced via the Cavendish method of immersing iron into sulphuric acid; this is a strongly exothermic reaction, resulting in the heating of the gas that was fed into the silk envelope through lead pipes. As the gas cooled during transport it decreased in volume and the envelope failed to completely inflate. The process was slow and laborious, but upon its release the balloon flew about 21 km downrange over the span of about 45 min, touching down in the village of Gonesse. The local population was evidently so terrified by the interloper that they destroyed the balloon with pitchforks and/or blades.

11.6 Louis Boilly (1761–1845), *The Montgolfier Brothers, Joseph-Michel (1740–1810) and Jacques-Étienne (1745–1799)*, mid-1790s. Black chalk and white gouache on paper; oval, 117 × 98 mm (4 5/8 × 3 7/8 in.). Pierpont Morgan Library 2007.125

The Montgolfier Brothers

The first successful aerostat designs to carry passengers were French balloons lifted by hot air rather than hydrogen, designed by brothers Joseph-Michel (1740–1810) and Jacques-Étienne (1745–1799) Montgolfier (Fig. 11.6). Born into a prosperous family of paper makers in Annonay, a town in the Ardèche department of southern France, the two brothers were not much alike. While Joseph—the twelfth child of sixteen born to Pierre Montgolfier (1700–1793) and Anne Duret (1701–1760)—was an idealistic inventor, fifteenth child Étienne was much more even-tempered and inclined to a businesslike manner. Pierre sent Étienne to Paris to train as an architect, but when eldest brother Raymond (1730–1772) died suddenly at a relatively young age, Étienne was summoned home to oversee the family business.

Paper making was a booming industry in the latter half of the eighteenth century, fueled by the ravenous appetite of an increasingly literate European population. Étienne spent the 1770s building up the Montgolfier business into a model throughout France by incorporating the latest manufacturing technologies into the family's mills. The other siblings, Joseph included, had roles at the factories, and it was in that environment that the first ideas culminating in the hot air balloon were formed. Around 1777, Joseph thought of building flying machines as he watched laundry drying on lines over a fire, occasionally billowing up due to vertical air currents rising off the flames.

He began carrying out experiments in 1782 while living at Avignon, having become convinced that smoke from fires was intermingled with what he called "Montgolfier gas," a substance with a special buoyant property. If that power could be harnessed, Joseph reasoned, it had a myriad of potential useful applications. Chief among these were martial strategies. Between 1779 and 1783, France and Great Britain sparred over Gibraltar, a key strategic position on the European side of the interface between the Atlantic Ocean and the Mediterranean Sea. The British fortress remained impregnable from both land and sea, and Joseph dreamed of a means of conveying soldiers and arms over its walls through the air. A successful demonstration of such a device could lead to a very lucrative contract from the state.

In November 1782, Joseph built a thin wooden frame enclosing a volume of a little over a cubic meter and covered the sides and top of the frame with taffeta cloth. Lighting a fire under the box, he was delighted to watch it lift off its stand and crash into the ceiling. He wrote Étienne at once: "Get in a supply of taffeta and of cordage, quickly, and you will see one of the most astonishing sights in the world." (Gillespie, 1983) The brothers set about building a similar device, increased in linear dimensions by a factor of three such that the volume was 27 times larger than Joseph's original design. While Étienne considered hydrogen as a more suitable lifting gas, he encountered many of the same challenges in its production as Jacques Charles and the Robert brothers would find mere months later. Joseph and Étienne conducted a test flight on 14 December during which they lost control of the craft; it drifted about 2 km from the launch site before crashing uncontrollably to the ground.

The brothers arranged a public demonstration of their balloon in the summer of 1783 in order to establish a claim to its invention before anyone else could. The design was scaled up again, this time resulting in an envelope with a volume of some 790 m^3; the globe-shaped envelope was fashioned out of three layers of thin paper backed by sackcloth that weighed a total of 225 kg. Sections of cloth-bound paper were formed into a rough sphere and held together by 1800 buttons and a reinforcing cord net around the exterior. Its maiden voyage was held on 5 June 1783 in their hometown before a delegation of officials from the États particuliers; within 10 min, the balloon reached an estimated altitude of nearly 2000 m and floated 2 km downrange. As word of the successful test spread, Étienne traveled to Paris to make preparations for further public demonstrations. Balloons of this design quickly became known as *Montgolfières*.

The Montgolfière was scaled up further. In September 1783, the brothers collaborated with the wallpaper manufacturer Jean-Baptiste Réveillon (1725–1811) to construct their largest balloon yet: a 1060 m^3 taffeta envelope varnished with a fireproof solution of alum.

FIGURE EXACTE
ET PROPORTIONS
DU GLOBE
AËROSTATIQUE,
Qui, le premier,
a enlevé
des Hommes
dans les Airs.

11.7 A contemporary engraving (1786) showing the Montgolfier Brothers' 1783 balloon. ("Figure and exact proportions of the 'Aerostatic Globe', which was the first to carry men through the air.") This type of balloon came to be known as a *Montgolfière*

The blue exterior was painted with elaborate Réveillon designs of suns, zodiac symbols, and various golden swirls and flourishes (Fig. 11.7). The caption accompanying the figure reads:

> The top portion was surrounded by fleurs-de-lys, with the twelve zodiac signs below. In the middle portion were images of the king's face, each surrounded by a sun. The bottom section was filled with mascarons and garlands; Several eagle's wings appear to support this powerful machine in the air. All of this ornamentation was gold on a beautiful blue background, so that this superb globe appeared to be gold and azure. The circular gallery, in which we see the Marquis D'Arlandes and Mr. Pilatre de Rozier, was covered in crimson draperies with gold fringes.

The balloon was test-flown on 11 September on the grounds of La Folie Titon in Paris and was deemed sufficiently airworthy to carry passengers, but concerns about the effects of the then-unknown upper atmosphere on living things prevented humans from climbing aboard. Instead, the inventors sent aloft a selection of barnyard animals in their place: a duck, a sheep and a rooster. Their survival during the test offered evidence that balloon travel was safe for people.

The first tethered flight with human passengers took place on 19 October with Réveillon, the scientist Jean-François Pilâtre de Rozier (1754–1785), and Giroud de Villette aboard. A manned free flight followed about a month later on 21 November. Louis XVI initially insisted that only condemned criminals could be among the passengers, but de Rozier and Marquis François d'Arlandes (1742–1809) successfully petitioned the king to travel on the flight instead. For this balloon, the hot air was supplied by an iron basket of coals secured below the neck of the envelope; the rendered the heat source controllable by the pilots so as to determine the rate of ascent and descent, and could be replenished easily by simply shoveling more fuel into the basket. The flight lasted slightly under a half hour, although there was sufficient fuel aboard to have flown for about 2 h. de Rozier and d'Arlandes were concerned about flying embers from the basket igniting the material of the envelope, so they gently set down as soon as they found open pasture land about five miles from the launch site.

Charles and the Robert brothers were not far behind. On 1 December, they launched a manned, 380 m^3 hydrogen balloon from the Jardin des Tuileries, adjacent to the Tuileries palace in Paris, before a crowd of several hundred thousand people. Among the VIPs watching the launch from a special viewing area was Benjamin Franklin, then the American diplomatic representative. There, too, was Joseph-Michel Montgolfier, given a place of honor in the proceedings wherein Charles had him release a small balloon to determine the suitability of conditions for the launch. Charles himself piloted the larger balloon, accompanied by Nicolas-Louis Robert; together they ascended to a peak altitude of about 500 m. The balloon's altitude was controlled through a combination of sandbags for ballast and a valve on the envelope allowing for venting of hydrogen.

After a free flight of a little over 2 h, they set down in Nesles-la-Vallée, some 36 km from the launch site. They touched down at sunset and Robert left the balloon's basket. Charles intended to continue flying and quickly ascended to an altitude of nearly 3000 m where he saw the sun rise over the western horizon. The altitude gain caused pain in his ears, and with night rapidly approaching he vented hydrogen through the envelope valve and began to descend again, landing a few kilometers away at Tour de Lay. It turned out to be his last flight as a pilot, but the balloon design featuring hydrogen as the lift gas soon became known a *Charlière* in honor of his achievements. The flight is famous for featuring the first scientific measurements from a balloon, consisting of readings taken from a barometer and thermometer at intervals in altitude; it inaugurated an era of scientific ballooning that persists to this day.

Ballooning milestones were rapidly established in the remainder of the eighteenth century. The first ascent in the British Isles was made in August 1784, followed shortly thereafter by the first crossing of the English Channel by balloon. Not far behind was history's first bona fide aviation disaster: in May 1785, a balloon flight gone awry resulted in a crash landing and ensuing fire that destroyed nearly 100 homes in the town of Tullamore in County Offaly, Ireland. The first flight in America took place on 10 January 1793 with President George Washington in attendance.

The following year, Joseph-Michel Montgolfier's vision of wartime ballooning was fulfilled in the first military use of balloons at the Battle of Fleurus during the French Revolutionary Wars; in that instance, spotters flew in a tethered hydrogen balloon to watch and report on the movements of Austrian army units. However, truly steerable balloons were decades away. The first dirigible, a steam-powered airship, was invented and flown by French engineer Henri Giffard (1825–1882) in 1852. By the end of the nineteenth century, dirigibles were employed as fully functional transport vehicles, and they experienced a golden age in the first three decades of the twentieth century. However, the age of dirigible flight, already dogged by the advent of the airplane, came to an abrupt end with the infamous destruction of LZ 129 *Hindenburg* at Lakehurst, New Jersey, on 6 May 1937.

DISAPPEARANCE

The deflation of Globus Aerostaticus seems to have taken place gradually in the last quarter of the nineteenth century. It appears in few books after Chambers (1877); a notable exception is the *Poole Bros. Celestial Handbook* (1892), which attributes it to "Lalande, 1798" with the caveat "These constellations, not being recognized by the B.A. [British Association] Catalogue,[6] are not described in this Handbook; the stars by which they were formed are inserted in the constellations from which they were taken." By the end of the century Allen declared that, like "most of his [Lalande's] stellar creations," it had "passed out of the recognition of science." Its stars were largely incorporated within the boundaries of the modern Piscis Austrinus by the IAU in 1928 (Fig. 11.8).

[6]Francis Baily, *The Catalogue Of Stars Of The British Association For The Advancement Of Science*, Richard & John E. Taylor, London (1845).

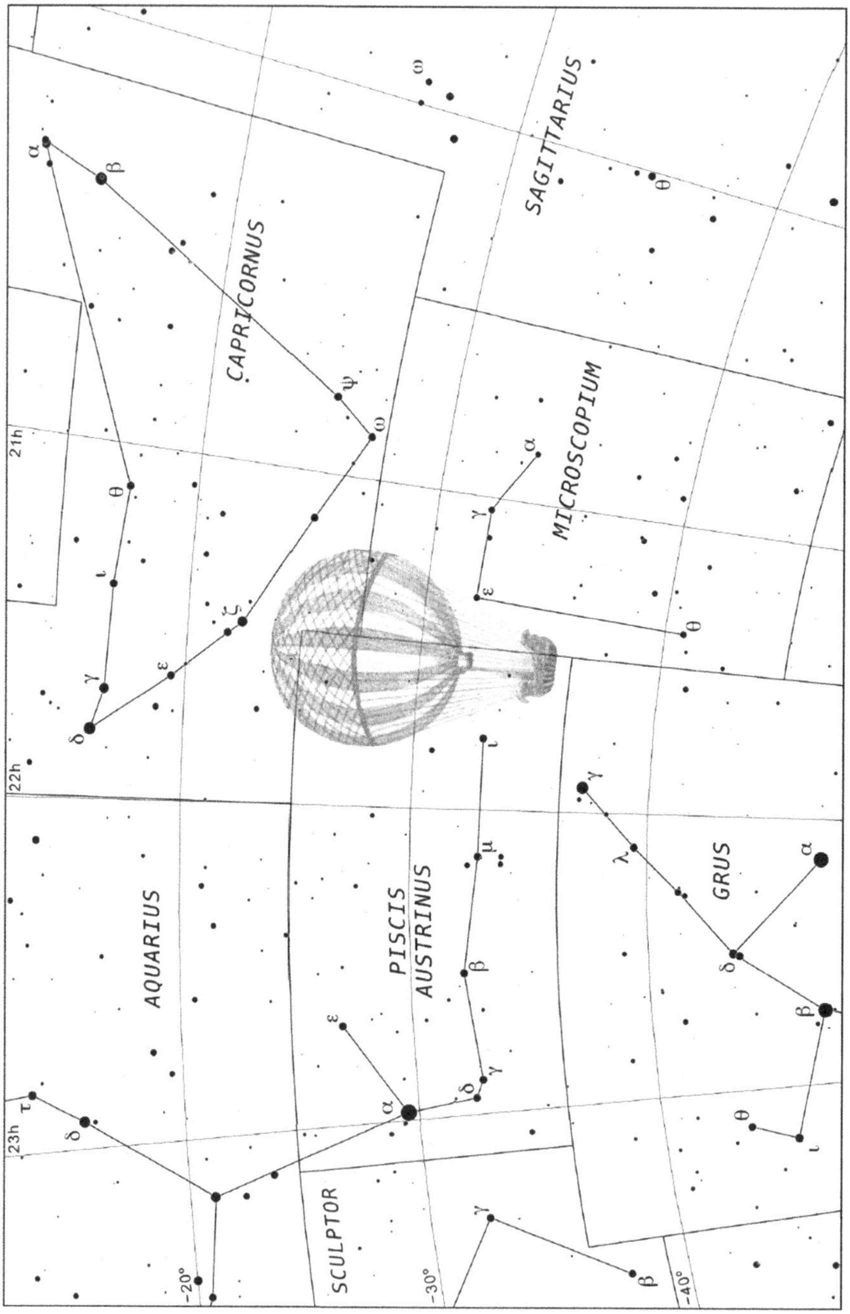

11.8 The figure of Globus Aerostaticus from Plate 16 of Bode's *Uranographia* (1801b) overlaid on a modern chart

12

Honores Frederici

Frederick's Glory

Genitive: Honoris Frederici
Abbreviation: HoF
Alternate names: "Gloria Frederici" (Agnes Clerke); "Friedrichs Ehre" (Bode); "Gloria Frederika"; "Gloria Frederica" (Burritt, 1833); "Frederici Honores" (referenced by Ridpath 1989); "Frederic's Trophy" (Young, 1807)
Location: Formed from stars along the present border of Andromeda and Lacerta[1]

ORIGIN AND HISTORY

Johann Elert Bode introduced this constellation in 1787 in commemoration of the recently deceased Frederick II of Prussia, first suggesting it in print[2] in his own *Astronomisches Jahrbuch* of that year as *Fredrichs Sternen Denkmal* ("Frederick's Star Memorial"):

> On 25 January this year, the local Royal Academy of Sciences held a meeting devoted to a public memorial and praise of Frederick, her serene founder. This celebration prompted me, in a short speech, to propose the introduction of a constellation consecrated to the memory of the immortal king under the designation

[1] *"Between Cassiopeia, Cepheus, Andromeda and the Swan"* (Bode, 1787); *"In the northern region of the heavens ... near ... Cepheus, Cassiopeia, Andromeda. Not far away are Pegasus and the Swan."* (Bode, 1792); *"A little to the north of Andromeda's hand"* (Green, 1824); *"Squeezed between the outstretched right arm of Andromeda and the Hevelius invention of Lacerta, the Lizard"* (Ridpath, 1989); *"Formed... out of the present stars of Lacerta and a few from Andromeda"* (Bakich, 1995).
[2] The original German text is reproduced at the end of this chapter.

J.C. Barentine, *The Lost Constellations*, Springer Praxis Books,
DOI 10.1007/978-3-319-22795-5_12

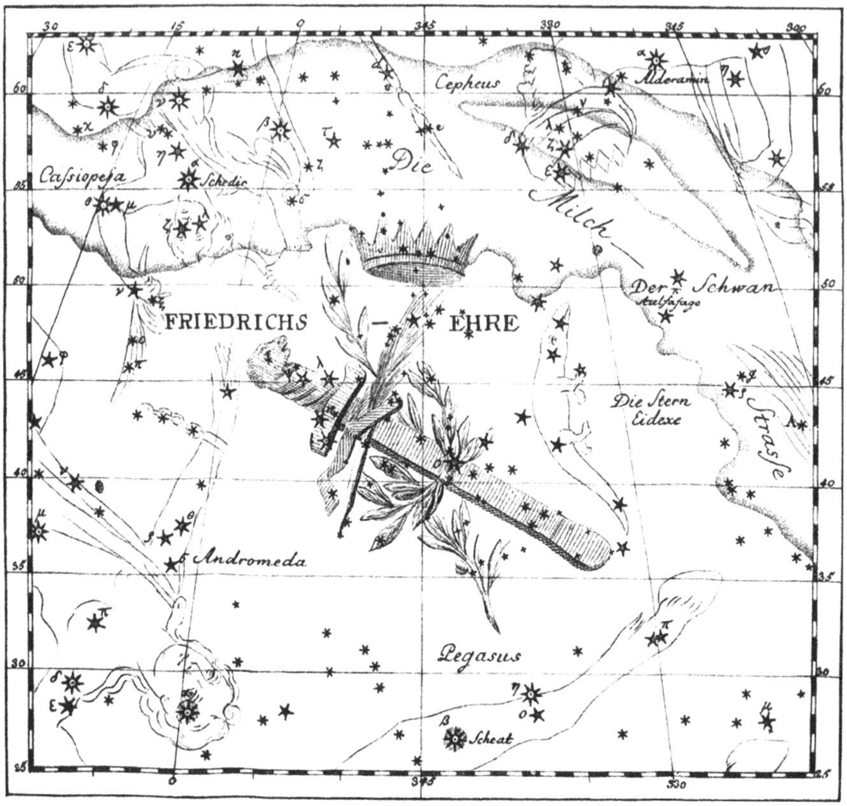

12.1 Honores Frederici, introduced by Bode in *Mémoires de l'Académie Royale des Sciences et Belles Lettres 1786–1787* (1792)

"Frederick's Honor," and I put to the academy the figure of the same in an engraved star chart, from which the figure listed in Table 2 has been taken.

I draw it between *Cassiopeia, Cepheus, Andromeda* and the *Swan* from 76 stars I have newly observed, and would require a halo ["rayed crown"], the sign of royal dignity, close to the southern edge of the Milky Way at the crown of Cepheus, because in Frederick's time the shimmering stars stood there above his earthly realm. Southward under the same, entwined with the unfading laurels of posthumous fame: A *sword*, a *pen* and an *olive branch* to indicate the eternal memory of this monarch as a *hero, sage*, and *peacemaker*.

Several astronomers of modern times honored famous princes with famous monuments in the sky, usually formed from the same stars that belonged mostly to neighboring constellations, and it was not uncommon that they were compelled to make minor changes to others, as I for example make in the formation of a new image from one hand of Andromeda and Hevelius' Lizard-stars, which the former I am moving a little eastward, and the latter in a more befitting location. The space, which I had chosen for my "Frederick's Glory," includes in this change 49 [stars] already in my star charts. My observations of January nights showed there were still around 27 stars and contributed their positions for an accurate sense of proportions in my marked chart. The new figure was now composed of 76 stars and two star clusters, among which included four stars of the fourth magnitude, three of the fifth magnitude, 24 of the sixth magnitude, 16 of the seventh magnitude, 24 of the eighth magnitude, and five of the ninth magnitude. To Pegasus belong five thereof, to Andromeda 26, to Cepheus six, to the Lizard nine, and Cassiopeia three stars.

The Russian Imperial Academy at St. Petersburg has my well-intentioned and disinterested proposal, the only ones to found a monument to Frederick in the heavens, and gave me approval in writing by Professor Euler on 24 April: "The Academy applauds this tribute Mr. Bode makes to the memory of the great Sovereign of the Century, and astronomers and gentlemen consented with pleasure to use the name of the new Constellation, every time they observe the stars that compose it." *

Mr. [Charles Peter] Layard,[3] secretary of the London Royal Society of Sciences, wrote to me on 7 June on behalf of the Society: *The Royal Society receiv'd your present of Fredrichs Sternen Denkmal, and I am directed to return to you their thanks for the fame. – Sir.*

Mr. [Thomas] Bugge[4] wrote this from Copenhagen to me on 26 June: *The Royal Society approved your high tribute that Danish astronomers might use your new constellation, its name, and its figure in their observations and writings ... She [the Society]* ordered me to indicate to you its full approval.

Mr. [Pierre] Mechain[5] wrote from Paris on 23 February: *I will present at the Academy an example of the new constellation you form as Frederick's Glory: The idea does you honor and all the philosophers cannot but applaud, the name of this great prince will live forever and well deserves to be writ upon the vault of heaven.*

---

[3]The Rev. Charles Peter Layard (1749–1803), Dean of Bristol, was Foreign Secretary of the Royal Society from 1784 to his death.

[4]Thomas Bugge (1740–1815) was the Danish Astronomer Royal and a Member of the International Commission on the Metric System (1798–1799).

[5]Pierre Méchain (1744–1804) was a French astronomer and surveyor who, along with Charles Messier, made major contributions to telescopic observations of comets and "deep sky objects" including nebulae, star clusters, and galaxies.

Mr. [Joseph Jérôme Lefrançois] de la Lande wrote me on the 7th of May: *I received with your letter of 12 February the design of your new constellation, I will communicate it to our astronomers: I will mention it in the new edition of my* Astronomy*; no one is more interested in the glory of this great prince for whom I was, so to speak, his handiwork, since I made my first trip to Berlin and observations in 1751 that began my career in astronomy.*

Professor Euler[6] wrote me on 24 July: The Board of local primary schools has already incorporated the new constellation "Frederick's Glory" on one of the same which appears on Mr. Gollovin's celestial globes.

Yet different astronomers have sent word of their approval of the proposed constellation in the letters below.

* On this occasion, I reported to Professor Euler that the astronomers of the Imperial Academy adopted my proposed designation and name of the new planet, and included about it a few pages from the St. Petersburg Calendar. -Bode

Bode further elaborated on the constellation and provided the first chart showing its location (Fig. 12.1) in an article titled "Monument astronomique consacré à Frédéric II" ("Astronomical Monument Dedicated to Frederick II") published[7] in *Mémoires de l'Académie Royale des Sciences et Belles Lettres* (1792):

Astronomers of the most remote antiquity usually dedicated to immortality the names and the merits of their illustrious contemporaries, their laws and their heroes, placing them after their death among the stars. The memory of these famous characters still lives among us after thousands of years gone by; and if later generations were able to alter their mythological mysteries and distort the primitive traditions, they were unable to dispense with them entirely. It is not unprecedented that modern astronomers have placed in the firmament also monuments of princes who distinguished themselves by their heroism and their charities.

Among all the princes who for centuries have disappeared from the world stage, is there more worthy a celestial monument than one to Frederick? he has surpassed everything we had ever dared expect from an inhabitant of this planet. Already poets, orators and historians have devoted their lyres and their quills to immortalize his exploits. The brush, the burin, and chisel have conveyed to posterity the features of his mortal remains. Do we not also allow the astronomer to search among the sublime objects in their care for the materials of an imperishable monument to the glory of such a king?

The brilliant assembly that is formed here to solemnly celebrate the memory of Frederick the august founder of the Academy, allow me to propose a new

[6]Leonhard Euler (1701–1783), a Swiss mathematician and physicist.

[7]The original French text is reproduced at the end of this chapter.

constellation that I dedicate to the glory of this prince. I place it in the northern region of the heavens, which is not occupied by any asterism (b), near which stars shine that are devoted to an ancient family of kings, to Cepheus, Cassiopeia, Andromeda. Not far away are Pegasus and the Swan. Making only a very slight change to the neighboring asterisms (c), I located seventy-six stars, many of which have been recently discovered by myself and which compose the new constellation of Frederick (d). I give this new constellation the name of Frederic's Glory (Friedrichs Ehre) (e). I place it along the southern edge of the Milky Way, and near the crown of Cepheus, a royal crown, because in the age of Frederick the stars in this location culminated above the center of his empire. Below the crown, leading toward the south, a sword, a pen, and an olive branch form a group surrounded by an immortal laurel, and perpetuate the glory of Frederick as a hero, philosopher and peacemaker.

If this meager expression of my enthusiasm and my gratitude is welcomed as I have reason to hope, I will draw my new constellation on appropriate segments on a globe of a foot in diameter that I intend to engrave soon. [I am] happy if I can contribute in some way to perpetuate the memory of Frederick.

May the most abundant measure of prosperity fall upon the days of the monarch who so gloriously succeeded Frederic! which indicates the reign of Frederick William the beloved, of the king adored by his nation, of this august patron of the Academy! That he may long be the father of his people! That the century that begins the long path to enjoy the splendor of his earthly crown, before he exchanges it for the glittering diadems of the firmament.

(a) Thus Halley placed in the firmament the oak and heart of Charles II,[8] the first in the southern hemisphere near the ship,[9] and the other near the hunting dogs.[10] Hevelius placed Sobieski's Shield[11] below the Eagle,[12] in memory of Polish King John III. There one hundred years ago Gottfried Kirch, the oldest astronomer of Berlin, placed near Eridanus the scepter of Brandenburg.[13] Finally Father Prezobut [Poczobut] of Wilna inserted Poniatowski's bull[14] near Ophiuchus, in honor of the reigning king of Poland. All these new constellations were created from unformed stars, most of which belonged to existing constellations.

(b) This region of the sky is of considerable extent. It offers stars that are easy to distinguish and we filled up this part of this space by tracing a rock and a long chain belonging to the figure of Andromeda.

[8]Robur Carolinum (Chap. 22) and Cor Caroli.

[9]Argo Navis (Chap. 5).

[10]Canes Venatici.

[11]Scutum.

[12]Aquila.

[13]Sceptrum Brandenburgicum (Chap. 24).

[14]Chap. 25.

(c) This change is made in order to meet the hand of Andromeda, which here consists of stars a little more to the east; and to recede westward toward the small lizard of Hevelius [Lacerta], receiving at the same time a more appropriate form. Many other astronomers before me have been allowed changes to the figures of the constellations as slight and essential, wherein they achieved a more interesting result.

(d) By means of the changes I have indicated I found forty-nine stars, which are all marked in my celestial maps, published in 1782.[15] I increased this number by my observations of January 1787, which gave me twenty seven stars and two clusters of stars scattered throughout space, and which had not yet appeared in any catalog or on any celestial map. These stars are in the representation I give the new constellation; their situation has been determined by glance. The new constellation is therefore then composed of seventy-six stars, four of which are of the fourth magnitude; three of the fifth; twenty-four of the sixth; sixteen of the seventh; twenty-four the eighth; and five of the ninth. Five of these stars belong to Pegasus; twenty-six to Andromeda; six to Cepheus; nine to the lizard[16]; and three to Cassiopeia. M. de la Lande noted to me, on 5 December 1789, that he has already seen the position of several of the stars belonging to the Frederick's Glory.

(e) Our worthy colleague, Professor Ramler,[17] is properly the inventor of the name of Frederick's Glory (Friedrichs Ehre). I found this name sufficiently appropriate that I did not hesitate to adopt it. Director Rode made the design of the new constellation following the instructions I gave him.

Bode continued to promote his invention, assuring its perpetuation for several decades by publishing it in his highly influential *Uranographia* (1801b). In the companion catalog *Allgemeine Beschreibung und Nachweisung der Gestirne* (1801a) he wrote[18]:

> It was in 1787 that I consecrated this astronomical monument to the memory of our immortal King Frederick II. I place it between Pegasus, the Swan, Cepheus, Cassopeia & Andromeda. Some of the principal stars of this constellation previously belonged to the chain and the northern hand of Andromeda.

Bode referred to the constellation using the German "Frederich's Ehre', meaning "Frederick's Honor" or "Frederick's Glory," appropriately Latinizing it as "Honores Frederici" when he showed the constellation in *Uranographia* (1801b; Fig. 12.2). Richard Hinckley Allen (1899) noted alternate names appearing in the nineteenth century literature: "Burritt's '*Gloria Frederica*, and Miss [Agnes] Clerke's '*Gloria Frederici*".

---

[15]The first edition of Bode's *Vorstellung der Gestirne.*

[16]Lacerta.

[17]The German poet Karl Wilhelm Ramler (1725–1798).

[18]"Dieses Sternendenkmal setzte ich im Jahre 1787, dem Andenken unsers unsterblichen Königs Friedrichs II zwischen dem Pegasus, Schwan, Cepheus, Cassiopeja und Andromeda. Einige der vornehmsten Sterne desselben gehörten sonst an der Kette und nordlichen Hand der Andromeda."

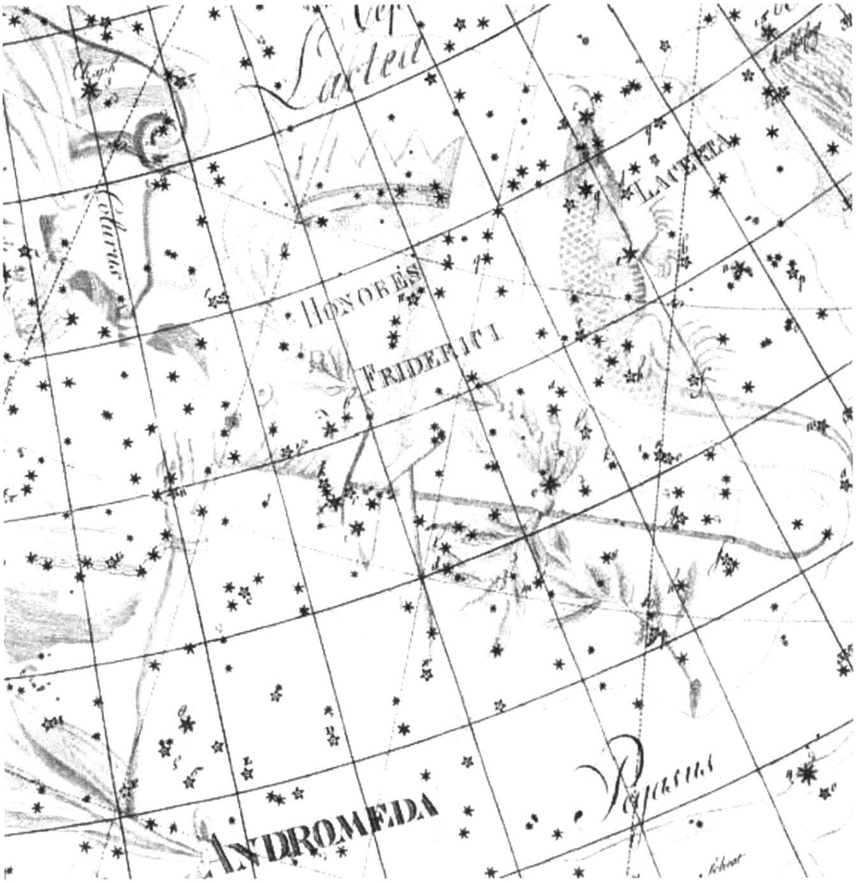

12.2 Honores Frederici as depicted in on Plate 4 of Bode's *Uranographia* (1801b)

Other mapmakers quickly took up Bode's suggestion. August Gottleib Meissner included the figure as '"Friedrichs Ehre" in *Astronomischer Hand-Atlas* (1805), while Thomas Young showed it as '"Frederic's Trophy" in the hemispheres reproduced in his *Course of Lectures on Natural Philosophy and the Mechanical Arts* (1807). Alexander Jamieson depicted Bode's figure in his *Celestial Atlas* (1822), and Jacob Green described the constellation briefly in *Astronomical Recreations* (1824), conveying his dislike for such contrived images:

> HONORES FREDERICI, or the Glory of Frederick, is a small asterism placed among the constellations by Bode in memory of Frederick II. king of Prussia. It is composed of a few stars, which remained in an unformed state, a little to the north of Andromeda's hand. Iota, Kappa and Lambda, stars of the fourth magnitude, placed by old astronomers in this hand, are now to be reckoned the principal stars of the

> Glory of Frederick. The emblem chosen to designate this group is a sword entwined with laurel: sometimes a crown is added. We freely confess that such symbols among the stars appear to us very much out of place.

Due to its location on the sky and his approach to projecting the sky onto the maps in his *Atlas Designed to Illustrate the Geography of the Heavens*, Elijah Hinsdale Burritt separated the crown from the rest of the figure, showing them separately on Plates 6 and 2, respectively (Fig. 12.3).

Despite the wide influence of *Uranographia*, Bode's figure was not universally adopted. In most cases cartographers who opted not to include Honores Frederici instead left its stars unformed and preserved Lacerta in place. Examples of this approach include Henry Brooke's *A Guide To The Stars* (Plate 1; 1820), Joseph Johann Littrow's *Atlas des Gestirnten Himmels* (Plates 1 and 2; 1839), and Friedrich Argelander's *Neue Uranometrie* (Plates 2 and 6; 1843). It does not appear on William Croswell's *Mercator Map of the Starry Heavens* (1810), but probably only because its position in the sky falls just off the edge of Croswell's map area. The trend of hit-and-miss appearances continued for several decades. By midcentury, it was already beginning to fall out of favor on both sides of the Atlantic; in Hanna Bouvier's *Familiar astronomy* (1858), Honores Frederici was declared "not generally recognised by astronomers, it having, by some, been suppressed."

In 1679 Augustin Royer introduced a constellation in nearly the same area using many of the same stars: Sceptrum et Manus Iustitiae (see Volume 2), or the Scepter and Hand of Justice, honoring the French King Louis XIV. Allen wrote, "Royer ... attempted to replace the earlier Lacerta of Hevelius by his Sceptre and Hand of Justice. But he borrowed for his new creation from the northern hand of Andromeda, which he moved to a more easterly position, entirely indifferent to the fact that it had been 'stretched out there for 3000 years.'" Royer's constellation was already long forgotten by the time Bode recycled the stars in honor of Frederick II.

ICONOGRAPHY

Frederick II of Prussia

The object of Bode's adoration was Frederick II, the Elector of Brandenburg and King of Bode's native Prussia from 1740 until his death (Fig. 12.4). He was the third monarch of the House of Hohenzollern, which reigned over much of Germany until the ultimate fall and abdication of Kaiser Wilhelm II (1888–1941) in 1918 at the conclusion of the First World War. Popularly known both as *Friedrich der Große* ("Frederick the Great") and *Der Alte Fritz* ("Old Fritz"), he remains best known as the brilliant military strategist who led Prussia to victory in the Seven Years' War (1754–1763).

Frederick was born in Berlin on 24 January 1712, the son of Frederick William I and Sophia Dorothea of Hanover. His father was King in Prussia and Elector of Brandenburg from 1713 until his death, and throughout was in personal union the sovereign prince of the Principality of Neuchâtel, a canton of western Switzerland. Frederick William ruled

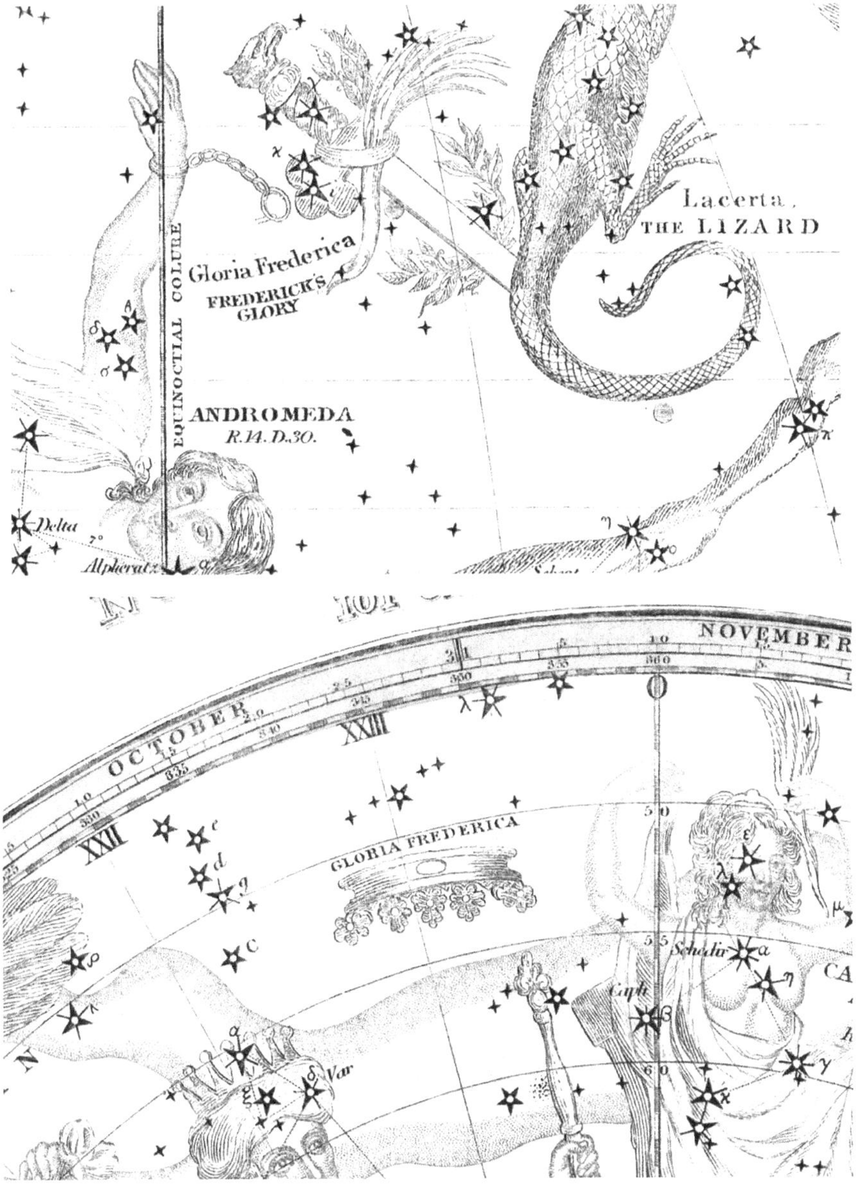

12.3 Honores Frederici ("Gloria Frederica") shown in two discontiguous pieces on Plate 2 (*top*) and Plate 6 (*bottom*) of an 1850 edition of Elijah Hinsdale Burritt's *Atlas Designed to Illustrate the Geography of the Heavens* (1835a)

12.4 *Portrait of Frederick II* (1781) by Anton Graff (1736–1813). Oil on canvas, 62 cm × 51 cm (24.4 in × 20.1 in); Charlottenburg Palace, Berlin

Brandenburg-Prussia with an iron fist; possessed of a violent and uncontrolled temper, historians have speculated that he suffered from porphyria (Pierach and Jennewein, 1999). In contrast, Sophia Dorothea, daughter of George I of Great Britain, was gentle and cultured, earning her the popular nickname "Olympia" on account of her manners and regal comportment. She was a source of refuge from his father's brutality, and to no surprise he became rather attached to her and would later deeply mourn her death.

Frederick William had great plans for his son and heir in the art of war, but young Frederick would have nothing of it, preferring to spend his time studying music and philosophy. He became enamored of the young nobleman Hans Hermann von Katte (1704–1730) who made fast friends with the Crown Prince. At age 18 he made plans to run away from his authoritarian father to the safety of Great Britain; von Katte counseled against such a move but came to accept the inevitable. Frederick made his move on 5 August 1730 by unsuccessfully escaping from his quarters while Frederick William's court visited the Rhenish Palatinate. von Katte was safely situated at Potsdam, but an intercepted communications between the two betrayed him as an accomplice.

Because both young men were commissioned officers in the Prussian army, Frederick William accused the pair of treason and considered forcing Frederick to renounce his right of succession in favor of his brother, Augustus William (1722–1758). The king ultimately spared Frederick's life and his future position, fearing the interference of the Imperial Diet of the Holy Roman Empire. Upon being found guilty of desertion at a court martial, von Katte was sentenced to life in prison; Frederick William was predictably furious at the sentence and ordered him executed. The order was carried out at the fortress of Küstrin on 6 November, and the younger Frederick was made to watch the execution by beheading. He despaired at his friend's death and rarely spoke of him for the remainder of his life. Frederick was pardoned and ordered to begin intensive training in military strategy at the Küstrin.

Marriage proposals involving potential brides from the British and Russian royal houses were dashed due to concerns over interference with military alliances. At one point, Frederick appeared ready to renounce his right to the succession in order to marry Maria Theresa of Austria, but Prince Eugene of Savoy successfully lobbied Frederick William to consent to his son's marriage to Elisabeth Christine of Brunswick-Bevern, a suitably Protestant relative of the Austrian Habsburgs. Frederick consented to the union after writing[19] his sister Wilhelmine "There can be neither love nor friendship between us." After marrying the intellectually vapid Elisabeth on 12 June 1733, he had her installed at her own palace in Berlin. Frederick kept himself strictly away from her at his own palace in Potsdam where he entertained himself among the strictly male members of his court, leading some historians to conclude that he was homosexual.

In the last years of his life, Frederick William restored his elder son to the Prussian army with the rank of colonel and granted him his own palace at Rheinsberg where Frederick entertained himself with a small retinue of actors, poets and musicians. It was the happiest time of his life, before he assumed the throne at age 28 upon the death of his father in 1740. At his accession, he took on his father's title of King *in* Prussia, because the territory he controlled was only a part of the historic Prussia; three decades later, after acquiring most of the rest of the historic territory, he styled himself King *of* Prussia.

As a ruler Frederick knew how to play both sides of an issue. He governed as a sort of enlightened absolutist, retaining the full spectrum of powers available to him as monarch but softened by an appeal to pragmatism. For example, he dealt with religious matters throughout his kingdom with approaches ranging from outright tolerance to abject oppression, especially in conquered realms in Poland. He reformed the Prussian civil service and opened high-level positions to men not born into the nobility while occasionally interfering in the minutiae of governance as he saw fit. And at the same time that he made decrees interfering with press freedom, he showered certain artists and philosophers with royal favor.

[19]Quoted by Louis Crompton in *Homosexuality and Civilization*, Harvard University Press (2009), p. 508.

Frederick was one of the most successful European military leaders of the eighteenth century. Shortly after coming to power, the Holy Roman Emperor Charles VI died on 29 October 1740 with the imperial crown going to his daughter, 23-year-old Maria Theresa of Austria. Frederick disputed her right of succession to the Empire in general and the Province of Silesia in particular. To try to force an outcome in his favor Prussia invaded Silesia on 16 December, leading to a quick victory and a military occupation; Frederick hoped the move would thwart the plans of Augustus III, King of Poland and Elector of Saxony, to consolidate his own lands via disputed territory in Silesia.

He struggled to retain control of his captured Silesian realms in 1741, facing real battle for the first time at Mollwitz on 10 April 1741. Incorrect intelligence led him to believe that the Austrian army had defeated his own, and in the interest of evading capture he fled the battlefield on horseback. It turned out instead that the very moment Frederick galloped away, his army in fact gained the upper hand and decidedly won the battle. Later recalling his grave mistake, he was forever humbled by his own personal humiliation during the battle and wrote that "Mollwitz was my school." He would go on to fight through four more years of the War of Austrian Succession, concluding the Treaty of Dresden with Austria on Christmas Day, 1745, and thus giving Silesia to Prussia.

Frederick again scored victories in the Seven Years' War (1756–1763) and the War of Bavarian Succession (1778–1779). In every conflict he personally led his men into battle, and had six horses shot out from under him during his military career. His battlefield innovations led him to be regarded by many as among the most brilliant military strategists in history. In 1807, some 20 years after Frederick's death, Napoleon Bonaparte (1769–1821) visited his tomb at Potsdam after the victory of the Fourth Coalition. After paying his respects, Napoleon is said to have turned to his officers and remarked,[20] "Gentlemen, if this man were still alive I would not be here." Napoleon studied Frederick's campaign narratives carefully, and even kept a small statuette of the Prussian king in his personal cabinet.

Frederick grew increasingly withdrawn toward the end of his life. Situated at Sanssouci Palace in Potdsam he saw his friends die off one by one as his age advanced, but he made no effort to replace them. The quality of his civil governance suffered, and despite continued popularity among his people, he spent progressively less time among both his subjects and figures in his government. Rather, in his final years he spent most of his time with his beloved Italian greyhounds, preferring their company to most people.

He died at Sanssouci on 17 August 1786 seated in an armchair in his study. It was his final wish to be buried next to his greyhounds in the palace vineyard, but his successor Frederick William II ensured that his body was interred next to his father at the Potsdam Garrison Church. Ultimately, his remains took a long route to their final rest. Adolf Hitler ordered the body moved into a salt mine to protect it from destruction near the end of World War II; after the Germans' defeat, American forces relocated it to Burg Hohenzollern. After German reunification, on 17 August 1991, the anniversary of his death, his coffin

[20]Quoted by Hannsjoachim Wolfgang Koch in *A History of Prussia*, Dorset Press, New York (1978), p. 160.

was disinterred and lay in state at Sanssouci covered by a Prussian flag. After dusk his body was quietly conveyed to his chosen final resting place.

Frederick the Great was lionized by later generations of historians, and held up as the very model of a proper German leader well beyond what the evidence from his life justified. Nineteenth century German historians praised him for raising the Prussian state up to the status of a major European power, and his reputation held firm even after the German Empire's humiliating surrender at the end of the First World War. While he was virtually deified by the Nazi regime as among the greatest leaders of German history, his image suffered in postwar Germany as the nation engaged in a generation of soul searching after suffering another catastrophic loss.

The Regalia

The figure devised by Bode is heavily laden with history and symbolism (Fig. 12.5). He intended it to represent a ceremonial sword, a pen and an olive branch, joined together with a laurel, the devices symbolizing Frederick II as a "hero, sage, and peacemaker." Perhaps to further solidify the regal reference, Bode added the figure of a crown offset from the other symbols.

The Sword

Swords in their modern sense are known from roughly the late Bronze Age as they evolved from shorter daggers. By the late fourth millennium BC, fabrication of long blades became possible for the first time using alloys first of arsenic-copper and then tin-bronze. Technology continued to improve during the Iron Age; although the techniques of the period continued to use hammered metal rather than quench-hardening, the use of iron made sword manufacturing easier from the standpoint of both obtaining raw materials and working the material.

From very ancient times, the sword has stood as a symbol of power, might and control. Swords are often included in the designs of heraldic arms to indicate descent from distinguished military and/or regal figures; in the Middle Ages, the sword became a representation of the word of God, and its occurrence in iconography was intended to reinforce the temporal authority of rulers as a consequence of the divine order. Along with the shield, the sword is the best known symbol of knighthood and chivalry. Since at least classical antiquity swords appear in mythology and folklore associated with magical properties, sometimes given to heroes by gods or spirits to enable the advancement of justice. A ruler of Frederick's era would have invoked the sword as a manifestation of both earthly dominion and divine favor.

12.5 Detail of Honores Frederici (*upper left*) depicted in *Urania's Mirror* (1825). A rather distressed Andromeda dominates the card, while both Triangula (including Triangulum Minus; Chap. 28) are shown below and to her left

The Pen

The pen is inextricably linked to language, which in turn serves as a manifestation of learning and the intellect, suggesting the mind of Frederick. In it Bode asserted that Frederick was a wise ruler whose actions on and off the battlefield were guided more by reason rather than emotion. However, the choice of the pen is distinct from other intellectual symbols such as the lamp of learning, which would indicate the role of scholar. Rather, in the case of a ruler, the pen evokes a sense of rhetoric and rational justification in an attempt to paint Frederick more in the sense of Plato's philosopher-king than a simple war-monger. Indeed, Frederick functioned in this role as an author on military subjects during his life.[21]

The Olive Branch

From classical antiquity, essentially every culture of the Mediterranean basin considered the olive branch symbolic of victory and/or peace[22]; its adoption by the ancient Greeks had a profound influence on Western culture generally. "To offer an olive branch" has become synonymous with the concept of warring parties suing for peace, as related[23] by Virgil in the *Aeneid*:

> But, when they saw the ships that stemmed the flood,
> And glittered through the covert of the wood,
> They rose with fear, and left the unfinished feast,
> Till dauntless Pallas reassured the rest
> To pay the rites. Himself without delay
> A javelin seized, and singly took his way;
> Then gained a rising ground, and called from far:
> "Resolve me, strangers, whence, and what you are;
> Your business here; and bring you peace or war?"
> High on the stern Aeneas his stand,
> And held a branch of olive in his hand,
> While thus he spoke: "The Phrygians' arms you see,
> Expelled from Troy, provoked in Italy
> By Latian foes, with war unjustly made;
> At first affianced, and at last betrayed.
> This message bear: The Trojans and their chief
> Bring holy peace, and beg the king's relief.

[21] For example, *Instruction to his Generals* (1797).

[22] See, e.g., Lucia Impelluso (2004), *Nature and its symbols*. Getty Publications.

[23] Book VIII, lines 107–123, trans. John Dryden (1697).

12.6 Detail of an early Christian marble funerary monument showing a dove bearing an olive branch, a person in the "orans" (prayer) position, and the chi-rho symbol. Terme di Diocleziano, Museo Nazionale Romano, Rome; provenance and date unknown

On first inspection there seems to exist an inherent conflict between the symbols of both war (the sword) and peace (the olive branch) appearing simultaneously among Frederick's regalia, as though the presence of one should contradict the other. There was no such conflict in the minds of the Romans who found an intimate connection between the notions of war and peace; for example, Mars, the god of war, had another aspect, Mars Pacifier, seen as the bringer of peace. The war god is shown in that aspect in the design of coins from the later Imperial era. Earlier, in the Pax Romana (*c.* AD 1–200), peace/war imagery in Roman art reinforced the power of the Emperors as protectors whose permanent war footing abroad ensured sustained prosperity at home. During this time, Roman envoys to foreign governments commonly used olive branches as tokens of peace. The Roman outlook was adopted by future leaders such as Oliver Cromwell (Chap. 22), Protector of England, who ably exploited his family's heraldic motto "Pax Quaeritur Bello" ("Peace is sought by war," or, "If you seek peace, prepare for war").

The olive branch is similarly tied as a symbol to the notion of divine blessing from its early appearance in Biblical version[24] of the Flood Myth:

> After forty days Noah opened a window he had made in the ark and sent out a raven, and it kept flying back and forth until the water had dried up from the earth. Then he

[24]Genesis 8:6–11 (New International Version).

> sent out a dove to see if the water had receded from the surface of the ground. But the dove could find nowhere to perch because there was water over all the surface of the earth; so it returned to Noah in the ark. He reached out his hand and took the dove and brought it back to himself in the ark. He waited seven more days and again sent out the dove from the ark. When the dove returned to him in the evening, there in its beak was a freshly plucked olive leaf! Then Noah knew that the water had receded from the earth.

Early Christians often incorporated the motif of the olive-branch-bearing dove in their art; the image frequently appears on Christian tombs and in meeting places such as catacombs (Fig. 12.6). Since the dove was also associated with the Holy Spirit, the combination of Judaeo-Christian and classical figures yielded the notion of earthly peace by divine command. In the context of Frederick's regalia the olive branch can be interpreted as a Christian symbol suggesting God's favor toward Frederick. The ensemble of olive branch and sword implies the notion of 'righteous' war as the key to earthly peace with Frederick as its agent.

The Laurel

In ancient Greece, the bay laurel tree (*Laurus nobles*) was sacred to the god Apollo who is often shown in art wearing a wreath of interlocking laurel branches on his head. Culturally, the Greeks viewed the laurel wreath as symbolic of honor, awarding them to their poets and heroes. This tradition has given us the English word "laureate," indicating a person of eminence recognized for his or her efforts often in association with various types of literary and academic awards. Too, the term "to rest on one's laurels" has an idiomatic meaning that refers to a person who relies entirely on past successes for continued fame and recognition. The Romans specifically associated laurels with martial victories; generals and emperors returning from successful military campaigns were celebrated with public "triumph" spectacles in which they were crowned with laurel wreathes (Fig. 12.7). As a component of Frederick's regalia, the laurel serves the dual function of symbolizing both military success and intellectual achievement, neatly bookended by both the sword and the pen.

DISAPPEARANCE

Bode's manifestly German expression of national pride waned in popularity through the last quarter of the nineteenth century and was functionally extinct before 1900. A number of influential atlases omitted the figure during this time, including Richard Anthony Proctor's *The Constellation-Seasons* (1876), William Henry Rosser's *The Stars and Constellations* (1879), and Charles Pritchard's catalog *Uranometria Nova Oxoniensis* (1885). Among its last appearances in a popular work is in the northern hemisphere chart in Eliza A. Bowen's *Astronomy By Observation* (1888). In 1899 Allen wrote, "It is now

12.7 Roman relief sculpture showing Agrippina (*right*) crowning her son Nero with a laurel wreath at his accession as emperor in AD 54. Agrippina, who Nero had murdered in AD 59, is shown holding a cornucopia symbolizing fortune and plenty, while the young emperor is clothed in the tunic and armor of a Roman military commander. Aphrodisias Museum, Geyre, Turkey

seldom mentioned, and has been discarded from the charts, while Lacerta maintains its position in this much occupied spot."

The last hurrah of Bode's tribute to the great Prussian king might possibly be in William Tyler Olcott's *Star Lore of All Ages*:

> The fourth magnitude stars λ, κ, ι Andromedæ and the fifth magnitude star ψ Andromedæ form a "Y"-shaped figure which bears the name of "Gloria Frederica" or Frederick's Glory, an asterism formed by Bode in 1787 in honour of the great Frederick II., of Prussia, who died in 1786. The figure is thus described: "Below a nimbus the sign of royal dignity hangs, wreathed with the imperishable laurel of fame, a sword, pen, and an olive branch, to distinguish this ever to be remembered monarch, as hero, sage, and peacemaker." This figure, with the exception of the

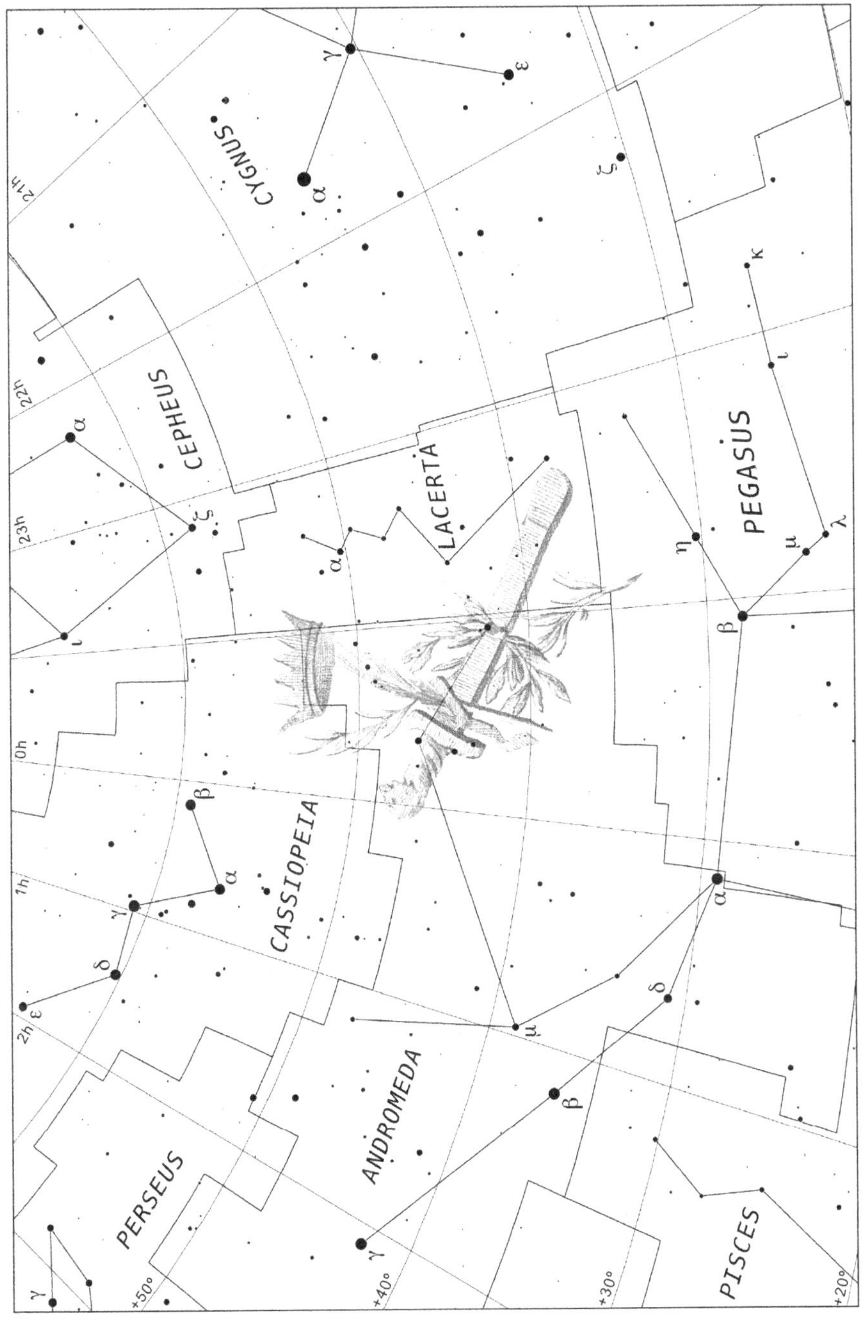

12.8 The figure of Honores Frederici from *Mémoires de l'Académie Royale des Sciences et Belles Lettres* (1792) overlaid on a modern chart

nimbus, appears on Burritt's Atlas, but later atlases omit the asterism entirely, and it is seldom mentioned.

In his *Atlas Céleste* (1930a), Eugène Delporte drew the boundary line between Andromeda and Lacerta directly through the middle of Bode's constellation, finally settling the disposition of its stars. A few stars from Bode's crown spill over Delporte's boundary into neighboring Cassiopeia (Fig. 12.8).

ORIGINAL BODE TEXTS

Astronomische Jahrbuch (1787)

Am 25 sten Januar d. J. hielt die hiesige Königl. Akademie des Wissenschaften eine öffentliche dem Gedächtniss und Lobe Friedrichs, ihres Durchlauchtigsten Stifters, gewidmeten Versammlung. Diese Feierlichkeit veranlasste mich, in einer kurzen Rede die Einführung eines neuen dem Andenken des unsterblichen Königs geweihten Sternbildes under der Benennung Friedrichsehre vorzuschlagen, und ich legte der Akademie die Figur desselben in einer gestochenen Himmelscharte vor, woraus die Fig. auf Tat. II. entlehnt worden.

Ich formire es zwischen der *Cassiopeja*, dem *Cepheus*, *Andromeda* und *Schwan* aus 76 zum Theil von mir neu beobachtete Sterne, und setze eine Stralenkrone, das Zeichen der Königl. Würde, an den Südl. Rande der Milchstr. nahe bey der Krone des Cepheus, weil im Jahrhundert Friedrichs die dort schimmernden Sterne, Seinem Erdenreich senkrecht standen. Südwärts under derselben hangen, mit dem unverwelklichen Lorbeer des Nachruhms umwunden: Ein *Schwerdt*, eine *Feder*und ein *Olzweig*, um diesen ewig unvergesslichen Monarchen als *Helden*, *Weisen* und *Friedensstifter* zu bezeichnen.

Es ist bekannt, dass mehrere Astronomen der neuern Zeit berühmten Fürsten dergleichen Ehren-Denkmäler am Himmel gegestistet, und selbige gewöhnlich aus Sterne bildeten, die grösstenteils zu benachbarten Gestirnen gehörten, auch dabey nicht selten genötigt waren, dergleichen unwesentliche Abänderungen vorzunehmen, wie ich z. B. bey Formirung des neuen Bildes mit der einen Hand der Andromede und dem Hevelschen Eidexengestirn, welche erstere ich durch etwas mehr östlich stehende Sterne lege, und letzteres in eine schicklichete Lage bringe. Der Raum, den ich für meine Fredischsehre gewält hatte, schliesst nach dieser Abänderung 49 schon in meinen Himmelscharten stehende Sterne ein. Hierzu suchte ich nun in einigen heitern Abenden des Januar-Monats noch 27 kleine dort herum stehende Sterne auf, und trug ihre Oerter nach einem genauen Augenmaass in meiner gezeichneten Charte ein. Das neue Bild war nun aus 76 Sternen und zwey Sternhäuflein zusammengesetzt, worunter sich 4 Sterne von der 4ten, 3 von der 5ten, 24 von der 6ten, 16 von der 7ten, 24 von der 8ten und 5 von der 9ten Grösse befinden. Zum Pegasus gehören hievon 5, zur Andromede 26, zum Cepheus 6, zur Bidexe 9, und zur Cassiopeja 3 Sterne.

Die Russisch-Kayserl. Akademie zu Petersburg hat meinen wohlgemeinten und uninteressirten Vorschlag, Friedrich den Einzigen ein Denkmal am Sterngefilde zu stiften, gebilligt, und mir ihre Approbation schriftlich durch Herrn. Prof. Euler unterm 24. April zugesandt. Es heisst am Schluss derselben: L'Academie applaudit à cet hommage que Mr. Bode rend à la mémoire d'un des grand Souverains de son Siècle, et Messieurs les Astronomes consentirent avec plaisir d'employer le nom de la nouvelle Constellation, toutes les fois qu'il s'agira dans leurs observations des etoiles qui la composent.*

Herr Layard, Secretair der Londoner Königl. Societät der Wissenschaften; schrieb unterm 7. Jun. im Namen der Societät an mich: *The Royal Society receiv'd your present of Fredrichs Sternen Denkmal, and I am directed to return to you their thanks for the same. – Herr.*

Herr Justizrath Bugge schrieb dieserwegen aus Kopenhagen an much unterm 26, Jun.: La Societe Royale a fort approuve votte homage, et elle ne doute pas, que les astronomes de Dannemark ferount Usage de votre nouvelle constellation, de son nom, et de sa figure dans leurs observations et dans ents ecrits Elle m'a ordonne de Vous marquer sa approbation entiere.

Herr Mechain schreibt aus Paris unterm 23. Febr.: *Je présenterai à l'Academie un de vos exemplaires de la nouvelle Constellation que vous aves forme à la gloire du grand frederic: Cette idée vous fait honneur et tous les philosophes ne peuvent qu'y applaudir, le nom de ce grand prince vivra éternellement et il mérite bien d'étre grave sur la vote des cieux.*

Herr de la Lande schreib unterm 7ten May: *J'ai eu avec votre Lettre du 12. Février la figure de votre nouvelle constellation, je l'ai communiquer à nos astronomes: j'en parlerai dans la nouvelle édition de mon Astronomie; personne ne s'interesse plus a la gloire de ce grand prince que moi qui fus pour ainsi dire Son ouvrage, puisque ce fut a Berlin en 1751 que je fis le premier voyage et les premieres observations qui m'ont ouvert la carriere de l'astronomie.*

Unterm 24. Jul. schrieb Herr Prof. Euler an mich: Das hiesige Directorium der Volkschulen hat bereits das neue Sternbild Friedrichsehre auf einer bey demselben von dem Herrn Hofr. Gollovin besorgten Himmelskugel ausragen lassen.

Noch haben verschiedene Astronom. mir ihren Beyfall über das vorgeschlagene Sternbild in Briefen su erkennen gegeben.

* Bey dieser Gelegenheit meldete mir noch Herr Prof. Euler, dass die Herren Astronomen der Kayserl. Akad. auch meine vorgeschlagene Benennung und Beze-

ichnung des neuen Planeten angenommen, und legte darüber einige Blätter aus dem Petersburget Kalender bey. -Bode

***Mémoires de l'Académie Royale des Sciences et Belles Lettres* (1792)**

Les astronomes de l'antiquité la plus reculée consacroient d'ordinaire à l'immortalité & les noms les mérites de leurs contemporains illustres, de leurs lois & de leurs héros, en les plaçant après leur mort parmi les astres. La mémoire de ces personnages fameux vit encore parmi nous après de milliers d'années révolues; & si les générations postérieures ont pu par leurs mystères mythologiques altérer & défigurer les traditions primitives, elles n'ont pu en faire perdre entièrement la trace. Il n'est pas sans exemple que des astronomes modernes aient placé au firmament les monuments des princes qui se distinguèrent par leur héroïsme & leur bienfesance.

Entre tous les princes qui depuis plusieurs siècles ont disparu de la scène du monde, en est-il un qui mérite un monument céleste à plus juste titre que Fréderic l'unique? lui qui a surpassé tout ce qu'on eût osé jamais attendre d'un habitant de cette planète. Déjà les poѐtes, les oratures & les historiens, ont consacré leurs lyres & leurs plumes à immortaliser ses exploits. Le pinceau, le burin, & le ciseau ont transmis à la postérité les traits de sa dépouille mortelle. Ne permettroit-on pas aussi à l'astronome de chercher parmi les objets sublimes dont il s'occupe les matériaux d'un monument impérissable à la gloire d'un tel roi?

La brillante assemblée qui s'est formée ici pour célébrer solennellement la mémoire de Fréderic, l'auguste fondateur de l'Académie, me permettra de lui proposer une constellation nouvelle que je consacre à la gloire de ce prince. Je la place dans une région boréale de firmament qui n'est occupée encore par aucun astérisme (b), & près de laquelle brillent les étoiles consacrées a une antique famille de rois, à Céphée, Cassiopée, Andromède. Non loin de là sont le Pégase, et le cygne. En n'apportant qu'un très-lèger changement aux astérismes voisins (c), je trouve soixante-seize étoiles, dont plusieurs ont été découvertes en dernier lieu par moi-mŘme & qui composent la nouvelle constellation de Fréderic (d). Je donne à cette nouvelle constellation la dénomination de gloire de Fréderic (Friedrichs Ehre) (e). Je place sur le bord méridional de la voie lactée, et près de la couronne de Céphée, une couronne royale, parce que dans le siècle de Fréderic les étoiles de ce lieu étoient perpendiculairement au dessus de centre de son empire. Au dessous de la couronne, en tirant vers le midi, un glaive, une plume, et un rameau d'olivier forment un groupe ceint d'un immortel laurier, et éternisent la gloire de Fréderic comme héros, philosophe et pacificateur.

Si cette foible expression de mon enthousiasme & de ma reconnoissance est accueillie comme j'ai lieu de l'espérer, je tracerai ma nouvelle constellation sur des segmens appropriés à un globe d'un pied de diamètre que je compte faire graver dans peu. Heureux si je puis contribuer en quelque manière a éterniser la mémoire de Fréderic. (f)

Que la plus abondante mesure de prospérité se répande sur les jours du monarque qui a si glorieusement succédé à Fréderic! qu'elle signale le régne de Fréderic Guillaume le bien-aimé, de ce roi adoré de sa nation, de ce protecteur auguste de l'Académie! Qu'il soit long-temps le pére de ses peuples! Que le siécle qui va commencer le voie long-temps jouir de l'éclat de sa couronne terrestre, avant qu'il l'échange contre les diadémes radieux du firmament.

(a) C'est ainsi que Halley plaça au firmament le chène & le coeur de Charles II, le premier dans l'hémisphère méridional près du vaisseau, & l'autre non loin des chiens de chasse. Hével plaça au dessous de l'aigle l'écu de Sobieski, en mémoire du roi de Pologne Jean III. Il y a cent ans que Godefroi Kirch, le plus ancien astronome de Berlin, plaça proche de l'ridan le sceptre de Brandenbourg. Enfin l'Abbé Prezobut à Wilna inséra près du serpentaire le taureau de Poniatowski, en l'honeur du roi de Pologne actuellement régnant. Toutes ces nouvelles constellations furent formées d'étoiles informes, dont la plupart appartenoient aux constellations ambiantes.

(b) Cette région du firmament est d'une étendue assez considérable. Elle offre des étoiles faciles à distinguer & l'on remplissait jusqu'a présent une partie de cet espace en y traçant un rocher & une longue chaîne appartenante à la figure d'Andromède.

(c) Ce changement consiste à faire répondre la main d'Andromede qui se trouve ici, à des étoiles un peu plus orientales; & à reculer vers l'occident le petit lézard d'Hevel, qui reçoit en meme temps une forme plus appropriée. Beaucoup d'autres astronomes se sont permis avant moi des changements aussi légers & aussi peu essentiels, dans les figures des constellations, lorsque par la ils ont obtenu un but plus intéressant.

(d) Moyennant le changements que je viens d'indiquer j'avois trouve quarante-neuf étoiles, qui se voient toute marquées dans mes cartes célestes, publiées en 1782. J'augmentai ce nombre par mes observations du mois de Janvier 1787, qui me donnerent vingt-sept etoiles & deux amas d'etoiles dispersées dans cet espace, & qui n'avoient encore paru dans aucun catalogue ni dans aucune carte céleste. Ces étoiles se trouvent dans la représentation que je donne de la nouvelle constellation; leur situation n'a ete déterminé que par le coup-d'oeil. La constellation nouvelle se trouva donc alors composee de soixante-seize etoiles, dont quatre sont de la quatrième grandeur; trois de la cinquieme; vingt-quatre de la sixième; seize de la septieme; vingt-quatre de la huitieme; & cinq de la neuvieme. De ces etoiles cinq appartiennent au Pegasé; vingt-six a Andromede; six a Cepheé; neuf au lézard; & trois a Cassiopeé. M. de la Lande me marque, sous date du 5 Décembre 1789, qu'il à déjà observe la position de plusieurs des étoiles appartiennent à la gloire de Frédéric.

(e) Notre digne Collégue, M. le Professeur Ramler, est proprement l'inventeur de la dénomination de gloire de Fréderic, (Friedrichs Ehre). J'ai trouvé cette dénomination si convenable, que je n'ai pas hésité de l'adopter. M. le Directeur Rode a fait le dessein de la nouvelle constellation d'après les instructions que je lui ai données.

13

Jordanis

The River Jordan

Genitive: Jordani
Abbreviation: Jor
Alternate names: "Fluvius Jordanis" (Allard, 1706)
Location: Followed a path roughly around (to the east, south, and west of) Ursa Major (Fig. 13.1).[1]

ORIGIN AND HISTORY

Jordanis was among the constellations formed by Petrus Plancius and published on his celestial globe of 1612. Prior to this, extending as far back as the time of Ptolemy, these stars were left "unformed"; as late as the time of Bayer (1603; Fig. 13.2) they were shown beyond the depicted extent of Ursa Major without any associated figure.

In *Star Names: Their Lore And Meaning* (1899) Richard Hinckley Allen attributed the invention of Jordanis to Jacob Bartsch:

> Bartschius drew on his map of this part of the sky the River Jordan, his Jordanis and Jordanus, not now recognized, indeed hardly remembered. Its course was from Cor Caroli, under the Bears and above Leo, Cancer, and Gemini, through the stars from which Hevelius afterwards formed Leo Minor and the Lynx, ending at Camelopardalis. But the outlines of his stream were left somewhat undetermined,

[1] *"Flows between the Bear and the Lion"* (Bartsch, 1624); *"Its course was from Cor Caroli, under the Bears and above Leo, Cancer, and Gemini"* (Allen, 1899); *"To the east, south, and west of Ursa Major."* (Bakich, 1995); *"Jordan had its source near the tail of the Great Bear in what is now the constellation of Canes Venatici.... From there it flowed between the Bear and Leo (an area now occupied by Leo Minor and Lynx) and ended near the head of the Bear next to Camelopardalis."* (Ridpath, 1989)

J.C. Barentine, *The Lost Constellations*, Springer Praxis Books,
DOI 10.1007/978-3-319-22795-5_13

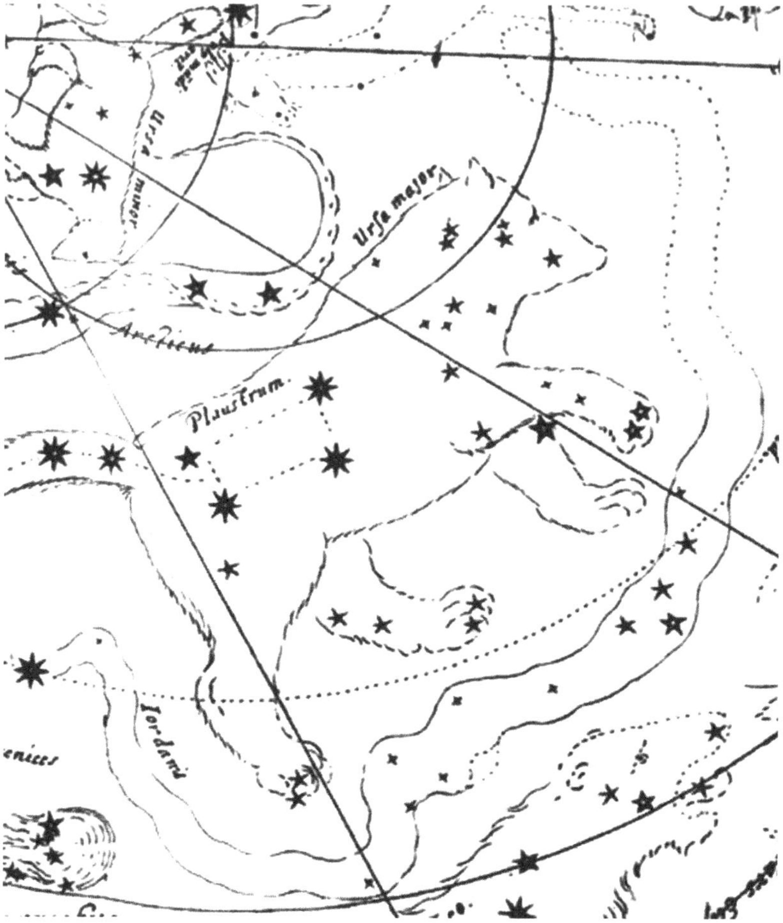

13.1 Jordanis shown winding below the feet of Ursa Major in Jacob Bartsch's *Planisphaerium stellatum*… (1661). Note the use of the label "Plaustrum" to refer to the Dipper/Plough asterism in Ursa Major

> much like those of Central African waters when guessed at by map-makers thirty years or more ago. This river, however, had already existed before his day on French star-maps and -globes.

Presumably the "French star-maps and -globes" to which Allen alludes are either a mistaken identification of Plancius' globe, or is meant to indicate French works produced in the roughly 10 years between Plancius and Bartsch that referenced the former. Allen also makes note of the portion of Jordanis that Bartsch indicates with a dotted, rather than solid, boundary as seen in Fig. 13.1 west of the head and forepaws of Ursa Major. He interprets this as the cartographer's prerogative to imaginatively fill in unknown reaches of natural

13.2 Johann Bayer's depiction of Ursa Major on Plate B of *Uranometria* (1603) leaves unformed the stars incorporated a few years later by Plancius into Jordanis

features (as the "Central African waters when guessed at by map-makers 30 years or more ago"), given that Bartsch indicates no stars in the western reaches of Jordanis.

Bartsch elaborated upon Plancius' invention in *Usus Astronomicus* (1624). On page 76 Bartsch described[2] it thusly: "It lies beneath the formless tails, of which Jordan is now the source, and which flows between the Bear and the Lion." He assigned thirteen stars to it. Later in the book, Bartsch wrote of it[3] more fully and claimed two sources for the river (stars labeled, conveniently, "Jor" and "Dan") (Figs. 13.3, 13.4, and 13.5):

> JORDANIS or the Jordan River of Judaea, [with] two sources east toward Lebanon, one called Jor and the other Dan, the most famous, sacred, lately fashioned from and adding to the Great Bear and Leo. Or, because the abundance of water in the Zodiac will be given the same, [it] is the second river of Paradise, the Euphrates, Genesis Chapter 2, verse 15.[4]

---

[2]"Subjacet informis caudae, quae IORDANIS est nunc Principium, quique Ursam inter fluit atque Leonum."

[3]"JORDANIS vel Jordanis fluvius Judae, oriens at Libani radices duobus fontibus, uno Jor, altero Dan vocato, in sacris celebratissimus, nuper ex informibus Helices, & Leonis adjectus. Vel quia profusioni aquae in Zodiaco idem tribuetur, sit secundus Paradisi fluvius Euphrates, Gen. cap. 2. v. 15."

[4]Bartsch evidently quoted the wrong verse number, which should instead be 14. Verses 10–14 of Genesis 2 describe the rivers issuing forth from the Biblical Eden: "A river watering the garden flowed from Eden; from there it was separated into four headwaters. The name of the first is the Pishon; it winds through the entire land of Havilah, where there is gold. (The gold of that land is good; aromatic resin and onyx are also there.) The name of the second river is the Gihon; it winds through the entire land of Cush. The name of the third river is the Tigris; it runs along the east side of Ashur. And the fourth river is the Euphrates." (New International Version)

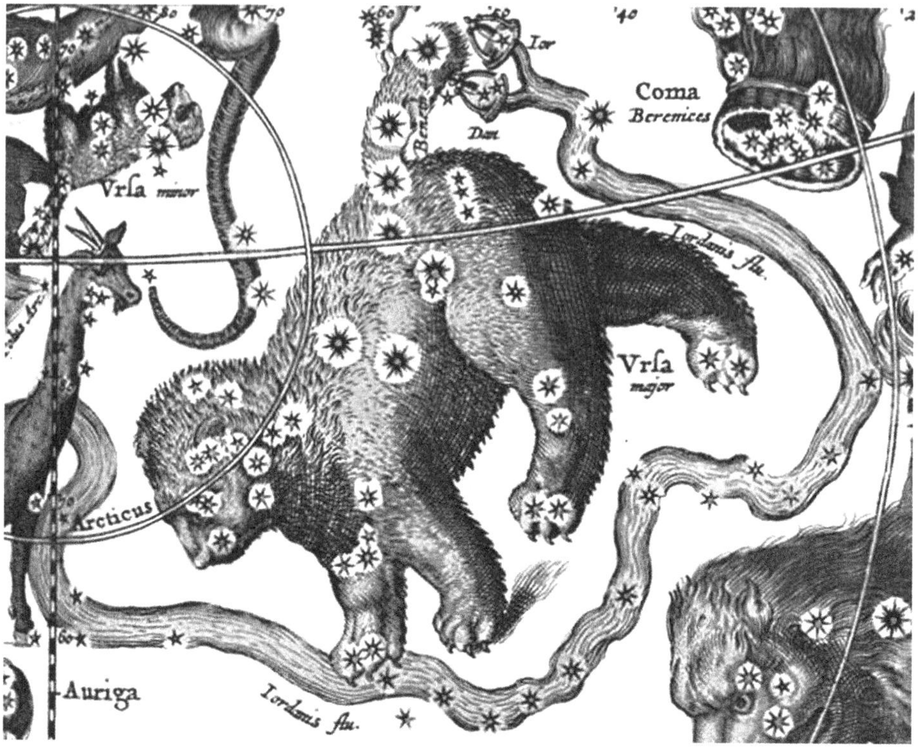

13.3 Detail of Jordanis shown on Plate 24 of Andreas Cellarius' *Harmonia Macrocosmica* (1661). The river's twin sources, "Ior" and "Dan," are depicted at top beneath Ursa Major's tail

However, Jor and Dan were not depicted on his 1661 chart (Fig. 13.1). In the same year, Andreas Cellarius showed Jordanis in *Harmonia Macrocosmica* (Fig. 13.5), following the course described by Bartsch, and explicitly including Jor and Dan.

Hevelius used *Prodromus Astronomiae* (1690) to introduce his own constellations in the space beneath the Great Bear: Leo Minor (the Lesser Lion), Lynx (the Lynx), and Canes Venatici (the Hunting Dogs), three constellations that became canonical in 1922 (Fig. 13.6). Unlike many of Hevelius' other inventions, the previous designation for these stars held on, and over a quarter-century passed before Jordanis disappeared from contemporary charts. The river appears on Carel Allard's map *Hemisphaerium meridionale et septentrionale planisphaerii coelestis* (1706) as "Fluvius Jordanis". Allard even added a little realism to the depiction by showing two islands in the river.

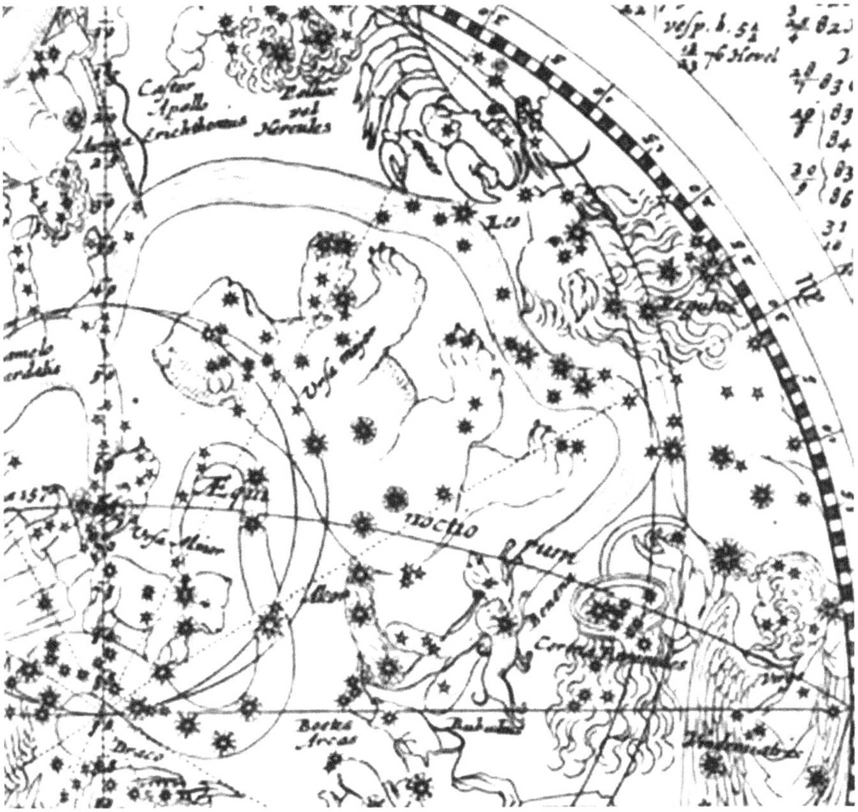

13.4 Jordanis, unlabeled, shown winding around Ursa Major in Volume 2, Table 83 of Stanislaw Lubieniecki's *Theatrum Cometicum* (1681), "Figura cometas ultimos A° Chr. 1664 et 1665 exhibens." ("Final figure showing the comets of the years 1664 and 1665")

Corbinianus Thomas depicted the constellation (Fig. 13.7) and described[5] it in his *Mercurii philosophici firmamentum firmianum* ... (1730) as:

> Palestine's well-known river, whose course arises at the sources *Jor* and *Dan*, which flow together into one channel. It flows around the Bear and the Lion and pours out stars, having a single ancient form, surrounded on all sides by its own banks

[5]"Fluvius Palestinae notissimus, cujus utì alveus, ita & nomenclatio ex fluviorum *Jor* & *Dan* in unum confluxu, enascitur. Ursam inter majorem & Leonem is incurvato tractu effusus stellas, antiquis *uniformes* seu *sporades* ripis suis coercet. Non ignoramus equidem in alias quoque ab aliis imagines hunc coeli tractum conformari, sed nobis, qui Jordanem reponunt, Gallorum nomenclaturam sequi placuit." (p. 175)

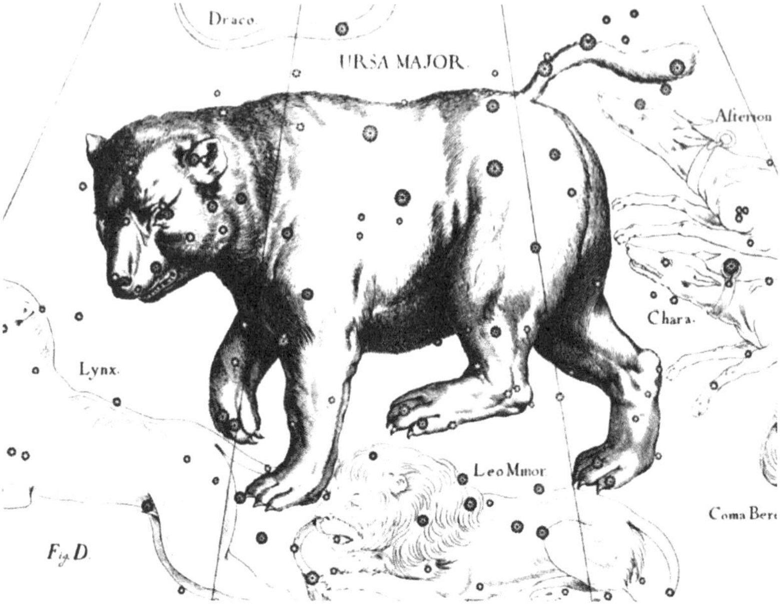

13.5 Ursa Major and surroundings as shown in Figure D of Hevelius' *Prodromus Astronomiae* (1690). In place of Jordanis, Hevelius introduced three new constellations: Lynx, Leo Minor, and Canes Venatici

or islands. Indeed, it is not known which other sources were used to compose *[this constellation]*, but we, who *[here]* restore the Jordan, are pleased to follow the naming convention of the French.

ICONOGRAPHY

Geography and Resources of the Jordan

Its relatively short 251-km run (Fig. 13.8) belies the lengthy course of the Jordan through the history of the Levant. Flowing from its headwaters above the Hula Valley of northern Israel to its terminus at the Dead Sea, the Jordan is formed from the confluence of four streams in its upper basin: the Ayun, the Hasbani, the Dan and the Banias. Along its course, the Jordan both supplies and empties the Sea of Galilee, south of which it receives the waters of tributaries the Yarmouk and the Zarqa. With the world's lowest average elevation for a river, the Jordan marks a rift valley that is part of a larger complex extending from southern Turkey to East Africa via the Red Sea.

The Jordan forms the border between both historical and contemporary nations. North of the Sea of Galilee, its course forms the western boundary of the Golan Heights, a disputed territory claimed by both Israel and Syria. South of the Sea, the remainder of

13.6 The course of Jordanis is shown on Carel Allard's map *Hemisphaerium meridionale et septentrionale planisphaerii coelestis* (1706)

its run establishes the boundary between Israel and the West Bank to the west and the Hashemite Kingdom of Jordan to the east. The West Bank's name refers to its situation along the river.

The Jordan is a source of fresh water in the region, but its natural resources are threatened by human activity. In the mid-nineteenth century the river carried so much water that visitors described many systems of rapids and waterfalls. Beginning in the 1960s, Israel, Jordan and Syria undertook diversion operations that have damaged the Jordan's ecosystem; in many places, the Jordan is now little more than a brackish stream only a few meters wide. Conflict over water rights in the Jordan contributed to the Six-Day War (1967) between Israel and its neighbors when Syria began diverting flow from its headwaters in collaboration with Lebanon and Jordan.

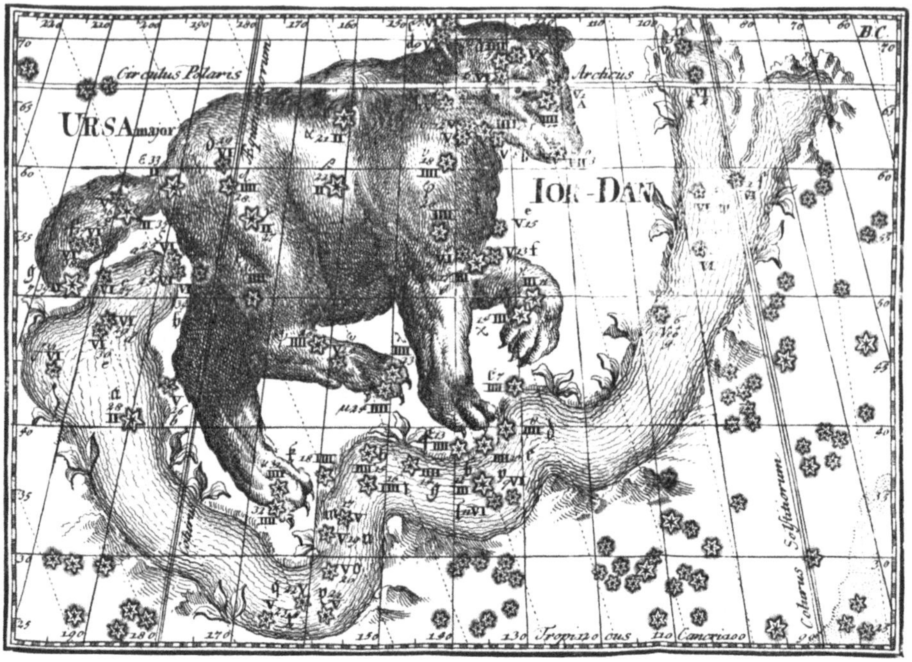

13.7 Jordanis ("Ior-Dan") depicted in Corbinianus Thomas' *Mercurii philosophici firmamentum firmianum*... (1730)

Another serious consequence of water removal from the Jordan is a reduced discharge into the Dead Sea; between this reduced flow, industrial extraction of salts from the Sea, and a high evaporation rate, the Sea is shrinking at an ever-increasing pace. Communities totaling a population of about 350,000 people along the lower reaches of the Jordan dump raw sewage directly into the river, threatening the wellbeing of various plant and animal species. Fouling of the waters further impacts tourism, given that thousands of religious pilgrims visit the region each year to be ritually baptized in its waters.

Biblical Significance of the Jordan

The Hebrew Bible contains no definitive description of the Jordan, and a picture of its significance to people of the time is gleaned only from scattered references, some of which are vague. In Genesis, Lot sees[6] that "the whole plain of the Jordan toward Zoar was well watered, like the garden of the Lord, like the land of Egypt." The river is referred to as a

[6]Genesis 13:10 (New International Version).

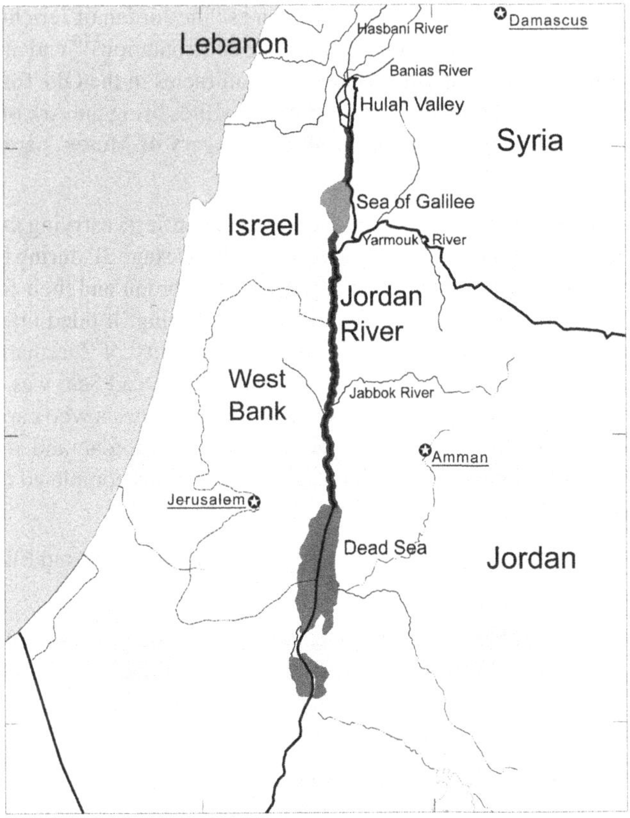

13.8 A political map of the Levant showing the course of the Jordan River (*heavy black line*)

source of life and fertility to the cities "of the plain" (*Kikkar ha-Yarden*), including Sodom, Gomorrah, Admah, Zeboiim and Zoar.

The river's role as a boundary line is mentioned in several places in the text. It is the line demarcating the dwelling places of the tribes of Reuben and Gad and the half-tribe of Manasseh to its east[7] and the "nine tribes and half of the tribe of Manasseh" led by Joshua

[7]Numbers 34:13–15; Joshua 13.

to its west.[8] In the vicinity of Jericho it was known as "the Jordan of Jericho."[9] A number of fords along the river are noted as sites of military confrontations[10] and ambushes.[11]

The waters of the Jordan bear witness to various miracles in the Old Testament. In the Book of Joshua,[12] the flow of the river halts as the Israelites bring the Ark of the Covenant to the water's edge in a scene borrowing from the imagery of Moses' parting of the Red Sea[13]:

> So when the people broke camp to cross the Jordan, the priests carrying the ark of the covenant went ahead of them. Now the Jordan is at flood stage all during harvest. Yet as soon as the priests who carried the ark reached the Jordan and their feet touched the water's edge, the water from upstream stopped flowing. It piled up in a heap a great distance away, at a town called Adam in the vicinity of Zarethan, while the water flowing down to the Sea of the Arabah (that is, the Dead Sea) was completely cut off. So the people crossed over opposite Jericho. The priests who carried the ark of the covenant of the Lord stopped in the middle of the Jordan and stood on dry ground, while all Israel passed by until the whole nation had completed the crossing on dry ground.

The motif of crossing a "dry" Jordan is repeated in the story of Elijah and Elisha in 2 Kings 2:7–14[14]:

> Fifty men from the company of the prophets went and stood at a distance, facing the place where Elijah and Elisha had stopped at the Jordan. Elijah took his cloak, rolled it up and struck the water with it. The water divided to the right and to the left, and the two of them crossed over on dry ground.

After Elijah was assumed bodily into heaven by way of a whirlwind,

> Elisha then picked up Elijah's cloak that had fallen from him and went back and stood on the bank of the Jordan. He took the cloak that had fallen from Elijah and struck the water with it. "Where now is the Lord, the God of Elijah?" he asked. When he struck the water, it divided to the right and to the left, and he crossed over.

---

[8] Joshua 13:6–7.

[9] Numbers 34:15; 35:1.

[10] Judges 12:5–6: "The Gileadites captured the fords of the Jordan leading to Ephraim, and whenever a survivor of Ephraim said, 'Let me cross over,' the men of Gilead asked him, 'Are you an Ephraimite?' If he replied, 'No,' they said, 'All right, say 'Shibboleth.' If he said, 'Sibboleth,' because he could not pronounce the word correctly, they seized him and killed him at the fords of the Jordan. Forty-two thousand Ephraimites were killed at that time." (NIV).

[11] Judges 7:24: "Gideon sent messengers throughout the hill country of Ephraim, saying, 'Come down against the Midianites and seize the waters of the Jordan ahead of them as far as Beth Barah.'" (NIV)

[12] Joshua 3:14–17 (NIV).

[13] Exodus 13–14.

[14] New International Version.

Elisha figures in two additional miracles at the river. In the first, God heals Naaman with its waters,[15] while in the second it is suggested that because Elisha is favored by God, he rightly uses divine power to retrieve an iron ax head lost in the river.[16]

The Jordan features prominently in the Jesus narrative of the New Testament from before the beginning of his ministry. John the Baptist used the waters of the river to baptize and urge repentance before the coming of the Messiah.[17] "Then Jesus came from Galilee to the Jordan to be baptized by John," Matthew relates,[18]

> But John tried to deter him, saying, 'I need to be baptized by you, and do you come to me?' Jesus replied, 'Let it be so now; it is proper for us to do this to fulfill all righteousness.' Then John consented. As soon as Jesus was baptized, he went up out of the water. At that moment heaven was opened, and he saw the Spirit of God descending like a dove and alighting on him. And a voice from heaven said, 'This is my Son, whom I love; with him I am well pleased.'

The baptism of Jesus (Fig. 13.9) was a popular subject of religious artwork from the Medieval period through the Renaissance.

When Jesus began to preach, he fulfilled a prophecy[19] about the Messiah indicating his future efforts lay 'beyond the Jordan':

> Leaving Nazareth, he went and lived in Capernaum, which was by the lake in the area of Zebulun and Naphtali – to fulfill what was said through the prophet Isaiah[20]:
>
> > 'Land of Zebulun and land of Naphtali,
> > the Way of the Sea, beyond the Jordan,
> > Galilee of the Gentiles'

---

[15] 2 Kings 5:13–14: "Naaman's servants went to him and said, 'My father, if the prophet had told you to do some great thing, would you not have done it? How much more, then, when he tells you, 'Wash and be cleansed!' So he went down and dipped himself in the Jordan seven times, as the man of God had told him, and his flesh was restored and became clean like that of a young boy." (NIV)

[16] 2 Kings 6:1–7: "The company of the prophets said to Elisha, 'Look, the place where we meet with you is too small for us. Let us go to the Jordan, where each of us can get a pole; and let us build a place there for us to meet.' And he said, 'Go.' Then one of them said, 'Won't you please come with your servants?' 'I will,' Elisha replied. And he went with them. They went to the Jordan and began to cut down trees. As one of them was cutting down a tree, the iron axhead fell into the water. 'Oh no, my lord!' he cried out. 'It was borrowed!' The man of God asked, 'Where did it fall?' When he showed him the place, Elisha cut a stick and threw it there, and made the iron float. 'Lift it out,' he said. Then the man reached out his hand and took it." (NIV)

[17] Matthew 3:1–6; Mark1:1–8; Luke 3:1–3; John 1:24–28.

[18] Matthew 3:13–17 (NIV). A similar version appears in Mark 1:9–10 and Luke 3:21–22, but is absent from the narrative in the Gospel of John.

[19] Matthew 4:13–15 (NIV).

[20] Isaiah 9:1–2.

13.9 "Baptism of Christ" (*c.* 1492) by Giovanni Battista Cima (Italian; 1460–1518). Oil on panel, 3.5 m × 2.1 m. San Giovanni in Bragora, Venice. Christ stands in the Jordan as he receives the sacrament from John the Baptist; unusually, Cima shows the river in cross-section around Christ's feet

Jesus returned to the river many times during his ministry, receiving believers from across the Jordan who came to hear him preach and to be healed[21] or transiting across it himself.[22] At the end of his life, he came to the waters of the Jordan seeking refuge from his enemies who sought his capture.[23]

The river became a feature of both sacred and secular music as well as works of art (Fig. 13.10) and literature. It is frequently invoked as a symbol of freedom and release from bondage in reference to the story of the Biblical exodus from Egypt to the Promised Land. The act of crossing the Jordan represents the final attainment of freedom and homecoming as expressed in the lyric of the early nineteenth century American folk song "The Wayfaring Stranger":

I'm going home to see my mother
I'm going home no more to roam
I'm just a-going over Jordan
I'm just a-going over home.

The Rivers of the Night Sky

Why put rivers in the sky? With few exceptions—namely, the extant Mensa and extinct Mons Maenalus (Chap. 15)—they are the only landform features so commemorated. There are neither mountain ranges nor oceans; no deserts, lakes or plains. But at one point around the mid-seventeenth century three rivers graced the Western night sky: Jordanis, Eridanus, and Tigris (Chap. 27).

Of these, only Eridanus survives in the modern canon of constellations. It is certainly the oldest of the three, and its origins trace back to the dawn of civilization. The Greeks originally knew it simply as ὁΠοταμός, the River; one idea explaining its proper name (White, 2008) is that it is a Greek adaptation of the Babylonian constellation MUL.NUN.KI ("Star of Eridu"). Eridu was a Sumerian city in a once-marshy region in the south of modern Iraq, sacred to the god Enki; "Eridanus" may then mean "River of Eridu." Sometimes identified with the Italian Po, the Danube of Hungary or even the Nile, Eridanus was probably always a purely mythical river said to flow through the far-northern land of Hyperborea. Its mythological basis is the story of Phaeton, who fell into its waters after losing control of the sun chariot of Apollo-Helios. Alternately, the river is named for the river god Eridanos, son of Oceanus and Tethys[24] and father of Zeuxippe.[25] In the

[21] Matthew 4:25; Mark 3:7–12.

[22] Matthew 19:1–2; Mark 10:1.

[23] John 10:39–41.

[24] "And Tethys bore to Ocean eddying rivers, Nilus, and Alpheus, and deep-swirling Eridanus..." Hesiod, *Theogony* line 338, trans. H.G. Evelyn-White.

[25] "Butes, son of Teleon and Zeuxippe, daughter of the river Eridanus, from Athens." Hyginus, *Fabulae* 14, trans. M. Grant.

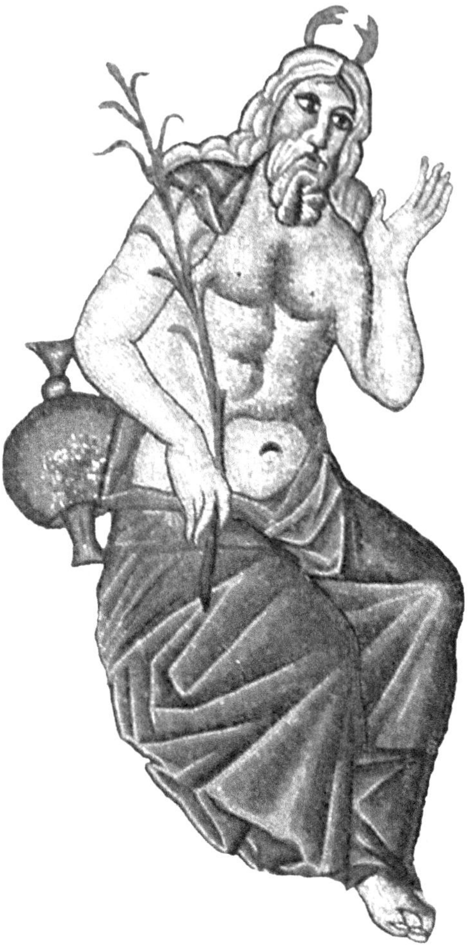

13.10 A personification of the River Jordan shown in a detail from the dome mosaic in the Arian Baptistry, Ravenna, Italy (ca. 490). The Jordan is depicted as a bare-chested, grey-bearded man with a pair of lobster claws atop his head. He holds a rush in his right hand, indicating he represents a river, and sits next to a vase or water-jar, symbolizing the source of the waters in which Christ was baptized

Georgics, Virgil called[26] Eridanus the "king of the rivers" whose mighty current "washed away forests in the whirl of his maddened vortex, and swept cattle and stables over the plains" as a portent of the death of Julius Caesar.

[26]Book 1, line 382, trans. A.S. Kline.

Already in antiquity Eridanus found a place among the stars. Ptolemy included it among the constellations of the *Almagest*, and four centuries later Nonnus of Panopolis (*c.* fifth century AD) noted[27] its presence in the heavens in Book 23 of his *Dionysiaca*: "I will drag down from heaven the fiery Eridanos whose course is among the stars, and bring him back to a new home in the Celtic land: he shall be water again, and the sky shall be bare of the river of fire."

Rivers serve many functions in human society: sources of fresh water, means of transportation, boundary markers. As constellations, they fill spaces where long, winding stretches among the stars suggest the bends of a lazy river. The river constellations might even reference the ultimate 'cosmic river:' our own galaxy. "There is much in the Euphratean records alluding to a stellar stream that may be our Eridanus," Richard Hinckley Allen wrote, "possibly the Milky Way, another sky river."

DISAPPEARANCE

The appearance of Jordanis on Allard's 1706 map and in Thomas' 1730 book were among its very last. Around the time of Allard's death, Johannes de Broen published the map *Hemelskaart voor de noordelijke en zuidelijke sterrenhemel uitgevoerd in Mercatorprojectie* (1709) on which he did not indicate a river flowing beneath the Great Bear, although he drew in some stars where Cellarius and Allard depicted Jordanis. Neither does it appear in Johann Gabriel Doppelmayr's *Atlas Coelestis* (1742). By midcentury, Hevelius' creations of Canes Venatici, Leo Minor and Lynx (1690) had achieved wide circulation, their figures displacing the river that once flowed through the same region of the sky. Johann Elert Bode adopted Hevelius' figures in his seminal atlases *Vorstellung der Gestirne* (1782; 1805) and *Uranographia* (1801b), ensuring the complete extinction of Jordanis by the dawn of the nineteenth century. Its traditional course (Fig. 13.11) follows a path through the boundaries of six constellations declared "official" by the International Astronomical Union in 1922: Camelopardalis, Lynx, Leo Minor, Leo, Ursa Major and Canes Venatici.

[27] *Dionysiaca* 23.380 ff, trans. W.H.D. Rouse.

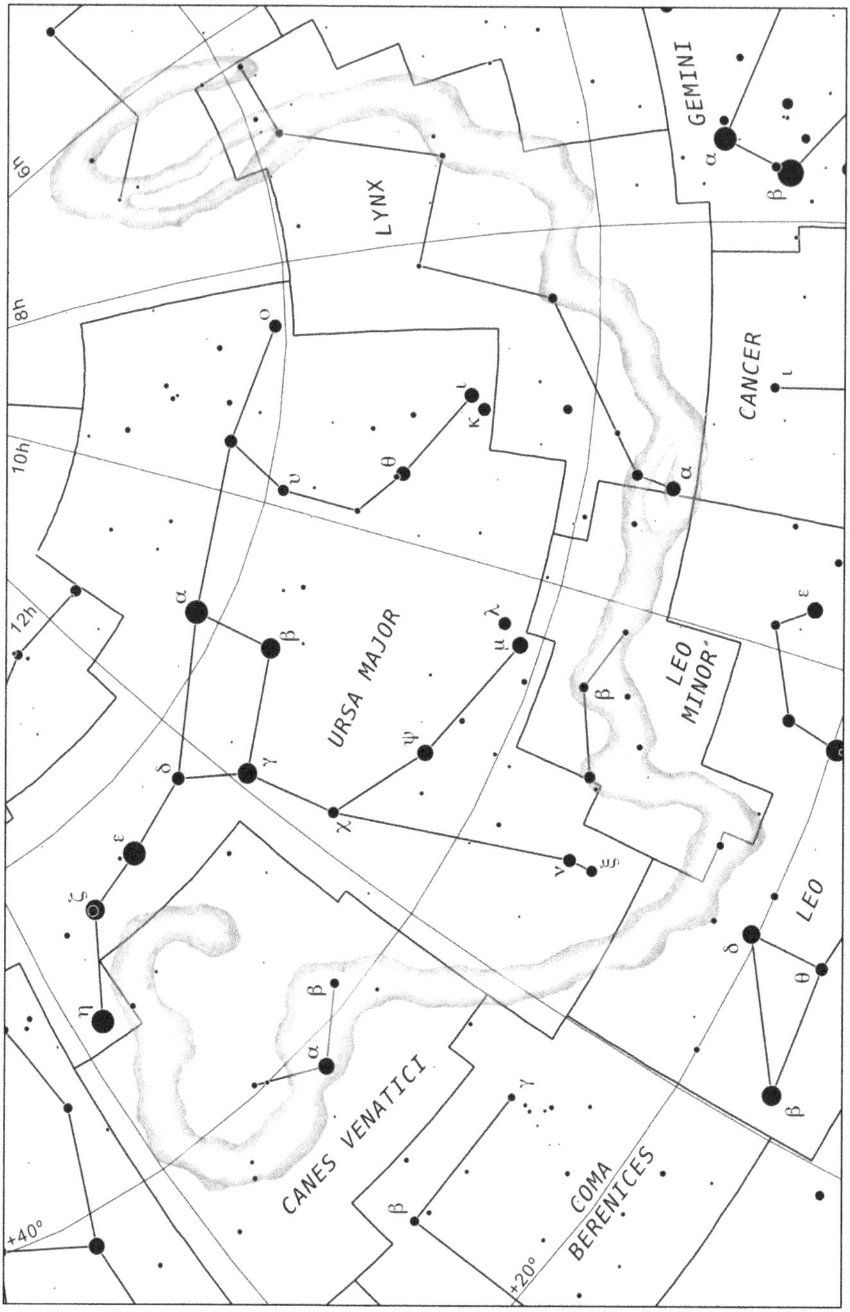

13.11 The figure of Fluvius Jordanis from Carel Allard's *Hemisphaerium meridionale et septentrionale . . .* (1706) overlaid on a modern chart

14

Machina Electrica

The Electric Generator

Genitive: Machinae Electricae
Abbreviation: MaE
Location: Below the south-central part of Cetus, between modern Fornax and Sculptor[1]

ORIGIN AND HISTORY

The Electrical Generator was first placed among the stars by Johann Elert Bode in *Uranographia* (1801b) (Fig. 14.1). In creating the new constellation, Bode merely borrowed a few stars from Sculptor, by his own account[2] in the companion catalog *Allgemeine Beschreibung und Nachweisung der Gestirne*:

> Since thus far no monument to the important invention of electricity was consecrated among the stars, I placed this new constellation east of the sculptor's workshop in the sky, and in the formation of the same made some changes to the latter.

These stars are left to Sculptor in his previous map, *Vorstellung der Gestirne* (1782; Fig. 14.2). According to Ridpath (1989), "Bode presumably was attempting to emulate the Frenchman Nicolas Louis de Lacaille who had introduced constellations representing

[1] *"Between the Phœnix and the Sea Monster"* (Green, 1824) *"Under the Whale, and west of the Chemical Apparatus"* (Kendall, 1845); *"Immediately south of Cetus, and north of Phoenix"* (Bouvier, 1858); *"South of the central portion of Cetus"* (Bakich, 1995); *"Between the modern Fornax and Sculptor"* (Ridpath, 1989).

[2] Da der wichtigen Erfindung der Electrizität bisher noch kein Sternen-Monument geweihet war, so habe ich deshalb dieses neue Sternbild ostwärts bey der Bildhauerwerkstadt an den Himmel gebracht, und zur Formirung deffelben mit letzterm einige Veränderung getroffen.

J.C. Barentine, *The Lost Constellations*, Springer Praxis Books,
DOI 10.1007/978-3-319-22795-5_14

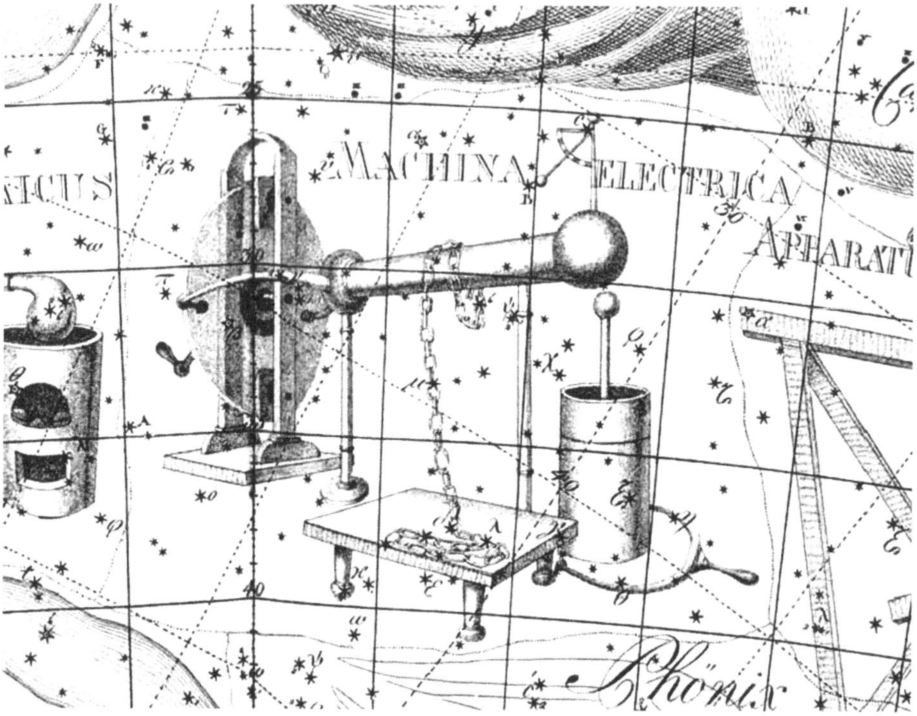

14.1 Machina Electrica depicted on Plate 17 of Bode's *Uranographia* (1801b)

scientific and technical inventions." Another of Bode's creations echoes this theme: Globus Aerostaticus, discussed previously in Chap. 11.

Following Bode's influential example, many nineteenth century cartographers included Machina Electric on their maps such as Jamieson (1822) and Hall (1825; Fig. 14.3). August Gottleib Meissner showed the constellation as "Elektrische Maschine" on Fig. XXIII, Table 31 of *Astronomischer Hand-Atlas* (1805), while Thomas Young called it "Electrical Machine" on his northern hemisphere chart in Plate XXXVII, Fig. 518 of *A Course of Lectures on Natural Philosophy and the Mechanical Arts* (1807). It appears by name only ("Machina Electrica") on Plate 15 of Jacob Green's *Astronomical Recreations* referring in his text to "Machina Electrica et Apparatus Chemicus:"

> MACHINA ELECTRICA, or the Electrical Machine, was introduced among the constellations by Bode. It is between the Phœnix and the Sea Monster. There is scarcely a star in the group visible to the naked eye. The Chemical Apparatus, or Apparatus Chemicus, is also south of Cetus, and may be found in one of the bends of the River. It contains two stars of the third magnitude, the others are much smaller. We owe this group to La Caille.[3]

[3]Nicolas Louis de La Caille (1713–1762), a French astronomer and priest.

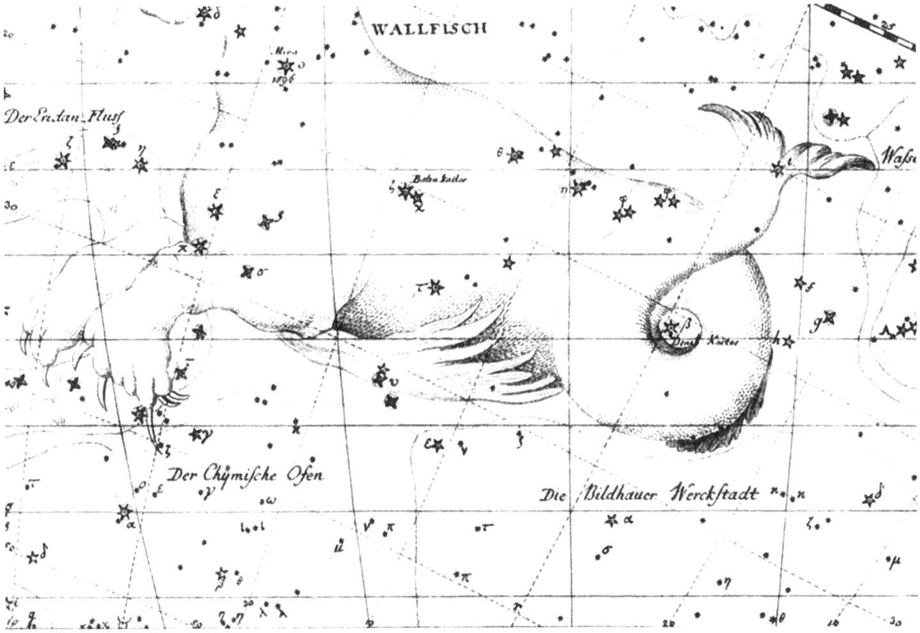

14.2 Detail of Plate 23 of Bode's *Vorstellung der Gestirne* (1782). The stars he would form 19 years later into Machina Electrica are shown in the space below Cetus ("WALLFISCH") and between Fornax ("Der Chÿmische Ofen", *lower left*) and Scupltor ("Die Bildhauer Werckstadt", *lower right*)

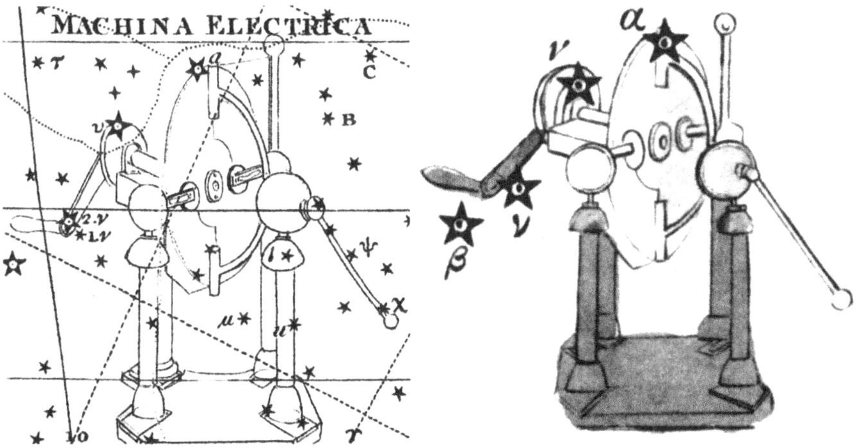

14.3 Two similar views of Machina Electrica as shown on Plate 23 of Alexander Jamieson's *Celestial Atlas* (1822; (left) and on Plate 28 of *Urania's Mirror* (1825; (right).

It also turns up by name on Plate XVII of Freidrich Argelander's *Neue Uranometrie* (1843), which is somewhat surprising considering that the composition of Argelander's charts is remarkably similar to that of the "modern" canon of constellations.

In the second half of the nineteenth century, Machina Electrica had about an even chance of appearing in any particular set of charts. More than one author commented on the fact that its stars were not particularly conspicuous; in her *Familiar Astronomy* (1858), Hannah Bouvier wrote that "It is situated immediately south of Cetus, and north of Phœnix, and contains no bright stars."

ICONOGRAPHY

A Brief History of Electricity

Human awareness of electricity extends as far back as the dawn of civilization. While earlier humans knew of natural electrical phenomena such as lighting, descriptions of electricity in nature are nearly as old as language itself. Egyptian texts from as early as the Second Dynasty (*c.* 2750 BC) refer to the electric catfish (family *Malapteruridae*) as the "thunderer of the Nile" that electrocuted other fish with shocks of up to 350 volts.[4] These and other electric fishes and eels were reported by naturalists and physicians around the Mediterranean basin from the Hellenistic to the Islamic periods.[5] In the first century AD Pliny the Elder described[6] the substantial jolt delivered by the electric ray, also commonly known as the torpedo fish:

> Would it not have been quite sufficient only to cite the instance of the torpedo, another habitant of the sea, as a manifestation of the mighty powers of Nature? From a distance, from a considerable distance even, and if only touched with the end of a spear or staff, this fish has the property of benumbing even the most vigorous arm, and of riveting the feet of the runner, however swift he may be in the race.

During the same period, the Roman physician Scribonius Largus noted the medical use of torpedo fish in the treatment of both headaches and gout in *Compositiones Medicae* (*c.* AD 47). From observations of these animals it seems that by the Medieval period scientists in the Islamic world had drawn a conceptual connection between terrestrial lighting and the phenomenon employed by the electric ray; echoing the Egyptian moniker, they applied a common Arabic term (*raad*) to both.

The earliest investigations into the nature of electostatics began in Greece around the seventh century BC. Thales of Miletus (*c.* 624–*c.* 546 BC) carried out simple experiments showing that pieces of amber when rubbed with cat fur would attract light objects such as feathers. With enough rubbing, one could even obtain a small, visible discharge spark to jump from the amber. The accumulation of charge on the amber was the result of the

---

[4]S. Finger and M. Piccolino, *The Shocking History of Electric Fishes: From Ancient Epochs to the Birth of Modern Neurophysiology* New York: Oxford University Press (2011).

[5]P. Moller and B. Kramer, "Review: Electric Fish". *BioScience* 41 (11): 794–796 (1991).

[6]*Naturalis history* Book XXXII, 2, trans. J. Bostock and H.T. Riley (1857).

triboelectric effect, in which mechanical friction redistributes electrons on the surface of a material. While Thales was incorrect in concluding that friction with the amber imparted the stones with magnetic qualities, he basically understood that some physical process involved in friction resulted in a force that could move objects. Otherwise, the phenomenon remained little more than a curiosity for some two millennia.

The English astronomer William Gilbert (1544–1603), appointed physician to the court of Elizabeth I, used the support of his royal pension to carry out a number of basic experiments like those of Thales in which he found additional properties of electricity (a word he coined). In addition to amber, he found that many other materials such as glass and wax could develop and hold an electrostatic charge. Gilbert seems to be the first to understand the concept of electrical insulation, noting that moisture seemed to prevent objects from becoming charged by friction. Most importantly for the future development of electromagnetic theory, he established that the force involved in electrostatics was indiscriminate in which substances it attracted whereas magnets seemed to only attract iron. Later in the seventeenth century, the Irish physicist Robert Boyle (1627–1691) found that electrostatic attraction and repulsion functioned in a vacuum as well as in the air, indicating that the phenomenon of electricity did not require air as a carrying medium.

In the following century, scientists began to think about ways to store and manipulate electric charge independent of the means of generating it. In 1745, two men—the German cleric Ewald Georg von Kleist and Dutch physicist Pieter van Musschenbroek—independently discovered that an electrostatic charge could be held fixed between two electrodes on the inside and outside of a glass jar using the same principle as the modern capacitor. Named for the city in which van Musschenbroek did his work, these "Leiden jars" became an important device for conducting experiments on the nature of electricity during the Enlightenment. The American polymath and statesman Benjamin Franklin (1706–1790) coined the term "battery" in 1749 to describe a set of Leiden jars wired together in series that could store relatively large amounts of electrical energy and release it for later use.

In 1780, the Italian physicist Luigi Galvani (1737–1798) inadvertently laid the foundation for the later development of the electrochemical battery during the laboratory dissection of a frog. The frog was held in place for the dissection using a brass hook; when Galvani happened to touch the frog's leg with his iron scalpel, the leg suddenly and unexpectedly twitched. From this simple observation, Galvani (incorrectly) deduced that the electricity that caused the dead frog's leg to move was intrinsic to the frog's body, and he called it "animal electricity." Galvani's colleague Alessandro Volta (1745–1827) offered the correct interpretation: noting that brass and iron were dissimilar metals, he concluded that when the metals were held near each other in the presence of a moist medium—namely, the frog's body—electricity flowed from one to the other. Carrying out further experiments of his own, he verified the effect and published his results in 1791. Based on this discovery, Volta invented the first true battery in 1800, now known as a voltaic pile (see the Battery of Volta in Volume 2).

Electricity came to be viewed as a physical phenomenon full of promise for new technologies, but the means of its production and storage remained limited. Attempts to build machines to convert mechanical energy to electrical energy were undertaken as

14.4 A glass globe friction generator design common in the eighteenth century, shown in Hubert-François Gravelot's *Die Elektrisierte* (1750). The caption reads in translation "I know where this great magical power is found, referred to by its electrical name; young beauties, it's in your eyes"

early as the mid-seventeenth century but did not ramp up significantly until the following century in which they became indispensable for generating currents for use in laboratory experiments. Some of the first designs used rotating glass globes (Fig. 14.4) rubbed at first by hand but later made to rub against a fixed piece of woolen fabric. The German physicist Georg Matthias Bose (1710–1761) further improved this design by adding a collector consisting of an insulated conducting cylinder suspended from silk strings. Another Bose innovation was the use of a "prime conductor," namely an iron rod held by an assistant whose body was prevented from grounding by standing on a block of insulating material such as resin.

In 1766, the Dutch physician Jan Ingenhousz (1730–1799) tested an electric generator that used flat glass plates instead of globes; by covering the plates in metal foil, he was able to create very large electrostatic discharges. This "frictional electric machine" design (Fig. 14.5) persisted well into the nineteenth century and was eventually superseded by the "influence machine" (Fig. 14.6), which made use of a small initial charge to start the electrostatic induction process. However, none of these designs proved sufficiently practical to drive large-scale electrification schemes.

Later developments in magnetic induction yielded generators that used magnetic induction rather than electrostatic methods to produce electric current. These resulted from the experiments of the English physicist Michael Faraday (1791–1867) described by Bachman (1918):

> A copper disk twelve inches in diameter and a fifth of an inch thick was fastened on a brass axle. This was so mounted that the disk could be turned rapidly. A powerful permanent horseshoe magnet was placed so that the disk revolved between its two ends. A metal collector was held against the edge of the disk, and a second collector

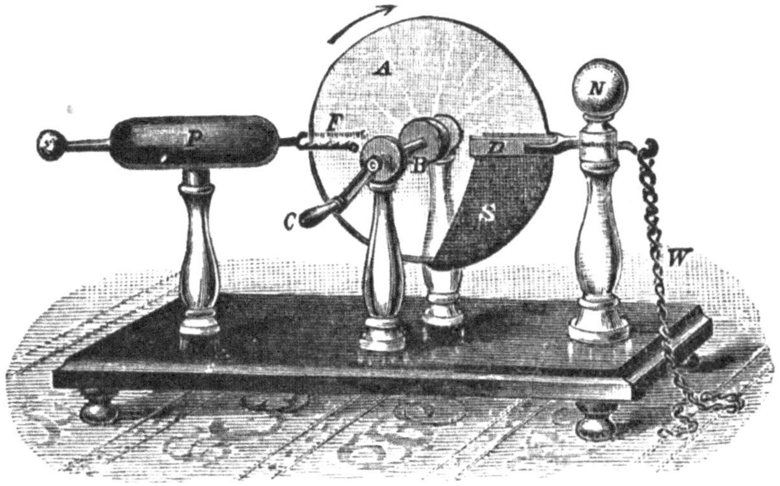

14.5 An electrostatic disk generator pictured on page 73 of Bachman (1918). The original caption read "Frictional electric machine; *A*, the glass disk; *P* and *N*, the prime and negative conductors"

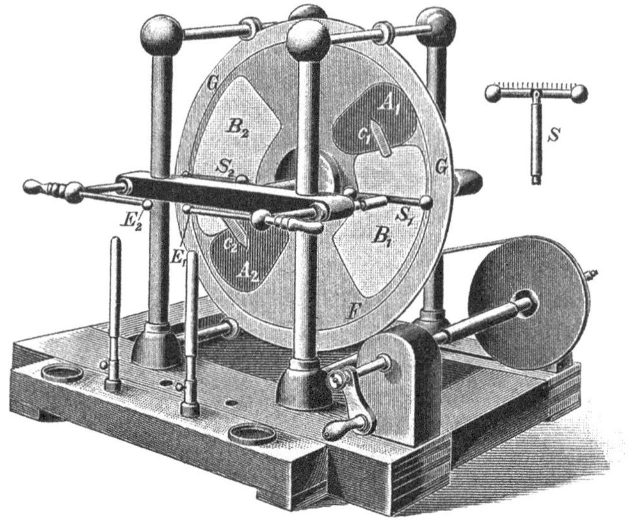

14.6 Engraving showing a "Holtz influence machine", a typical mid-nineteenth century electrostatic generator introduced by German physicist Wilhelm Holtz (1836–1913) in 1865. E. Mueller-Baden, *Bibliothek allgemeinen und praktischen Wissens für Militäranwärter*, Band III, Deutsches Verlaghaus Bong & Co., Berlin (1905)

> was fastened to the axle. Faraday turned the disk, and a steady current of electricity was produced. This was the first dynamo ever made.

As the rotating conductor experienced a time-varying magnetic field, an electromotive force was generated in the disk that produced a small DC voltage. This approach required no charge to "start" and ultimately yielded a more reliable supply of electricity than any electrostatic device. The principles elucidated by Faraday remain the standard means of generating electricity from mechanical energy to this day.

However, Bode drew his maps before Faraday's time when electrostatic generators retained state-of-the-art status in the world of electricity. The device shown in *Uranographia* (Fig. 14.1) appears to be a simple disk generator, although later cartographers "improved" its design (Fig. 14.3) to more closely resemble influence machines of the first quarter of the nineteenth century.

DISAPPEARANCE

Machina Electrica had a fairly brief run of less than a century; the degree to which it persisted at all likely had to do with the influence of *Uranographia*. One of the last instances in which a popular astronomy text of the day treated it as canonical is found in George Chambers' *A Handbook of Descriptive Astronomy* (1877); Chambers names this constellation and attributes it to "Bode's maps" without citing a date. It merited a brief mention in the table "Old And New Constellations In Chronological Order" in *Poole Bros. Celestial Handbook* (1892), attributed to "Bode, 1798, A.D."

At end of the century it had been discarded, and according to Richard Hinckley (1899) was by then "generally omitted from the maps and catalogues." When the modern constellation boundaries were finally established in 1930, the stars of Machina Electrica were distributed more or less evenly between Fornax and Sculptor (Fig. 14.7).

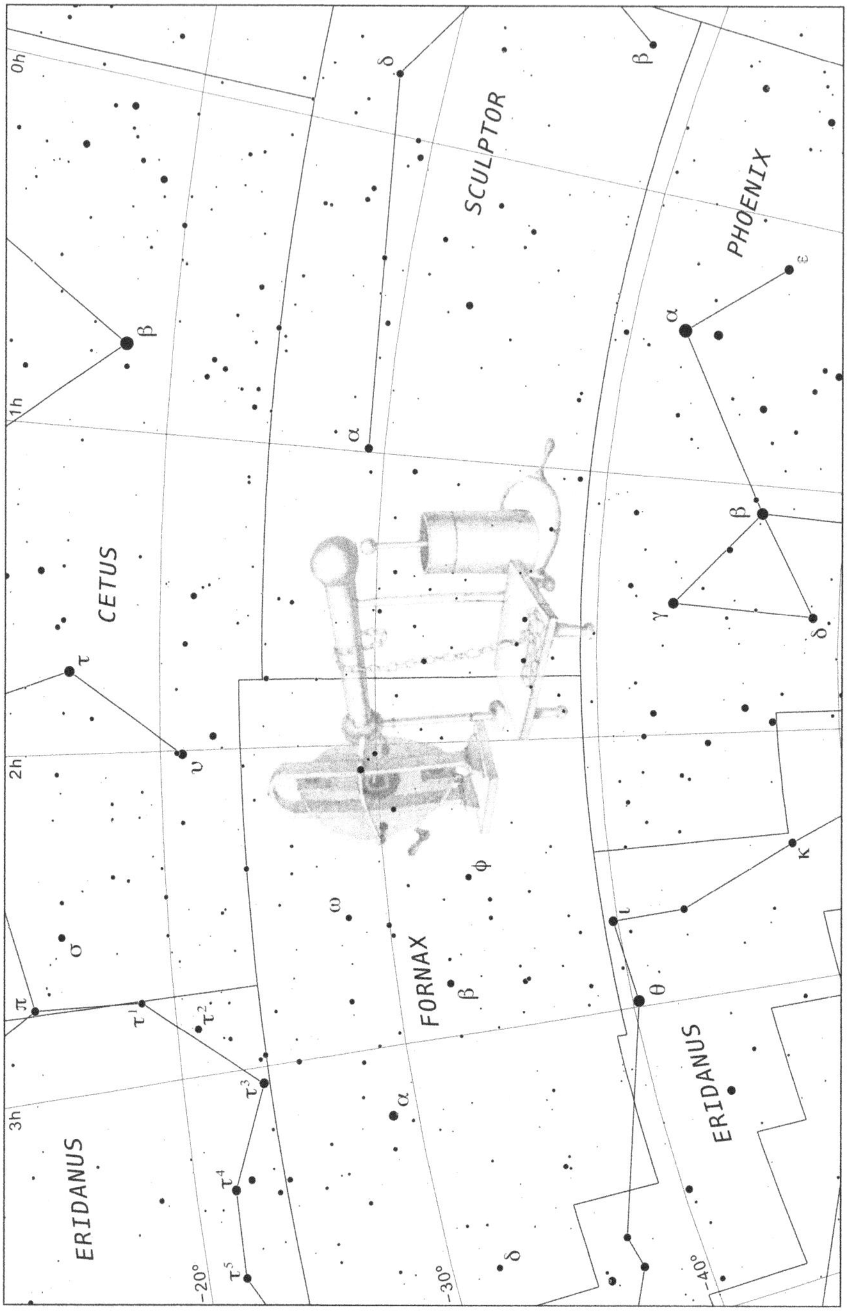

14.7 The figure of Machina Electrica from Plate 17 of Bode's *Uranographia* (1801b) overlaid on a modern chart

15

Mons Maenalus

Mount Mainalo

Genitive: Montis Maenali
Abbreviation: MoM
Location: At the feet of Boötes, between its stars and those of Virgo and Libra[1]

ORIGIN AND HISTORY

Mons Maenalus was introduced by Johannes Hevelius in *Prodromus Astronomiae* (1690), created from unformed stars in the space between the southern end of Boötes and the eastern extent of Virgo (Fig. 15.1). While Hevelius' work was highly influential, few other cartographers rushed to adopt the new figure perhaps because the included stars weren't especially bright. Barely a decade after the posthumous publication of Hevelius' star list, Phillipe de la Hire left Mons Maenalus off his *Planisphère céléste septentrionale* (1702), choosing to simply leave the space below Boötes empty (Fig. 15.2). Carel Allard similarly drew Boötes without a mountain to climb on his 1706 map *Hemisphaerium meridionale et septentrionale planisphaerii coelestis*. But within a generation of Hevelius mapmakers gradually began to include Mons Maenalus as a matter of routine, as on the maps of Johann Leonhardt Rost (Fig. 121; 1723), Christian Doppelmayr (1742) and Bode (1801) (Figs. 15.3, 15.4, and 15.5).

[1] *"At the feet of Boötes"* (Bouvier, 1858; Allen, 1899; Bakich, 1995).

J.C. Barentine, *The Lost Constellations*, Springer Praxis Books,
DOI 10.1007/978-3-319-22795-5_15

15.1 Detail of Mons Maenalus (*bottom*) shown on Figure F of Johannes Hevelius' *Prodromus Astronomiae* (1690)

15.2 Boötes as depicted on de La Hire's map *Planisphère céléste septentrionale* (1702)

Johann Elert Bode wrote[2] in *Allgemeine Beschreibung und Nachweisung der Gestirne* (1801a):

[2]"Nach den Erzählungen der Dichter war dieser Bärenhüter oder Hirte Ikarus, der Vater der Erigone (Jungfrau im Thierkriese). Er hatte vom Bacchus die Kunst Wein zu keltern gelernt, um solche die Menschen zu lehren. Dies veranlasste, dass er todt geschlagen wurde, weswegen man ihm under die Sterne versetzte. Er folgt südostwärts auf dem grossen Bären. Ein Stern erster Grösse, Arcturus, glänzt mit einem röthlichen Lichte in diesem Gestiern, rechts und links von demselben, so wie nordwärts und mach den dreyen am Schwanz des grossen Bären hin, sind Sterne 3ter und 4ter Grösse an den Fussen, dem Gürtel, den Schultern, Arm und Kopf des Bootes sehr kenntlich. Den Berg Maenel hat Hevel eigentlich zuerst unter die Fusse des Bootes gesetzt. Auf einem in Arkadien gelegenen Berge dieses Namens erbauete Maenal, ein Sohn des Königs Lykaon, eine Stadt."

15.3 Boötes, Canes Venatici, Coma Berenices, and Mons Maenalus as shown on Fig. XXVI of Johann Leonhardt Rost's *Atlas Portatilis Coelestis* (1723)

> According to the tales of the poets this was Boötes or the shepherd [Icarius of Athens], the father of Erigone (the Virgin of the zodiac). He had learned grape pressing from Bacchus in order to teach humanity the art of wine[making]. As a result, he was beaten to death, which is why he was placed among the stars. He follows southeastward from the Great Bear. A star of the first magnitude, Arcturus, shines with a reddish light, right and left of it, as northwards and makes three of them toward the tail of the Great Bear, the stars are of the 3rd and 4th [magnitude]. Mons Maenalus was originally placed by Hevelius under the feet of Boötes. Built on a mountain located in Arcadia by that name Maenal, a son of King Lycaon, a city.

In 1827, James Ryan described the figure in *The New American Grammar of the Elements of Astronomy*:

> The Mountain Maenalus in Arcadia was sacred to the god Pan, and frequented by shepherds: it received its name from Maenalus, a son of Lycaon, king of Arcadia. It was made a constellation and placed by Hevelius under the feet of Boötes.

ICONOGRAPHY

The mythological origins of the name "Maenalus" are unclear. It may refer to either of two mythical individuals. One is a son of Lycaon who founded the eponymous Arcadian

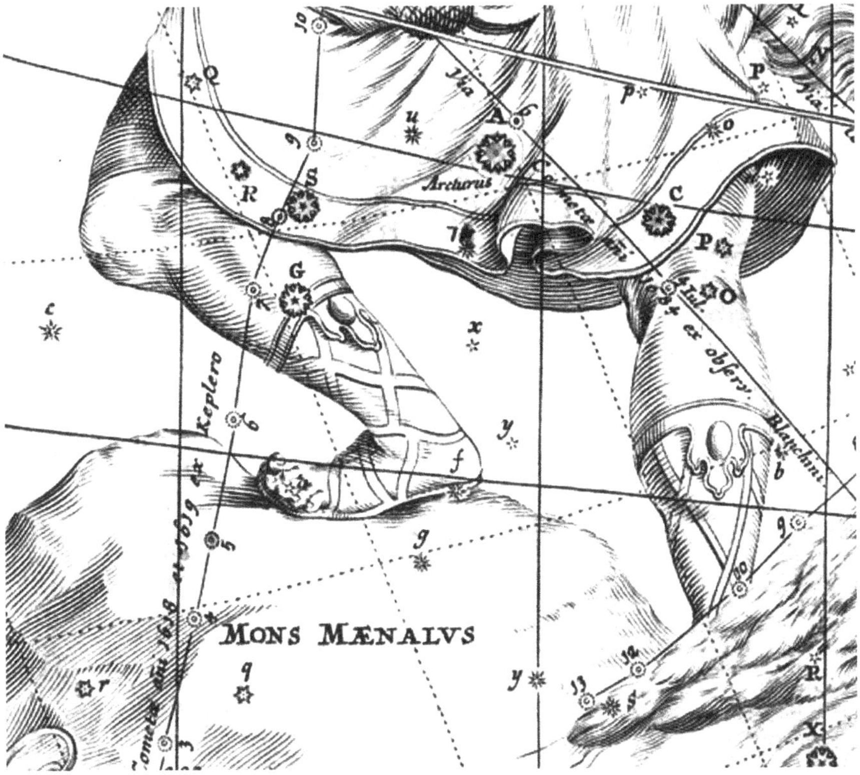

15.4 Detail of Mons Maenalus shown in Christian Doppelmayr's *Atlas Coelestis* (1742). Doppelmayr also notes here the paths of the comet observed by Johannes Kepler in 1618–1619 (*left*) and that seen by Francisci Blanchini in 1684 (*right*)

town.[3] The other is the father of Atalanta,[4] the virgin huntress beloved of Meleager.[5] On the other hand, Ridpath (1989) indicated who Maenalus was *not*:

> Mons Maenalus was sacred to the god Pan who frequented it. Ovid in his Metamorphoses said that Mons Maenalus bristled with the lairs of wild beasts[6] and was a favourite hunting ground of Diana and her entourage, including Callisto.[7] In saying

[3]Pausanias, *Description of Greece* VIII, 3. §1 (second century AD)

[4]Pseudo-Apollodorus, *Bibliotheca* III, 9, fin.

[5]Atalanta has a curious, possible connection to Argo Navis (Chap. 5). In some versions of the quest for the Golden Fleece, Atalanta sailed with the Argonauts as the only woman aboard the ship.

[6]Ovid, *Metamorphoses* I, 216.

[7]Ovid, *Metamorphoses* II, 415, 442.

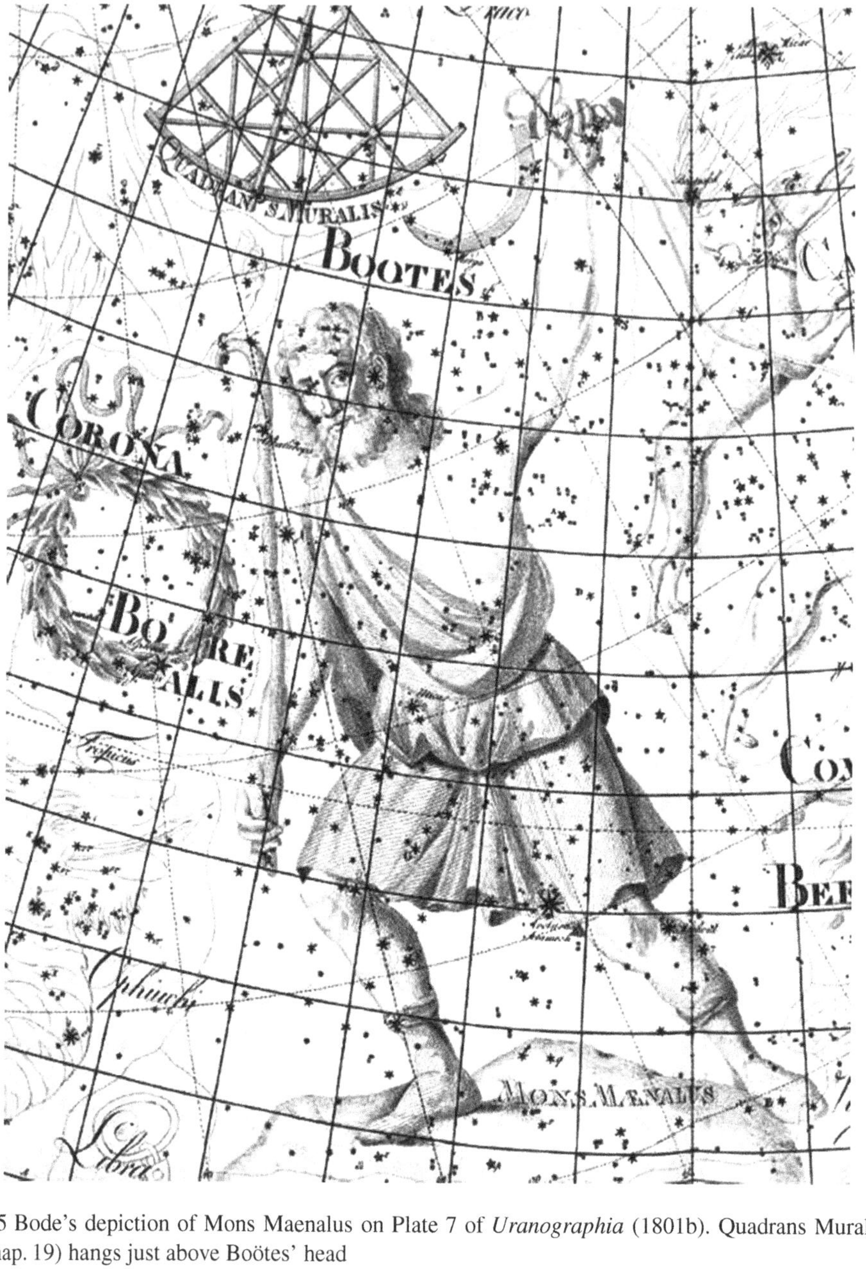

15.5 Bode's depiction of Mons Maenalus on Plate 7 of *Uranographia* (1801b). Quadrans Muralis (Chap. 19) hangs just above Boötes' head

15.6 Boötes and Canes Venatici as shown on Plate 10 of *Urania's Mirror* (1825). Mons Maenalus (*bottom*) appears to hold the status of an asterism

> this, Ovid clearly rejects the story that Maenalus was Callisto's grandson, as the mountain would not yet have got its name.

In the first century AD, the Roman poet Virgil personified the idyllic slopes of Maenalus in the *Eclogues*[8]:

> Maenalus has ever tuneful groves and speaking pines; ever does he listen to shepherds' loves and to Pan, who first awoke the idle reeds.

[8] Virgil, *Eclogae* VII, 22 ("Damon") trans. H.R. Fairclough.

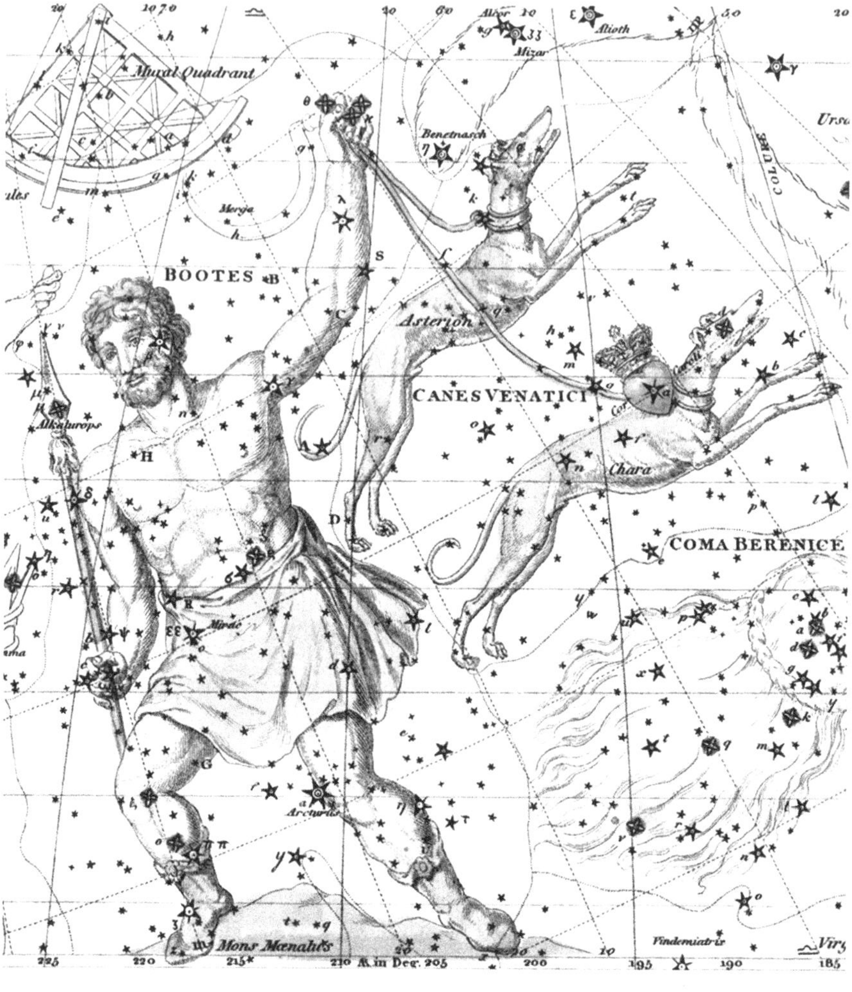

15.7 A typical presentation of the "Boötes Group" (Boötes and Canes Venatici) from Plate VII in Alexander Jamieson's *Celestial Atlas* (1822). Boötes is shown standing on Mons Maenalo (*bottom-center*) and holding the leashes of the Hunting Dogs. Note that the bright star α CVn in the dog, traditionally named "Chara," is shown as a literal depiction of its formal name, "Cor Caroli" (Charles' Heart, *center-right*). By the end of the nineteenth century, Cor Caroli would cease to be shown as a kind of asterism in atlases, and Mons Maenalus would disappear completely

Richard Hinckley Allen (1899) offered an intriguing theory connecting the Greek representation of Mons Maenalus to ancient Indian astronomy:

> Landseer has a striking representation of the Husbandman, as he styles Boötes, with sickle and staff, standing on this constellation figure. A possible explanation of its origin may be found in what Hewitt writes in his Essays on the Ruling Races of Prehistoric Times:
>
> > "The Sun-god thence climbed up the mother-mountain of the Kuṣhika race as the constellation Hercules, who is depicted in the old traditional pictorial astronomy as climbing painfully up the hill to reach the constellation of the Tortoise, now called Lyra, and thus attain the polar star Vega, which was the polar star from 10000 to 8000 B.C."
>
> May not this modern companion constellation, Mons Maenalus, be from a recollection of this early Hindu conception of our Hercules transferred to the adjacent Boötes?

Allen didn't give sufficient context to the passage quoting J.F. Hewitt. The full paragraph of Hewitt's text is as follows, with emphasis added to the sentence quoted by Allen, along with Hewitt's original footnotes:

> In the mythological astronomy of this year the sun-god, which is the ship Argo, steered by the Hindu father-god Agastya, the star Canopus, who, in the Rigveda, brought from out the lightning Vasiṣhṭha, the most creating *(vasu)* fire, the perpetual fire burning on the altar of the fire-worshippers, the son of Mitra, the moon-god, and Varuna, the god of rain *(var)* and the dark night.[9] This steersman star was, by his wife Lopā Mudra, the moon-fox *(lopā)* , the father of the three Dasya or Dravidian Tamil races,[10] the Cheroos, sons of the bird (chirya); the Cholas, sons of the mountam *(kol)* ; and the Paṇḍyas, the yellow sons of the sun-antelope Pandu, the father of the fair *(paṇḍu)* race. He steered the sun-ship to the house of the bird of winter, Corvus, and **the sun-god thence climbed up the mother-mountain of the Kuṣhika race as the constellation Hercules, who is depicted in the old traditional pictorial astronomy as climbing painfully up the hill to reach the constellation of the Tortoise, now called Lyra, and thus attain the polar star Vega, which was the polar star from 10,000 to 8000 B.C**. On the other side of the polar constellation of the Tortoise, the sun-god became transformed into the rain-bird, the constellation of the Bird Ornis of the Greeks, and Cygnus, the Swan of Latin astronomical mythology, who brought Soma, the seed of life, to earth; while the mother-mountain bird, who laid the egg whence the hundred parent sons of the Kushite race were born, the hundred *(satā)* creators *(vaēsa)*, who formed the crew of the ship Argo in the Zendavesta, was the Vulture, the Gridhra, or sacred bird

---

[9]"Rigveda, VII. 33, io, 11."

[10]"Mahabharata Vana (*Tirtha-Yatra*) Parva, XCVI.–XCVIII. pp. 307, 314."

of the Rigveda, who, in Egyptian mythology, ruled the year and gave her name to this constellation.[11] This pictorial astronomy telling us the history of the sun-god, the polar star, and the mother-bird, must, as we know from its agreement with the *Phainomena* of Aratus, have come down to us from the traditional picture-writing of the Akkadians, from whom Aratus got his facts through Eudoxus.

Hewitt implies a tradition at least predating the classical Indian texts such as the *Rigveda*, composed in the early Vedic period roughly between 1700 and 1100 BC[12] It seems possible then, that the story of a hero-figure (Hercules) climbing a mountain—later, either from possible confusion or mere convenience, transferred to Boötes—originated long before the Greeks. As they adopted a story transmitted from the east, the identity of the mountain became associated with one of their own. In Book IV of *Elegies* (first century AD), the Latin elegiac poet Aurelius Propertius[13] suggested a similar connection between Hercules and Mons Maenalus, by way of his club (*ramus*), which Propertius says originated on the slopes of the famed mountain. The episode concerns Hercules' appearance on the Palatine Hill wherein a mythic origin of the Roman Forum was outlined:

> Cacus lay low, thi'ice smitten on the brow with
> the Maenalian club, and thus spake Alcides : "Go,
> ye oxen, go, oxen of Hercules, the last labour of my
> club. Twice, oxen, did I seek ye, and twice ye were
> my prey. Go ye and Avith your long-drawn lowing
> hallow the Place of Oxen ; your pasture shall in times
> to come be the far-famed Forum of Rome." He
> spake, and thirst tortured his parched palate, while
> teeming earth supplied no water.[14]

Somewhere along the way between antiquity and Hevelius, the association between Hercules and the Mainalo was lost. It seems likely to have been a matter of convenience to Hevelius, who found unformed stars in the right place beneath Boötes from which to make a mountain, whereas Hercules was traditionally depicted on charts as all but stepping on the head of Draco, leaving little room to insert his mountain.

---

[11]"Professor Romieu, *Sur an Decan*, etc., p. 39, has identified the Egyptian star of the Vulture with the constellation Lyra, the star of the goddess Ma-at, the mother of law and order; and in Egyptian mythology the vulture ruled the year. In the Rigveda the vulture Gridhra is represented as a rival ruler of time with the Ashvins, or twins, who are invoked to come and drink the Soma cup early in the morning before the greedy vulture (Rigveda, v. 77, 1), to whom the Marka or Soma cup of the dead (*Mahrka*) was offered."

[12]Oberlies (1998) estimated a composition date of c. 1100 BC for the youngest hymns in Book 10 of the *Rigveda*. He cited "cumulative evidence" for a terminus post quem of the earliest hymns in the wide range of 1700–1100 BC In any case, the core of the *Rigveda* material clearly dates to the late Bronze Age.

[13]Sextus Aurelius Propertius (*c*. 50 BC—aft. 15 BC) was a Latin elegiac poet of the Augustan era.

[14]Propertius, *Elegies* IV(5), 9, 15, trans. H.E. Butler.

DISAPPEARANCE

Mons Maenalus was left off charts with increasing frequency in the last half of the nineteenth century. It was by then viewed as an anachronism. In *Poole Bros. Celestial Handbook* (1892), Mons Maenalus appears in the table "Old And New Constellations In Chronological Order," in which its formation is erroneously attributed to "Flamsteed, 1725, A.D." The table comes with the caveat that "These constellations, not being recognized by the B.A. Catalogue, are not described in this Handbook; the stars by which they were formed are inserted in the constellations from which they were taken." While recognizing that the stars comprising Mons Maenalus were derived from those formerly associated with Boötes, the *Celestial Handbook* grants them back to the Herdsman, consigning the former constellation to the dustbin of history.

As usual, Allen (1899) included a mention of Mons Maenalus at the end of the century although curiously he did not dismiss it as a relic, instead noting only that "It culminates in June, due south from β Boötis and north of β Librae." By the time of the first IAU General Assembly, it appears that astronomers considered Mons Maenalus completely subsumed back into Boötes. Accordingly, Eugéne Delporte drew the modern boundary between Boötes and Virgo such that all of the stars formerly belonging to Mons Maenalus are now decidedly considered part of Boötes (Fig. 15.8).

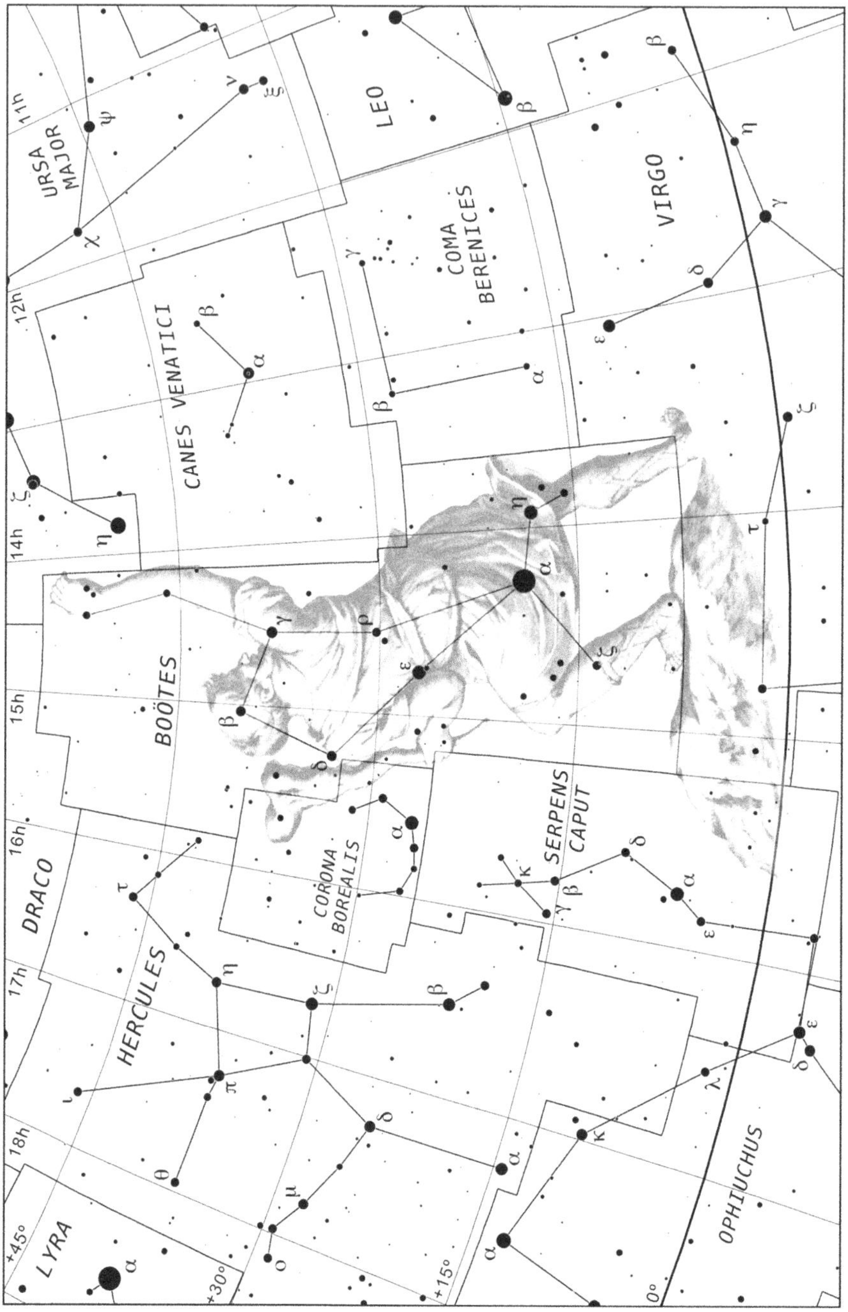

15.8 The figures of Boötes (*center*) and Mons Maenalus (*bottom*) from Figure F of Hevelius' *Prodromus Astronomiae* (1690) overlaid on a modern chart

16

Musca Borealis

The Northern Fly

Genitive: Muscae Borealis
Abbreviation: MuB
Alternate names: "Musca" (Ryan, 1827); "Lilium" (Pardies 1674, Allard 1706); "Apis" (Plancius 1612); "Vespa" (Bartsch, 1624); Fleur-de-lys (Pardies 1674)
Location: North of Aries and east of Triangulum, comprising the stars now designated 33 Arietis, 35 Arietis, 39 Arietis, and 41 Arietis[1]

ORIGIN AND HISTORY

The group of four bright stars most commonly identified on historic charts as Musca Borealis has a complex history in which both of two competing representations were ultimately discarded. The representations are here considered separately (Fig. 16.1).

As Musca

There is some confusion over when the first figure inclusive of these stars was introduced. Kanas (2007) identifies its origin as "Apes" (the Bee) on Petrus Plancius' 1613 globe, created from four stars left unformed by Ptolemy described as lying "over the rump" of Aries in the *Almagest*. This name immediately created the possibility of confusion with the southern constellation "Apis" (later Musca) identified by Plancius and Dutch explorers

[1] *"Over the back of the Ram"* (Allen, 1899); *"Between Aries and Perseus"* (Kendall, 1845); *"Directly north of Aries"* (Bouvier, 1858); *"Four stars north of Aries"* (Bakich, 1995).

J.C. Barentine, *The Lost Constellations*, Springer Praxis Books,
DOI 10.1007/978-3-319-22795-5_16

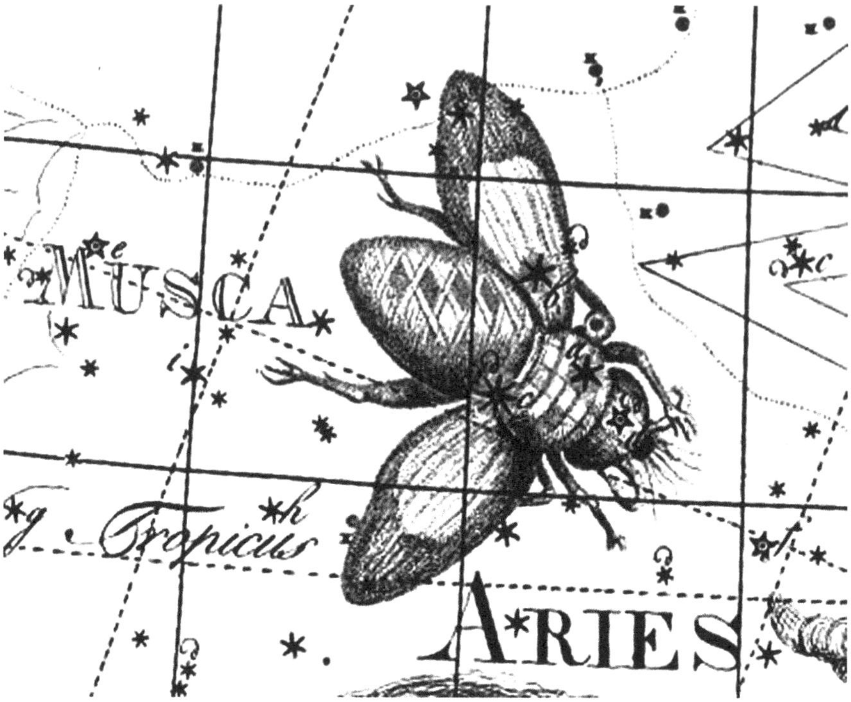

16.1 Musca Borealis shown in a detail of Plate 11 from Johann Elert Bode's *Uranographia* (1801b)

Pieter Dirkszoon Keyser and Frederik de Hutman around 1598. The northern counterpart certainly does not seem to have been in existence by the time of Bayer, who drew its stars in *Uranometria* (1603, Fig. 16.2) but left it without any sort of figure to itself. Richard Hinckley Allen (1899) thought the figure of Musca was originally identified by Jakob Bartsch in the first quarter of the seventeenth century:

> Houzeau[2] attributed its formation to Habrecht,[3] but others to Bartschius,[4] who called it Vespa, the Wasp, although also Apis, the Bee; and, still further changing the figure, wrote that it represented Beel-zebul, the god of flies, the Phoenician Baal-zebub; this insect being the ideograph of that heathen divinity, varied at times by the Scarabaeus. La Lande's Apes probably is a typographical error. To whom we owe its present title I cannot learn; but it is thus given in the Flamsteed *Atlas* of 1781,

[2]Jean-Charles Houzeau de Lehaie (1820–1888) was a Belgian astronomer and journalist known for his travels in the United States during the mid-nineteenth century.

[3]Isaac (II) Habrecht (1589-1633) was an astronomer, physician and cartographer at Strasbourg.

[4]Jakob Bartsch (c. 1600–1633) was a German astronomer and cartographer from Lubań, now in Poland.

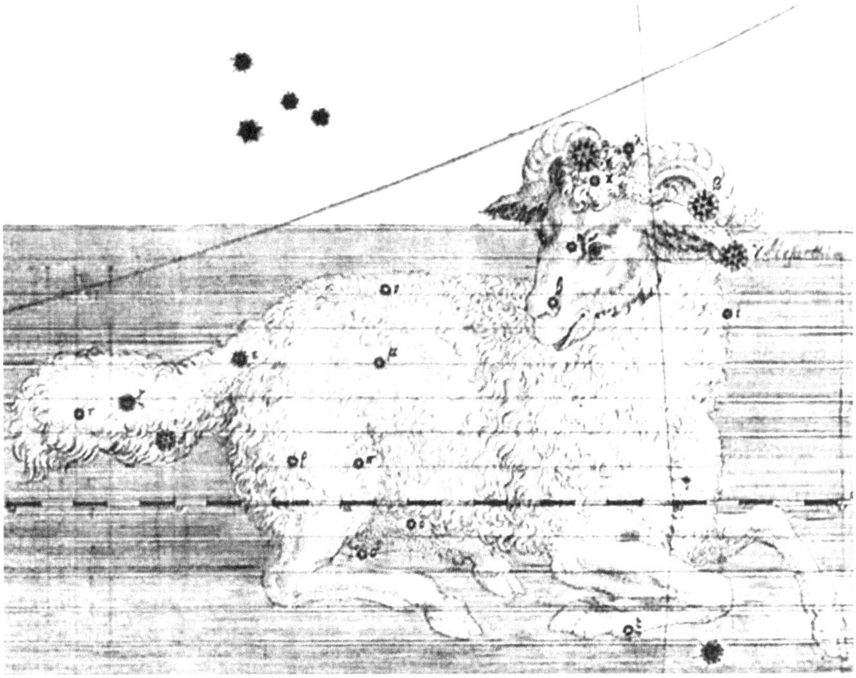

16.2 The four stars of Musca Borealis were drawn by Bayer above and to the left of Aries on Plate X of Johannes Hevelius' *Uranometria* (1603), but were not shown with any particular figural depiction. The same is true for the identical stars on Plate L (Perseus)

attributing that version of the name to a mistake made by Joseph Jérôme Lefrançois de Lalande nearly two centuries later. Bartsch certainly seems to have been the first popularizer of this constellation, referring to it as "Vespa" (the Wasp) it in *Usus Astronomicus* (1624; Fig. 16.3).

In the decades after Bartsch, most atlases included the insect, labeling it either "Apis" or "Apes." It was marked the former on a set of celestial globe gores engraved by Matthäus Greuter (1556?–1638) in 1636 (Fig. 16.4), though which were not printed on paper until 1695 by Domenico de Rossi in Rome. Cellarius used the latter label on Plate 24 of *Harmonia Macrocosmica* (1661; Fig. 16.5). Hevelius labeled the four stars as "Musca" in Figure BB of *Firmamentum Sobiescanum* (1687), (i.e., add here (1687)) but listed them under the constellation Aries in *Prodromus Astronomiae* (1690). This suggests that Musca Borealis rose only to the status of an asterism in Hevelius' mind (Fig. 16.6). However, "Musca" was bound to cause trouble because of the southern constellation of the same name; later, in the eighteenth century, the southern Fly would sometimes appear labeled "Musca Australis" in charts in order to avoid confusion. In the early nineteenth century, the picture became even more complicated: Alexander Jamieson (1822) labeled the constellation devised by Plancius and promoted by Hevelius "Musca Borealis" to

16.3 Musca Borealis shown as "Vespa" in a detail of Jakob Bartsch's *Planisphaerium stellatum* … (1661)

distinguish it from the southern constellation, which in turn resulted in mapmakers dropping the "Australis" from the latter.

Musca Borealis was even turned into a subtle piece of Roman Catholic propaganda in its earliest years. When Julius Schiller tried to redefine the constellations in terms of Judaeo-Christian symbolism in *Coelum Stellatum Christianum* (1627), Aries became the apostle Peter, seated on a throne. Retaining the four stars of Musca Borealis as unformed within Aries, according to Ptolemy, Schiller came up with a clever means of incorporating them into his new figure. As seen in Fig. 16.7, the stars became the ends of Peter's keys to the church; in the era of the European Wars of Religion, this clearly signaled support for the notion of Catholic supremacy through the Petrine succession. However, like every other constellation devised by Schiller, none entered wide circulation.

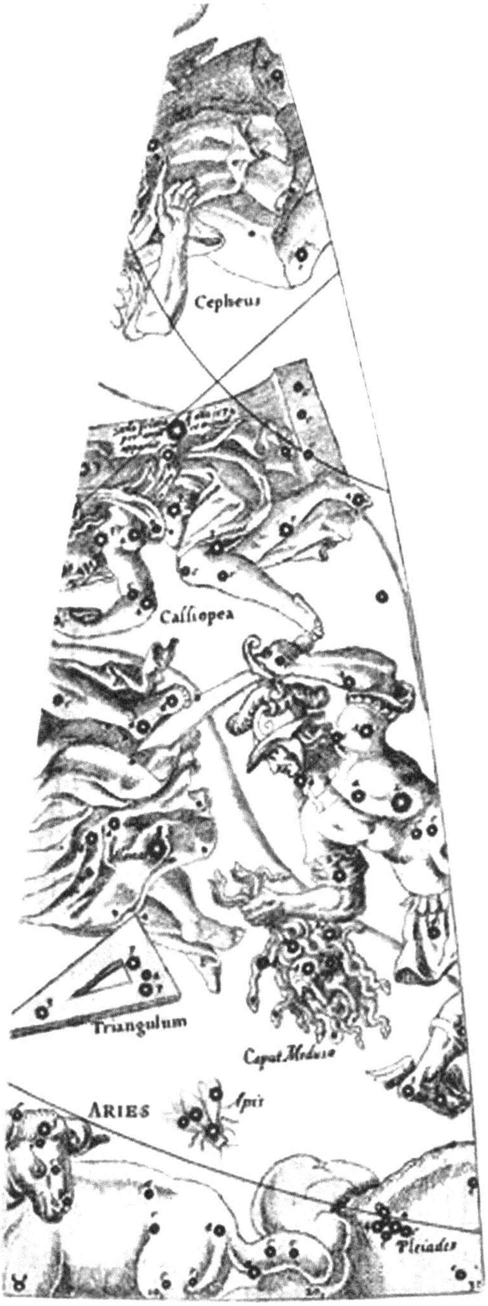

16.4 Gore 11 of the set of celestial globe gores engraved by Matthäus Greuter in 1636 and printed on paper in 1695. Musca Borealis is shown as "Apis" near the *bottom* of the figure. Overall dimensions of the gore are approximately 44 × 19 cm

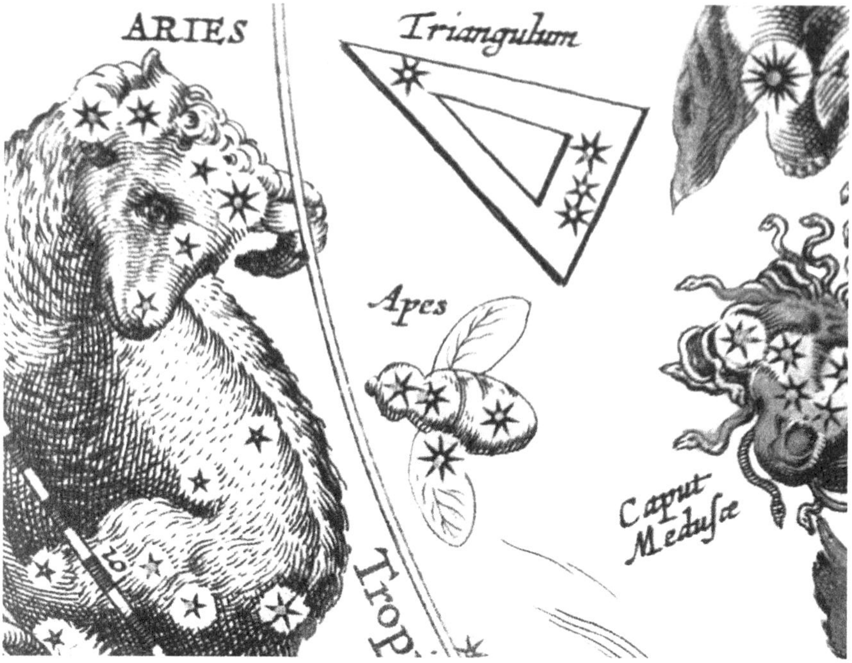

16.5 Andreas Cellarius depicted Musca Borealis as "Apes" on Plate 24 of *Harmonia Macrocosmica* (1661); Caput Medusa (see Volume 2) appears at right

As Lilium

The French architect and cartographer Ignace-Gaston Pardies (1636–1673) introduced a different version of this constellation on Plate 2 of his map *Globi coelestis in tabulas planas redacti descriptio*, first published posthumously in 1674. However, he did not label the figure on his chart, as seen in Fig. 16.8. Nick Kanas notes that, since Pardies was a Jesuit, his maps enjoyed wide popularity among Jesuit astronomers and remained influential, seeing two additional additions published in 1693 and ca. 1700. "The name Lilium," Ian Ridpath writes in *Star Tales* (1989), "first seems to have appeared five years [after Pardies] in a chart and catalogue published by another Frenchman, Augustin Royer, although he made the lily much bigger." This is Royer's series of four celestial maps entitled *Cartes du Ciel Réduites en Quatre Tables* and first published in 1679. In the same work, Royer introduces three other new constellations: Crux (the Cross, as completely distinct from Centaurus) and Columba (the Dove), which later became canonical, and Sceptrum et Manus Iustitiae, which did not (see Volume 2) (Figs. 16.9 and 16.10).

ICONOGRAPHY

Tracing the symbolism of this constellation is somewhat convoluted because it received more than one complete makeover in its ~250-year history.

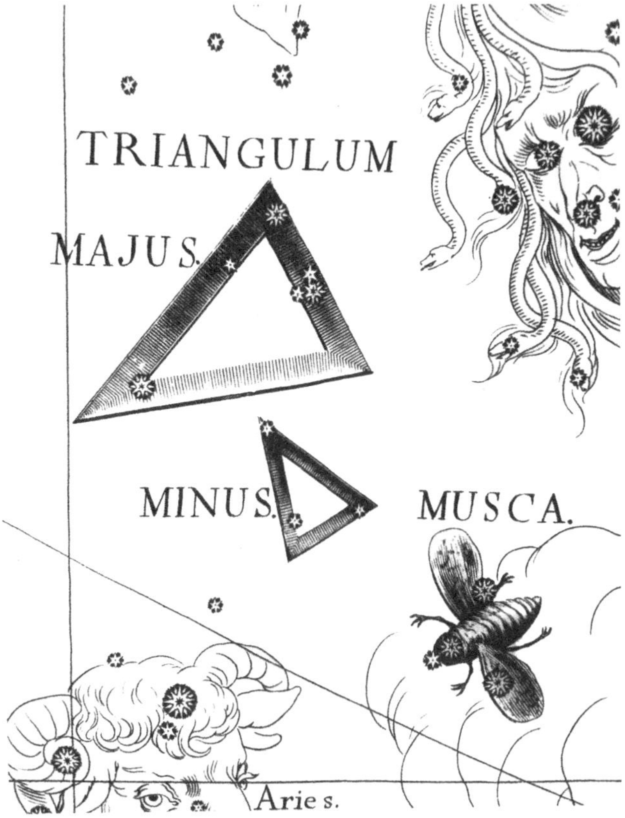

16.6 Musca shown on Fig. BB of Hevelius' *Firmamentum Sobiescanum* (1690) along with Caput Medusae (unlabeled at *upper right*; see Volume 2), Triangulum, Triangulum Minus (Chap. 28), and the head of Aries

The Bee, or Fly

Bartsch found a distinct Biblical meaning in this constellation. In *Usus Astronomicus* (1624) he wrote[5]:

> The Bee, once belongning to Aries. To me it is Vespa Beelzebub, god of the flies, noting Luke 11:15[6] etc. Or Apes, meaning those that appeared from the carcass of the lion slain by Samson, Judges 14:8.[7]

[5]XXVII. APES olium ad (Aries) informes pertinebant: jam informam redactae. *Mihi vel Vespa Beelzebub, deum muscarum* notet, Luc. II. v. 15. &c. vel *Apes* significent eas, quae ex Leonis a Simsone trucidati cadavere, Jud 14. v. 8. provenere (p. 88).

[6]"But some of them said, 'By Beelzebul, the prince of demons, he is driving out demons"'. (NIV)

[7]"Some time later, when he went back to marry her, he turned aside to look at the lion's carcass, and in it he saw a swarm of bees and some honey." (NIV)

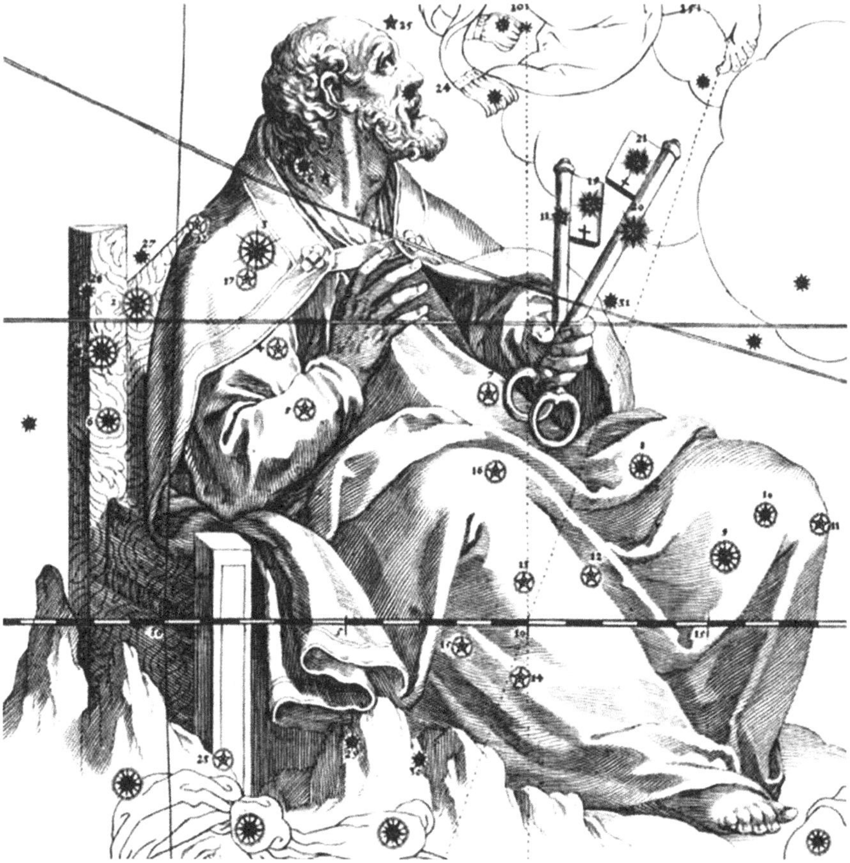

16.7 Aries rebranded as Saint Peter by Julius Schiller in *Coelum Stellatum Christianum* (1627). The four stars depicted on contemporary maps as Musca Borealis are here repurposed as part of Peter's emblematic crossed keys (*upper right*)

It appears that Bartsch essentially threw everything against the wall in looking for a Biblical interpretation of the Bee. While the reference to Judges 14 is likely the example most Biblically-literate individuals would come up with at first glance, the Luke 11 verse is decidedly darker—unless Bartsch deliberately sought a dichotomy of dark and light. "Beelzebul" is probably a corruption of Ba'al Zebûb, a Semitic god appropriated as the "lord" (taking the literal sense of "ba'al") of the Philistine city of Ekron. Ba'al Zebûb famously figures in the incident related 2 Kings, in which

> After Ahab's death, Moab rebelled against Israel. Now Ahaziah had fallen through the lattice of his upper room in Samaria and injured himself. So he sent messengers, saying to them, "Go and consult Baal-Zebub, the god of Ekron, to see if I will recover from this injury." But the angel of the Lord said to Elijah the Tishbite, "Go up and

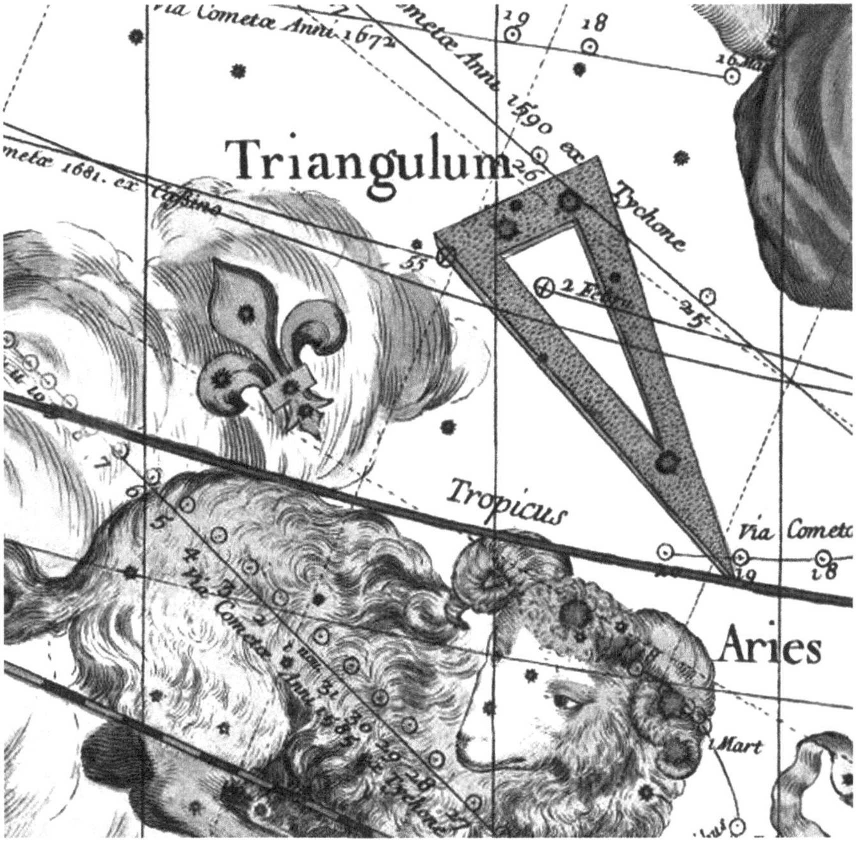

16.8 The stars formerly identified as Musca Borealis were arbitrarily redrawn as "Lilium," representing the royal symbol of France, by Ignace-Gaston Pardies on Plate 2 of his map *Globi coelestis …* (1674). The version shown here is from the second edition of his map, printed in Paris in 1693

> meet the messengers of the king of Samaria and ask them, 'Is it because there is no God in Israel that you are going off to consult Baal-Zebub, the god of Ekron?' Therefore this is what the Lord says: 'You will not leave the bed you are lying on. You will certainly die!'" So Elijah went.[8]

Based on the angel's words, Elijah condemned Ahaziah to die because Ahaziah sought the counsel of Ba'al Zebûb rather than the God of Israel.[9]

[8]2 Kings 1:1–4 (NIV).
[9]2 Kings 1:16–17 (NIV).

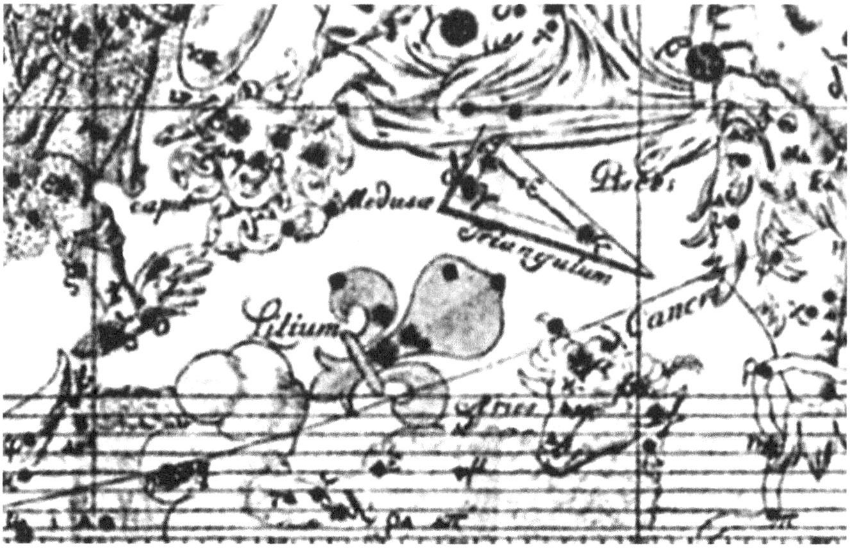

16.9 Lilium received its first proper label, in Latin, in Augustin Royer's *Cartes du Ciel Reduites en Quatre Tables* (1679a)

The association of Ba'al Zebûb with flies probably derives from historical Jewish scholarship, which characterized Ba'al as a pile of dung, comparing Ba'al's adherents to flies. Therefore, to refer to Ba'al as the "Lord of the Flies" is to mock the Ba'al religion, considered by the ancient Israelites as nothing more than idol worship.

The Luke 11 episode is echoed in parallel instances in Mark 3:22–29 and Matthew 12:24–28. Beelzebul also appears as prince of the demons in the *Testament of Solomon*, a pseudepigraphical Old Testament work composed sometime between the first and fifth centuries AD. In it, Beelzebul claims he was formerly the highest ranking angel in Heaven[10] and was associated with the star Hesperus,[11] the common Greek term for the planet Venus in its aspect as the Evening Star. We can therefore understand Bartsch's reference to Beelzebul thus as one to Satan—a personification of evil—more specifically than to the Ba'al religion, even though the latter can be thought of as symbolizing a form of moral decay. In contrast to the "good" bees that provide the life-sustaining sweetness of honey, the alternative Biblical view of Apes is its association with flies, rotting, and death. Bartsch sought both in his 1624 description of Apes.

[10] *Testament of Solomon*, 6.2.
[11] *Testament of Solomon*, 6.7.

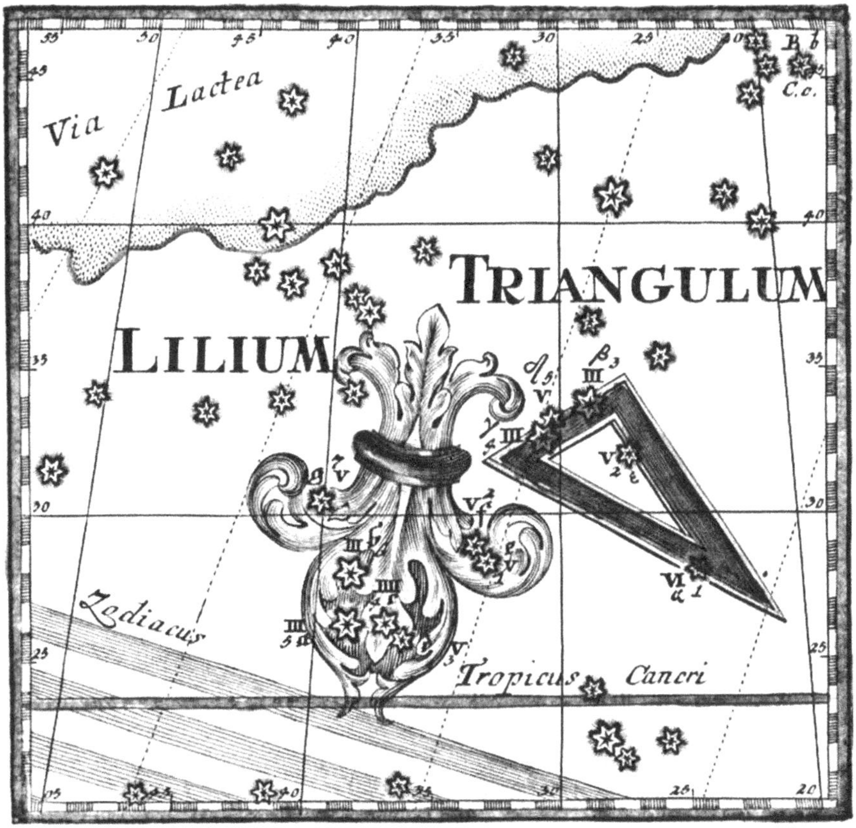

16.10 "Lilium" depicted in Corbinianus Thomas' *Mercurii philosophici firmamentum firmianum* (1730)

The Fleur-de-Lis

Based on contemporary depictions in late seventeenth century atlases, Lilium was clearly intended as a fleur-de-lis, an element of French royal heraldry of ancient origin. The lily symbolized by the fleur-de-lis became associated with French royalty from the reign of Clovis I (AD 509–511), widely considered the first king of what would later become the French nation represented by the original union of Frankish tribes. Clovis converted Catholicism at the best of his wife, Clotilde, a princess of Burgundy who retained her Catholic allegiance in the face of the dominant Arianism of her court. According to tradition, he was baptized by Saint Remigius[12] on Christmas Day, AD 496, at a small church near the present Abbey of Saint-Remi in Reims. Legend holds that the Virgin Mary

[12]Saint Remigius of Reims (AD 437–533) was Bishop of Reims and Apostle of the Franks.

gave a single lily to Clovis at the baptism, implying the divine institution of the monarchy that issued from him; this point was raised by French bishops at the Council of Trent (1545–1563) in support of their arguments favoring the precedence of their king, Francis I.

Long before Christianity, the lily was associated with the idea of purity; the Church adopted it from pagan uses and remade it into a symbol of Mary's special position in the Catholic worldview. Legend, then, imbued Clovis (and his descendants) with both a holy purity and temporal supremacy over their realms. On Christmas Day, AD 800, Pope Leo III crowned Charlemagne as the first Holy Roman Emperor in Old St. Peter's Basilica in Rome; tradition holds that the Pope presented the new emperor with a blue banner embroidered with golden fleurs-de-lis (*Azure semé of fleurs-de-lis Or*). This design persisted as the personal arms of the French king through the end of the monarchy in 1792 (Fig. 16.11). Its history as a heraldic charge dates to at least the twelfth century when it first appeared on a field of Philippe II (1180–1214). The arms are certainly associated with French kings from 1200 on. Charles V reduced the number of fleurs-de-lis to three in 1376, reportedly in reference to the Holy Trinity; Henry IV of England soon emulated the new pattern in the first quarter of his arms to symbolize the English claim to the throne of France.

The Scottish heraldry historian Alexander Nisbet (1657–1725) wrote in his comprehensive survey of the heraldry of Scottish families *A System of Heraldry, Speculative and Practical: With the True Art of Blazon* (1722):

> The other lilies, as those of France, so well known .... having only but three leaves, is by the Latins called *flos iridis*, and by the French *fleur de l'iris; being always called the flower of the rainbow or iridis, which the French call fleur-de-lis*, from the river Lis, as some will; and anciently *flams* or *flambs*, which signifies the same: Whence the Royal Standard of France was called the *oriflam* or *oriflambe*, being a blue banner, charged with golden flower-de-luces, a suitable figure, say some, for the Franks, who come from the marshes of Friezland.

Indeed, the stylized fleur-de-lis more closely resembles the iris than an actual lily. Historically, there was much confusion on this point, and only in the nineteenth century did the iris gain an identity separate from the family of true lilies. Nisbet references the river "Lis", which today is known in France as *Lys*, and its continuation into Flanders is known in Flemish as *Leie*. The golden iris remains common on banks of the Flemish portion of the river. Abstracted by the herald, the upper pair of the five petals of the iris is joined to the center petal, while the outer two curve downward.

DISAPPEARANCE

Musca Borealis persisted on charts and in the descriptions of popular books on astronomy throughout the nineteenth century. Thomas Young showed it as, simply, "Fly" on Plate XXXVI, Fig. 517 of his *Course of Lectures on Natural Philosophy and the Mechanical Arts* (1807), while William Croswell included it as "Musca" on his 1810 *Mercator Map*

16.11 The royal standard of the Kindgom of France, combining the *Pavillon royal de France* (a flag representing the Bourbon monarchy) with the royal arms. The fleur-de-lis device features prominently in both elements. Image licensed under Creative Commons Attribution-Share Alike 3.0 (Unported) from Wikimedia Commons user Oren neu dag

of the Starry Heavens, among the first star charts produced and printed in America. As mentioned previously, Alexander Jamieson depicted it on his 1822 chart, as did Jacob Green in his *Astronomical Recreations* (1824). The Fly was given perhaps its most detailed figural rendering, along with Aries, on Plate 16 of *Urania's Mirror* (1825; Fig. 16.12); this version appears to be unique among charts in that the insect is shown with two pairs of wings instead of the traditional one. Ezra Otis Kendall described it in *Uranography* (1845) as "A small constellation, containing one star of the 3d and two of the 4th magnitudes," while Hannah Bouvier wrote in her *Familiar astronomy* (1858) that it "was formed by Bartschius, the son-in-law of Kepler, for which reason Hevelius retained it in his catalogue."

The constellation slowly began to drop out of charts and books in the second half of the century. One of the first prominent omissions was in Richard Anthony Proctor's *Constellation-Seasons* (1876), although George Chambers, in *A Handbook of Descriptive Astronomy* (1877), mentioned this constellation as "Fleur-de-lys" and attributed its introduction to Royer. William Henry Rosser left Musca off his constellation lists in *The Stars and Constellations* 1879, and it does not appear in Charles Pritchard's catalog of stars

16.12 Aries and "Musca Borealis" depicted on Plate 16 of *Urania's Mirror* (1825). Note also on this plate labels for "Psalterium Georgii" (Chap. 18) and "Triangula", containing Triangulum Minus (Chap. 28)

Uranometria Nova Oxoniensis from 1885. Similarly, it was left out of the table "Old And New Constellations In Chronological Order" in *Poole Bros. Celestial Handbook* (1892); "Apis" is listed, but is presumably the Musca of southern skies.

Writing at the close of the nineteenth century, Allen declared the Northern Fly extinct: "The constellation has been retained in some popular astronomical works, although not figured by the scientific Argelander, Heis, nor Klein, nor recognized in the British Association Catalogue." Still, it persisted on a few charts through the turn of the century. Ernest Lebon drew it as "Mouche" in *Histoire Abrégée de l'astronomie* (1899). In perhaps its last significant appearance in print, William Tyler Olcott gave it more than passing mention in his discussion of Aries in *Star Lore of All Ages* (1911), but it is not clear whether he considered it an asterism or a proper constellation:

> The faint stars east of Hamal on the back of the Ram form a little group known as "Musca Borealis," the Northern Fly. The figure appears in Burritt's Atlas. According to Allen the inventor of this asterism is unknown. Musca has been also styled "the Wasp" and "the Bee."

Due to its late survival, Musca Borealis might well have become canonical, but its fate was sealed at Rome during the first IAU General Assembly in 1922 when it was left off the list of constellations for which three-letter abbreviations were decided. In his *Atlas Céleste* (1930a), Eugéne Delporte drew the boundary between Aries and Triangulum to carefully include the stars of Musca within the former, suggesting that he intended to revert to the ancient status quo ante, in which the stars were left unformed but properly considered part of Aries (Figs. 16.13 and 16.14).

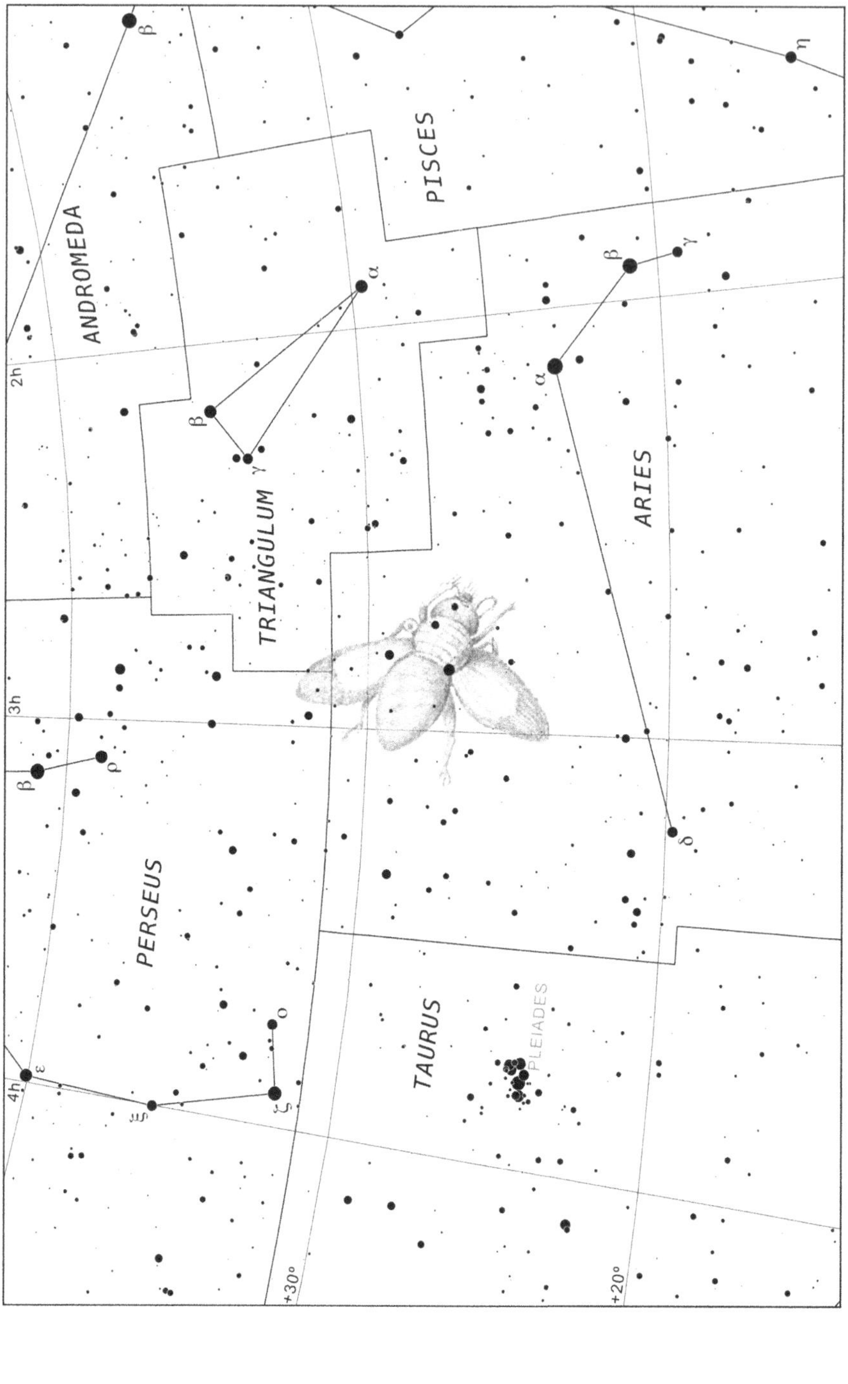

16.13 The figure of Musca Borealis from Plate 11 of Bode's *Uranographia* (1801b) overlaid on a modern chart

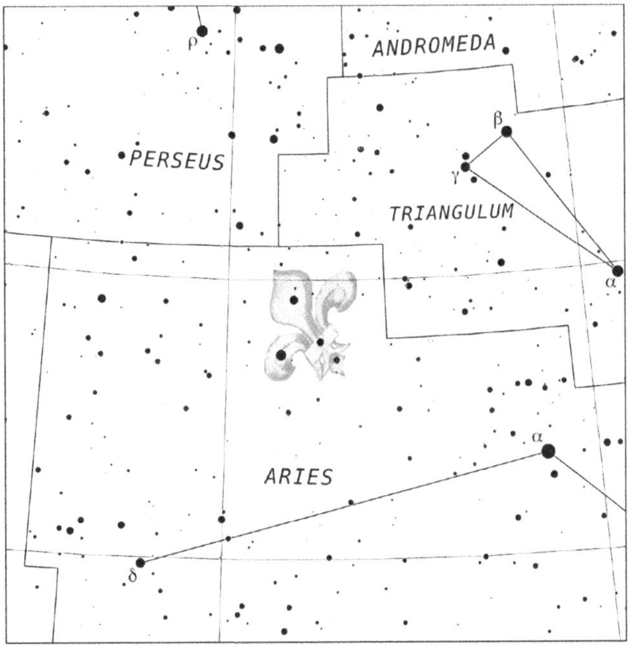

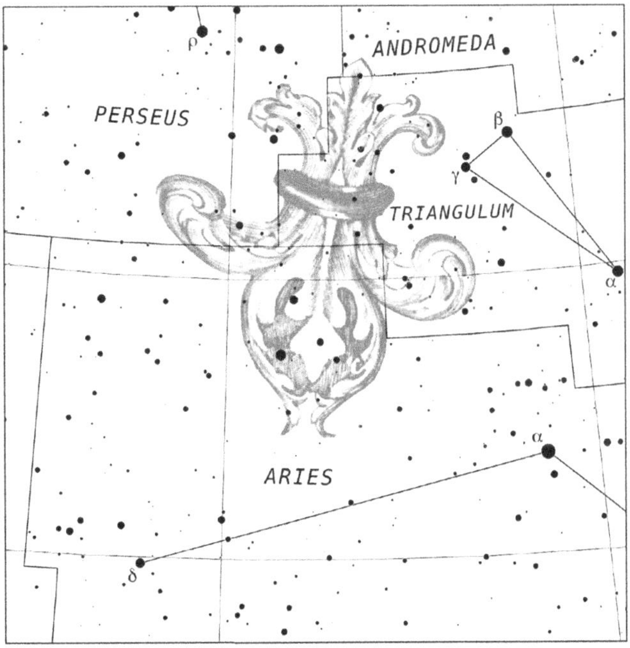

16.14 Two interpretations of Lilium overlaid on a modern chart. *Top*: Ignace-Gaston Pardies, Plate 2 of *Globi coelestis in tabulas planas redacti descriptio* (1674). *Bottom*: Corbinianus Thomas, *Mercurii philosophici firmamentum firmianum* (1730)

17

Officina Typographica

The Printer's Workshop[1]

Genitive: Officinae Typographicae
Abbreviation: OfT
Alternate names: "Atelier Typographique" (*Urania's Mirror*, 1825); "Atelier de l'Imprimeur" (Preyssinger, 1862); "Antlia Typographiae" (Bowen, 1888)
Location: East of Canis Major and immediately north of Puppis[2]

ORIGIN AND HISTORY

This constellation was first introduced by Johann Elert Bode in *Uranographia* (1801b) as "Buchdrucker-Werkstatt". In the accompanying catalog, *Allgemeine Beschreibung und Nachweisung der Gestirne*, he described[3] it thusly:

[1]Most authors render Bode's contrived New Latin name too literally as "Printing Office." A closer match to the original intended sense is obtained from the French translation "atelier typographique" (printing workshop), so I have here rendered it as the "Printer's Workshop". A shorter English version that retains the spirit of the original is "Printshop."

[2] *"Between the head of the Great Dog, the Unicorn, and the Cat"* (Kendall, 1845); *"Directly east of Canis Major, and south of Monoceros"* (Bouvier, 1858); *"[S]tars immediately east of Sirius"* (Allen, 1899); *"It lay in what is now the northern part of Puppis, the stern of the ship Argo, between Canis Major and the hind legs of Monoceros"* (Ridpath, 1989); *"Directly east of Sirius"*Sirius" (Bakich, 1995).

[3]"Dieses Gestirn erscheint zuerst in diesen Charten, gerade links vom Sirius und dem Kopf des grossen Hundes, und macht sich an verschiedenen daselbst in und bey der Milchstrasse stehenden Sternen kenntlich. Ich habe erst ganz neuerlich in Vorschlag gebracht, um damit das Andenken einer äusserst wichtigen über 350 Jahr alten Erfindung eines Deutschen, am Sterngewölbe zu erhalten. Es besteht aus einem Theirl der Druckerpresse, dem Schriftkasten, Ballen &c."

J.C. Barentine, *The Lost Constellations*, Springer Praxis Books,
DOI 10.1007/978-3-319-22795-5_17

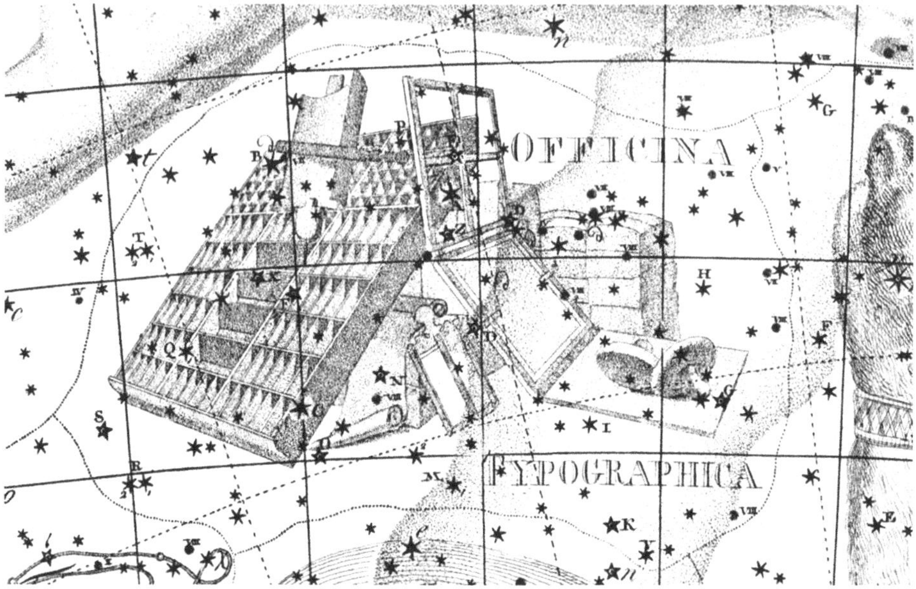

17.1 Officina Typographica depicted on Plate 18 of Bode's *Uranographia* (1801b). It is here reproduced at near-full size to show the fine quality of the engraving that characterizes Bode's masterwork

> This constellation first appears in these charts, just to the left of Sirius and the head of the Greater Dog, and is composed of various stars situated in and along the Milky Way. I have only very recently proposed it in order to preserve in the starry vault the memory of a very important invention by a German over 350 years ago. It consists of a portion of the printing press, the font box, the bale, etc.

Officina Typographica was one of two constellations devised by Joseph Jérôme Lefrançois de Lalande in 1798 (*"la Presse de Gutemberg* [sic] *et la Globe de Montgolfier"*) and suggested to Bode along with Globus Aerostaticus, discussed in Chap. 11, to commemorate two great technological advances of Germany and France respectively. Figure 17.2 shows the part of the sky in which he formed the constellation in the editions of his *Vorstellung der Gestirne* preceding (1782) and following (1805) the publication of *Uranographia*. The comparison is instructive as it demonstrates how Bode often identified stars that didn't seem to be properly claimed by any existing constellation after the first edition of *Vorstellung* and helped himself to fill in the otherwise empty spaces between figures in the sky. In that the pattern of available stars is not reminiscent of any particular figure, it seems most likely that Bode had in mind the figure in the first place and simply placed it without much regard to whether the stars involved formed a pattern resembling the figure. That the stars involved tend to be relatively faint, the prominence of any pattern is reduced.

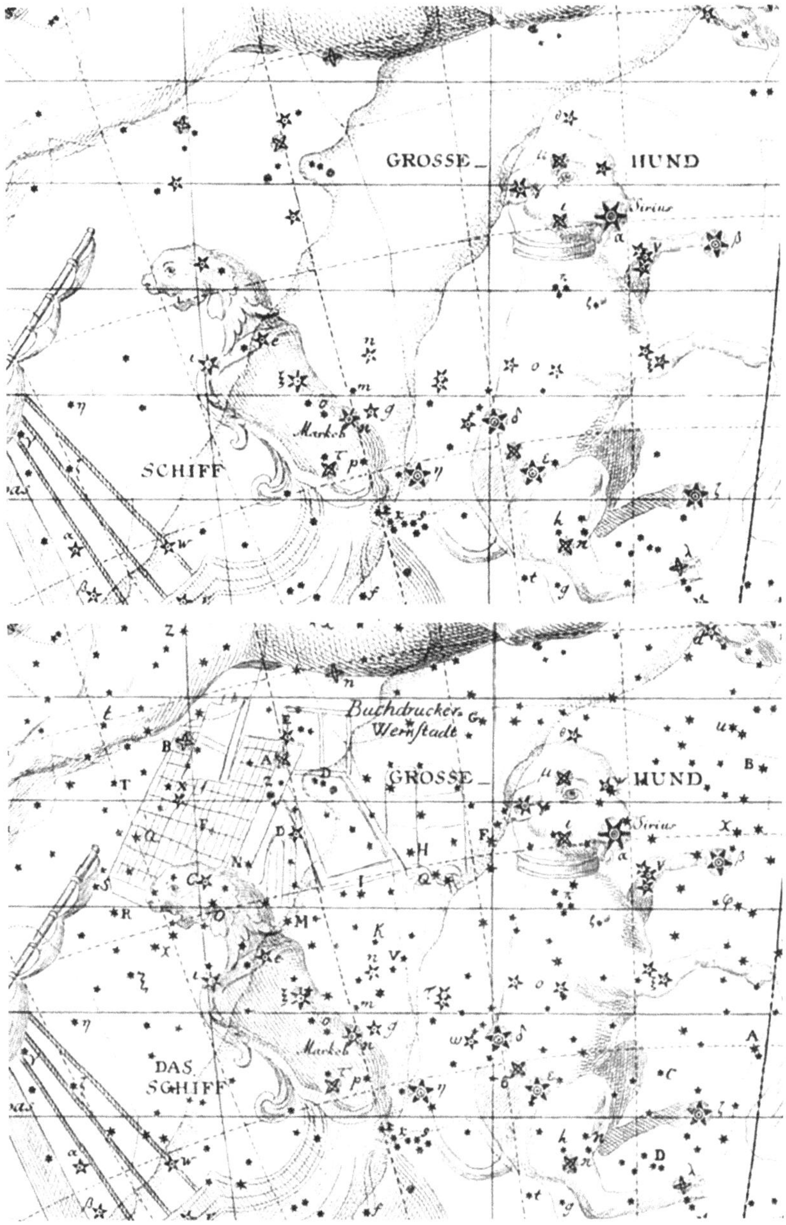

17.2 Before and after: Excerpts of Plate 25 from two editions of Bode's *Vorstellung der Gestirne*, 1782 (*top*) and 1805 (*bottom*). In addition to showing the marked increase in stars plotted from one version to the other, the two figures show the addition of Officina Typographica ("Buchdrucker Werkstadt") between editions. Note also that Argo Navis ("Schiff" in 1782) gained a definite article ("Das Schiff" in 1805) in the interim

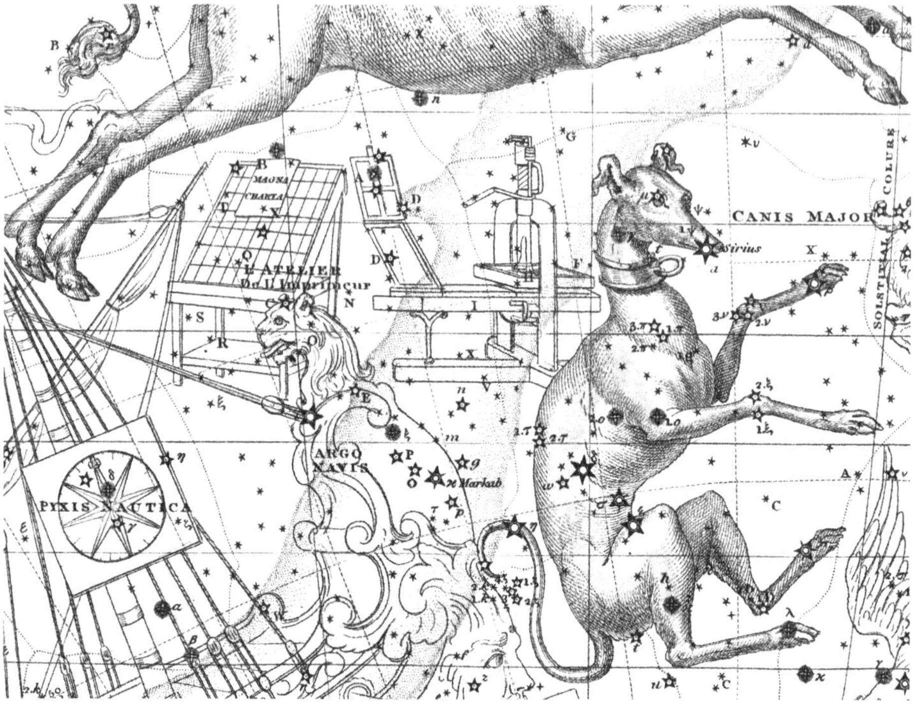

17.3 Officina Typographica as shown on Plate 25 of Alexander Jamieson's *A Celestial Atlas* (1822). Jamieson's innovation in the figural representation of the Printshop: the entire press is shown, whereas Bode and others only included certain elements of the press such as the decker. This version, however, leaves ambiguity as to exactly where Puppis (here, Argo Navis) ended and Officina began

Adoption of the new constellation was hit or miss after the publication of *Uranographia.* August Gottleib Meissner showed the figure as "Buchdrucker Werkstatt" on Fig. XXXIV, Table 32 and Fig. XXXV, Table 33 of *Astronomischer Hand-Atlas* (1805). Two years later, Thomas Young included the "Printer's Workshop" on his northern hemisphere chart (Plate XXXVII, Fig. 518) in his *Course of Lectures on Natural Philosophy and the Mechanical Arts.* William Croswell did not show it on his 1810 *Mercator map of the starry heavens*, but it was depicted on Plate 25 of Alexander Jamieson's atlas (1822; Fig. 17.3) and on Plate 31 of *Urania's Mirror* (1825; Fig. 17.4) in the following decade.

Jacob Green showed the constellation by name only on Plate 16 of *Astronomical Recreations* (1824), but his passing mention of it in the text shows that he held it and other similar "modern asterisms" in low regard:

> Besides the seven constellations figures on this plate, there are five small modern asterisms mentioned, viz. Officina Typographica, Cæla Sculptoris, Horologium, Plateum Pictoris, and Pyxis Nautica. We shall not however enter into any detail

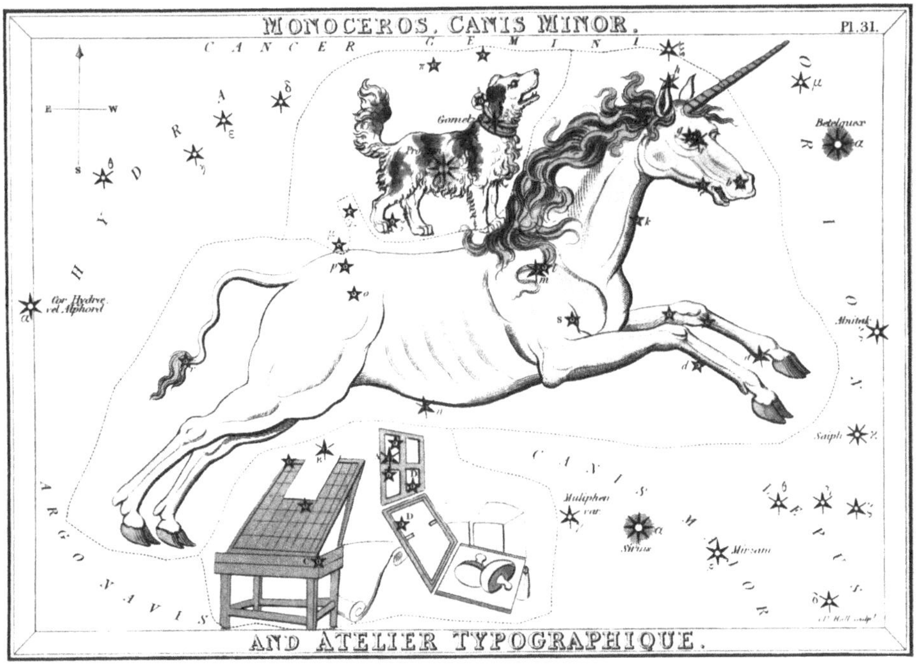

17.4 Officina Typographica ("Atelier Typographique") shown beneath Monoceros on Plate 31 of *Urania's Mirror* (1825). The elements of the figural depiction borrow heavily from Bode (1801b)

concerning them. For the most part they are composed of very minute stars, and those who become familiar with all the ancient groups will readily find these, which are of modern date, filling up the space left unnoticed by the old astronomers.

Ezra Otis Kendall accorded full constellation status to Officina Typographica in his *Uranography* (1845), rendering its name in English as "The Printing Press." Hannah Bouvier counted Officina Typographica among "the new constellations formed by Bode" in her *Familiar Astronomy* (1858), noting that while it included no bright stars its position along the Milky Way made it remarkable for other reasons:

Q. Does it contain any stars of the FIRST or SECOND magnitude?
A. No ; it is composed of small stars, and contains some rich clusters not visible to the naked eye.

Officina Typographica made noteworthy appearances during roughly the second half of the nineteenth century in the atlases of Riedig (1849; Plate II, as "Buchdruck. Werkst."), Preyssinger (1862; Plate II, as "Atelier de l'Imprimeur") and Bowen (1888, on the all-sky map of page 108 as "Antlia Typographiae").

ICONOGRAPHY

A Brief History of Printing

The idea of forming symbols in a way that decoupled the production of written material from the hands of individual scribes is an ancient one, extending back to the dawn of human civilization. Printing as a concept in which multiple impressions of the same writting could be easily reproduced many times emerged around 3000 BC in ancient Sumeria in the form of cylinder seals and similar objects. These objects contained incised designs that were pressed into or rolled across slabs of wet clay, leaving a lasting design once the clay hardened. Stamps on clay bricks found in the southern Sumerian cities of Uruk and Larsa and dating to the second millennium BC show uneven spacing among the character impressions, suggesting that they may have been made with a primitive form of movable type.[4] This technology is distinguished from the other preserved records of ancient Near East cultures made by pressing a stylus into wet clay, which still required on-the-fly composition and was not mass-reproducible.

Another early example of a "printed" work is the Phaistos Disk (Fig. 17.5), a small (15 cm diameter), flat, circular clay object found during excavation of the Minoan palace of Phaistos on Crete in 1908. The archaeological context of the object remains in dispute, but an origin in the middle to late Minoan Bronze Age (*c.* second millennium BC) is likely. Both sides of the Disk bear a spiral inscription of symbols evidently stamped into the wet clay. The language represented by the symbols is unknown and has been the subject of various decipherment efforts for over a century. The striking similarity in the form of individual symbols on the Disk strongly suggests production via some kind of movable type. As the German typographer Herbert Brekle put it,[5]

> If the disc is, as assumed, a textual representation, we are really dealing with a "printed" text, which fulfills all definitional criteria of the typographic principle. The spiral sequencing of the graphematical units, the fact that they are impressed in a clay disc (blind printing!) and not imprinted are merely possible technological variants of textual representation. The decisive factor is that the material "types" are proven to be repeatedly instantiated on the clay disc.

Similarly, the first coins, introduced simultaneously in Iron Age Anatolia, Archaic Greece, India and China in the seventh century BC, could be thought of as the first "printed" mass media. Master designs were incised into pieces of stone or metal, which served as either molds for casting coins or punches for minting them in the traditional sense. They yielded products that were highly uniform in appearance—a hallmark of printing.

The earliest mass printing technologies involving flexible cloth and paper substrates originated in East Asia. Woodblock printing appeared in Han Dynasty China. In this relief

---

[4]J. Marzahn, *Aramaic and Figural Stamp Impressions on Bricks of the Sixth Century* BC *from Babylon*. Harrassowitz Verlag: Wiesbaden (2010), p. 11, 20, and 160.

[5]"Das typographische Prinzip. Versuch einer Begriffsklärung". *Gutenberg-Jahrbuch*, Vol. 72, pp. 58–63 (1997).

17.5 Side "B" of the Phaistos Disk, a Bronze Age Minoan ceramic object imprinted with characters in a process described as an early form of movable type. Heraklion Archaeological Museum. Greece

printing technique the design is cut into a flat wooden surface, which is inked and pressed directly onto the printing medium; the parts of the wood that are not cut away receive the ink and constitute the design. The Chinese also invented movable type in the modern sense in the mid-eleventh century AD. Early type was made of porcelain but broke easily under repeated impressions; the method was improved in the following two centuries by making type first from carved wood and then hammered copper. Still, woodblock printing remained the dominant printing technology in China for many centuries.

The development of cast metal type later made famous in Europe occurred in early thirteenth century Korea using type cast in bronze using techniques originally invented to cast coins. Characters were cut as positive reliefs into slabs of beech wood, which were pressed into wet clay to make a mold; molten bronze was poured into the molds and allowed to cool and harden before the mold matrix was removed and the metal polished. Using this technique, the Koreans made the world's oldest extant work printed using movable metal type, a version of the Jikjisim Sutra produced in 1377. Knowledge of this process slowly diffused westward in the late fourteenth and early fifteenth centuries.

The woodcut process arrived in Europe first, around the thirteenth century. Early European block printing was done on fabric before paper became widely available after about 1400. The cheap printing medium fueled a strong market for small religious images and playing cards printed from blocks in the early fifteenth century. The technology also awakened an increasingly literate European audience to books, which for the first time could be mass-produced rather than limited strictly to bespoke, hand-lettered editions. "Block books" consisting of both text and images printed from woodcuts became highly popular bestsellers of their day in the half-century before the European invention of cast metal moveable type.

The ultimate limitation of woodblock printing paved the way for moveable type: despite their suitability for mass printing, woodblocks were still custom individual works whose elements could not be repurposed once a block was cut. They were durable but absolutely fixed once carved, and their production was a highly labor-intensive process. The next major advance in European printing involved mass producing the printing elements themselves, a step that would prove to change the trajectory of world history.

Johannes Gutenberg

It should be noted that Bode's constellation celebrates not simply the process of printing, but specifically of an innovation that permitted the rise of both typography and the ability to mass-produce ideas. The invention of moveable type in Europe is attributed to Johannes Gensfleisch zur Laden zum Gutenberg[6] (*c.* 1398–1468), the son of a German goldsmith who adapted metal smithing techniques to the production of cast metal type in the mid-fifteenth century. Gutenberg was born at Mainz just before 1400 into the house of Gensfleisch, an established patrician family. He is now more famously known by the toponym "zum Gutenberg," indicating the name of the family home in Mainz.

Little is known of Gutenberg's formative years. His father worked in the archepiscopal mint, a job that led his descendants in following centuries to claim hereditary positions as mint masters. The Gensfleisch were driven out of Mainz in 1411 as the result of a local uprising against the patricians, resulting in the flight of over a hundred ancient families. Gutenberg's family relocated to Eltville am Rhein, across the Rhine border in Hesse, where Johannes may have studied at the University of Erfurt. By 1434 he was living in Strasbourg where lived some relatives on his mother's side of the family; there he plied his trade as a goldsmith and was enrolled in the local militia.

Gutenberg seems to have come up with the idea for moveable, cast metal type in around 1439–1440 as part of a scheme to bilk religious pilgrims visiting Aachen, an important city in the early history of the Holy Roman Empire. In 1439 the city planned to exhibit relics of Charlemagne there, but a severe flood that year caused the exhibit to be called off.

[6]It has been suggested that Laurens Janszoon Coster (*c.* 1370–*c.* 1440) may have invented cast metal movable type in Holland as early as the 1420s, but the documentary evidence is much less well established than that favoring Gutenberg.

Gutenberg was at the time engaged in a financial venture that involved manufacturing and selling polished metal mirrors to pilgrims coming to the Aachen exhibition; these mirrors were believed to capture "holy" light from religious relics. After the Charlemagne exhibit was canceled, the city was left to deal with the matter of repaying investors who had fronted the planning costs. Gutenberg intimated that he had a revolutionary moneymaking secret to share, which may have been the financial promise of mass-produced books enabled by moveable type printing.

Around this time Gutenberg was evidently working on the technology in Strasbourg, although there is no specific, surviving evidence for this in the way of printing trials. Regardless of when he perfected the process, the historical record is silent for several years until 1448 when Gutenberg was back in Mainz. There he obtained a loan from his brother-in-law, Arnold Gelthus, which may well have been intended to finance the purchase of a press or associated implements.

Gutenberg's idea was evidently the first instance in Europe in which pieces of moveable, reusable type were cast from metal, yielding durable materials for producing inked impressions of letters on vellum and paper. His process involved making duplicates of master characters carved from hard metal *punches*; the punches were hammered into a soft copper substrate, creating a *matrix*. The copper matrix became a mold into which was poured molten type metal, a material made of an alloy of lead, tin and antimony. Varying the alloy contents yielded type with different physical properties, each suitable for one or more typecasting methods. When the liquid type metal cooled, the resulting *sort* was easily removed from the matrix. Eventually sorts became deformed by metal flow under the pressure of repeated impressions on the press, but with a saved matrix for each they could be recast on demand. The sorts could be reused in any desired combination, resulting in literally "moveable" type. Type cast in this way results in printed works that are highly uniform in their appearance relative to hand-lettered manuscripts where one often finds considerable variation from page to page and copy to copy. The uniformity of type produced using Gutenberg's process resulted in the emergence of distinct typefaces enabling better identification of manuscripts that bore no printer's marks—like Gutenberg's.

While Gutenberg is most famous for his 42-line Bible (Fig. 17.6), considered to be the first complete book printed with the moveable type method, his first endeavor was the printing of a simple German poem in 1450. He used this as a proof of concept to entice potential investors, and soon attracted the interest of the wealthy local financier Johann Fust who funded Gutenberg to the tune of 800 guilders. Shortly thereafter Gutenberg took on Peter Schöffer, later Fust's son-in-law, a former scribe in Paris who likely designed Gutenberg's first typefaces.

The idea for the monumental Bible project came about around 1450 and it was complete by 1455. Gutenberg borrowed a further 800 guilders from Fust to finance the Bible, hoping for a very large return on investment given the value of hand-lettered Bibles for ecclesiastical use. Work on the project began in 1452 while Gutenberg continued printing other, more popular (and less-expensive texts) like Latin grammars. During the same time it is thought that Gutenberg supplemented his book-publishing income by printing indulgences for the Church.

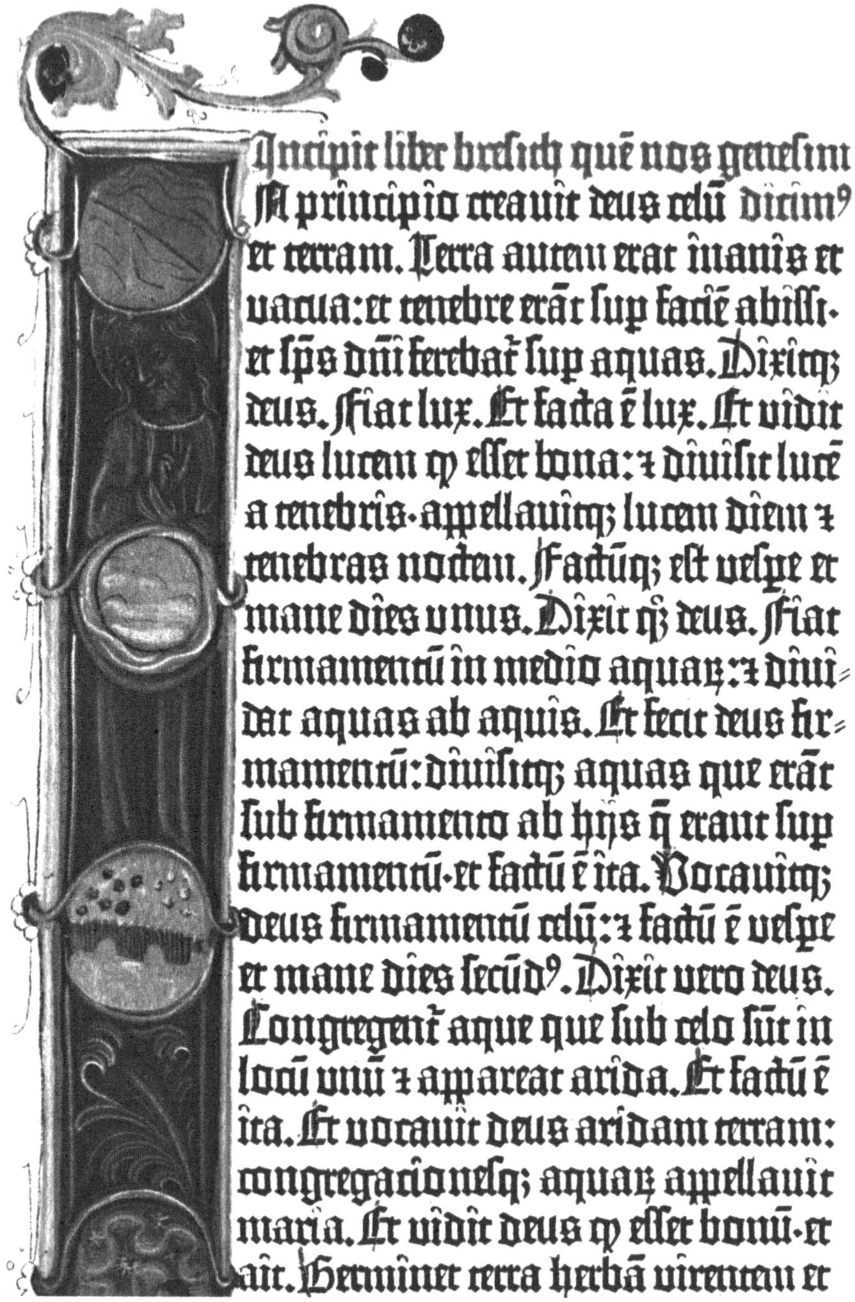

Incipit liber bresith quē nos genesim
In principio creauit deus celū dicim⁹
et terram. Terra autem erat inanis et
uacua: et tenebre erāt sup faciē abissi·
et sp̄s dn̄i ferebat̄ sup aquas. Dixitq;
deus. Fiat lux. Et facta ē lux. Et uidit
deus lucem q̄d esset bona: ⁊ diuisit lucē
a tenebris· appellauitq; lucem diem ⁊
tenebras noctem. Factūq; est uespe et
mane dies unus. Dixit q̄ȝ deus. Fiat
firmamentū in medio aquaꝝ: ⁊ diui=
dat aquas ab aquis. Et fecit deus fir=
mamentū: diuisitq; aquas que erāt
sub firmamento ab hiis q̄ erant sup
firmamentū· et factū ē ita. Vocauitq;
deus firmamentū celū: ⁊ factū ē uespe
et mane dies secūd⁹. Dixit uero deus.
Congregent̄ aque que sub celo sūt in
locū unū ⁊ appareat arida. Et factū ē
ita. Et uocauit deus aridam terram:
congregacionesq; aquaꝝ appellauit
maria. Et uidit deus q̄d esset bonū· et
ait. Germinet terra herbā uirentem et

17.6 The incipit of the Book of Genesis from Johannes Gutenberg's 42-line Bible of *c.* 1455 (Folio 5 recto, volume 1; Staatsbibliothek, Berlin), showing the uniformity of the type produced by Gutenberg's method. This copy was hand-illuminated after printing

The Bible project resulted in roughly 180 complete copies of the 42-line Bible; most copies were printed on paper, but a few received vellum editions. Some copies were illuminated and/or rubricated by hand, such as the example in Fig. 17.6, giving them a look more like their hand-lettered predecessors. The completed Bibles sold for 30 florins each, several times the average salary of a typical clerk of the era, but far less than the cost of a manuscript Bible.

The printing enterprise ultimately landed Gutenberg in legal hot water. Fust accused Gutenberg of misusing funds from the sale of the Bibles and demanded his money back, but Gutenberg was by then in a terrible financial bind due to the high cost of producing them. His debt then exceeded 20,000 guilders, far in excess of what profits from sale of the books could have yielded. Fust sued Gutenberg in the archbishop's court and won a judgment granting him half the Bibles printed to date and control over Gutenberg's printing workshop.

Absent a stream of income from his printing work, Gutenberg was left essentially bankrupt. However, he took the knowledge of his process to a shop in the town of Bamberg, where he participated in the printing of another Bible around 1459. Authenticating printed works said to have come from his shops is difficult because of a lack of printer's mark, but various editions of books are attributed to his hand based on similarity of production and typeface style. Meanwhile, his former shop—now controlled by Fust and Scöffer—pioneered printer's marks with their Mainz Psalter (1457).

Gutenberg continued to publish but tumult and controversy followed him for the rest of his life. He was exiled from Mainz in 1462 in the aftermath of the city's defeat at the hands of forces loyal to archbishop Adolph von Nassau. By then in his early sixties, Gutenberg returned to Eltville where he was eventually reconciled to von Nassau and awarded the title of Hofmann in 1465 for his achievements in printing. The honor came with a stipend and other benefits, but they were insufficient to account for Gutenberg's outstanding debts. He may have ultimately returned to Mainz at the end of his life, where he was buried after his death in 1468. The church and cemetery in which he was buried were later destroyed, and Gutenberg's grave is now lost.

Gutenberg's type casting process was utterly revolutionary, enabling the manufacture and distribution of vast numbers of sacred and secular works. In the century after he lived, moveable type would prove indispensable in dissemination of ideas during the Protestant Reformation. Knowledge and learning were now easily communicable, beginning changes whose effects persist in human culture to this day. It is worth remembering, however, that his method was incredibly manually intensive; setting and printing a single page of the 42-line Bible could take half a day of work in a shop that may have employed as many as 25 workmen. It was only through later innovations that the technological challenges associated with large-scale printing were addressed and the power of the printed word fully unleashed.

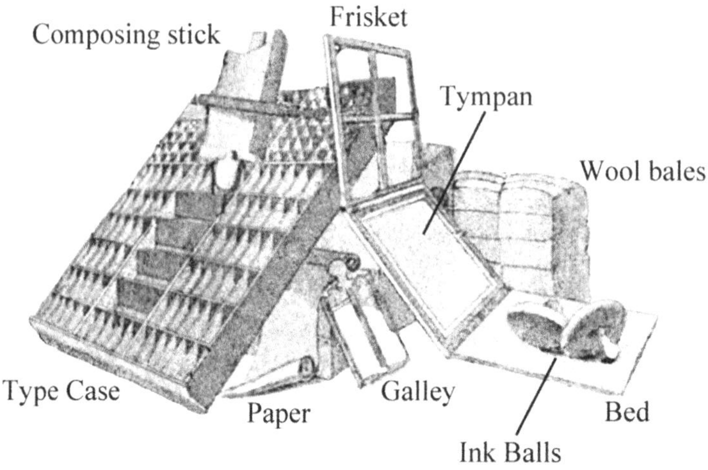

17.7 The figure of Officina Typographica from Plate 18 of Johann Elert Bode's *Uranographia* (1801b) with the constituent elements labeled

Printing Elements Shown by Bode

The composition of Bode's figure includes items dictated by the process of letterpress printing, the dominant paper printing technology in Bode's time. The elements of the printing process were relatively unchanged since Gutenberg's time, and would have been familiar to anyone who visited a printer's shop until the nineteenth century, when mechanical processes gradually displaced manual methods. To recognize the elements in Bode's design, labeled in Fig. 17.7, one must first understand how the manual letterpress process worked. The method consisted of four distinct steps: composition, imposition and lock-up, and printing.

Composition

Also known as typesetting, composition is the step in which individual pieces of moveable type are "composed," or assembled, to form the text to be printed. Manual composition is the responsibility of the *compositor*, who selects individual letters from a **type case** where the cast metal type is stored. The traditional arrangement of type cases kept capital letters in a separate tray above the minuscule letters, leading to the English terms "upper case" and "lower case" referring to the two letter sizes. The compositor uses a **composing stick**, usually a wooden frame with a sliding guide that held the pieces of type in place during typesetting. Working from left to right and placing the letters upside-down, the compositor set several lines of text before transferring those lines to a type **galley**.

Imposition

When an entire galley corresponding to one complete page of text is formed, its contents are "imposed," or rendered safely fixed, for transfer to the press. Imposition is performed by a *stoneman*, so named because this work is traditionally carried out atop a large *imposition stone*. The stoneman's task is to arrange the individual galleys of type relative to each other in such as way as to make most efficient use of printing **paper**. Flat pieces of wood or metal *furniture* are added to the blank spaces to prevent the type blocks from moving during printing. The stoneman makes use of a mallet to wedge the type and furniture in place, assuring a uniformly flat printing surface, and binds the pieces into place using a set of strings or cords into a *chase*, or frame. Depending on its size, several galleys' worth of text could be printed at once onto a single sheet, so the layout process of imposition is crucial so as to not waste materials.

Lock-Up

The final step before printing fixes all the pieces (type, blocks, furniture and chase) into place, creating a *forme*. This is done by turning a set of *quoins*, expandable wooden or metal blocks, using keys or levers to "lock" the forme. The printer checks the forme for typesetting errors before preparing it for printing, and ensures that the pages are oriented to face the correct direction and retain the proper margins.

Printing

A manual letterpress generally requires two operators: one person to ink the type and the other to physically actuate the press itself (Fig. 17.8). Inking is performed using a set of **ink balls** attached to handles. The ink balls were traditionally made of sheep's wool stuffed into a thin leather casing; the wool retained the ink like a sponge, keeping the leather surface wet. The inker places ink on one of the two balls, then presses it against the other and works the ink around until the desired degree of consistency and uniformity is achieved. The ink is then applied to the type by "beating" the ink balls against it, taking care to avoid either too little ink, leading to an incomplete impression, or too much, which resulted in bleeding on the paper.

A sheet of paper is loaded into a three-part device used to hold it fixed during printing. Slightly damp paper is used as it is more flexible than dry paper; this allows the inked type to more efficiently "bite" into the paper and makes for better impressions. The paper is carefully pinned in place on the **tympan** and is covered by a windowed frame called a **frisket**, consisting of one opening per page being printed to the sheet. The printing **bed** holds the formes to be printed. The frisket and tympan are folded over into place atop the bed, and the entire assembly is fed into the press. The press operator turns a wheel or lever, actuating a large screw that pushes a heavy wooden or metal *platen* against the frisket-tympan-bed assembly, creating the inked impression on the paper.

17.8 Printers operating a manual letterpress in a mid-sixteenth century woodcut. At *center-right*, a "beater" applies ink to type set in a forme, while to his right a "puller" removes a freshly-printed sheet from the tympan. In the background, two compositors work before angled type cases

The assembly is removed from the press and disassembled; the "puller" carefully peels the paper away from the type in order to prevent smearing of the wet ink and hangs it up to dry. The dry sheets are then folded and/or cut and trimmed to the size appropriate for the desired dimensions of the final printed product.

Bode shows all of the important parts of this process in his figure of Officina Typographica except for one essential piece: the press itself! It is unclear why he left it out, although in *Allgemeine Beschreibung und Nachweisung der Gestirne* he wrote that it consists of "a portion of the printing press," evidently referring to the frisket, tympan and bed. Instead of a complete representation of the contents of a traditional printer's workshop Bode gives us a still life of the printer's accessories, as if one might find all these pieces casually arranged in the corner somewhere.

Bode's design influenced many other nineteenth century mapmakers, most of who simply adopted his design down to the details. There are a few exceptions; Alexander Jamieson added the press itself in his *Celestial Atlas* (1822), positioning the bed on the press and showing the tympan and frisket unfolded above it in space. He further provided a more realistic depiction of a type case as they were typically used, suspended at an angle on a wooden frame. In this way, the compositor could sit in front of the type case and more readily see the entirety of its contents while composing. However, Jamieson dispenses with the other printing accoutrements shown in Bode's 1801 figure.

DISAPPEARANCE

As with many of Bode's suggestions, the popularity of Officina Typographica peaked in the mid-nineteenth century and began a slow decline. Kendall (1845) gave a one-sentence description of Officina, writing merely that it was "introduced by Bode," while Chambers (1877) attributed it to "Bode's maps." Neither cited a date. Among the last mentions in a popular astronomy text of the era was in the *Poole Bros. Celestial Handbook* (1892) in which it was attributed to "Bode, 1798, A.D." Richard Hinckley Allen wrote that it appeared on a contemporary map, but the reference is lost:

> Officina Typographica, the Printing Office, was formed by Bode—at all events, first published by him—from stars immediately east of Sirius; but it is seldom found on the maps of our day, nor recognized by astronomers, although Father Secchi[7] inserted it on his planisphere of 1878.

Ridpath (1989) suggested that the chief reason for the eventual disuse of Officina Typographica and Globus Aerostaticus had to do with their identity with specific nations: "Although subsequently shown on many popular maps they were not eventually accepted, perhaps because the motives for their invention were too overtly nationalistic." Most authors and cartographers who elected to not include Officina Typographica on their maps simply left its stars among those of northern Puppis, as they had been considered as long ago as the time of the *Almagest*. At the adoption of the modern constellation boundaries in 1928 this definition was made permanent (Fig. 17.9).

---

[7]Father Pietro Angelo Secchi, S.J. (1818–1878) was an Italian astronomer who served as Director of the Observatory at the Pontifical Gregorian University in Rome from 1850 until his death.

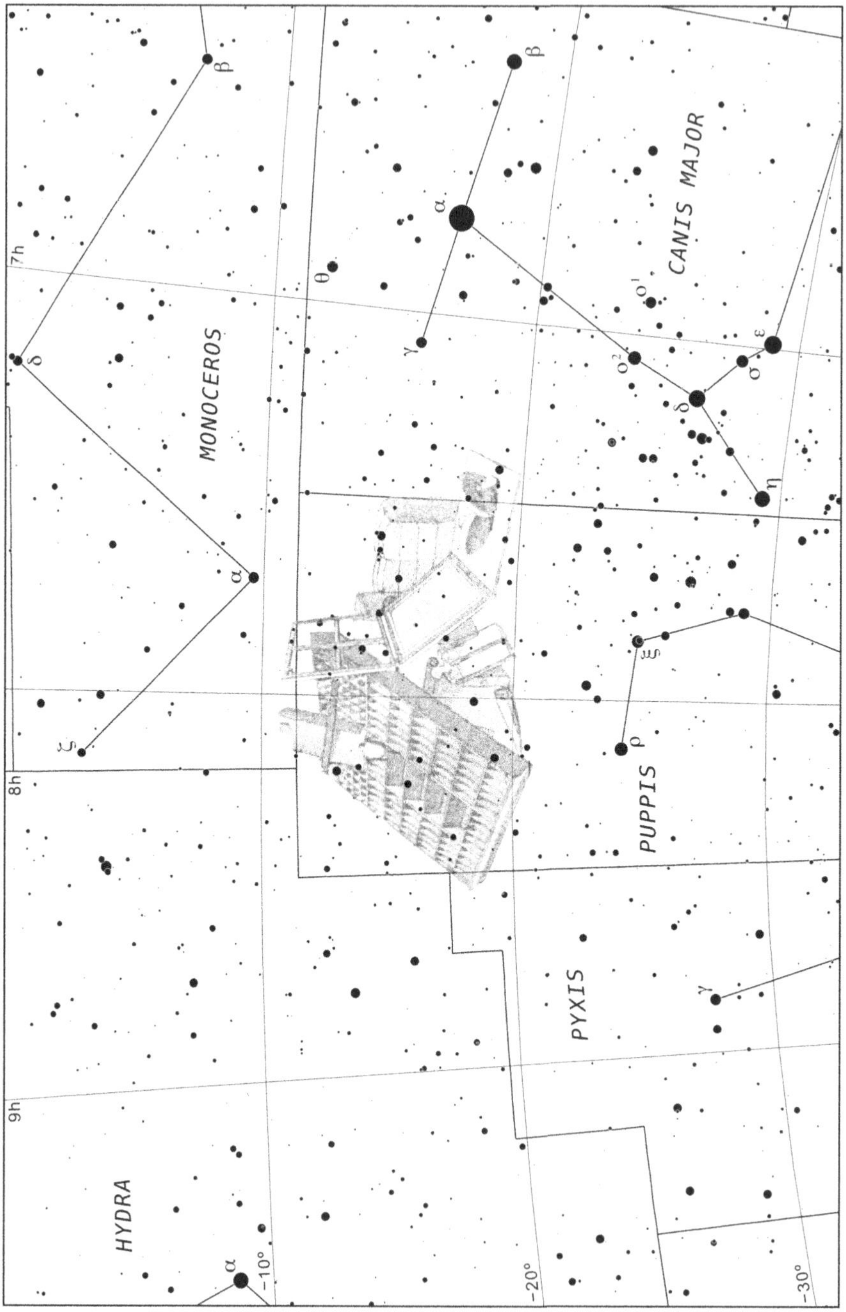

17.9 The figure of Officina Typographica from Plate 18 of Bode's *Uranographia* (1801b) overlaid on a modern chart

18

Psalterium Georgianum

King George's Harp

Genitive: Psalterii Georgiani
Abbreviation: PsG
Alternate names: "Psalterium Georgii" (Jamieson, 1822; Burritt, 1835a; Middleton, 1842); "Harpa Georgii" (Bode, 1801b; Colas, 1892); "Georgsharfe" (Meissner, 1805); "K. George's Harp" (Young, 1807); "Harfe" (Riedig, 1849); "Harpe de George" (Preyssinger, 1862)
Location: Several fourth-magnitude stars in now in northern Eridanus and extreme southern Taurus, principally defined by 10 Tau, 17 Eri, 24 Eri, 30 Eri, 32 Eri, and 35 Eri[1]

ORIGIN AND HISTORY

This constellation was introduced in 1789 by the Hungarian abbot Maximilian Hell, then the director of the Vienna Observatory, and published that year in the annual *Ephemerides astronomicae* for 1790 (Fig. 18.1). Hell intended to honor King George III of Great Britain, patron of the musician and astronomer William Herschel who discovered the planet Uranus in 1781 (see Chap. 26). Hell made use of some unformed stars north of a bend in Eridanus which are sufficiently south of the classical figure of Taurus to avoid being considered part of the Bull. Richard Hinckley Allen summarized its origin in *Star Names: Their Lore And Meaning* (1899), incorrectly identifying both the year of the invention and the intended recipient of its glory:

[1] *"[S]outh of the Bull and east of the Whale"* (Bode, 1801a); *"Bounded by Taurus, Cetus, and Eridanus"* (Green, 1824); *"Between Taurus and Eridanus, eastwardly from the head of the Whale"* (Kendall, 1845); *"Between the paws of the Whale and the feet of Taurus"* (Bouvier, 1858); *"Between the fore feet of Taurus and the River Eridanus"* (Allen, 1899); *"Between the forefeet of Taurus and Eridanus"* (Bakich, 1995).

J.C. Barentine, *The Lost Constellations*, Springer Praxis Books,
DOI 10.1007/978-3-319-22795-5_18

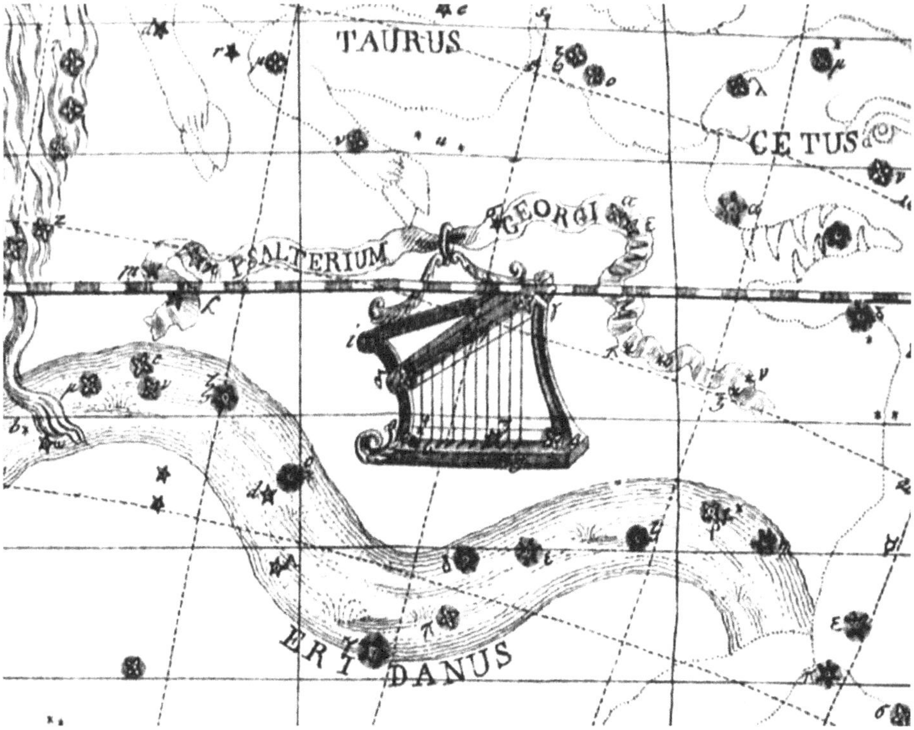

18.1 Psalterium Georgianum as introduced by Maximilian Hell in *Ephemerides astronomicae* for 1790

> Psalterium Georgii or Georgianum, sometimes Harpa Georgii, was formed in 1781 by the Abbé Maximilian Hell, and named in honor of King George II of England. On the Stieler Planisphere it is Georg's Harfe, from Bode's Georgs Harffe. It lies between the fore feet of Taurus and the River Eridanus, its stars all very inconspicuous, unless it be the $4\frac{1}{2}$-magnitude o^2 Eridani, which was borrowed for its formation.[2]

Johann Elert Bode called it "Harpa Georgii" in *Allgemeine Beschreibung und Nachweisung der Gestirne* (1801a) and described[3] Hell's motivation to create the new figure:

[2]This is an erroneous claim. See later in the text for an explanation.

[3]Diese Harse hat der Abt Hell dem jetzigen König von Grossbritannien Georg III zu Ehren den Himmel gebracht. Sie steht sudlich underhalb dem Stier östlich vor dem Wallfisch, und ist aus Sternen vierter Grösse zufammengesetzt, die sonst zum Eridanfluss gehörten.

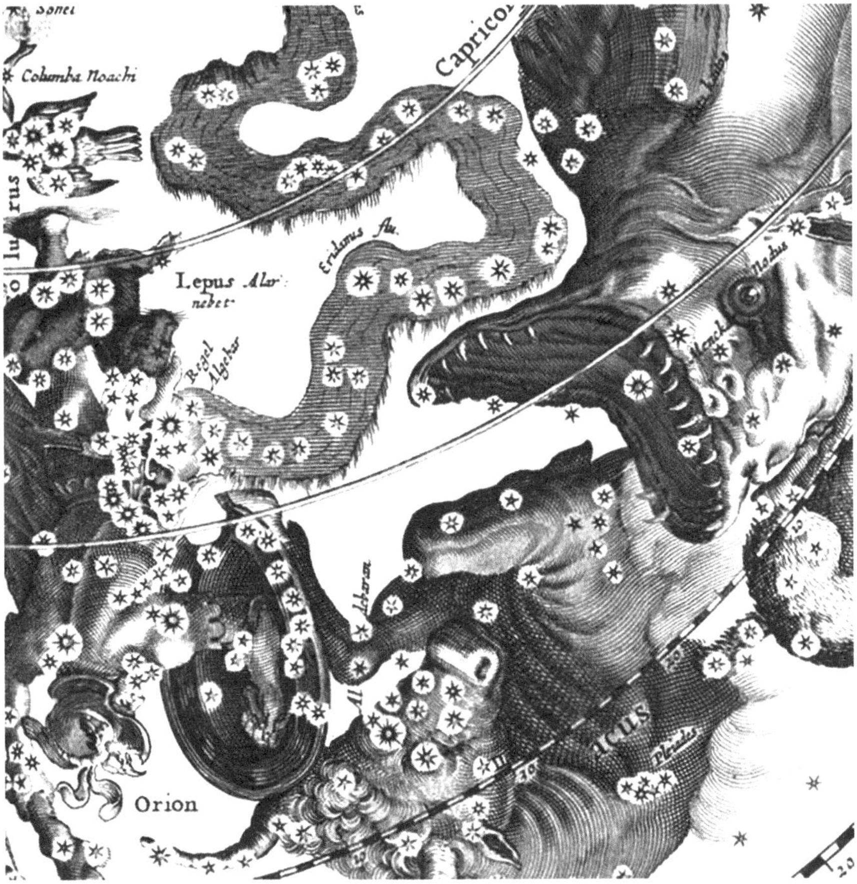

18.2 Stars that would later become part of Psalterium Georgianum are shown here on Plate 27 ("Hæmisphærium Australe") in Andrea Cellarius' *Harmonia Macrocosmica* (1661). The stars define the lower part of the Whale's gaping maw, slightly above center in this excerpted view

> This harp has been placed in the sky to honor the present King of Great Britain, *George III.*, by Abbé Hell. It is south of the Bull and east of the Whale, and is composed of fourth magnitude stars that otherwise belonged to Eridanus.

Prior to Hell, the stars were shown by various cartographers as either part of Cetus (Cellarius, 1661; Fig. 18.2), incorporated into decorative flourishes in the figure of Eridanus (Bayer, 1603; Fig. 18.3), or simply left unformed (Hevelius, 1690; Flamsteed, 1729; Bode, in the first edition of *Vorstellung der*, Gestirne, 1972, top image in Fig. 18.5).

Exactly how the figure of Psalterium Georgianum was drawn is important to understand which cartographers considered which stars part of it. There are three distinct representations of the Harp: (1) the psaltery shown by Hell (1789); (2) the upright harp

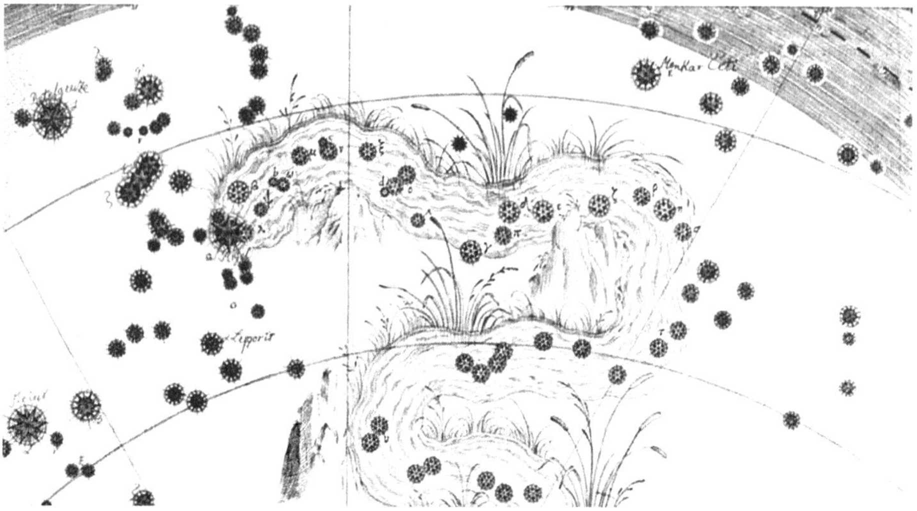

18.3 The northern reaches of Eridanus shown on Plate Mm of Johannes Bayer's *Uranometria* (1603). Two stars later incorporated into Psalterium Georgianum by Abbot Hell appear at top-center among some decorative reeds along the River's bank; they are now designated 10 Tauri (*rightmost of the two*) and 32 Eridani (*leftmost*)

without a figurehead ornament first drawn in Christian Goldbach's *Neuester Himmels* (1799); and (3) the upright harp with the figurehead ornament that first appeared in Jamieson (1822). Hell's concept of the harp (Fig. 18.1) most closely matches the ancient psaltery (see Iconography) and bears little resemblance to the upright harps depicted by Bode and other later cartographers. His design included a ribbon device bearing the constellation's name from which the psaltery was suspended, another feature disregarded by others.

Goldbach drew an upright harp similar in design to those that emerged during the Middle Ages. In *Neuester Himmels* (1799) it is tilted heavily forward, leaning east (Fig. 18.4, top); Goldbach anchored it to the River by borrowing ζ Eridani to make one of its stars. On the other hand, Bode deliberately drew the course taken by Eridanus around the end of his harp, carefully avoiding the incorporation of either ϵ or ζ Eridani into its design. As a result, Bode somewhat artificially carved out part of the River's traditional course into which to fit the base of the Harp; this seems to have been done with the intent of preserving certain proportions of the figure as a matter of convenience, since no bright star happens to be found in the gap between δ and ϵ Eridani. However, when Psalterium Georgianum was added to the 1805 edition of *Vorstellung der Gestirne*, as seen in the bottom panel of Fig. 18.5, the result is quite different. Here the Harp tilts somewhat less, but its base clearly overlaps Eridanus and *both* figures appear to lay claim to ϵ Eridani. This seems to have been done to take advantage of a few faint stars south of ϵ that curve gently to the east, tracing perfectly the ornate, curving base of the Harp as drawn.

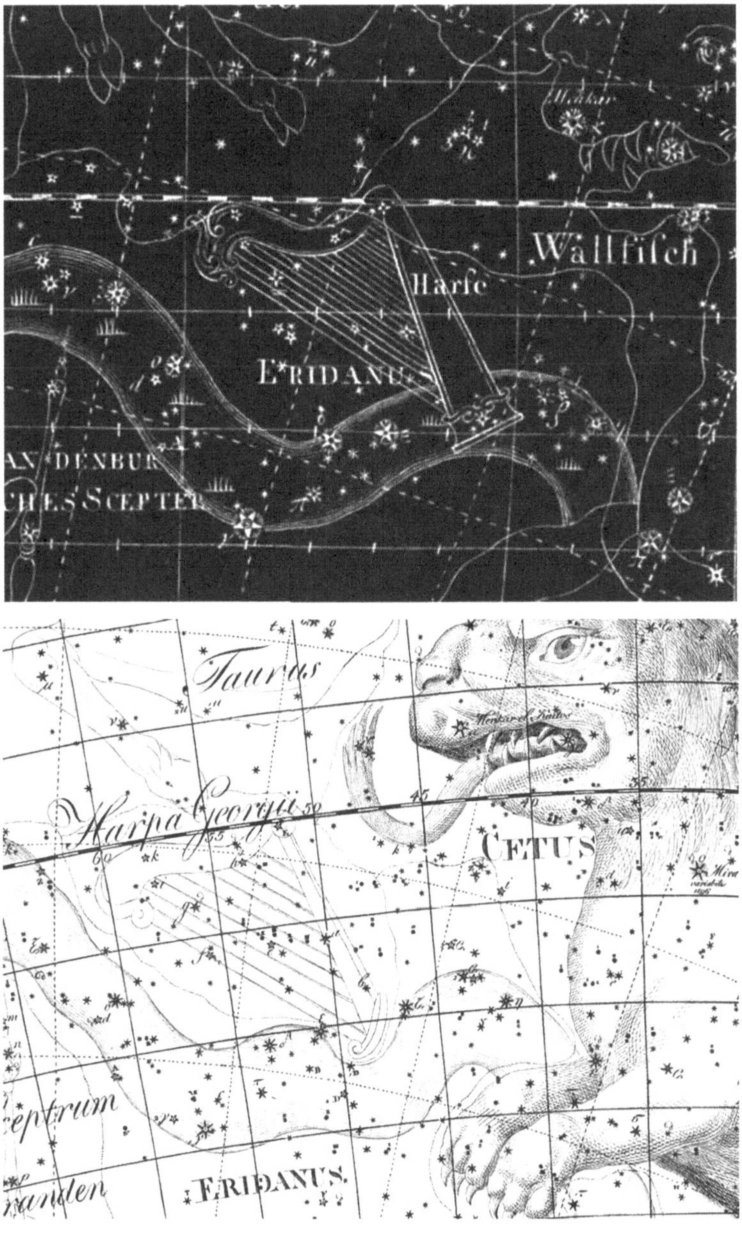

18.4 Two views of the Psalterium Georgianum design as an upright harp without figurehead ornament. *Top*: Plate 24 of Christian Goldbach's *Neuester Himmels* (1799) as "Harfe." *Bottom*: Plate 17 of Johann Elert Bode's *Uanographia* (1801b) as "Harpa Georgii". Note that Goldbach included ζ Eridani in the base of his figure, whereas Bode deliberately drew the course of Eridanus around the end of his harp so as to not incorporate any of its stars

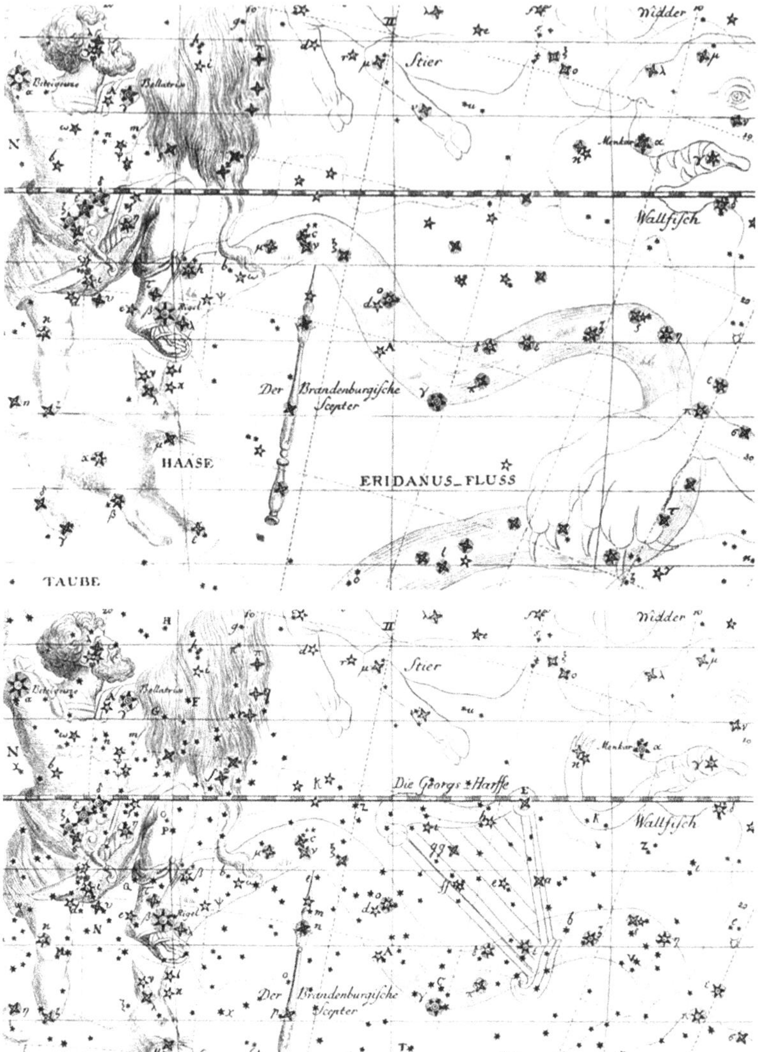

18.5 Before and after: the region between Orion (*left*) and Cetus ("Wallfisch," *upper right*) shown on Plate 24 in two editions of Bode's *Vorstellung der Gestirne*: 1782 (*top*) and 1805 (*bottom*). In the intervening 23 years Hell created Psalterium Georgianum, which Bode adopted in *Uranographia*. Note that the latter edition of *Vorstellung* included considerably more, fainter stars. The extinct constellation Sceptrum Brandenburgicum (Chap. 24) appears below center-left as "Die Brandenburgische Scepter"

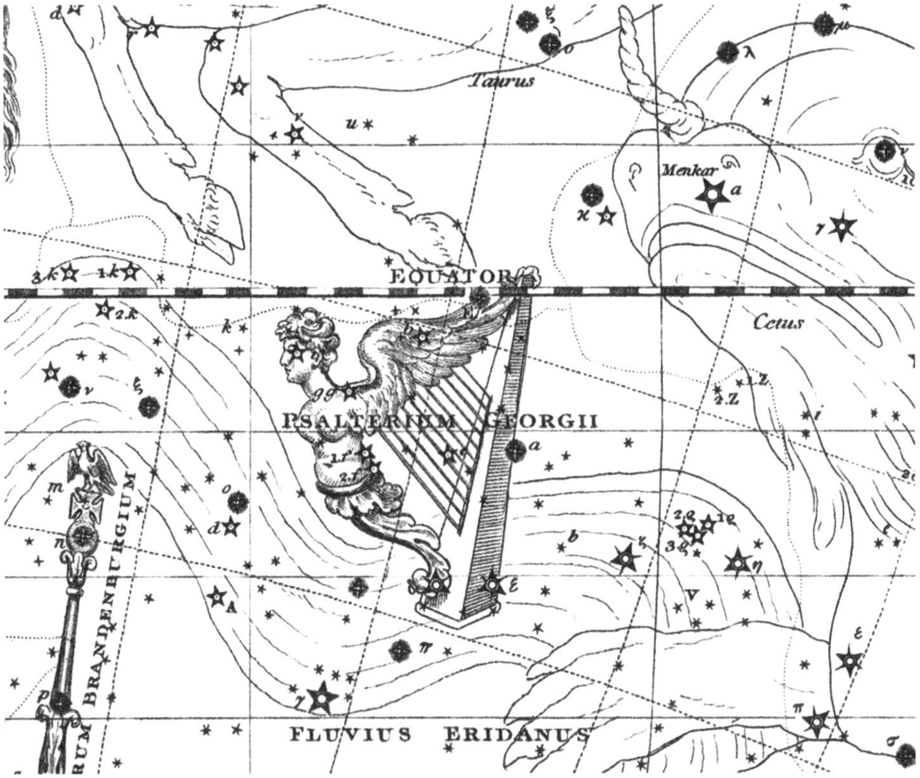

18.6 Psalterium Georgianum received a more artistic treatment on Plate 24 of Alexander Jamieson's *Celestial Atlas* (1822). Sceptrum Brandenburgicum (Chap. 24) appears at lower left as "Sceptrum Brandenburgium"

Alexander Jamieson took a distinctly different approach to depicting Psalterium Georgianum as "Psalterium Georgii" on Plate 24 of his *Celestial atlas* (1822), shown here in Fig. 18.6. His rendition of the harp placed a winged anthropomorphic figure like a ship's figurehead suspending the strings above the soundbox, and he oriented the whole device in a more upright manner with respect to the cardinal directions on the sky. The base of his figure distinctly took in both δ *and* ϵ Eridani, while otherwise he identified essentially the same stars as Bode. The artist responsible for *Urania's Mirror* (1825) followed this prescription exactly on Plate 28 of the card set, drawing a Jamieson-esque figure that also clearly took δ Eridani and ϵ Eridani for the Harp's base (Fig. 18.7). Bakich (1995) wrote that Psalterium's "brightest star is o^2 Eri," but this is not correct. There are no maps in which the figural depiction of the Harp extends sufficiently far enough east to include either o^1 or o^2 Eridani.

The three broad design types led to different identifications of stars belonging to Psalterium Georgianum, as listed in Table 18.1. For the harp without figurehead, Bode's

Table 18.1 Comparison of principal stars in Psalterium Georgianum identified by three cartographers (Hell 1789, Bode 1801 and Jamieson 1822) along with the Bayer, Flamsteed, or HD catalog equivalent. In each source, stars are listed in order of increasing right ascension.

Hell	*Equivalent*	*Bode*	*Equivalent*	*Jamieson*	*Equivalent*
ν	5 Eri	b	HD 21019	a	17 Eri
ξ	7 Eri	a	17 Eri	ϵ	ϵ Eri (shared)
o	94 Cet	E	10 Tau	E	10 Tau
π	95 Cet	e	22 Eri	δ	δ Eri (shared)
ϵ	κ^1 Cet	h	24 Eri	b	24 Eri
α	κ^2 Cet	f	30 Eri	1.f	HD 24371?
β	17 Eri	g	32 Eri	2.f	30 Eri?
γ	10 Tau	i	35 Eri	gg	32 Eri
σ	12 Tau	k	HD 25457	i	35 Eri
ζ	22 Eri			k	?
η	24 Eri				
ϕ	HD 24371				
δ	32 Eri				
ι	35 Eri				
κ	HD 28375				
λ	45 Eri				
μ	HD 29335				

Uranographia is used as the source because Goldbach did not label any of its stars. Of the listed sources, five stars (10 Tau, 17 Eri, 24 Eri, 32 Eri and 35 Eri) received labels in all three charts. Both Allen and Backich claim that Hell included o^2 Eridani among the stars of Psalterium Georgianum, but this is clearly not correct; see Fig. 18.1.

The American cartographer Jacob Green had little time for recently-invented constellations like Psalterium Georgianum, considering them contrivances for the greater glorification of their creators rather than useful markers for delineating the construction of the night sky. In this case, he didn't even mention the inventor by name in his *Astronomical Recreations* (1824):

> HARPA GEORGII, or the Harp of George, is a new constellation introduced on the maps by one of the German astronomers, in honor of the late king of England, George III. It is composed of a few very small stars, taken from Eridanus the neighbouring group. It is bounded by Taurus, Cetus, and Eridanus.

To decades later, Ezra Otis Kendall mentioned the constellation in *Uranography* (1845) but got the year of its introduction wrong:

> *The Harp*, between Taurus and Eridanus, eastwardly from the head of the Whale, was established by Hell in the year 1791, in honour of King George III. of England.

In her *Familiar Astronomy* (1858), Hannah M. Bouvier attributed the dedication of the Harp to the wrong King George:

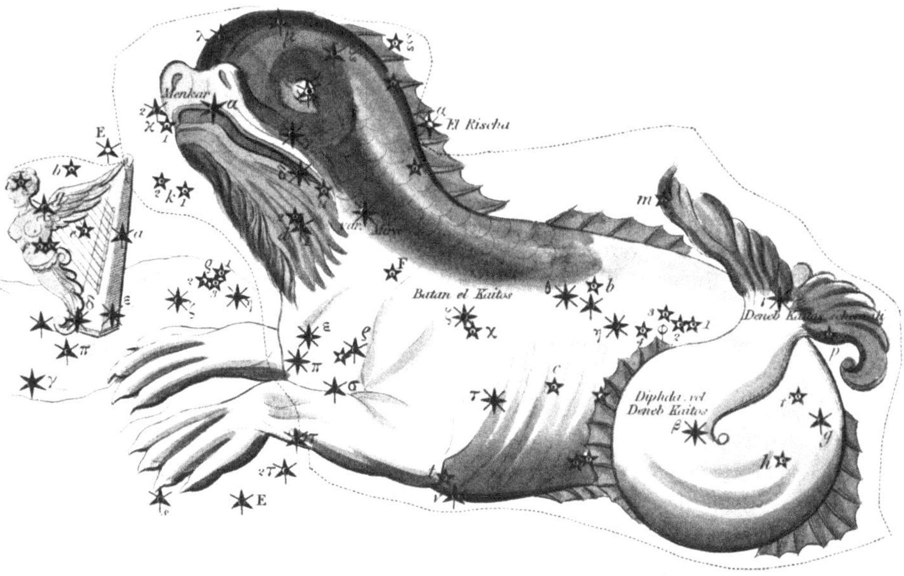

18.7 Psalterium Georgianum (*left*) shown in proximity to Cetus (the Whale) on Plate 28 of *Urania's Mirror* (1825)

Q. How is the constellation Psalterium Georgianum REPRESENTED*?*
A. By a harp, which is situated between the paws of the Whale and the feet of Taurus.

Cetus, or the Whale, is represented on the map as a monster with a tail like a fish, but having *paws* somewhat resembling those of a dog.

Q. Does this constellation contain any CONSPICUOUS *stars?*
A. It does *not*. There is one star in this constellation, called *Reid,*[4] which comes to the meridian with the Pleiades ; it is a star of the fifth magnitude.

This is one of the new constellations, and is not acknowledge by all astronomers. It was formed by Maximilian Hell, a Hungarian astronomer, in honor of George II. of England.

[4]"Reid" appears to be a corruption of either Beid or Keid, proper names for o^1 and o^2 Eridani, respectively. In any case, neither star was a constituent of any variation on Hell's constellation published in the nineteenth century.

ICONOGRAPHY

Hell's Psaltery

Hell's choice of both label and imagery for his invention clearly indicates that he had a specific type of stringed instrument in mind: a psaltery. It is an instrument whose history extends back as far as ancient Greece where it was called ψαλτήριον ("stringed instrument"), a word that derives from the verb ψαλλω ("to pluck"). In the context of musical instruments, this verb specifically indicated a style of playing characterized by plucking with the fingers rather than a plectrum. As an example of the later use of the term, the translators of the Authorised (King James) Version of the Bible used the word "psaltery" to translate several ancient Hebrew and Aramaic words whose meanings are otherwise imprecisely known.[5]

The ancient psaltery had no soundboard or other resonant chamber, rather consisting of several strings suspended within a frame and is more closely related to the lyre than the modern harp (Fig. 18.8). In the early Christian era, the basic design of the psaltery was modified by adding a soundboard and standardizing the tuning of the strings. This version was also known as a canon from the Greek κανών, ("rule" or "principle"); another way of translating the Greek is "mode," referring to a set of mathematical principles that determine the note value of a particular string. This tradition is said to extend back to the time of Pythagoras, whose followers applied the idea of musical consonance derived from various ratios of string lengths such as 2:1 (an octave), 3:2 (a fifth), 4:3 (a fourth), and 9:8 (a whole tone). They built versions of the canon called a monochord, consisting of a soundboard with two bridges between which is stretched a string; a third bridge divides the strings into two parts corresponding to different ratios to achieve the intonation of their system of music.

The psaltery remained in wide use through the Medieval period, seen frequently in art throughout Europe (e.g., Fig. 18.9). Their design and details vary considerably, including their shape, size and number of strings.

Later Versions of the Harp

In his second edition re-issue of John Flamsteed's *Atlas Coelestis* (1795), Jean Nicolas Fortin ignored Psalterium Georgianum altogether and left its stars unformed. When Christian Goldbach published his *Neuester Himmels* at Weimar in 1799, his evidently became the first major atlas to show Hell's new figure (Fig. 18.4, top). However, his figure bears no resemblance to Hell's as shown in *Ephemerides astronomicae*: Goldbach's harp looks like a modern concert harp, consistent with the type of instrument that would have been familiar to him.

---

[5]These instances include the Hebrew words transliterated as *keli* (Psalms 71:22; I Chronicles 16:5) and *nevel* (I Samuel 10:5; 2 Samuel 6:5; I Kings 10:12; I Chronicles 13:8; 15:16, 20, 28; 25:1, 6; II Chronicles 5:12; 9:11; 20:28; 29:25; Nehemiah 12:27; Psalms 33:2; 57:6; 81:2; 92:3; 108:2; 144:9; and 150:3), and the Aramaic *pesanterin* (Daniel 3:5, 7, 10, and 15).

18.8 A seated woman playing a version of the psalterion called a τρίγωνον (*trigonon*; "three-cornered"). Red figure pelike from Anzi, Apulia, *c.* 320–310 BC

European harps developed proper sound boxes in the Medieval period, with strings strung directly from the angled box to the "arch," a structural element protruding from the top of the box. The arch is usually curved, as shown in Goldbach's figure, in order to assure that the strings remain equidistant despite their varying lengths. By Goldbach's time, the design of European harps typically included a "pillar," a third structural element spanning the ends of the arch and sound box to provide additional rigidity and prevent detuning of the strings under their own tension; however, Goldbach's figure appears to show no such element.

Goldbach's harp takes into account a largely different set of stars than those indicated by Hell, suggesting that Goldbach probably was unfamiliar with Hell's figure and knew of the constellation only on the basis of written descriptions. In order to avoid taking stars largely from Taurus and Cetus, he tipped the harp forward to the east, the curve of its arch

18.9 "King David with writers and musicians," a scene from the *Chronicle of the World* (*c.* 1240) by Rudolf von Ems (*c.* 1200–1254), an Austrian epic poet. King David (*left*) plays a harp, while the musician at lower right plays a psaltery. Tempera on parchment, 16.5 × 19.5 cm; Zentralbibliothek, Zürich

following more or less exactly the sinuous boundary between Taurus and Eridanus. Despite this placement, the point at which his harp's arch joins the sound box is marked by one star borrowed from Taurus (10 Tauri). Despite Goldbach's explanation of his sources in the preface of *Neuester Himmels*, he did not specifically mention his source for Psalterium Georgianum. Goldbach's design completely supplanted Hell's proper psaltery for at least the first decade of the nineteenth century and was shown in the atlases of Bode (1801b), Meissner (1805), and Young (1807); in a few cases, such as Riedig (1849), this simple design persisted as late as midcentury.

In his *Celestial Atlas* (1822), Alexander Jamieson showed the harp composed of nearly the same set of stars as in Goldbach and Bode, but he changed the aesthetic appearance of

18.10 *Winged Victory of Samothrace* (*c.* 200–190 BC). Marble; 2.4 m height. Louvre, Paris. Photo by Marie-Lan Nguyen (2007), used with permission

the harp itself (Fig. 18.6). He added a pillar consisting of woman's head and bare-breasted torso, adapting the arch to be a pair of wings, a design referring directly to traditional nautical figureheads.

Bow ornaments on ships date back at least as far as ancient Egypt and throughout history often took the form of animals whose attributes of speed, acute vision, grace and ferocity were conjured with carved or painted images. Female figureheads baring one or both breasts became popular in about the seventeenth century and are thought to reflect the superstitions of sailors who believed that their images could calm stormy seas. When shown with wings, the figureheads were thought to reference the Greek goddess Nike, a divine charioteer who flew around battlefields rewarding victorious forces with glory and fame.

The best known classical image of the goddess is the Winged Victory of Samothrace (Fig. 18.10), found on the Aegean island in 1863 and now on display at the Louvre in Paris. A masterpiece of Hellenistic sculpture dating approximately to the decade between 200 and 190 BC, the archaeological context in which it was found suggests it occupied a site in the Samothrace temple complex within view of a ship monument erected by the Macedonian king Demetrius I Poliorcetes (337–283 BC). It originally stood on a marble pedestal meant to symbolize the prow of a ship and represents Nike descending from the heavens to meet the returning, triumphant fleet. In later renditions as a figurehead, the winged figure was intended to lead seafaring forces to victory, her wings fluttering in the breeze ahead of the fast-moving ships of the era.

From Jamieson's time until cartographers began to leave the figure of Psalterium Georgianum off their charts around midcentury, most depictions took the form of his figurehead harp including those found in Burritt (1835a) and Middleton (1842). Curiously, Jamieson's design bears a remarkable resemblance to the coat of arms of Ireland (Fig. 18.11) in use during the period of the island's status as a client state of the Kingdom of England (1542–1707) and the Kingdom of Great Britain (1707–1800). Jamieson left no explanation for why he chose the symbol appearing on the royal standard of Ireland for his version of Hell's constellation; he himself had no obvious Irish connections, having been born on the Isle of Bute in Rothesay, western Scotland. However, as the constellation was understood to honor George III, it is no stretch to conclude that Jamieson took a straightforward approach to portraying a harp in specific connection to the nominal head of the Irish state.

As a subject of that monarch, and in consideration of the times, it may be the Jamieson had in mind to lend support for English hegemony over Ireland in the person of George III who was himself of German extraction. The roughly century and a half before the publication of Jamieson's atlas was a time of Protestant ascendancy in Irish history, setting the stage for the civil war that would engulf the island in the early twentieth century. By placing a symbol of royal power over Ireland in the sky, Jamieson seems to have clarified Hell's intent in the context of nineteenth century world politics. But did Hell himself have politics in mind when he first put his psaltery on a celestial map?

1789, the year in which Hell published Psalterium Georgianum, marked the centennial of the Glorious Revolution and the ultimate consolidation of English power in Ireland after the victory of the Protestant William III over the Catholic James II at the Battle of the Boyne in 1690. 1788–1789 also marked a regency crisis in England when George III first fell ill. In order to secure their kingdom in Ireland, the English allowed the island some degree of legislative independence in the form of the Constitution of 1782; however, executive control of Ireland remained firmly situated in London. The leader of the Irish parliament at the time, Henry Grattan (1746–1820), wanted to appoint the Prince of Wales (later George IV) as Regent of Ireland, but the King recovered from his illness before the regency could be enacted.

18.11 A representation of the armorial achievement of the Kingdom of Ireland (1542–1800) from Hugo Gerard Ströhl's *Heraldischer Atlas* (Stuttgart, 1899)

Another possibility is that in choosing a musical instrument Hell indirectly honored the very musical William Herschel; the figure may have served double duty being both a direct musical reference as well as a political one. Furthermore, George was a patron of science beyond his support for Herschel, perhaps earning him a place in the stars separate from his other achievements. In the end is left only speculation, because Hell himself evidently did not explain the reasons for his choice.

18.12 *King George III In Coronation Robes* (*c*. 1765) by Allan Ramsay (1713–1784), oil on canvas, 2.36 m × 1.59 m. Art Gallery of South Australia, Adelaide

George III of the United Kingdom

George William Frederick (Fig. 18.12) was King of Great Britain and Ireland from his accession on 25 October 1760 until the union of the two nations on 1 January 1801, at which point he became King of the United Kingdom of Great Britain and Ireland until his death. He was the third British monarch of the German House of Hanover, and the first in that line whose first language was English, and who never visited or resided in Germany. Despite spending the entirety of his life in southeast England, he was both a Prince-Elector of the Holy Roman Empire and Duke of Hanover (properly, the Duchy of Brunswick and Lüneburg) until he was proclaimed King of Hanover on 12 October 1814.

George was born second in line to the throne, the son of Frederick, Prince of Wales (1707–1751), and Princess Augusta of Saxe-Gotha (1719–1772), and grandson of King George II (1683–1760). He was two months premature at the time of his arrival at Norfolk House in London on 4 June 1738 and not expected to survive, but he lived and thrived. However, his grandfather disliked Prince Frederick and took little interest in Frederick's children until the Prince died of a lung injury in 1751, leaving 12-year-old George as heir apparent. The child automatically inherited Frederick's title as Duke of Edinburgh, but 3 weeks after his death the King created George Prince of Wales. The young prince received a comprehensive education, reading and writing both English and German, and was the first British monarch to systematically study science in his youth.

George became infatuated with Lady Sarah Lennox, the sister of the Duke of Richmond, in 1759 and was prepared to propose marriage to her when his mother's confidant and later prime minister, Lord Bute, quashed the idea. George II tried to marry him to Duchess Sophie Caroline Marie of Brunswick-Wolfenbüttel but both the Prince and his mother resisted the idea. Shortly thereafter the King died suddenly on 25 October 1760, making the younger George king at age 22; the need for a wife became considerably more acute. The King married Princess Charlotte of Mecklenburg-Strelitz in the Chapel Royal at St. James' Palace on 8 September 1761—the same day the he met her in person for the first time. Despite their initial unfamiliarity, they went on to enjoy a happy marriage that produced nine sons and six daughters.

George's long reign was defined by a series of military conflicts involving his kingdoms, colonies and the rest of Europe in which he saw his world powerfully transformed into something resembling the modern geopolitical scene. His early years as King saw the defeat of France in the Seven Years' War (1754–1763), followed a decade later by the loss of his American colonies in the Revolutionary War. Involved in the Napoleonic wars from 1793, his forces eventually secured the defeat of Napoleon at Waterloo in 1815. While losing control of the economic powerhouse of America, his other successes set the stage for the emergence of Britain as the dominant world power of the nineteenth century under his granddaughter, Queen Victoria.

George III is remembered, fairly or otherwise, principally for his 'madness,' a chronic mental illness that eventually became permanent and displaced him from power late in life. Various recurrent spells of the illness left him in poor physical condition around the turn of the century, and by 1810 despite the peak of his popularity as monarch, he was left nearly blind from cataracts and in near-constant pain from rheumatism. George fell dangerously ill late in the year, and eventually accepted the need for a Regency Act passed the following year that named the Prince of Wales regent for the rest of the King's life. By the end of 1811 he was clearly insane and unfit to participate in government; he retired permanently to Windsor Castle where he lived in seclusion until his death on 29 January 1820. The Prince Regent succeeded his father as George IV.

History's assessment of George has evolved almost continuously since his death, his various biographers' views of him seen through lenses heavily colored by available sources and public sentiment. Viewed poorly on account of his illness, the King's other accomplishments in governance were usually displaced by his reputation as the tyrant of the American Revolution and in Britain as the "scapegoat for the failure of imperialism" (Brooke, 1972). Around the time of the Revolution's bicentennial in 1975–1981, an historical reassessment of George's life and acts rehabilitated his image, recasting him as a crucial figure in the emergence of the modern world political order.

DISAPPEARANCE

Psalterium Georgianum is an eighteenth century invention whose persistence into the nineteenth century is certainly attributable to its adoption by Bode and its inclusion in *Uranographia.* It seems unlikely that it would have become popular without Bode's considerable influence. However, its heyday was limited to the first half of the new century, after which it was resorbed into Eridanus. Argelander left it out of *Uranometria Nova* (1843); on Plate VII, Argelander simply drew the northern boundary of Eridanus to roughly match that shown by Bode in the first edition of *Vorstellung der Gestirne.* Despite mentions by Kendall (1845) and Colas (1892, ascribed to "Hell, 1789, A.D."), it had vanished from charts by the start of the twentieth century. Allen remarked in 1899, "Psalterium is not now recognized by astronomers." When the modern constellation boundaries were set in 1928, the northern edge of Eridanus was drawn to completely encompass the stars of Psalterium Georgianum (Fig. 18.13).

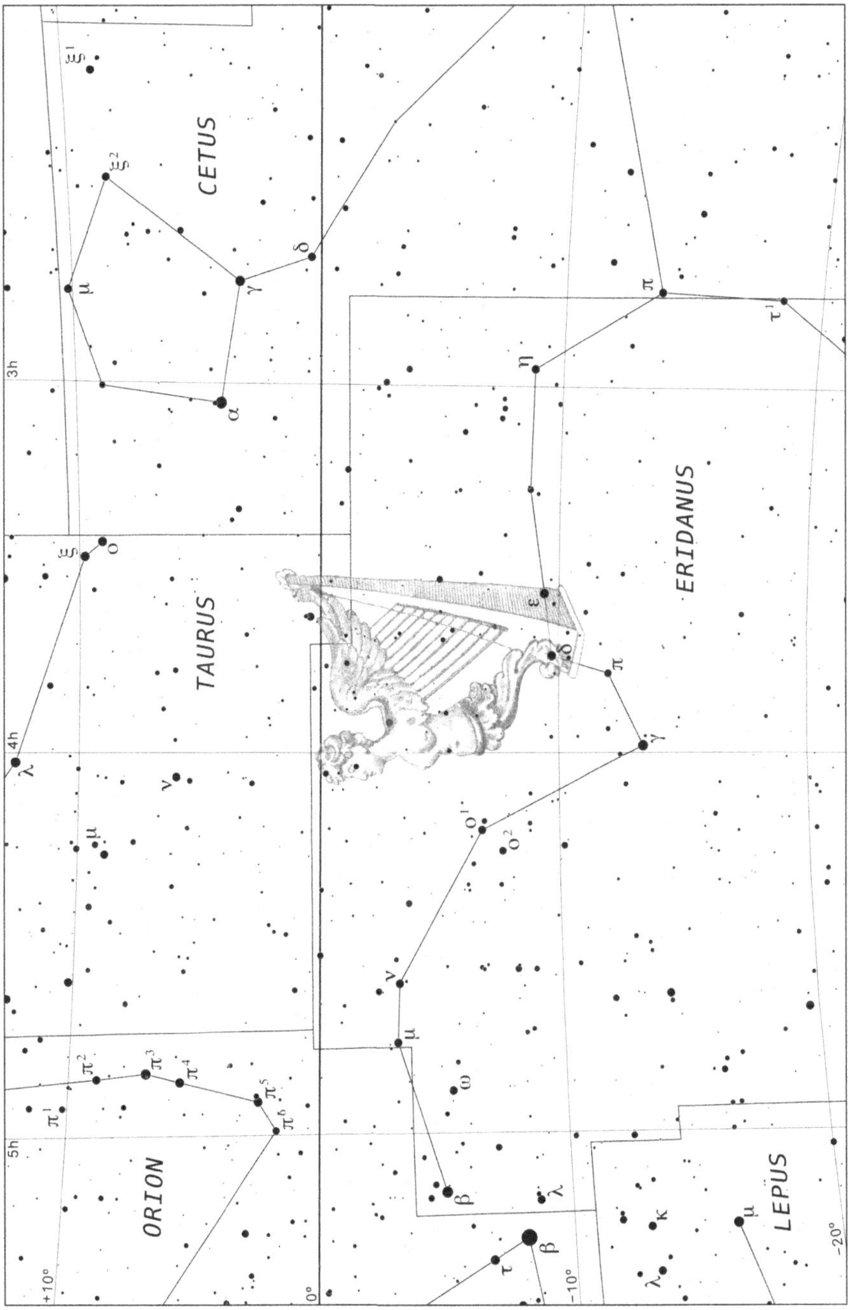

18.13 The figure of Psalterium Georgianum from Plate 24 of Alexander Jamieson's *Celestial Atlas* (1822) overlaid on a modern chart

19

Quadrans Muralis

The Mural Quadrant

Genitive: Quadrantis Muralis
Abbreviation: QuM
Alternate names: "Mural Quadrant" (Young, 1807; Burritt, 1835a; Middleton, 1842); "Quadrant" (Riedig, 1849)
Location: Between the head of Boötes and the body of Draco[1]

ORIGIN AND HISTORY

Quadrans Muralis (Fig. 19.1) was introduced by Joseph Jérôme Lefrançois Lalande and first appeared as "Le Mural" in the second edition of the reissue of Flamsteed's *Atlas Coelestis* published by Jean Fortin in 1795. In the text accompanying the *Atlas Céleste de Flamstéed*, Fortin argued[2] that the new figure was justified in the tradition of Lacaille's naming of southern constellations after items of scientific apparatus:

[1] *"Between the Dragon, the Herdsman and Hercules"* (Fortin, 1795); *"Between the Dragon, Hercules, and Boötes"* (Bode, 1801b); *"Bounded by Draco, Ursa Major, Boötes, and Hercules"* (Green, 1824); *"Immediately north of Boötes"* (Bouvier, 1858); *"Between the right foot of Hercules, the left hand of Boötes, and the constellation Draco"* (Allen, 1899; Bakich, 1995).

[2] "LE MURAL, ou Quart-de-cercle Mural (Pl. 2), est encore une nouvelle Constellation que le citoyen Lalande a placée dans une espace vide, entre le Dragon, le Bouvier & Hercule. Lacaille, après avoir observé les étoiles australes, forma des Constellations nouvelles avec les instruments de la Physique & des Arts (Mém. 1752). A son example, on a cru pouvoir consacrer dans l'Hemisphere Boréal l'instrument précieux qui a servi aux observations de 30 mille étoiles, c'est à-dire au plus grand monument de l'Astronomie; & les Astronomes à venir, profitant de cet immense travail, conserveront sans doute une Constellation propre à en rappeller la mémoire. Toutes ces nouvelles Constellations sont une richesse de plus pour notre Atlas." (pp. v–vi).

J.C. Barentine, *The Lost Constellations*, Springer Praxis Books,
DOI 10.1007/978-3-319-22795-5_19

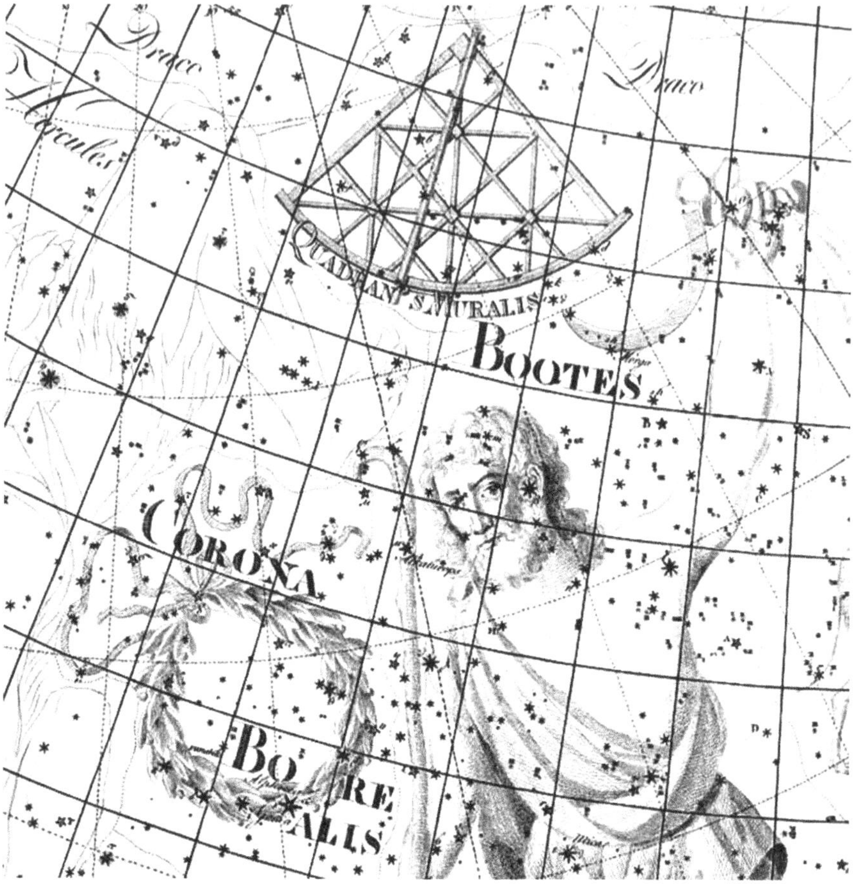

19.1 Quadrans Muralis, Boötes, and Corona Borealis shown on Plate 7 of Bode's *Uranographia* (1801b)

> THE MURAL, or Mural Quadrant (Plate 2), is yet a new constellation that the citizen Lalande placed in the empty space between the Dragon, the Herdsman and Hercules. Lacaille, after observing the southern stars, formed new constellations out of instruments of the arts and sciences (*Mémoires* 1752). By his example, we thought to devote *[space]* in the northern hemisphere to the costly instrument used in the observations of 30,000 stars, that is to say the greatest monument of astronomy; and future astronomers, taking advantage of this tremendous work, without a doubt will retain this constellation of their own to recall its memory. All of these new constellations add even greater value to our atlas.

The following year he published a table of circumpolar stars in *Connaissance des Temps* that included ten stars between the fifth and seventh magnitudes he attributed to the new constellation, although more were shown in the *Atlas*. Richard Hinckley Allen wrote that it was "[F]ormed by La Lande in 1795, as a souvenir of the instrument with which he and his nephew, Michel Le Français, observed the stars subsequently incorporated under this title into the latter's *Histoire Celeste Française*." Together with his nephew Michel Lefrançois de Lalande (1766–1839), the elder Lalande noted positions of some 50,000 stars while at the Collège de France.

The "instrument" is the mural quadrant, an astronomical measurement device; in placing it among the stars, Lalande took a cue from his predecessors who formed new constellations in the seventeenth and eighteenth centuries to honor scientific devices: Johannes Hevelius (Sextans) and Nicolas Louis de Lacaille (Antlia, Fornax, Horologium, Microscopium, Octans, Reticulum, Telescopium).

In *Allgemeine Beschreibung und Nachweisung der Gestirne*, the companion catalog to his *Uranographia* (1801b), Johann Elert Bode similarly described[3] the "new constellation" as having been

> only recently introduced in 1795 by Mr. de la Lande between the Dragon, Hercules, and Boötes in commemoration of the mural quadrant, with which Mr. de la Lande and his nephew Mr. le François, for some years, have observed consecutively at Paris a great number of stars of the sixth magnitude and fainter, whose positions until that time had not been determined.

Bode took two liberties with his inclusion of Lalande's constellation. First, he slightly reduced the sky area of the constellation, as compared to the manner in which it was shown by Lalande, by adjusting its boundaries to exclude some stars from its more traditional neighbors. And second, he Latinized "Le Mural" as "Quadrans Muralis," more directly referring to the measurement instrument and bringing its name in line with the Latin nomenclature of other constellations shown on his maps. Given that its stars were faint, most cartographers prior to Lalande dealt with this part of the sky by simply plotting no stars (see Fig. 19.2).

Due to the influence of *Uranographia*, Quadrans Muralis remained a staple of most charts and books during the first half of the nineteenth century with a few notable exceptions such as Middleton (1842) and Argelander (1843). However, not all authors were impressed by a constellation formed entirely from very faint stars. In *Astronomical Recreations* (1824), Jacob Green noted that "there are about nine very small stars in it, most of which are of the seventh and eighth magnitudes," and lamented that it was ever charted in the first place: "It is perhaps to be regretted that this little constellation was

[3]"Dies neue Gestirn ist erst im Jahr 1795 von dem Herrn *de la Lande* zwischen dem *Drachen, Herkules*, und *Bootes* zum Andenken des Mauerquadranten eingeführt, womit derselbe und sein Neveu Herr *le François*, seit einigen Jahren, eise sehr grosse Anzahl bisher noch nicht nach ihrer Stellung bekkanten Sterne sechster und geringerer Grösse, zu Paris beobachtet haben." (p. 8)

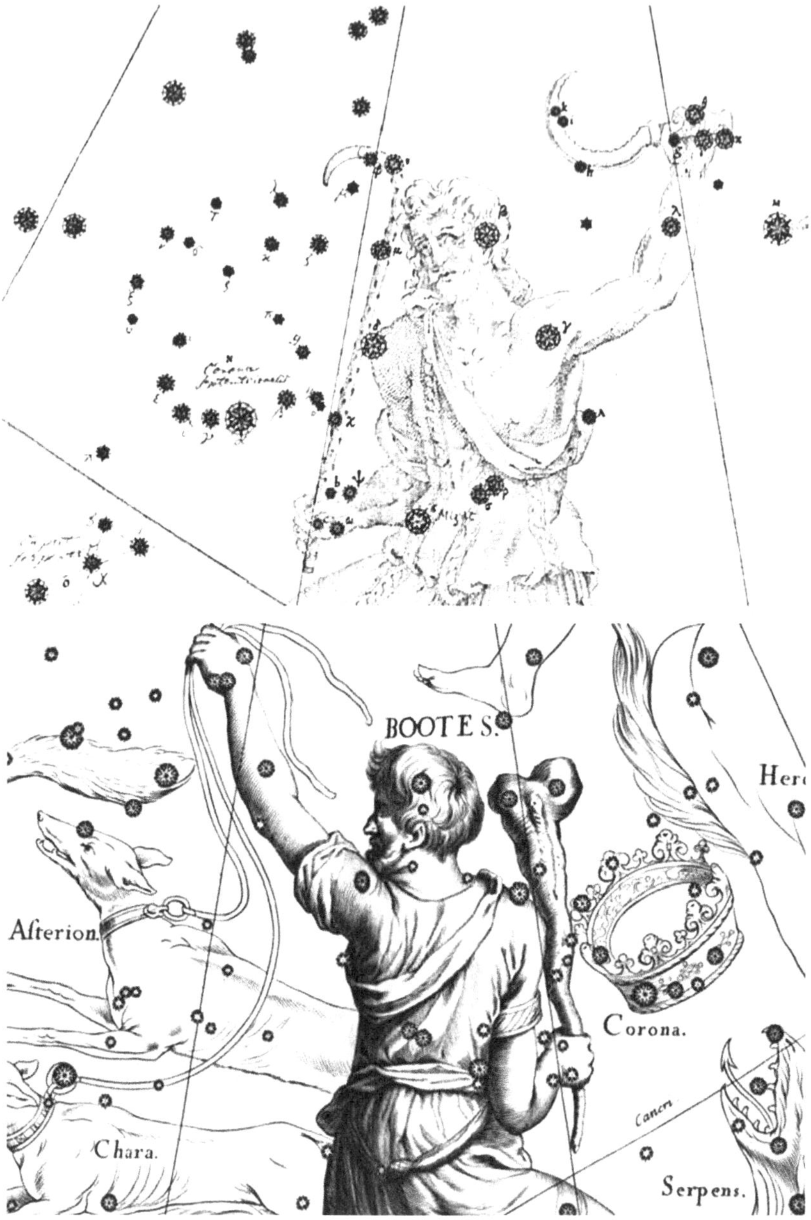

19.2 Two views of northern Boötes and neighbors from Plate E of Bayer's *Uranometria* (1603; *top*) and Figure F of Hevelius' *Prodromus Astronomiae* (1687; *bottom*). Note that the two depictions are reversed right-to-left (east-west), as the two cartographers drew the sky from points of view inside (*top*) and outside (*bottom*) the celestial sphere defined by the plane of the sky

19.3 Quadrans Muralis (*top*) is shown together with Boötes, Canes Venatici, and Coma Berenices on Plate 10 of *Urania's Mirror* (1825). Note at this date, α Canes Venaticorum is still shown as the quasi-asterism "Cor Caroli"

formed, as there is no particular advantage to be derived from it, and as it encumbers our catalogues with an additional name." Hannah Bouvier put it most succinctly in her *Familiar Astronomy* (1858) (Figs. 19.3):

> *Q*. Are there any CONSPICUOUS stars in the constellation Quadrans Muralis?
> *A*. There are *not*.

ICONOGRAPHY

A quadrant is one of the very simplest tools devised to make measurements in positional astronomy. As its name implies, the quadrant consists of one-quarter of a complete circular disc, its curved outer edge subtending an angle of 90° as seen from the opposite point on the right angle. A plumb line is suspended from the point of the right angle; to find the elevation of the Sun, Moon, or other object above the local horizon, a sighting is taken along one of the straight-line sides of the quadrant and the elevation read from a graduated scale printed or incised along the curved edge.

Its invention is usually attributed to Ptolemy as an improvement on the astrolabe, a device for measuring the elevation angles of various astronomical objects; Ptolemy is

19.4 Ptolemy shown using a quadrant in a woodcut from Giordano Ziletti's *Clavdio Tolomeo Principe De Gli Astrologi...*, published in Venice in 1564 (11.4 cm × 10.9 cm). His status as an astrologer is suggested by the armillary sphere at *lower left*, while he indicates that he is a geographer by pointing toward the ground with his right hand. Given the manner in which Ptolemy holds the quadrant, it is evident that the anonymous artist did not know how the instrument was actually used

often depicted in Medieval and Renaissance texts holding the device as a token of his status as an astronomer (Fig. 19.4). However, the historical introduction of the quadrant probably pre-dates Ptolemy; since knowledge of one's longitude is especially valuable in navigation at sea, it is likely that the first people to employ it practically were mariners in antiquity. Much as the telescope would later be adapted to astronomical studies from distinctly terrestrial uses, the quadrant probably passed from the hands of navigators to those of astronomers.

The quadrant could also be used as an instrument for reckoning the local solar time, which differs from civil time according to the location of an observer within a given

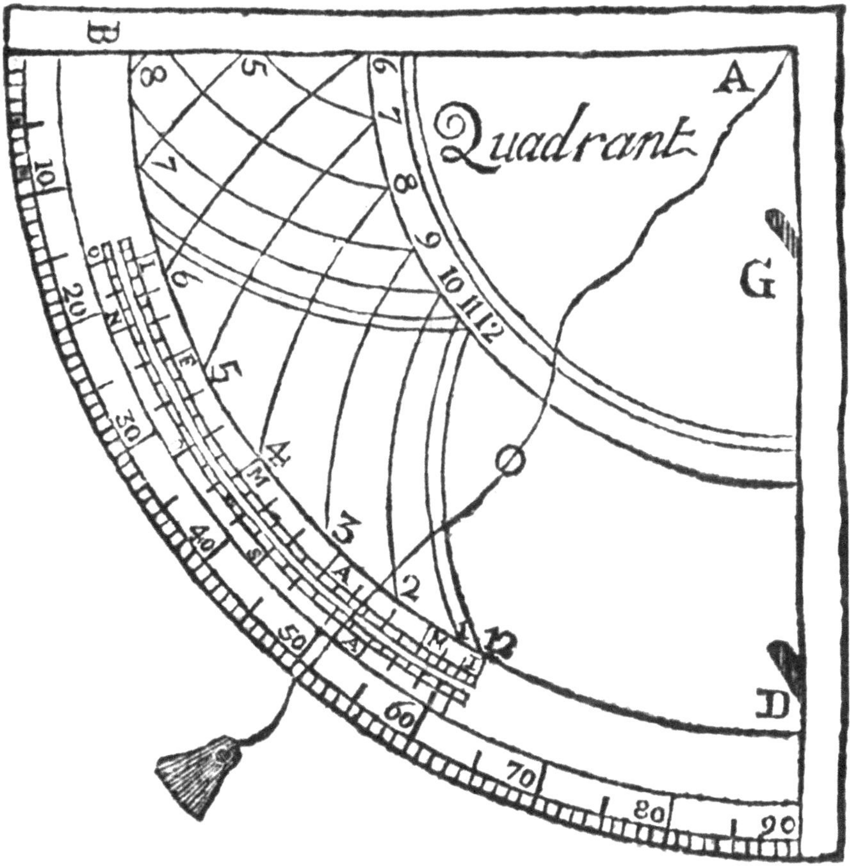

19.5 An horary quadrant, used to determine the local solar time, shown in George Fisher's *The Instructor: or, Young Man's Best Companion* (7th edition, 1744). This particular quadrant as drawn is calibrated for a latitude of about 51.5°. The figure original caption read, "This Quadrant, or Quarter of a Circle is variously useful, on sundry Accounts, *viz.* to take Heights and Distances, whether accessible or inaccessible: to find the Hour of the Day, *&c.*"

time zone; it can be thought of, to use a familiar device as an example, the time that a sundial would indicate. And example of such an "horary" quadrant is shown in Fig. 19.5 from an eighteenth century instructional text called *The Instructor: or, Young Man's Best Companion*. Its function is described thusly:

> To find the Hour of the Day: Lay the thread just upon the Day of the Month, then hold it till you slip the small Bead or Pin-head [along the thread] to rest on one of the 12 o'Clock Lines; then let the Sun shine from the Sight G to the other at D, the Plummet hanging at liberty, the Bead will rest on the Hour of the Day.

Here, like a sundial, the quadrant must be constructed in a particular way to account for the observer's latitude; as described and shown in the figure, the calibration is built into the design of the instrument.

The angular resolution of a quadrant is proportional to its size, given the relationship between the length of an arc on a circle, s, the radius of the circle, r, and the angle subtended by the arc, θ, via the relation

$$s = r\theta$$

As s grows in size proportional to r, progressively finer graduations can be inscribed along the arc. The positional errors of measurements made in antiquity suggest that ancient quadrants were small, handheld objects; Renaissance astronomers required better precision and quickly realized that the corresponding circular radii needed to achieve such precision would result in quadrants far too large to wield by individual observers. This resulted in the introduction of the *mural quadrant*, so named for the Latin word "mura" ("wall"): a very large device fixed to a wall for mechanical support and stability. An example of such a mural quadrant is shown in Fig. 19.6, a plate from the massive Enlightenment reference work *Encyclopédie, ou dictionnaire raisonné des sciences, des arts et des métiers* published in 32 volumes over 1751–1777. To further increase the precision of the instrument, the plumb line was replaced by a pair of sights fixed to a rotating pivot at the right-angle corner of the quadrant; the viewing end of the sight was constrained to move only along the graduated arc. Mechanical sights were later replaced with telescopes upon their introduction in the seventeenth century. Mural quadrants were typically oriented such that the plane of the quadrant aligned parallel to the meridian, an imaginary arc across the sky passing through both north and south celestial poles and dividing the sky into equal eastern and western halves. As such, these quadrants were also *transit devices*, used to time the instant at which a given object crossed the meridian and thus reached its highest elevation angle above the local horizon in a night.

Lalande may have sought to honor the role of the mural quadrant in astronomy in particular because of the crucial role it played in the formulation of Johannes Kepler's laws of planetary motion in the early seventeenth century. Kepler was hired as an assistant by Tycho Brahe (1546–1601), a Dane among the most famous figures in the history of astronomy. For the last quarter of the sixteenth century, Tycho enjoyed the patronage of the Danish king, Frederick II, during which time he built a sprawling compound on Ven, a small island in the Öresund strait between Scania and Sjælland, in modern Sweden and Denmark, respectively. Uraniborg ("the Castle of Urania") and its neighboring, underground observatory complex Stjerneborg (built *c.* 1581) were a center of learning at which Tycho hosted nearly one hundred artists and students between 1576 and 1597.

Tycho's work entirely predated the introduction of the telescope, and was limited to the non-optical astronomical instruments of the day. These included a large (1.94 m-radius) brass mural quadrant he had installed at Uraniborg. Tycho is shown using the quadrant in his 1598 book *Astronomiae instauratae mechanica* (Fig. 19.7). His version of the instrument is different than the previously described versions. The brass arc of a circle is shown set into a curved wall along which the observer would sight toward a

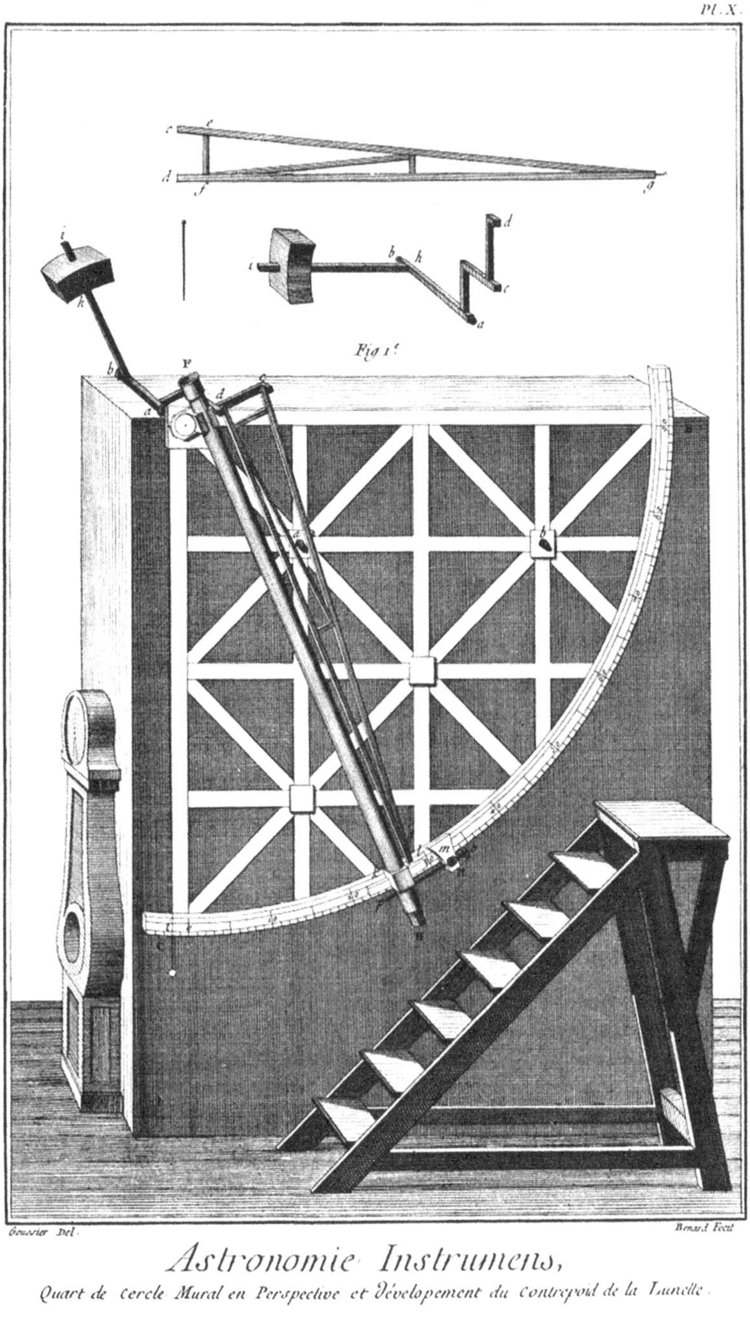

19.6 A mural quadrant shown on Plate X, Volume 5 ("Sciences—Mathématiques—Astronomie") of *Encyclopédie*, co-edited by Denis Diderot and Jean le Rond d'Alembert (1767)

19.7 An engraving of Tycho Brahe's mural quadrant at Uraniborg in his *Astronomiae instauratae mechanica* (1598), from a portrait by Thobias Gemperlin (*c.* 1550–*c.* 1602). Tycho points toward the slit, high on the opposite wall, through which the night sky was viewed and measurements made with the quadrant; he called out his observations aloud to an assistant (seated figure at *lower left*) who recorded them. A second assistant at lower right reads the time of observation from clocks. Behind the scene, on the far wall, is a mural painting featuring vignettes of Tycho's other scientific instruments

narrow slit high up on the facing wall; when an object was seen along the line of sight through the slit, the user noted the number next to a mechanical pointer on the brass arc. Tycho and his assistants observed many thousands of stars this way to the highest precision of any such measurements made to date.

Commenting on Tycho's "Star Catalog D" (1598), Rawlins (1993) noted that "Tycho achieved, on a mass scale, a precision far beyond that of earlier catalogers," and that his achievement was the result of "an unprecedented confluence of skills: instrumental, observational, & computational." Tycho aspired to reach an angular precision of one minute of arc, which is one-sixtieth of an angular degree; for purposes of comparison, this is about one-thirtieth of the apparent angular diameter of the full Moon. He claimed to have accomplished this goal, although many of his measured stellar positions were only reliable to within about two arcminutes. Wesley (1978) found on comparing Tycho's published positions to those recorded in his observing ledgers that the latter were actually, on average, good to *better* than one arcminute. However, systematic errors due to problems with calibrations and incorrect transcriptions sometimes rendered the published positions of stars off by several degrees.

Frederick II died in 1588 and was succeeded by his 11-year-old son, Christian IV; Tycho's relations with the new court soured and in 1597 he left Denmark permanently. At the invitation of the Holy Roman Emperor Rudolf II, he built a new observatory in a castle some 50 km from Prague at Benátky nad Jizerou in the modern Czech Republic. After working there for the balance of 1600, he returned on Rudolf's orders to Prague where he lived to his death. His sponsors allowed him to continue gathering astronomical measurements in return for casting astrological charts and advising them with predictions of various civil and military events. He died following an episode in which he refused to leave the table during a banquet in Prague on 13 October 1601. According to Kepler's firsthand account of the event, Tycho's refusal to leave the proceedings long enough to relieve himself, on grounds of etiquette, cost him his life when his bladder burst, leading to acute septicemia. On the night before he died (23 October), falling into a fever-fueled delirium, Tycho repeatedly exclaimed "Ne frustra vixisse videar!" ("Let me not seem to have lived in vain!")

Kepler was left with the voluminous positional data Tycho had acquired over the prior 30 years. After trying repeatedly to fit circles to the orbit of Mars gleaned from Tycho's positions, he came to realize that the Red Planet's orbit could only be described by an ellipse in which the Sun occupied one focus. No matter how many epicycles upon deferents he added to the model, he could not force the orbit of Mars within the Aristotelian ideal of perfectly circular planetary orbits centered on the Earth. Without a physical understanding of the root cause, Kepler was left to conclude—at least phenomenologically—that some set of natural laws governed the orbital motions of the planets and that Nicolaus Copernicus was right after all. Only many decades after Kepler's death did Isaac Newton first propose a Universal Law of Gravitation from which Kepler's Laws of planetary motion necessarily followed. Even after the introduction of the telescope, enabling considerably more precise positional measurements than the mural quadrant made possible, later observations vindicated Kepler's work.

DISAPPEARANCE

Lalande's constellation enjoyed a run of less than one hundred years and was already on the road to extinction by the last quarter of the nineteenth century. Its brief popularity on charts of the earlier half of the century is certainly attributable to its depiction in *Uranographia*. Having captured neither the attention of Ptolemy originally, or that of his nineteenth century successors, the Mural Quadrant soon fell into disuse. In 1899 Allen wrote that it was by then "[N]ot recognized by modern astronomers." At the time the modern constellation boundaries were finalized in 1928, the former constellation was split in two by the re-drawn boundary between Draco and Boötes (Fig. 19.8). Of its former stars the most well-known is BP Boötis, an RS Canum Venaticorum-type variable.

Nevertheless, Quadrans Muralis lives on, albeit in an oblique way. The first annual meteor shower of the calendar year, peaking on 3 January, remains known as the "Quadrantids," its radiant located among the stars of this former constellation. At the time the shower was first noticed, in 1825, the constellation was still commonly shown on charts of the day.

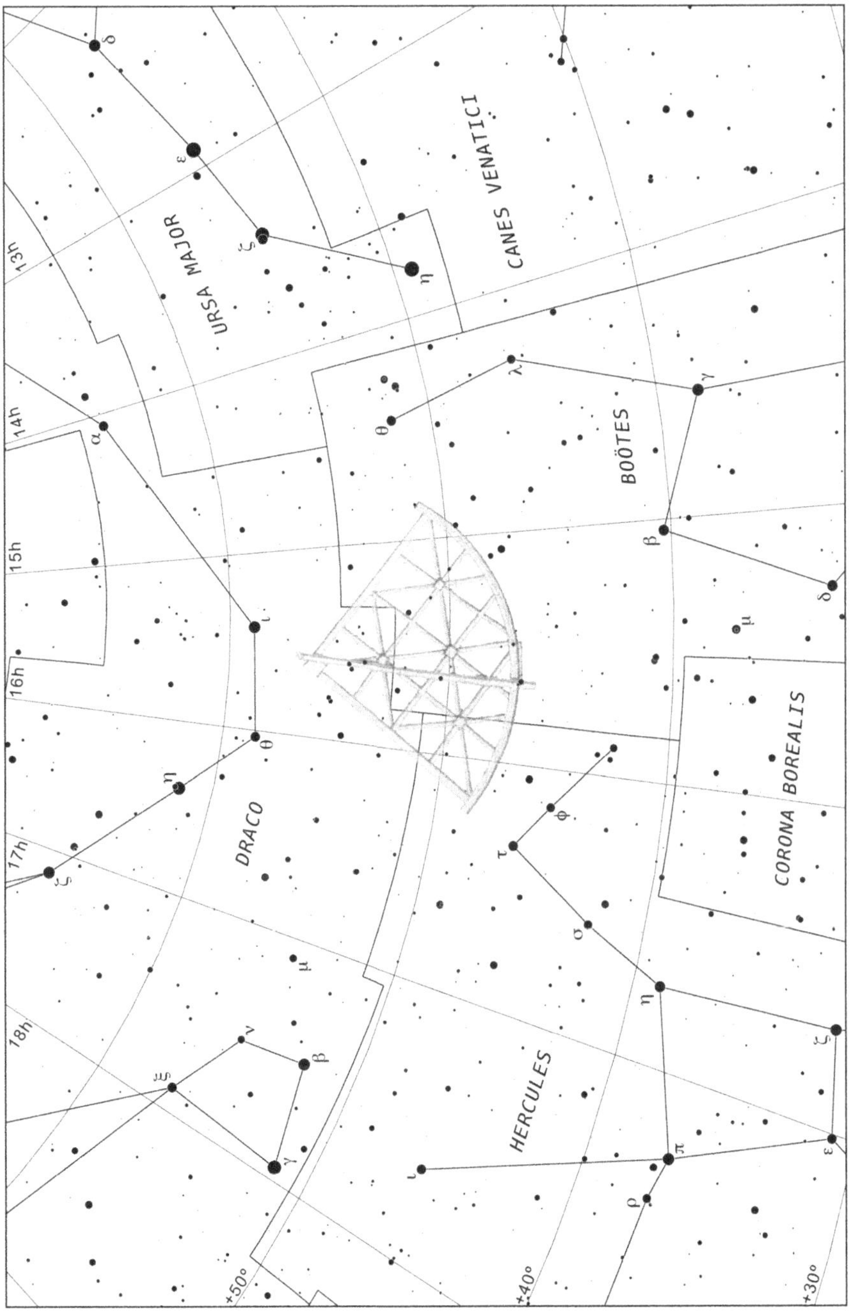

19.8 The figure of Quadrans Muralis from Plate 7 of Bode's *Uranographia* (1801b) overlaid on a modern chart

20

Rangifer

The Reindeer

Genitive: Rangiferi
Abbreviation: Ran
Alternate names: "Le Reene" (Le Monnier, 1743); "Tarandus" (*Urania's Mirror* 1825, Ryan 1827, Kendall 1845, Chambers 1877); "le Réne" (Le Monnier 1746, Fortin 1776); "La Renne" (Fortin 1795); "Renna"; "Tarandus vel Rangifer"
Location: Between Cassiopeia and Camelopardalis[1]

ORIGIN AND HISTORY

The story of Rangifer begins more than a century before it first appeared on any chart in the designation of a group of northern stars left unformed by Ptolemy as "Camelopardalis," the Giraffe. Richard Hinckley Allen described its stars as "long, faint, and straggling like its namesake. ... [stretching] from the pole-star to Perseus, Auriga, and the Lynx, the hind quarters within the Milky Way." Allen thought that it was originally introduced by Jacob Bartsch, "who published it, in outline only, in 1614, and wrote that it represented to him the Camel that brought Rebecca to Isaac." None of the early depictions of this constellation show it as a camel, so this origin story seems unlikely. Rather, Allen asked whether the Bartsch story explained why Richard Anthony Proctor "changed its title to Camelus," although he pointed out that the only nineteenth century work to adopt the name change

[1] *"Formed from various other faint stars belonging to Cassiopeia and Cepheus."* (Bode, 1801a); *"Bounded by Ursa Minor, Cepheus, Cassiopeia and Camelopardalis."* (Green, 1824); *"Situated between the middle of the Shephed and the Pole-Star."* (Kendall, 1845); *"Directly north of Cassiopeia, and east of Cepheus."* (Bouvier, 1858); *"A small and faint asterism between Cassiopeia and Camelopardalis"* (Allen, 1899); *"Formed of faint stars between Cassiopeia and Camelopardalis"* (Bakich, 1995).

J.C. Barentine, *The Lost Constellations*, Springer Praxis Books,
DOI 10.1007/978-3-319-22795-5_20

was the 1894 edition of Flammarion's *Astronomie Populaire*. Bartsch did indeed describe a camel in *Usus Astronomicus* (1624), but in turn incorrectly attributed its introduction to Isaac Habrecht (1621).

Ridpath (1989) correctly identified its introduction with Petrus Plancius on his 1612 globe, the same on which he also gave the world another fanciful animal, Monoceros (the Unicorn). In the early seventeenth century, European explorers' descriptions of African giraffes, while straightforward, were taken at home with a grain of salt because a giraffe, as described, seemed too improbable. Its Latin name translates as "cameleopard," as the animal was thought by some to be a literal (if unlikely) hybrid of two more familiar creatures. Thus the "cameleopard" was chosen to represent a set of fourth magnitude and fainter stars near the north celestial pole overlooked by the Greeks. Yet even fainter stars lay unformed further north than Camelopardalis.

The symbolic value of these northerly stars was important in the introduction of Rangifer by Pierre-Charles Le Monnier on a star chart in his book *La Théorie des Comètes* (1743; Fig. 20.1), a work he dedicated to the French mathematician Pierre-Louis Moreau de Maupertuis (1698–1759). Maupertuis was commissioned to lead an expedition to Lapland in the 1730s as part of a scientific study of the shape of the Earth; in honor of his efforts, Le Monnier found a suitably northern animal to place in the heavens—the reindeer.

The Maupertuis Expedition

Aside from his scholarly accomplishments, Maupertuis is best known for leading the northern of two expeditions in the mid-eighteenth century that had as their object determining the overall shape of the Earth. While the notion of a spherical Earth originated in antiquity, the determination that the Earth is not described exactly as a sphere, but rather an ellipsoid, was made in his time as geodesy emerged as a modern science. Such determinations relied upon precise measurements of *meridian arcs*, parts of great circles drawn along the Earth's surface along lines of constant longitude or latitude. The first modern meridian arc measurements were made by Jean Picard (1620–1682), a French astronomer and priest; it was a "modern" determination in the sense that Picard laid out a precise baseline with the use of wooden rods, and used a telescope, attached to a quadrant, to make careful measurements of angles. To increase the precision of his angular measurements, he outfitted his telescope with a reticle made of crosswires and used micrometer screws to make very fine adjustments to the quadrant's position. In 1669–1670, Picard measured 1° of latitude along the Paris Meridian with this method. He defined thirteen triangles along a line between Paris and the clock tower of Sourdon, near Amiens, finding that 1° of latitude subtended 110.46 km; for a spherical Earth, this implied a radius of 6328.9 km or a mere 0.44 % less than the modern value.

Jacques Cassini (1677–1756), son of the Italian astronomer Giovanni Domenico Cassini, extended Picard's arc north to Dunkirk and south to the border with Spain and divided the arc into two pieces that met in Paris. Computing the length of a degree

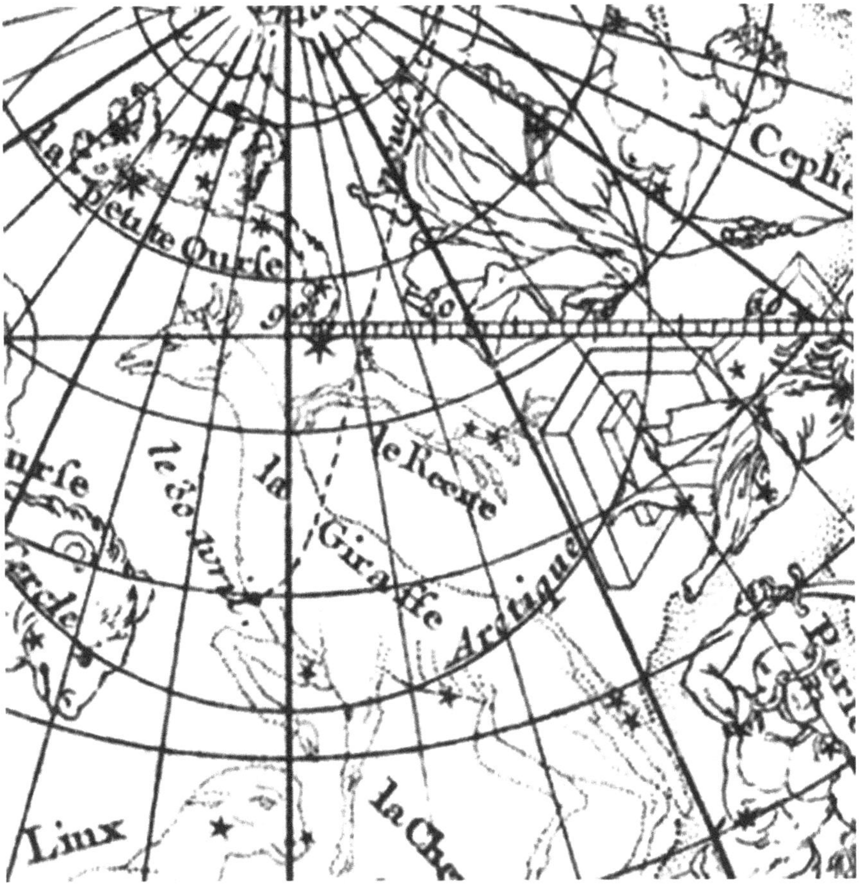

20.1 An excerpt of the chart "Hemisphere Boreal" in Pierre Charles Le Monnier's *La Théorie des Comètes* (1743) on which he first introduced Rangifer as "le Reene"

independently from measurements made on both arcs, Cassini found that the lengths disagreed with each other, with the length along the northern piece coming in slightly shorter than that on the southern piece. After carefully ruling out measurement errors, the data clearly contradicted the underlying assumption of a perfectly spherical Earth, and provided the first observational evidence that the true shape of the Earth was described by some other figure of rotation. Specifically, Cassini's measurements implied that the shape of the Earth was that of a prolate spheroid, an egg-shape in which the long axis of the figure is parallel to the Earth's rotation axis.

However, both Isaac Newton and Christiaan Huygens predicted exactly the *opposite* shape—an *oblate* spheroid, flattened rather than extended along the poles. The reasoning was that the rotating Earth would tend to bulge out at the equator where the tangential

20.2 A portrait of Pierre-Louis Moreau de Maupertuis (1698–1759) in the frontispiece of his *La Figure de la Terre* (1738). Maupertuis wears a warm fur hat and stole, indicating his 1736–1737 expedition to the Scandinavian far north, and his left hand pushes down on a globe, referring to his measurements demonstrating the oblate shape of the Earth

speed of rotation is highest. Since the Earth obviously held together as a single, quasi-spherical body, it must compensate for this outward bulging by drawing in along a smaller radius at the poles. From his calculations, Newton predicted that the Earth was oblate with a flattening of one part in 230.

Cassini's claim of a prolate Earth against the theoretical prediction of a flattened, oblate Earth could only be settled by determining the lengths of a meridian arc along the equator and comparing the results to similar measurements made along any great circle of longitude. Because all such great circles are perpendicular to the rotation direction of the planet, any well-determined line of longitude would suffice, but latitude was different. There are two extrema in a spherical system, one at the equator and the other at the poles. The length of a degree of longitude measured along the equator is a maximum, decreasing in size with latitude until it vanishes to zero at either pole. To measure the greatest possible range of values of the length of a degree of longitude, the Swedish physicist Anders Celsius (1701–1744) proposed a measuring expedition to resolve the ongoing debate during his visit to Paris in 1735. King Louis XV and the French Académie des Sciences actually commissioned two expeditions—the French Geodesic Missions—to the greatest extremes of latitude then reachable. The first (1735–1744), led by the French astronomers Charles Marie de La Condamine, Pierre Bouguer, Louis Godin and Spanish geographers Jorge Juan and Antonio de Ulloa, was sent to modern-day Ecuador to measure a meridian arc on the equator. The second (1736–1737) was dispatched to Lapland, the furthest point north reachable from Europe lying largely above the Arctic Circle. The Académie chose Pierre-Louis Moreau de Maupertuis (Fig. 20.2) to lead the expedition.

Maupertuis was born at Saint-Malo, France, and was educated privately in mathematics. On completion of his formal education, his father arranged for him a commission in the cavalry of the Armée de Terre; the commission was largely ceremonial and permitted him time to build his reputation as a mathematician. After 3 years in this position, he relocated to Paris where he moved in fashionable social and scholarly circles and was admitted to the Académie in 1723. His early work in mathematics largely supported and extended the Newtonian theory of mechanics and universal gravitation, as opposed to the kinetic theory of gravitation of René Descartes and others. At the time of Newton's death in 1727, his ideas about physics were still largely contested outside England, but Maupertuis positioned himself as a strong defender on the Continent. Filled with this zeal, he was a natural choice for the Académie to help settle the issue of the Earth's shape definitively.

The Lapland expedition included mathematicians Charles Etienne Louis Camus and Alexis Claude Clairaut, astronomer Pierre-Charles Le Monnier, draughtsman M. d'Herbelot, secretary M. Sommereuxthe, abbot Réginald Outhier, and Anders Hellant (1717–1789), an astronomer from Tornio, then in the Kingdom of Sweden, who served as their interpreter and guide. It left France in April 1736, stopping briefly in Stockholm before continuing north and arriving in Tornio on 19 June. Maupertuis originally intended to make the measurements in the archipelago of the Gulf of Bothnia, but quickly realized that the islands were too low in elevation for accurate triangulation. The headmaster of the school in Tornio suggested the River Tornionjoki, which would be frozen over in the winter months. A baseline of 14.3 km was established and accurately measured along the river in the winter of 1736–1737. The crew spent a year gathering data, leaving on their return journey on 10 June 1737. Upon the expedition's return to France, Maupertuis calculated the length of the meridian degree along the Tornio to be 57,437.9 toises, or 111.95 km. Comparing this to the figure of 57,060 toises (111.21 km) near Paris indicated that the Earth in fact was flattened at the poles, with a degree of flattening of about one part in 150. Gross measurement errors were later found in the data, but the conclusion of an oblate Earth remained intact. Louis XV rewarded Celsius for his suggestion with a life pension of 1000 livres per year. Maupertuis published the results in *La Figure de la Terre* (1738). The book included an adventure narrative of the expedition, and an account of the Käymäjärvi Inscriptions.[2] On his return home he became a member of almost all the scientific societies of Europe.

Le Monnier again published his new constellation in *Institutions astronomiques* (1746) without providing any additional information about it. In the second half of the eighteenth century, the figure slowly gained acceptance by virtue of its inclusion in several key atlases. Jean Nicolas Fortin added it to Map 2 of his reissue of Flamsteed's *Atlas Coelestis* with a name that evolved between the two editions; rendered first as "le Réene" in 1776, it became "la Renne" in 1795. Following the example of Fortin, Johann Elert Bode included the figure under the German spelling "Rennthier" on Plate 2 of the first edition of *Vorstellung der Gestirne* (1782). It was similarly shown by Christian Goldbach on Map 2 of *Neuester Himmels* (1799).

As in the case of several other constellations that flourished and died out in the nineteenth century, Rangifer experienced a heyday after appearing in Bode's influential *Uranographia* (1801b; Fig. 20.3) under the Latinized spelling of its name. In the accompanying *Allgemeine Beschreibung und Nachweisung der Gestirne*, Bode described[3] its origin as "from Le Monnier, in remembrance of the meridian degree measured in Lapland at the North Pole in 1736, placed in the sky, and formed from various other faint stars belonging to Cassiopeia and Cepheus." Bode again showed the figure, labeled in German, on Plate 2 of the second edition of *Vorstellung der Gestirne* (1805). In the same year August Gottleib Meissner showed the figure in Fig. XX, Table 18 of *Astronomischer Hand-Atlas* as "Renthier" (Fig. 20.4). Shortly thereafter, in 1807, the constellation made its English-language debut as "ReinDeer" on Plate XXXVI, Fig. 517 of Thomas Young's *A Course of Lectures on Natural Philosophy and the Mechanical Arts* (Fig. 8.7). It appeared in name only, without a corresponding figure, as "Rennthier" on Tables 1–2 of Joseph Johann Littrow's *Atlas des Gestirnten Himmels* (1839).

Alexander Jamieson took a decidedly different tack with respect to nomenclature by labeling it "Tarandus" on Plate 2 of his *Celestial Atlas* (1822); *tarandus* is the species name for the reindeer or caribou, properly *Rangifer tarandus* (see Iconography, below). Jamieson set a trend followed by a number of other mapmakers in the next generation who also referred to it as "Tarandus." It was shown under this label on Plate 2 of *Urania's Mirror* (1825; Figs. 8.8 and 20.5) alongside neighboring Custos Messium (Chap. 8). James Ryan (1827) noted that "the Rein Deer was made out of the unformed stars between Cassiopeia and the north pole". Jacob Green (1824) described the figure as an assembly of "some

[2]Also known as the "Käymäjärvi Inscriptions," an ancient stone with undeciphered characters of disputed origin in Finnish Lapland.

[3]"Ist von le Monnier, zum Andenken des in Lappland beim Nordpol im Jahre 1736 gemessenen Meridiangrades, an den Himmel gesetzt, und aus verschiedenen sonst zur Kassiopeia und dem Cepheus gehörigen kleinen Sternen, formirt." (Bode, 1801a)

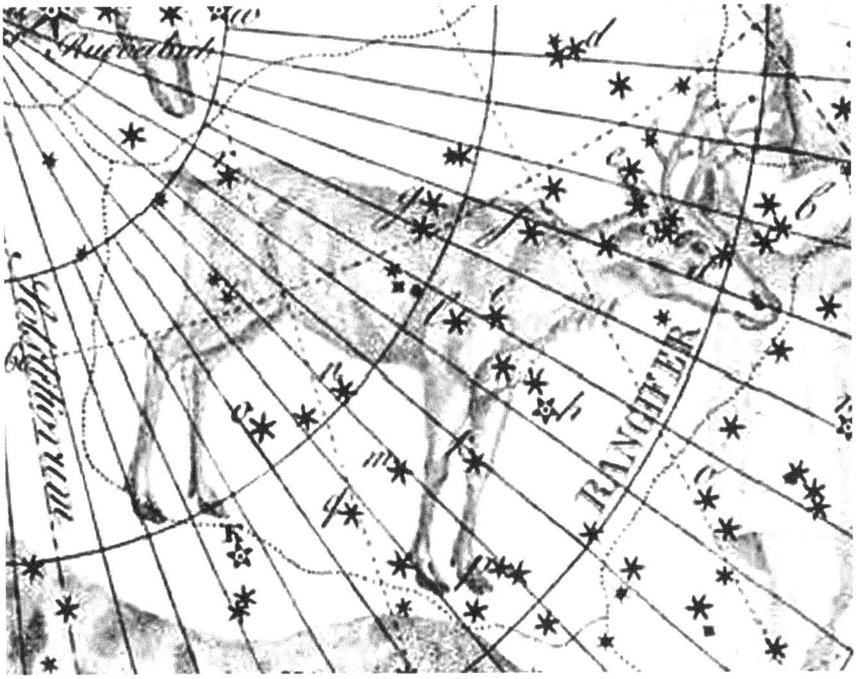

20.3 One of the most artistically detailed renderings of Rangifer, from Plate 3 of Bode's *Uranographia* (1801b)

remote and telescopic stars belonging to two of the ancient neighboring groups." Ezra Otis Kendall counted "Tarandus—The Reindeer" on a list of constellations in *Uranography* (1845); the figure was "situated between the middle of the Shepherd and the Pole-Star, and contains only small stars." Similarly, Hannah Bouvier wrote in her *Familiar astronomy* (1858) that Tarandus "contains no conspicuous stars."

ICONOGRAPHY

The reindeer (*Rangifer tarandus*), known as the caribou in North America, is a species of deer found in circumpolar regions of Siberia, Europe and North America. Reindeer inhabit a variety of habitats, from arctic regions to mountains and boreal forests. There are three major haplogroups of reindeer, identified with different geographic regions isolated during the Pleistocene[4]: (1) Beringia and northern and eastern Eurasia; (2) western Eurasia; and (3) North America. The last two are refugia cut off from the main Eurasian population due

[4]F. Oystein and K.H. Roed, "Refugial origins of reindeer (Rangifer tarandus L.) inferred from mitochondrial DNA sequences". *Evolution* 57(3), pp. 658–670 (2003).

20.4 Rangifer (as "Renthier") depicted on Fig. XX, Table 18 of August Gottleib Meissner *Astronomischer Hand-Atlas* (1805). Nearby stands Custos Messium (Chap. 8) as "Erndtehuter"

to advances of the ice sheet in both areas. Fourteen recognized subspecies exist, two of which are extinct.[5]

The taxonomic name for the reindeer is a compound of two late Latin words. *Rangifer*, chosen by the Swedish biologist Carl Linnaeus (1707–1778) in 1758 for the reindeer genus, was first used by Albert Magnus (bef. 1200–1280) in Book XXII, Chap. 268 of *De Animalibus* (*"Dicitur Rangyfer quasi ramifer"*). It is thought that the origin of the word

[5]D.E. Wilson and D.M. Reeder, *Mammal Species of the World—A Taxonomic and Geographic Reference*. Baltimore, Maryland: Johns Hopkins University Press/Bucknell University (2005).

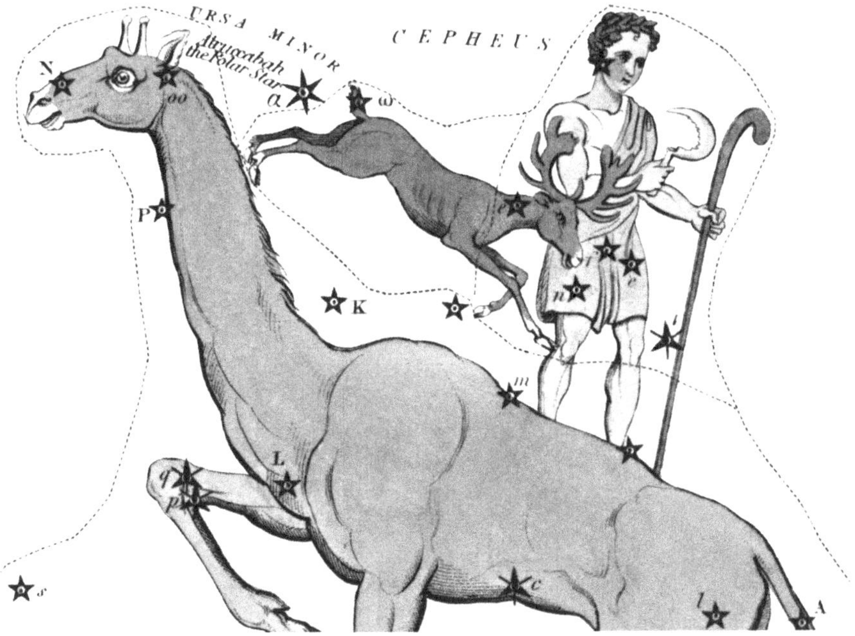

20.5 Detail of Plate 2 from *Urania's Mirror* (1825) showing Camelopardalis, Rangifer, and Custos Messium (Chap. 8). The full plate is reproduced in Fig. 8.8

is *raingo*, the Saami word for the animal. For the species *tarandus*, Linnaeus referenced[6] Chap. XXX of Ulisse Aldrovandi's *Quadrupedum omnium bisulcorum historia*... (1621). The English name "reindeer" derives from the Old Norse *hreinn*, meaning "horned animal," itself descending from the Proto-Indo-European **kroinos* via the Proto-Germanic **hrainaz*. "Caribou" is a French loanword, derived from the word *qualipu* in the Eastern Algonquian Mi'kmaq language. The Mi'kmaq word means "snow shoveler," which refers to the behavior of the animal in which it digs through the snow with its front hooves in the search for food.

Anatomically modern reindeer emerged during the Middle Pleistocene; the oldest known reindeer fossils date to roughly 650,000 years before present in northern Europe. In *Pleistocene Mammals of Europe*, Björn Kurtén wrote[7] "That the species is so old may seem surprising as it is a highly specialized form, but of course there is no definitive proof that the reindeer had acquired all of its modern adaptive characters at that early date." By the latest stage of the Pleistocene, the reindeer ranged across much of Europe from Spain

[6] Aldrovandi used *tarandus* and *rangifer* to refer to two distinct animals, the reindeer and the deer, respectively. Both were illustrated in *Quadrupedum omnium bisulcorum historia* ..., reproduced in Fig. 20.6.

[7] London: Weidenfeld & Nicolson (1968), p. 170.

20.6 Engravings of two animals depicted in Ulisse Aldrovandi's *Quadrupedum omnium bisulcorum historia* ... published at Bologna in 1621. *Left*: "Tarandus" (Chap. XXX, page 861). *Right*: "Rangifer" (Chap. XXI, page 863)

to southern Russia. It was a favored game animal of humans until its numbers dwindled in the Mesolithic as the population moved north with the retreating ice sheet. In much of the reindeer's modern range, humans are its main predator.

Reindeer are among the largest mammals in their native range. Males are about 2 m in length, stand 0.8–1.5 m tall at the shoulder and weigh between 160 and 180 kg, although weights in excess of 300 kg have been measured; females are slightly shorter (1.6–2 m) and weigh less (80–120 kg) than males. Newborn calves average about 6 kg and grow rapidly, often doubling their weight in 10–15 days.

Reindeer are the only cervid species in which both genders grow antlers. While there is significant variation in antler size among populations, the typical bull reindeer's antlers are second in size among extant deer species only to those of the moose; bull antlers can be over a meter in width. Their fur color varies considerably, among individuals within a population, subspecies, and even season. Northern subspecies tend to be both smaller on average and lighter in color than southern subspecies.

A ruminant animal, reindeer possess a four-chambered stomach to break down plant matter. The favored winter diet of reindeer is lichen, particularly reindeer moss (*Cladonia rangiferina*). To digest this tough material, they are the only known mammal species to make use of the digestive enzyme lichenase, which converts the complex glucan lichenin into simple glucose. In other seasons, their lichen diet is augmented by leaves from deciduous trees and native species of grasses and sedges. In times of extreme nutritional stress, reindeer are known to have eaten small rodents, bird eggs and fish. In order to find continuous sources of food, herd animals like reindeer must stay on the move. While some subspecies are more sedentary, most reindeer are migratory animals that make long travels of several hundred kilometers per year between their preferred winter and summer ranges. If a herd is sufficiently small and food plentiful, it may migrate either slowly, infrequently, or not at all.

Toward the end of summer, in early September, bull reindeer begin shedding the velvet covering of their antlers and begin their annual autumn migration. Their necks swell considerably in September as their testosterone levels surge in anticipation of the rut, or breeding season. Males fight for females with increasing frequency as the month proceeds, with breeding taking place in mid-to-late October. After a seven-month gestation, females calve between mid-May and early June; most births are single, and the typical adult cow reindeer is pregnant each year. The calves are highly vulnerable to predators such as bears and wolves. Reindeer evolved a strategy for dealing with the threat by overwhelming predators with a large numbers of births over a relatively short period of time, swamping the predator demand for food.

Reindeer are extremely well-adapted to the bitter cold characterizing much of the year over their natural range. They cope with the cold air by warming it through specialized nasal turbinate bones that have the effect of significantly increasing the surface area of the nasal cavity; water vapor produced in the animals' respiration is captured before exhalation and used to moisten dry air before it is inhaled into the lungs. Their hooves dynamically adapt to the conditions of each season. In summer the land is relatively soft and wet from snowmelt, so the footpads soften and become spongy to provide extra traction on slippery surfaces. As the land hardens during the transition to winter, the footpads shrink and toughen, retracting beneath the rim of the hoof; the durable hoof material then cuts into ice for better traction. Reindeer also use their hooves to dig into the snow in search of reindeer moss. Their fur has also evolved to protect them from the cold, consisting of a dense undercoat topped by a longer overcoat whose hairs consist of hollow strands, providing some thermal insulation against heat loss due to winds.

Maupertuis made some rather honest observations[8] about reindeer and their various uses by the Lapps in "Relation d'Un Voyage Fait Dans La Lapponie Septentrionale, Pour Trouver Un Ancien Monument" ("Relation Of A Journey Made In Northern Lapland To Find An Ancient Monument") from *Les oeuvres de Mr. de Maupertuis* (1752):

---

[8]"Les Rennes méritent que nous en disions ici quelque chose. Ce sont des especes de Cerfs, dont les cornes fort rameuses jettent leurs branches sur le front. Ces animaux semblent destinés par la Nature, à remplir tous les besoins des Lappons. Ils leur servent de Chevaux, de Vaches, & de Brebis.

"On attache le Renne à un petit Bateau, appellé Pulka, pointu par devant pour fendre la neige; & un homme, moitié assis, moitié couché dans cette voiture, peut faire la plus grande diligence, pourvu qu'il ne craigne, ni de verser, ni d'être à tous moments submergé dans la neige.

"La Chair des Rennes est excellente à manger, fraîche, ou sêchée. Le lait des femelles est un peu acre, mais aussi gras que la crême du lait des Vaches; il se conserve longtems gelé, & les Lappons eu font des fromages, qui seroient meilleurs, s'ils étoient faits avec un peu plus d'art & de propreté." (pp. 322–323).

> We should say something here about the reindeer. These are a species of deer whose strong, branching horns project toward the front. These animals seem intended by Nature to fulfill all the needs of the Lapps. They functioned as their horses, cows and sheep.
>
> The Lapps harness the reindeer to a small boat, called a *pulka*, with a pointed nose to plow through the snow; and a man, half sitting, half lying in the vehicle, can make great headway, provided he is neither afraid to be fall out nor to be at all times buried in snow.
>
> The meat of the reindeer makes for excellent dining, whether fresh or dried. Their milk is slightly pungent, but also is more fatty than cream from the milk of cows; it keeps frozen for a long time, and from it the Lapps made cheeses, that would be better if they were made with a little more art and cleanliness.

DISAPPEARANCE

While Rangifer saw wide circulation throughout the bulk of the nineteenth century, its undoing began relatively early when it was left out of Elijah Hinsdale Burritt's popular *Atlas Designed to Illustrate the Geography of the Heavens* (1835; Fig. 20.7). Burritt showed Rangifer's stars but labeled the figure of neighboring Camelopardalis seemingly to indicate he felt the stars belonged to the Giraffe. It suffered another blow upon being left out of Friedrich Argelander's *Neue Uranometrie* (1843); Argelander drew a boundary between Cassiopeia and Cepheus on his circumpolar Plate 1 to more or less evenly distribute Rangifer's stars between them.

The Reindeer was increasingly left off of charts in the 1850s and 1860s; notable examples include Table II of Otto Möllinger's *Himmels-Atlas* (1851); Plate 17, Map 5 of Alexander Keith Johnston's *Atlas of Astronomy* (1855), and Plate II of Ludwig Preyssinger's *Astronomie Populaire ou Description des Corps Célestes* (1862). It was left out of the charts in Richard Anthony Proctor's *The Constellation-Seasons* (1876), and does not appear anywhere in either William Henry Rosser's *The Stars and Constellations* (1879) or Charles Pritchard's catalog of stars in *Uranometria Nova Oxoniensis* (1885).

Among its last appearances on a map is as "Tarandus" in Eliza A. Bowen's *Astronomy By Observation* (1888); it was mentioned briefly in *Poole Bros. Celestial Handbook* (1892) as "Rangifer," and attributed to "Lemonnier, 1776, A.D." By the end of the century, it was functionally extinct; in 1899, Allen noted "It has seldom been figured, and now is never mentioned." Along with neighboring Custos Messium (Chap. 8), its stars were split more or less evenly between Cassiopeia and Cepheus when the modern constellation boundaries were drawn in 1928. Even then, some memory of the former figure was retained; the furthest corner of the boundary between Cepheus and Camelopardalis was drawn such that it neatly contained the Reindeer's hindquarters (Fig. 20.8).

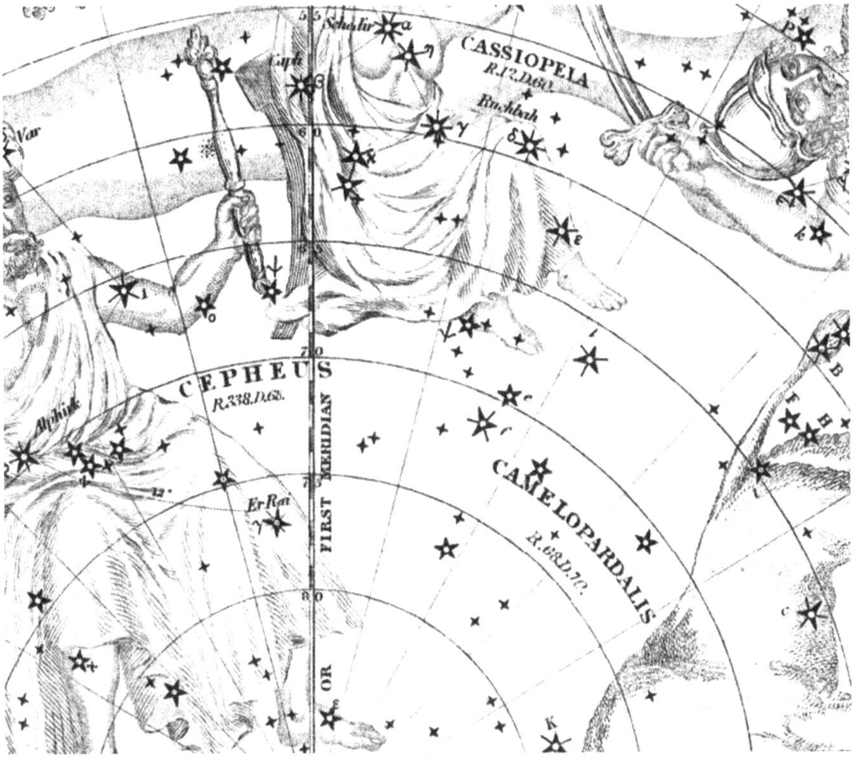

20.7 The component stars of Rangifer shown unlabeled on Plate VI of Elijah Hinsdale Burritt's *Atlas Designed to Illustrate the Geography of the Heavens* (1835)

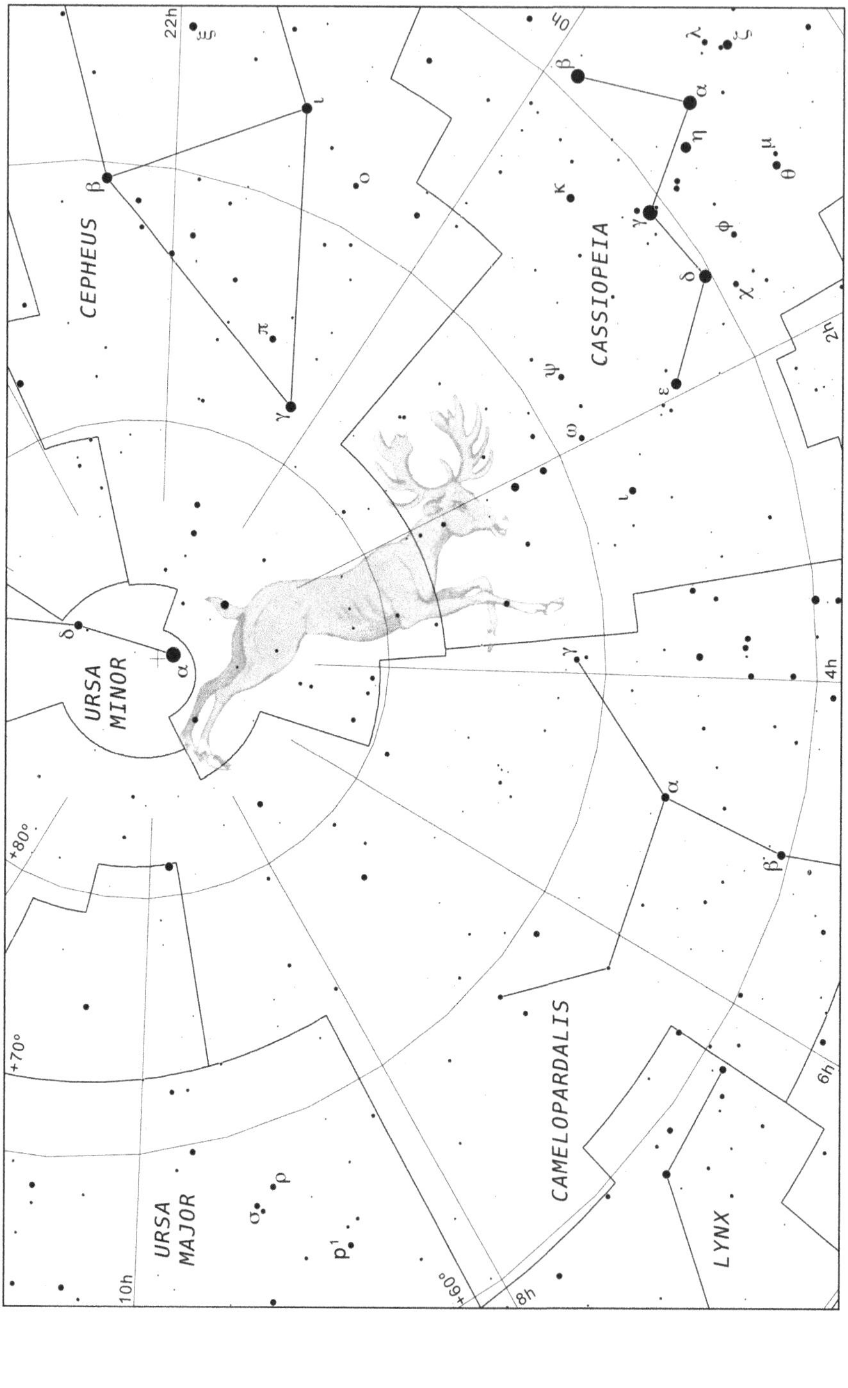

20.8 The figure of Rangifer from Plate 2 of *Urania's Mirror* (1825) overlaid on a modern chart

21

Rhombus

The Bullroarer

Genitive: Rhombi
Abbreviation: Rho
Alternate names: "Rhombo" (Habrecht, 1621); "Rhombois" (Royer, 1679a); "Quadratum" (Allard, 1706)
Location: The region bounded by α and β Reticulum, and γ and ν Hydri

ORIGIN AND HISTORY

This constellation first appeared on a 1621 globe called *Globus Coelestis* published by Isaac Habrecht II (1589–1633). Habrecht was a professor of astronomy and mathematics at the University of Strasbourg and a doctor of both medicine and philosophy.[1] He was the son of another Isaac Habrecht (1544–1620), who with his brother Josias (1552–1575) was a clockmaker in Schaffhausen, Switzerland. The Habrecht brothers were commissioned to build Strasbourg's second astronomical clock, completed in 1574. The younger Isaac took up the skill of fine clockmaking from his father, producing several 'clockwork' globes, some of which featured maps of the night sky.

[1] Habrecht is remembered in the history of science for having made the first correct determination by geometric triangulation of the altitude of a meteor while in flight through the Earth's atmosphere. Working in collaboration with Wilhelm Shickard (1592–1635) he concluded in 1623 that the glowing trail of a meteor is produced while it is roughly 20 German miles, or about 150 km, above the surface of the Earth. The result correctly contradicted the Aristotelian proposition that meteors were a phenomenon specifically of the *lower* atmosphere.

J.C. Barentine, *The Lost Constellations*, Springer Praxis Books,
DOI 10.1007/978-3-319-22795-5_21

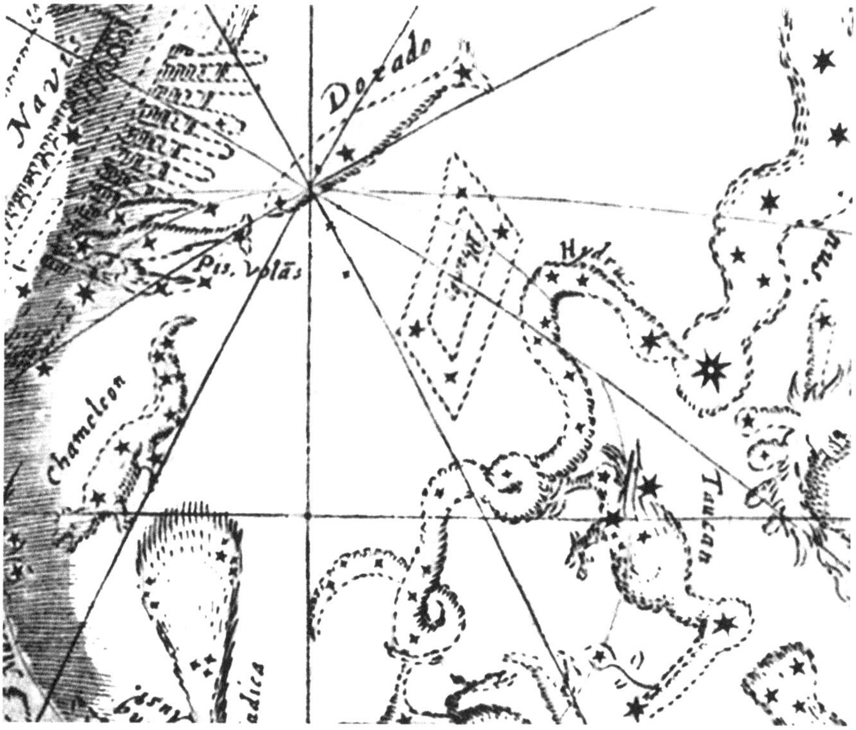

21.1 Rhombus shown in its original incarnation as "Rhombo" on Isaac (II) Habrecht's 1621 *Globus Coelestis*. Part of Argo Navis (Chap. 5) appears in the *upper left corner*

Habrecht named the figure "Rhombo," and as shown on his map (Fig. 21.1) it seems to be a square or rhomboidal framework with an open center.[2] It is not clear from early charts whether the figure was truly intended to be a square, with all sides at right angles to each other, or a parallelogram with sides at two pairs of different angles. In *Mercurii philosophici firmamentum firmianum* (1730), Corbinianus Thomas wrote[3] that its creation was wholly obvious, being "The result of the geometric shape of the constituent stars." Its name provides a clue as to the significance of the figure represented (see Iconography).

Moore and Rees (2011) simply referred to it as "the Square" and cited "Allard 1700" for its origin, although its mention by Bartsch in *Usus Astronomicus* (1624) makes clear[4] that Habrecht's reference preceded Carel Allard's map by some 85 years:

[2]The figure appeared again in Habrecht's *Planiglobium Coeleste, et Terrestre* (1628).

[3]"Ob formam stellarum in hanc fig[urum] Geometricam coëuntium," p. 210.

[4]"RHOMBUS inter 2 nubeculas, primum in Cl[arissimus] Dn. D[octor] Habrechti Globo formatus, ex 4. stell. angulos totidem constituentibus," p. 66.

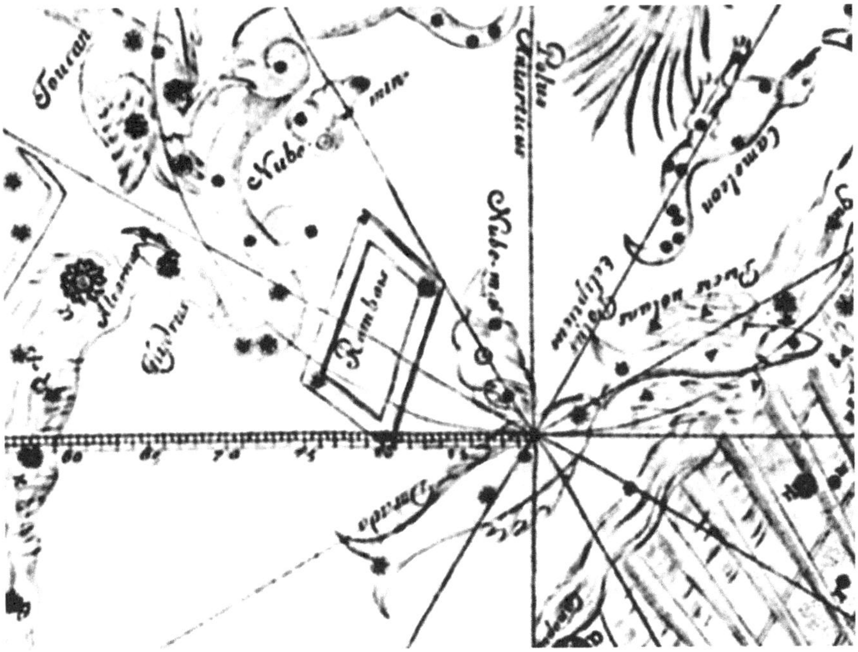

21.2 "Rombois" appears just to the *left of center* in Augustin Royer's *Cartes du Ciel* (1679a)

> RHOMBUS between *[the]* two *[Magellanic]* clouds, first formed on the famous scholar Dr. Habrecht's globe, *[consisting]* of four stars and as many angles.[5]

It was similarly noted by Edward Sherburne in *The Sphere of Marcus Manilius* (1675):

> RHOMBUS, which Habrechtus hath formed in his Globe out of four stars constituting each Angle thereof.

Augustin Royer showed Habrecht's figure, labeled "Rombois," in *Cartes du Ciel* (1679a; Fig. 21.2), and listed its stars in the companion tables *Tabula Universalis Longitudinum et Latitudinum Stellarum.* His label for the constellation almost certainly indicates the word for "rhombus," which in modern French would be rendered *rhombi.*

Vincenzo Coronelli may have followed Royer directly in adding "Romboide" ("parallelogram" in Italian) to his southern hemisphere map in *Epitome cosmografica* (1693; Fig. 21.3, top). Coronelli's depiction of the figure shows some kind of ornament within the parallelogram-shaped frame, suggesting a fully solid object unlike the manner in which it was shown by previous cartographers. Future maps that included the figure

[5]The same text is repeated in *Planisphaerium stellatum* (1661). Even though this edition of Bartsch's work included a northern hemisphere chart, there is no equivalent southern hemisphere chart on which Rhombus might have been depicted.

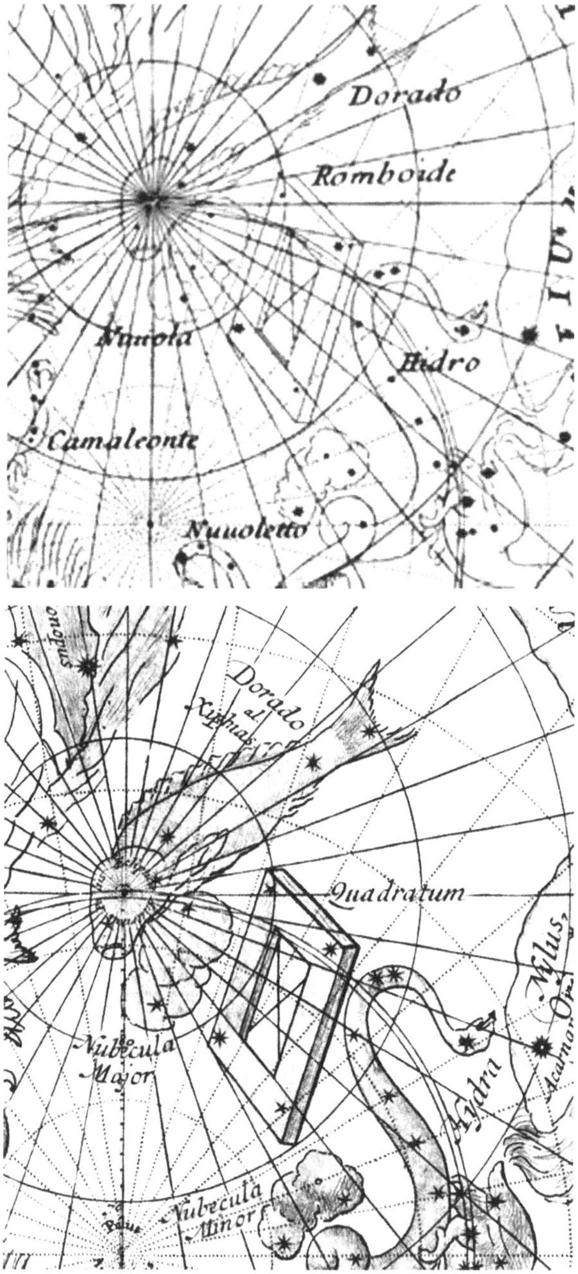

21.3 Two strikingly similar depictions of Rhombus showing clear influence of one cartographer on another. *Top*: “Romboide” shown on Plate 18 of Vincenzo Coronelli’s *Epitome cosmografica* (1693). *Bottom*: “Quadratum” from Carel Allard’s 1706 map *Hemisphaerium meridionale et septentrionale planisphaerii coelestis*

invariably involved a design substantially identical to Coronelli's. Carel Allard copied almost every detail in Coronelli's map in *Hemisphaerium meridionale et septentrionale planisphaerii coelestis* (1706; Fig. 21.3, bottom) in which he Latinized the constellation's name to "Quadratum" (from *quadrātum*, "square"). It appears that his was the only widely-distributed chart to ever show that name.

The constellation dwindled in popularity among mapmakers after 1700. In perhaps its last appearance, Corbinianus Thomas labeled it "Rhombus" in *Mercurii philosophici firmamentum firmianum* (1730; Fig. 21.4).

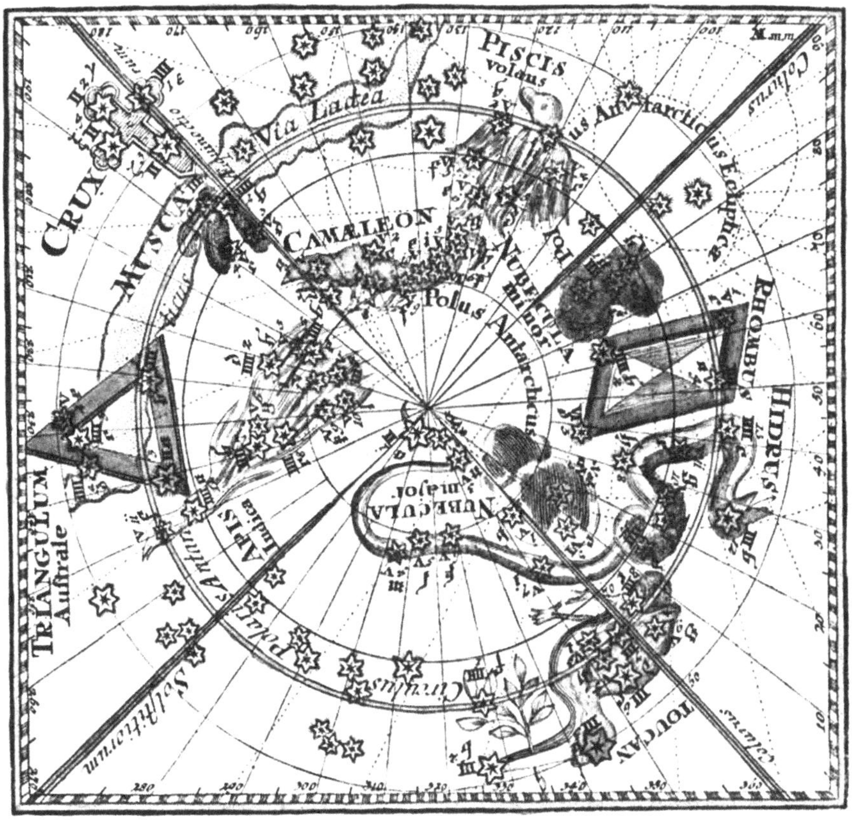

21.4 A map showing the constellations near the south celestial pole (*center*) from Corbinianus Thomas' *Mercurii philosophici firmamentum firmianum* (1730). Rhombus is indicated to the *right of center*

Where Was Habrecht's Constellation Actually Located?

Various authors from the nineteenth century on have generally stated as fact that "Rhombus" was repurposed in toto by Nicolas Louis de Lacaille in 1752 to form his constellation Reticulum (the Reticle; Fig. 21.7), which remains recognized by the International Astronomical Union as canonical. Richard Hinckley Allen (1899) wrote as much in his discussion of Reticulum:

> Reticulum Rhomboidalis, the Rhomboidal Net, is generally supposed to be of La Caille's formation as a memorial of the reticle which he used in making his celebrated southern observations; but it was first drawn by Isaak Habrecht, of Strassburg, as the Rhombus, and so probably only adopted by its reputed inventor.

But while the modern boundaries of Reticulum certainly contain some of the stars that originally defined Rhombus, it seems impossible that the *figure* implied by Reticulum is identical in shape and extent to the that of Rhombus depicted on charts of the seventeenth and early eighteenth centuries.

Lacaille's figure was indeed a rhombus with corners consisting of α, β, δ, and ϵ Reticuli, but that shape is oriented at a distinct angle relative to the figure shown on the Coronelli/Allard map (Fig. 21.3), two of whose sides are oriented much more nearly parallel to hour lines of right ascension. Contemporary charts all show a quadrilateral figure nearly as large in extent as nearby Dorado, which spans more than 10° on the sky. Based on those charts, it appears more likely that the brighter stars of modern Reticulum form only one of the sides of Rhombus (α and β Reticuli). γ Hydri is almost certainly another corner.

That the fourth corner of the figure is a star in Hydrus, not Mensa, is suggested by superimposing part of Allard's map on a modern star chart as shown in the top panel of Fig. 21.5. Here, geometric distortion between the maps' different projections has been accounted for by aligning the stars α, β, and γ Doradus (defining the body of Dorado), and β Hydri (the head of the Water Snake). Given the time in which Allard's map was produced—before the precise positional measurements of Lacaille a half-century later—the match is reasonably good. The superimposed maps affirm that α and β Reticuli form one side of Rhombus, and that γ Hydri forms one corner. The fourth corner must therefore be ν Hydri. It is clear, in any case, that the small quadrilateral defining the figure of modern Reticulum cannot possibly be the same as envisioned by cartographers back to Isaac Habrecht himself.

ICONOGRAPHY

Isaac Habrecht evidently left no written description of Rhombus or his understanding of what the figure represented, but some useful insight into his motives is indicated by its shape. The constellation at first appears to be some sort of tool, much like a modern carpenter's square, perhaps with adjustable corners so as to permit various angles to be formed with the sides. However, examination of contemporary engravings and woodcuts

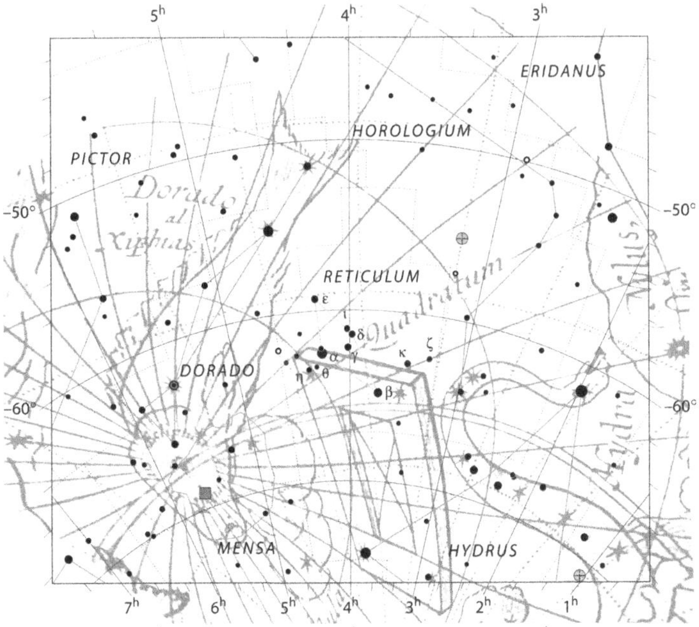

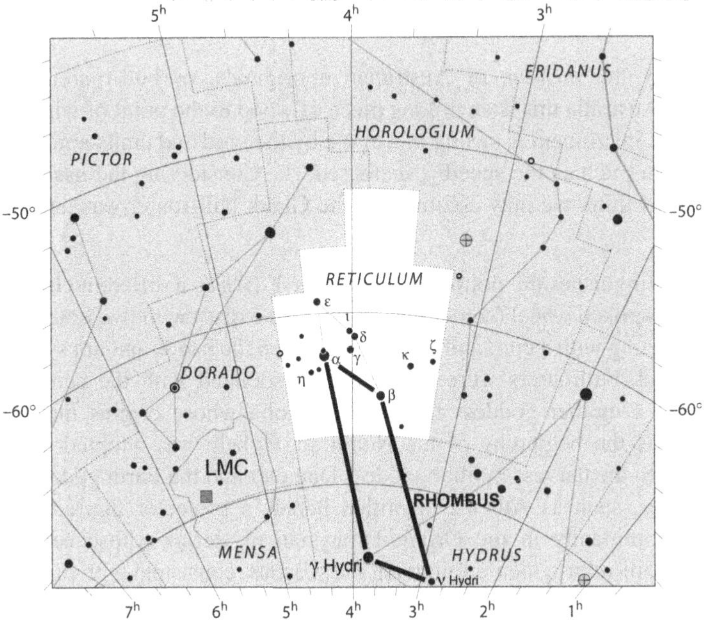

21.5 *Top*: an excerpt of Carel Allard's (1706) map, corrected for geometric distortion, superimposed on an International Astronomical Union/*Sky And Telescope* Magazine map centered on the modern constellation Reticulum. *Bottom*: a proposed construction of Rhombus (*heavy black lines*) on the same modern map. IAU/*S&T* map used with permission

of scenes such as carpenters' and stonemasons' shops does not reveal any similarly-shaped tool; neither does the figure seem to represent some type of measurement device.

The answer to understanding what Rhombus represents may be found on Habrecht's original map. His label, "Rhombo," corresponds to the Latin word *rhombō* and indicates a wooden object which would now be properly called a "bullroarer." The device almost certainly predates the rise of civilization. Christopher Partridge notes[6] that bullroarers are used among various modern, primal religions in different parts of the world in imitation of the voice of a supreme deity, or of the voices of dead ancestors, and that stone objects resembling bullroarers have been found in Upper Paleolithic caves, but their use "symbolizing the presence of the Supreme God or other supernatural beings ... is far from certain." Rather than transmitted knowledge, it appears that the invention of the bullroarer occurred spontaneously among cultures widely separated by geography, but in essentially all cases it is found particularly within religious and ceremonial contexts (Fig. 21.6).

The Latin *rhombō* turn derived from the ancient Greek ῥόμβος (*rhombos*), 'something that spins', in turn derived from ῥέμβω, meaning 'to turn around repeatedly'. When attached to a length of rope and swung in a circle overheard the flow of air over the object yields a loud hissing or growling sound whose volume grows louder or softer depending on the force of its motion. The Cambridge classicist Andrew S.F. Gow drew the connection between the Greek word in literature and the noisemaking instrument, noting[7] that the object labeled ῥόμβος is certainly the origin of the term for the geometrical figure "rhombus." It is, Gow argues,

> pretty plainly the *turndun* of Australian aboriginals, or bull-roarer of modern England. In Australia this is an oblong piece of wood to the point of which a cord is attached. The instrument is swung in a circle by the cord and emits a muttering roar which rises in pitch as the speed is increased. ... Considering the use of the word in geometry, I think we may assume that the Greek bull-roarer was usually of this pattern.

Gow clearly distinguishes the ῥόμβος from the ἴυγξ (*iynx*), a different Greek instrument consisting of "a spoked wheel (sometimes it might be a disc) with two holes on either side of the centre" strung with a cord and operated between the hands and not swung overhead.

Ancient Greek bullroarers were particularly associated with the ceremonies of the cult of Cybele, a mother goddess figure of Anatolia whose origins may extend back as far in time as the beginning of the Neolithic. Cybele was adopted by the Greeks, lending aspects to the harvest-mother goddess Demeter and the Earth goddess Gaia; some Greek city-states, such as Athens, identified her as a protector figure. The bullroarer also figured prominently in the Orphic-Dionysian mysteries whose ceremonies were founded on a death-rebirth theme following the seasons, consistent with practices in many

[6] *Introduction to World Religions* (2nd ed.); Minneapolis: Fortress Press (2013).

[7] "ΙΥΓΞ, ΡΟΜΒΟΣ Rhombus, Turbo". *Journal of Hellenic Studies*, 54(1), pp. 1–3 (January 1934).

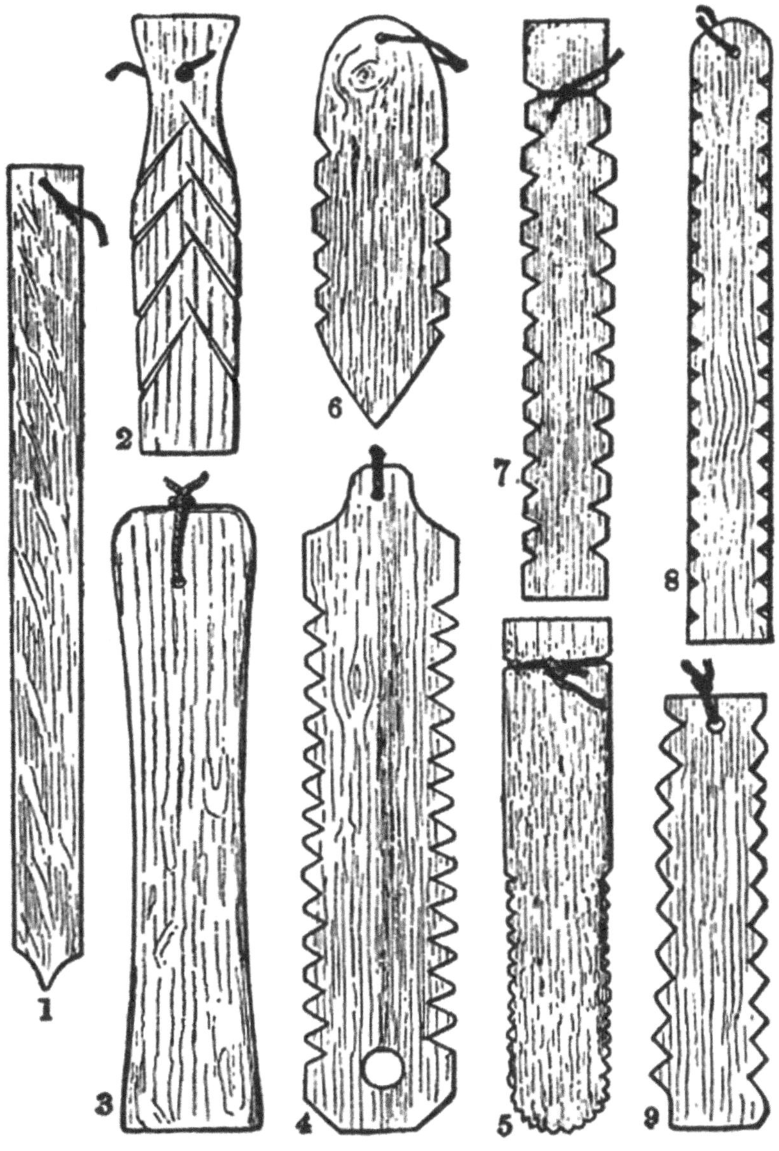

21.6 "Bull-Roarers from the British Islands" in Alfred C. Haddon's *The Study of Man*, New York: G.P. Putnam's Sons (1898), p. 221. The sources are: (*1*) "Ballycastle, County Antrim"; (*2*) "Warwickshire, Staffordshire and Shropshire"; (*3*) "Warwickshire"; (*4*) "Montgomeryshire"; (*5*) "Derbyshire"; (*6*), (*7*), (*8*) "Norfolk"; and (*9*) "Balham (Surrey)". Edge serrations on some examples were used to modify the air flow over the bullroarers in order to change their acoustic properties

ancient agricultural cults. Aeschylus described[8] the sound of the bullroarer in the Orphic-Dionysian rites in a fragment of one of this lost plays:

> And bull-voices roar thereto from somewhere out of the unseen, there are fearful semblances... From an image as it were the sound of thunder underground is borne on the air heavy with dread.

In a classical context, the bullroarer was clearly intended as a means of commanding silence and reverence in conjuring the presence of a god. In the Age of Exploration, travelers to the New World and the southern hemisphere encountered similar devices among the indigenous populations of faraway lands. It is possible, but unproven, that Isaac Habrecht placed a bullroarer in the southern hemisphere sky as a nod to the tales returned from voyages of discovery. In this, he would have been following the lead of the travelers themselves, such as the exotic animals turned into constellations by Pieter Dirkszoon Keyser and Frederick de Houtman while in the employ of the cartographer Petrus Plancius during the generation before Habrecht.

A closer match may be found in the figure of Triangulum Australe, a constellation that remains part of the modern canon. It was introduced as "Triangulus Antarcticus" by Plancius on a celestial globe published by the Dutch cartographer Jacob Floris van Langren at Amsterdam in 1589, although Plancius positioned the figure south of Argo Navis (Chap. 5) in an area of the sky now occupied by the constellations Volans and Chamaeleon. On Figure Aaa of his *Uranometria* (1603), Johann Bayer placed the Triangle in its modern position in the space bounded by Ara, Apus and Circinus, and gave the figure its current name. Bakich (1995) suggests that Plancius' figure was previously noted by the Italian explorer Amerigo Vespucci (1454–1512) "as early as 1503." According to Kanas (2007), Vespucci described a number of southern stars used as navigational aids including "a textual description of three stars that probably are in Apus or Triangulum Australe" published in *Mundus Novus* (1504), a pamphlet in the form of alleged correspondence between Vespucci and his Florentine patron, Lorenzo di Pierfrancesco de' Medici (1463–1503).

By Habrecht's time, the southern sky had already been initially charted, but there were many gaps that would not be filled until the work of Nicolas de Lacaille was published more than a century later. Furthermore, the best maps available to Habrecht still contained significant positional errors; the case of Triangulum Australe shows that at the turn of the seventeenth century, the locations of southern hemisphere constellations were still subject to major revisions. It is possible that the quadrilateral figure of Rhombus was some navigational convenience to southern sailors communicated to Habrecht as if it were a constellation in its own right. Short of a description in Habrecht's own words surfacing at some future date, the story of this figure will probably never be fully known.

[8]Fragment 57, translated by Jane Ellen Harrison in *Themis: A Study of the Social Origins of Greek Religion*, Cambridge: Cambridge University Press (2010), from August Nauck's *Tragicorum graecorum fragmenta*, Leipzig: Teubner (1889).

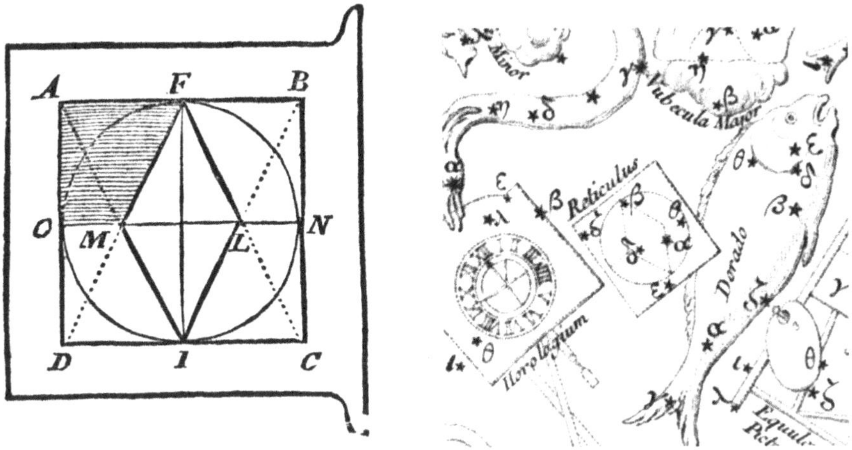

21.7 *Left*: drawing of the 'third-style' reticle used by Nicolas Louis de Lacaille in his southern hemisphere observations of the 1750s, from page 34 of *Coelum Australe Stelliferum* (1763). *Right*: Lacaille's constellation Reticulus, from the chart "Coelum Australe" in same work, partially retained the parallelogram shape that characterized its predecessor Rhombus

(Incorrect) Identification with Reticulum

Isaac Habrecht's constellation of 1621 was almost certainly misidentified by later authors as either identical or substantially similar to Lacaille's Reticulum (Fig. 21.7, right), as demonstrated in the previous section. When Lacaille introduced the constellation in *Coelum Australe Stelliferum* (1763) he drew the figure of a reticle, an optical device used in visual telescopic observations to gauge angular distances and separations. He described[9] how he came to form the figure in a 1752 report to the Académie Royale des Sciences:

> To fill the large empty spaces between the ancient constellations, I have assumed new ones: I put among them the figures of the main instruments of the arts. … The rhomboidal *Reticle [is]* a small astronomical instrument used to compile this catalog: it is made from the intersection of four lines drawn from each corner of a square to the middle of the two opposing sides.

As drawn by Lacaille (Fig. 21.7, left), the reticle is a glass slide onto which is ground a pattern of fine lines. When inserted in the near field of a telescope eyepiece, the lines are in focus as are stars and other objects at infinity. To facilitate easier reading of the pattern under dark conditions, reticles are sometimes faintly illuminated from the side; most of

[9]"Pour remplir les grandes intervalles vides entre les constellations anciennes, j'en ai supposé de nouvelles: j'y ai mis les figures des principaux instruments des arts. Le *Reticule* rhomboïde, petit instrument astronomique qui a servi à dresser ce catalogue: on le construit par l'intersection de quatre droites tirées de chaque angle d'un carré au milieu de deux côtés opposés." From *Histoire de l'Académie Royale des Sciences* for the year 1752, page 588.

the light passes through the glass, but some is scattered at angles by the etched lines and reaches the observer's eye. Carefully inscribed patterns on the glass help the observer estimate angular distances to a precision of arcseconds or better.

Exactly how Reticulum came to be improperly associated with Rhombus is evidently lost to history. Certainly Lacaille was aware of earlier maps such as those of Corbinianus Thomas, but in neither his Académie Royale des Sciences report nor *Coelum Australe Stelliferum* did he suggest that his invention was intended to replace or supersede Habrecht's constellation. It may be simply coincidence that Lacaille's reticle sported a rhomboidal lozenge whose shape echoed that of a nearby constellations that had by that time appeared on charts for a century. As the association between the two sources came to be understood as received knowledge the confusion was visually echoed in the figure shown by most cartographers for Reticulum, which retained the rhomboidal shape originally drawn by Lacaille himself.

DISAPPEARANCE

Rhombus seems to have vanished from charts by the time of Lacaille, who may have taken this as license to use some of its stars in creating Reticulum. While it is found on Carel Allard's 1706 and in Corbinianus Thomas' 1730 charts, it was omitted entirely by de La Hire (1702), de Broen (1709) and Doppelmayr (1742). It seems safe to conclude that the invention of Reticulum was simply not a wholesale rebranding of Rhombus but rather the formation of an entirely new constellation. However, his reference to "le Romboide" seems to indicate that he was at least aware of what he was doing, and may well have been a reference for the former constellation he disassembled and part of which he repurposed. Today, the stars of Isaac Habrecht's Rhombus are distributed more or less evenly between the modern constellations of Reticulum and Hydrus (Fig. 21.8).

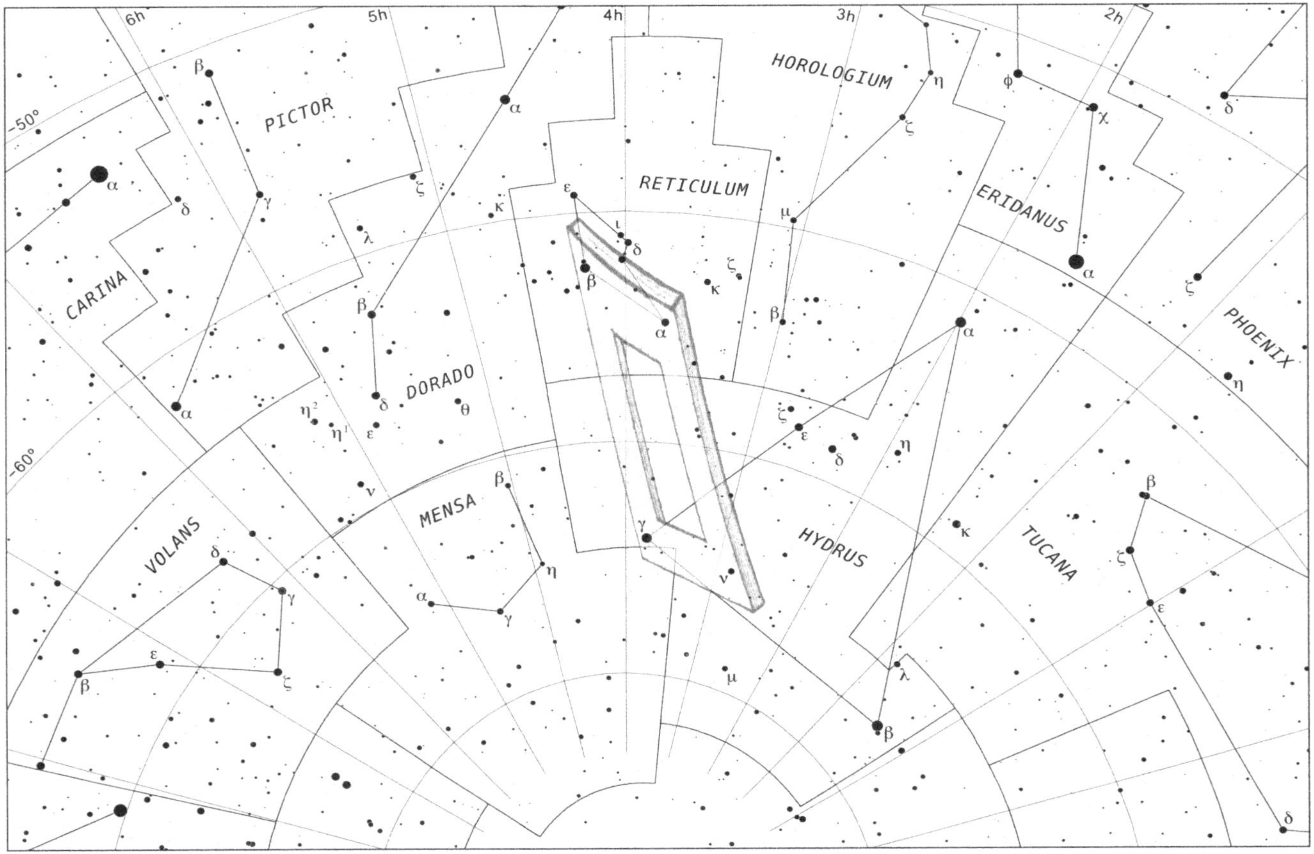

21.8 The figure of Rhombus (as "Quadratum") from Carel Allard's *Hemisphaerium meridionale et septentrionale planisphaerii coelestis* (1706) overlaid on a modern chart. Note that this is a suggested reconstruction; see main text for details

22

Robur Carolinum

The Royal Oak

Genitive: Roboris Carolini[1]
Abbreviation: RbC
Alternate names: "Rober Caroli", "Robur Caroli II" (Green, 1824); "Robur Caroli" (Doppelmayr 1742; Bode 1801b; Ryan 1827; Kendall 1845; Chambers 1877; Colas 1892)
Location: Between Crux and Carina[2]

ORIGIN AND HISTORY

The introduction of this constellation (Fig. 22.1) is attributed to Edmond Halley in honor of his patron, King Charles II of England. At the time of its introduction, Halley was recently returned to England from a trip to the South Atlantic island of St. Helena in 1676 to observe and catalog stars of the southern hemisphere. He presented the results of the survey to the Royal Society in 1678 and published them the following year in *Catalogus Stellarum Australium*. His citation read, "In memory of the hiding place that saved Charles II of Great Britain etc., deservedly translated to the heavens forever." His flattery of the King paid off: as Richard Hinckley Allen (1899) put it, "This invention secured for Halley his master's degree from Oxford, in 1678, by the king's express command."

Prior to Halley's innovation, given the ancient understanding of the stars composing Argo, mapmakers struggled with an inconvenient truth: Argo, formally, had no prow. Nature did not exactly oblige in giving astronomers a complete set of stars with which to define the Ship, so those seeking a figural depiction of Argo were left to come up with

[1]Following Allen's example: *"Halley's 2d-magnitude α Roboris was changed to β Argūs, now in Carina."*
[2]*"To the east of Argo Navis"* (Green, 1824); *"Between the Centaur and the northern part of Argo Navis"* (Kendall, 1845); *"25 stars, including Beta Carinae"* (Bakich, 1995).

J.C. Barentine, *The Lost Constellations*, Springer Praxis Books,
DOI 10.1007/978-3-319-22795-5_22

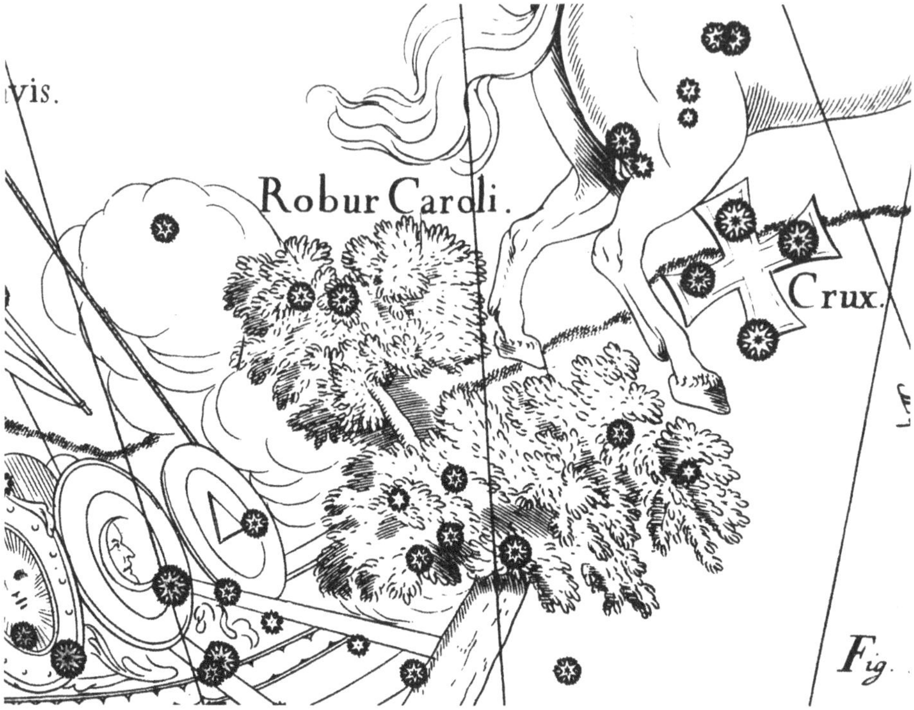

22.1 Detail of Robur Caroli depicted on Figure Eee of Hevelius' *Prodromus Astronomiae* (1690)

ad hoc ways to hide the fact that part of the boat was literally missing. The most common way cartographers dealt with this problem was to show the front of the Ship disappearing either into a cloud of mist or behind a conveniently-placed rock; Bayer, for example, opted for the latter means in *Uranometria* (Fig. 5.4). Halley simply reinvented part of the stars defining this rock as Charles' Oak, but that still left a need for disguising the front of Argo with clouds, as in Johann Rost's *Atlas Portatilis Coelestis* (1723; Fig. 22.2).

Halley listed 12 stars in the new constellation, among which was the star now known as η Carinae, an usual type of erupting variable. While he characterized it as of the fourth magnitude, it brightened considerably in the next century. When Nicolas Louis de Lacaille produced his star catalog in the mid-eighteenth century and gave Bayer-style Greek letter designations to the bright stars, it was then near the second magnitude. After this episode it returned to fourth magnitude before enduring another eruptive episode in the mid-nineteenth century. It peaked in brightness near magnitude −0.8 in April 1843, making it the second-brightest star in the sky after Sirius. It later faded to below eighth magnitude but in the early years of the twenty-first century has again brightened to naked-eye visibility at about magnitude 5.

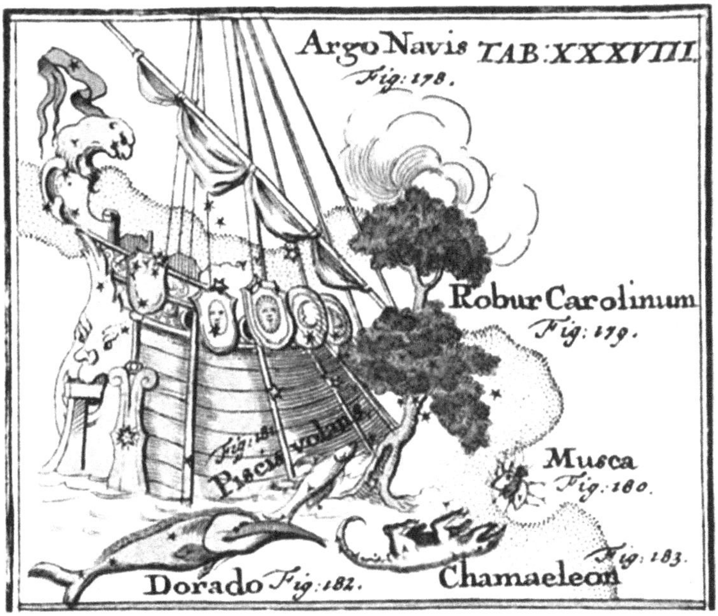

22.2 Johann Rost's depiction of Robur Carolinum in Plate 38 of *Atlas Portatilis Coelestis* (1723). The previous assignment of Robur's stars as part of a mist into which it appears Argo is sailing is kept in reference by the billowing clouds above the upper branches of the Oak

Among the first mapmakers to adopt the new figure was Ignace-Gaston Pardies in the second edition of *Globi coelestis* (1693; Fig. 22.3), although he did not label it. Not all cartographers rushed to include Halley's tree; for example, Philippe de La Hire left the rock intact on *Planisphère céleste septentrionale* 1702. In de La Hire's case, there was probably some patriotic chauvinism involved since to include the Oak would have honored the king of a foreign country that still claimed ceremonial sovereignty over France at the turn of the eighteenth century (Fig. 22.4).

In *Mercurii philosophici firmamentum firmianum* ... (1730) Corbinianus Thomas described[3] the constellation originating "From the most celebrated English astronomer, Edmond Halley, since it goes below the ecliptic, in order to devote himself freely to the observations of southern stars, on the island of St. Helena, dedicated to King Charles II of England." He included Robur Carolinum in the same plate on which he depicted Argo Navis (Fig. 5.2). As with many recently-invented constellations of the day, Jacob Green grudgingly mentioned Robur Carolinum in his *Astronomical Recreations* (1824):

[3]"A Celeberrimo *Angliæ* Astronomo, *Edmundo Hallejo*, cùm ultra æquatorem, ut meridionalium stellarum vacaret observationibus, in insula S. Helenæ dicata ageret, Carolo II. tunc Angliæ Regio consecratrum." (p. 209)

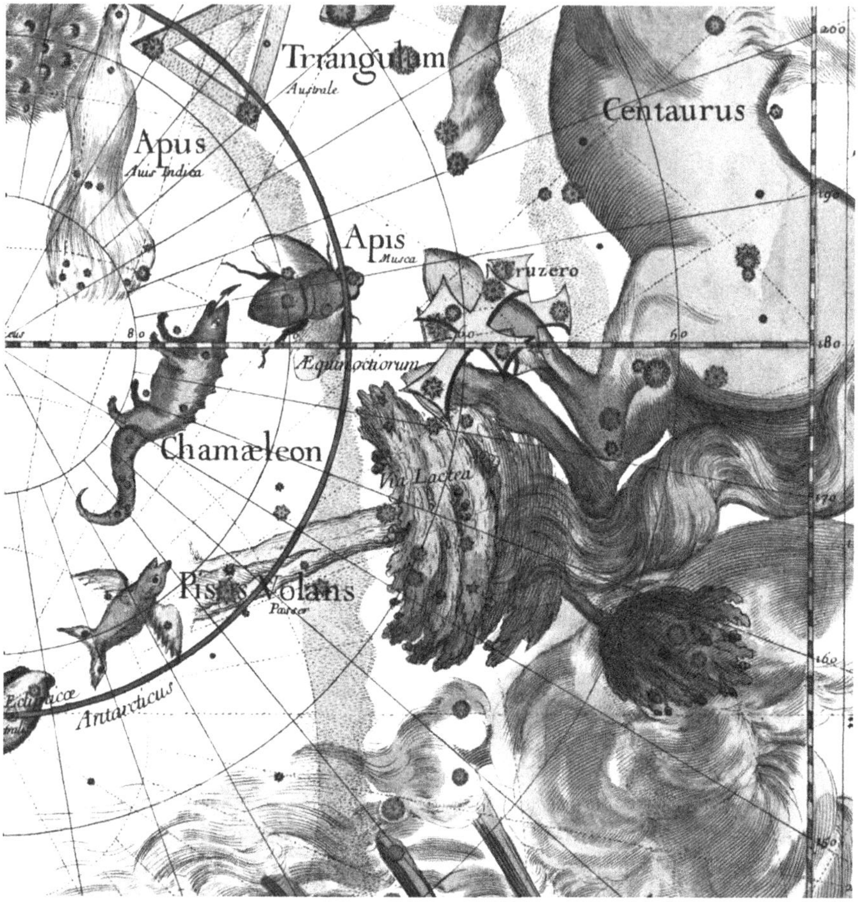

22.3 Robur Carolinum shown (unlabeled) on Plate 6 of the second edition of Ignace–Gaston Pardies' *Globi coelestis* (1693)

> We shall say but a few words concerning the small modern asterisms mentioned on this Plate.
> Rober Caroli, or Charles's Oak, is to the east of Argo Navis, and contains one star of the first magnitude. Dr. Halley introduced this asterism on the maps in 1676, when he was in the Island of St. Helena. He named it in honour of the oak whose branches concealed Charles II. after the battle of Worcester.

On the corresponding plate (17; Fig. 22.5), he only showed one star belonging to the constellation, that "of the first magnitude:" α Roboris, later β Argūs and now β Carinae.

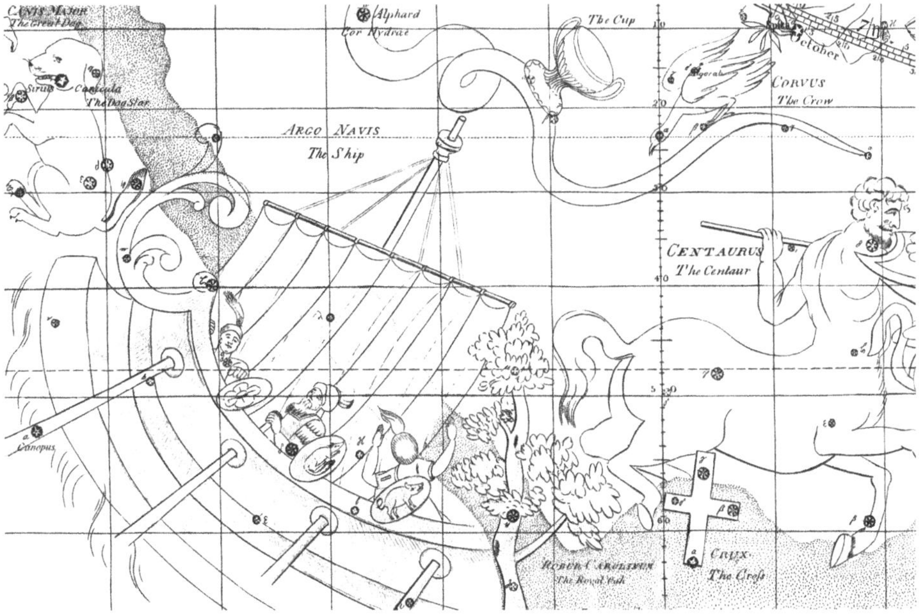

22.4 Part of William Croswell's *Mercator map of the starry heavens* (1810) from Argo Navis (*left*; Chap. 5) eastward to Centaurus; Robur Carolinum is at *bottom-center*

ICONOGRAPHY

Charles' Oak occupies a position of infamy in the story of English history. The episode occurred during the English Civil War (1642–1651), a series of political struggles and armed conflicts over the nature of government itself that pitted the anti-monarchial Parliamentarians (known as "Roundheads") against the Royalists (known as "Cavaliers"). The War was a key event in the century and a half that separated the English Reformation from the Glorious Revolution and set the stage for the emergence of the United Kingdom as a modern world power. It was against the backdrop of these events that a tree literally preserved the English monarchy.

Events of the War

The political machinations and power struggles that brought Britain to the edge of the abyss began a century before the War erupted, as Henry VIII pulled England away from Rome during the Protestant Reformation and founded the Church of England, of which he conveniently appointed himself "Supreme Governor". Thus began a consolidation of power in the monarch not seen since the days of King John (1166–1216), whose authority was abruptly curtailed when English barons cornered him in a meadow at Runnymede,

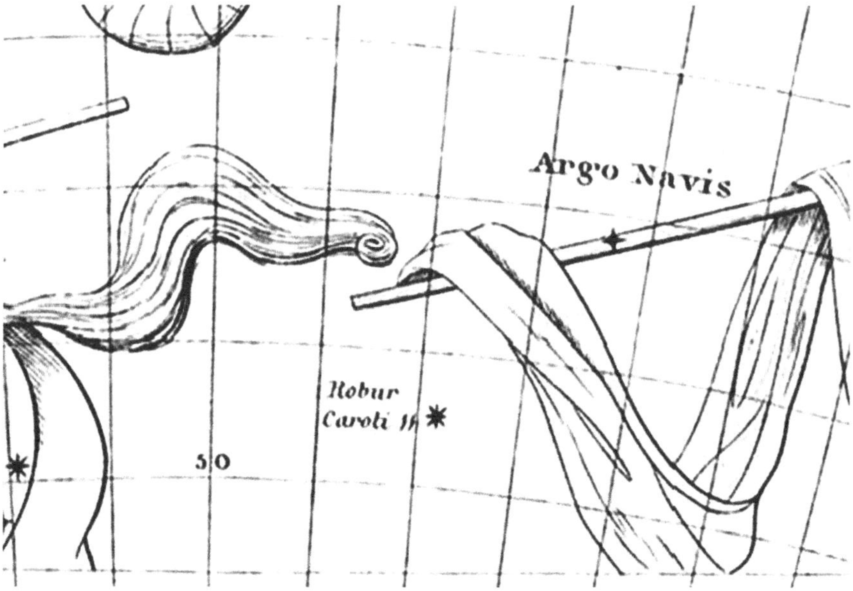

22.5 Robur Carolinum (as "Robur Caroli II") is indicated by a single star (β Carinae) on Plate 17 of Jacob Green's *Astronomical Recreations* (1824). The hind end of Centaurus is seen at *lower left*

Surrey, during the First Baron's War (1215–1217) and forced his assent to the Magna Carta. In addition to asserting new spiritual power over his subjects, Henry VIII attempted to extend his temporal realm through a series of ill-fated expeditions intended to make him the de jure King of France. On his death, the throne of England passed to his son Edward VI (1537–1553) whose untimely death at age 15 set off a fight between rival Protestant and Catholic claimants until Henry's daughter Elizabeth I emerged victorious.

After her long reign ending in 1603, the childless Elizabeth was succeeded by her cousin James VI of Scotland, who became James I of England. Consistent with the religious beliefs of most of the Scottish ruling class, James was raised in the Protestant Church of Scotland. England seemed safe for Protestantism in the hands of James, whose personal union with the English throne literally birthed the notion of a United Kingdom. But after the events of the Gunpowder Plot on the night of 4–5 November 1605, in which Catholic conspirators attempted to disrupt the state opening ceremony of the second session of James's first English Parliament with a catastrophic explosion (intended, in part, to kill the monarch), James took a hard line on English Catholics whose first loyalty was to the Pope. In May 1606, Parliament passed the Popish Recusants Act,[4] requiring an oath of allegiance denying the Pope's authority over the King and making it an act of high treason to refuse to obey the King in all matters, spiritual and temporal.

[4] 3 Jac.1, c. 4.

Despite the marriage of James's son Charles I (1600–1649) to the Bourbon princess Henrietta Maria of France (1609–1669), a Catholic, the United Kingdom remained firmly Protestant. Charles, however, took upon himself and amplified his father's notion of rule by divine right, answerable to no one but the Almighty. Throughout his early reign, he became increasingly autocratic and acted without the approval of Parliament to, among other things, levy taxes on his subjects to fund an ongoing siege of the French coast from 1627 with the goal of protecting the Huguenots (French Protestants) at La Rochelle against the reprisals of the Catholic French government. Resistance from Parliament over the tax and enactment, which the members of Parliament felt had stretched Charles' power to its legitimate limit, led to its dissolution in 1629.

Thus began the eleven years of Charles' personal reign over England in which he answered to no one other than God. While there was precedent for such a move, his power to tax remained limited under English common law absent Parliamentary approval. At last he was forced to summon a new Parliament in November 1640, which later became known as the Long Parliament. Attempts by this Parliament to rein in the powers of the monarch proved fruitless, and by 1642 England suffered unrest in neighboring Ireland and deteriorating conditions at home. After Charles' plan to arrest members of the Long Parliament he felt were conspiring against him failed spectacularly on 4 January 1642, Parliament moved to seize London and the King fled, first to Hampton Court and then to Windsor Castle. Both sides began to arm, and by mid-1642 the Kingdom was in a state of civil war.

Despite some early battlefield successes, in the ensuing 3 years the King suffered one defeat after another. The balance of power finally tipped in his loss at the Battle of Naseby on 14 June 1645. It became clear the Royalist cause was lost, but Charles was unwilling to surrender to the Parliamentarians; instead, he beseeched the Scottish Presbyterian army for safety, and turned himself over to its authority while the army negotiated with the Parliamentarians. After nine months, the Scots reached an agreement with the English Parliament to hand over Charles to face justice in exchange for £100,000 and the promise of further monies in the future. The Scots withdrew from Newcastle and the commissioners of the Long Parliament took custody of Charles in January 1647. As many as 300,000 people, or upward of 6 % of England's population, are thought to have died during the War.

The Trial and Execution of Charles I

Charles was far from contrite after his detention, and attempted to stir the pot in 1647 as he simultaneously appeared to negotiate in good faith with Parliament; the resulting Second Civil War (May–August 1648) brought fresh violence, but the Royalists again suffered defeat at the hands of the Parliamentarian New Model Army.

Having lost for a second time, Charles saw his only recourse was to attempt to bargain for his life, but the Army, led by Oliver Cromwell (1599–1658), opposed any further negotiations with someone they viewed as a tyrant incapable of being rehabilitated. Along with radical MPs, Cromwell and other Army officers decided that the kingdom would never resume peace while Charles remained alive. A schism among the MPs resulted in a purge of non-hardliners on 6–7 December 1648. The reconstituted "Rump Parliament" indicted Charles for treason early the following month. The charge against Charles asserted that the King

> for accomplishment of such his designs, and for the protecting of himself and his adherents in his and their wicked practices, to the same ends hath traitorously and maliciously levied war against the present Parliament, and the people therein represented ... [and that the] *wicked designs, wars, and evil practices of him, the said Charles Stuart, have been, and are carried on for the advancement and upholding of a personal interest of will, power, and pretended prerogative to himself and his family, against the public interest, common right, liberty, justice, and peace of the people of this nation.*

The House of Lords failed to pass the bill and it naturally lacked royal assent; Charles therefore refused to acknowledge the authority of the High Court of Justice assembled to try him. After an interim of 2 weeks while preparations were made, the trial opened on 20 January 1649 in Westminster Hall. Charles began his attempts to interfere with the proceedings immediately; once proceedings were declared open, Solicitor General John Cook (1608?–1660) stood to the right of the King to read the indictment. As Cook began to speak, Charles tapped him sharply on his shoulder with his cane and ordered him to "hold." Ignoring this, Cook continued to read. Charles tapped him with the cane again and rose to speak, but Cook ignored him. Finally, furious at being ignored, Charles struck Cook across the shoulder so hard that the silver tip of his cane broke off and fell to the floor between the men. Charles demanded to know "by what power I am called hither. I would know by what authority, I mean lawful." Insisting that the Commons held no power to try the monarch, he refused to plead. This was taken under common law as a plea of guilty, but evidence was presented nonetheless as part of the historical record against Charles. Still, he was denied basic due process; held away from the proceedings while evidence was presented, he was afforded no opportunity to question witnesses.

The Court rejected the doctrine of sovereign immunity, and Charles was declared guilty of treason at a public session on 27 January 1649 and sentenced to death. All 67 Parliamentary Commissioners present at the hearing rose to their feet to show their agreement with the sentence, and 59 signed their names to the death warrant; after the Restoration of the monarchy in 1660, this warrant was used to identify the Commissioners who signed it—by then known as the "Regicides"—and prosecute them for treason.

The Parliamentarians permitted Charles to receive his two children then living England under their control, Elizabeth Stuart (1635–1650) and Henry Stuart, Duke of Gloucester (1640–1660), who said their tearful goodbyes to him on 29 January. The following morning, before his execution, Charles asked for two shirts to insulate him against the cold of the winter morning such that the assembled crowd would not mistake his shivers for an expression of fear. He was led to a scaffold erected for the purpose before the Banqueting House of the Palace of Whitehall (Fig. 22.6), where he made a short speech maintaining his innocence of all charges and reiterating his vision of the relationship between King and his Subjects:

22.6 Detail of *A View of the Place and Manner of K. Charles the First's Execution.* Copperplate engraving; artist and date unknown

> I must tell you that their liberty and freedom consists in having government. ... It is not their having a share in the government; that is nothing appertaining unto them. A subject and a sovereign are clean different things.

Shortly after two o'clock in the afternoon, Charles put his head on the block and stretched forth his hands to indicate his readiness. The executioner, who was never later positively identified, then beheaded the King with a single stroke of the ax.

Charles II's Conquest of England

Charles II (Fig. 22.7) very nearly lost his life during his attempted conquest of the English throne, and perhaps only by a stroke of luck involving an oak tree managed to escape to exile. In June 1650, some eighteen months after his father's execution, Charles was ready to begin the conquest, but at merely 20 years old, he lacked resources, funds, and significant followers. While he was formally recognized by Scotland as his father's legitimate successor, his cause appeared fully lost in his English realms. However, he remained convinced that they devolved upon him by right, and he resolved to take back what was properly his. The Scots received him after he signed a covenant containing certain limits on his power to which they felt he must agree in order to exercise the

22.7 Portrait of Charles II of England at approximately age 40 by Simon Pietersz. Verelst (1644–bef. 1717). Royal Collection object 409151; oil on canvas, 219.1 × 135.8 cm

authority vested in him; further, as his only friends in the British Isles, he knew the support of Scotland was key to his restoration in England. Applying the same resolve as he had to ridding the country of the elder Charles as a precondition for peace, Cromwell marshaled the Army against Charles II and drove north in the summer of 1651.

On being pursued, Charles' forces drove south into the heart of England, but stopped to regroup at Worcester. Here Charles found a base of support for his efforts and locals sympathetic to the Royalist cause who agreed to shelter and defend him from Cromwell. While Charles could begin to think of himself as an actual king in a very small-scale sense, his sense of ease vaporized inside a week when Cromwell's forces arrived. The attempted Royalist invasion ended abruptly. Charles and his men conferred on a bridge a half-mile outside Worcester at dusk on 3 September and all agreed that they faced certain death if they did not quickly engineer an escape back to Scotland. Doing so required that they disband and take separate escape routes from the city, or else they would surely be captured traveling together in an identifiable column. Charles went forth with a very small group of his closest advisors, endeavoring to make themselves as unremarkable in appearance as possible so as to not gather attention. In the wake of the Civil War, the degree of lawlessness remaining in the far-flung corners of England made it possible for even the heir to the throne to move about unnoticed, provided that he was not leading an army.

Charles and his advisors traveled steadily into the night of 3–4 September. Some twenty miles from Worcester, in a secluded forest, they arrived at a pair of houses separated by a half-mile from each other at a place called Bocobel. The place was away from all the major high roads nearby, and was so isolated that in the previous century it had attracted "popish recusants," English Catholics who maintained loyalty to Rome despite official persecution. The mansion at Boscobel contained a number of "priest holes," small hiding places built into homes to hide Catholic clergy in the event of sudden intrusion and inspection by Protestant authorities. It seemed like a secure location to hold Charles in the event Cromwell closed in. Charles and company first arrived at the other house, once a convent in the Catholic era known as the "White Ladies Priory" (officially, the Priory of St. Leonard at Brewood), where the sympathetic housekeepers took to disguising Charles, including cutting his hair short to make him more closely resemble a Roundhead than a Cavalier and rubbing his face with ash from the hearth to make him look like a peasant.

Among the tenants on the Bocobel estate were several brothers by the name of Penderel. Richard Penderel (*c.* 1606–1672) arrived at the invitation of the mansion owner; he told the King that he must immediately leave the place given security concerns, and led him about a half mile from the Priory into the forest to wait while Penderel attempted to obtain intelligence on the movement of Parliamentarians in the area. However, being exceptionally sensitive to the delicacy of the situation, Penderel deliberately failed to inform the advisors in Charles' company, who rode out in fear when they discovered him missing. They were quickly found and captured by Cromwell's forces, branded as traitors, and either killed or imprisoned for nearly the following decade. One of the King's followers, Lord Wilmot, managed to escape Cromwell and make his way back to Charles the following day.

Charles was left in the forest around sunrise on 4 September when it began to rain. The King tried to find shelter among the trees but was drenched in the downpour. Richard Penderel was given a blanket by the residents of a nearby cottage and brought it to Charles, who rather turned it into a cushion on account of the hard ride of many hours he endured the night before. He remained in the forest all day as it continued to rain. A nearby road was visible from the woods and Charles passed the hours keeping careful watch for Cromwell's troops approaching. He also considered carefully his circumstances and options, deciding by the day's end that the best choice was to escape west to Wales where he could find secure hideouts among the mountains. From there, he would make his way to the coast and then onto a ship to take him to France where he could regroup and plan his next moves. As night fell, Penderel returned to fetch the King and bring him back to the house.

Explaining his decision, Penderel agreed to make the necessary arrangements. Charles would be outfitted with the tools of a woodman and adopt the alias "Will Jones" for the journey. After a brief rest with time to dine and remake Charles' appearance again, the two set out into the darkness and rain.

Penderel and Charles continued on foot for some hours before arriving in the town of Madeley, Shropshire, harried by a miller as they attempted to cross a bridge over a tributary stream of the River Severn, who very nearly blew their cover. In Madeley, Penderel sought a Mr. Woolf who he knew could help them into Wales; he did not identify his companion but said only that he could not travel safely by the light of day. Woolf was skeptical and pressed Penderel for further information; the latter said that his companion was an Army officer who had deserted at Worcester. This did not encourage Woolf, who allegedly proclaimed that he would not "hazard my life by concealing him, which I should not be willing to do for any body, unless it were the king."

Penderel, realizing his options were few, then positively identified his companion, and Woolf at once agreed to harbor him in his home. Fearing for their safety in the main house, he put them up for the moment in his barn, concealing them among the hay. Woolf promised to find safe passage for them and would report back later in the day; he returned as evening fell with unfavorable news about the prospects for travel into Wales. Republican forces were stationed at every conceivable crossing along the Severn with strict orders to allow none to pass without a thorough examination. A price was on Charles' head, he told them, and severe penalties prescribed for anyone caught harboring the fugitive King. Woolf declined to house them any further and suggested they return to Boscobel.

The Royal Oak

Charles, left with no alternative, departed Woolf's house at midnight on 5 September with Penderel after changing his appearance yet again. Returning to the bridge over the tributary where they met the miller, they found there was no choice but to ford the river. Penderel was terrified to get in the water because he could not swim, and feared drowning in the swift current if he lost his footing. Charles, however, could swim, and set about to test the depth. Reporting back favorably, he led Penderel into the water and guided him across in the dark by hand. They arrived at Boscobel before dawn, where Penderel again left the King in the forest while he ascertained the safety of the house. At the mansion he found Colonel William Careless, one of Charles' generals, whom he brought into the woods to see the King. Worn out by fatigue, Charles was led to the house for food and warmth by the fire.

As the new day dawned, Penderel began to consider where to conceal the King, for the danger existed that soldiers could descend at any moment and demand unfettered access to search the premises; likewise, the forest was ruled off-limits on the presumption that Cromwell's men would search the woods as surely as the house. As they mulled various options, Careless spotted a cluster of oaks standing apart in a field near the house. One of

22.8 Detail of *King Charles II and Colonel William Carlos in the Royal Oak*, by Isaac Fuller (d. 1672). National Portrait Gallery NPG 5249; oil on canvas, 212.7 cm 4 × 315.6 cm, 1660s(?)

the trees was especially dense with foliage and could potentially conceal the King. Careless thought that while the woods would be an attractive search target, troops would be unlikely to go literally tree to tree. Presented with this possibility, Charles readily consented.

As soon as it was light enough to see outside, Careless, Richard Penderel and his brother William accompanied the King to the tree where they established rests in its uppermost branches. Finding the result acceptable, the Penderels left Careless alone with the King for the day; the Colonel served as watch while allowing Charles to sleep, assuring him that he would make certain that the King did not fall out of the tree. In the relative safety of the oak's branches, Charles managed to sleep for more than an hour for the first time in days. Despite the expected searches of the woods by Republican soldiers, Careless and the King remained safely ensconced high above ground (Fig. 22.8).[5]

At nightfall, they descended from the oak and carefully returned to the house as Charles declared he could not endure the privations of the tree for another day and that they should make alternate arrangements for the next day. By 7 September, the immediate threat of Cromwell's search had passed, and the King's associates were able to concoct a successful plan to spirit him out of England. Penderel took Charles to the nearby house of a man called Lane, whose sister was about to travel to Bristol. Charles accompanied her in the guide of her servant, and set sail at Bristol first for Brighton on the south coast of England, and then quickly on to the coast of France.

He would bide his time in exile for 9 years before returning to England. With few supporters left and no money, he could not obtain the financing to attempt another challenge to Cromwell's government. France and the Dutch Republic formed an alliance with England in 1654, forcing Charles to try to obtain assistance from the Spanish crown. With a favorable response, Charles raised a crude fighting force among his loyal, exiled subjects on the Continent, forming the basis for what became his post-Restoration army.

The Interregnum

A new republican regime, the "Commonwealth," filled the power vacuum in England after Charles' execution led by Cromwell. The period from 1649 until 1653 was one of military rule in which the executive powers of government lay with the Council of State while legislative authority remained with the Rump Parliament. Cromwell became the de facto dictator of England in this time, leading a brutal conquest of Ireland in 1649–1650 and a campaign against the Scots, who proclaimed Charles II their king after the elder's death, the following year. Both Ireland and Scotland were ruled under military occupations for the duration of the Commonwealth.

In 1653, a constitution was promulgated that established a form of government called the Protectorate, making Cromwell the unelected leader for life. Cromwell styled himself "Lord Protector" but functioned in every respect as a king, even signing his name thereafter as "Oliver P." where the *P* indicated "Protector" much as previous English monarchs had signed a postnominal *R* for "Rex" (king) (Fig. 22.9). Cromwell was formally offered the crown by Parliament in 1657 as part of a proposed constitutional settlement, and while he agonized over the decision for weeks he ultimately rejected it, saying that he "would not seek to set up that which Providence hath destroyed and laid in the dust." Nevertheless, Cromwell was installed as Lord Protector in a 26 June 1657 ceremony at Westminster Hall, complete with the regalia and symbols of royal power including King Edward's Chair, purple robes, a sword of justice and a scepter.

---

[5]Charles later told the diarist Samuel Pepys in 1680 that while he was hiding in the oak, a Republican soldier walked directly beneath it.

22.9 A half-crown issued by Oliver Cromwell in 1658. The observe shows the Lord Protector in classical style, wreathed with a laurel branch, and bears the circumscription OLIVAR[IVS] D[EI] G[RATIA] R[EI]P[VBLICAE] ANG[LIAE] SCO[TIAE] ET HIB[ERNIAE] &c PRO[TECTOR] (*"Oliver, by the grace of God and of the Republic, of England, Scotland, and Ireland, etc., Protector"*)

During the early Interregnum, the political influence of the Puritans (constituting a majority of Parliament) and their supporters rose, and their austere beliefs were imposed on the country. The Puritans suppressed celebrations of holidays such as Christmas and Easter and banned various pastimes as "vice," much to the chagrin of many English people. As Cromwell consolidated power, it was clear he would be a liberal influence to counter the Puritan perspective, and attitudes gradually relaxed. The Church of England was later formally disestablished by the Commonwealth government, but the question of with what to replace it was never formally answered. Cromwell proved himself largely an effective administrator, but his popularity suffered by his insistence on levying high taxes to maintain a large standing army thought necessary to contain threats of rebellion in Ireland and Scotland.

Although the office of Lord Protector was not explicitly hereditary, Cromwell was allowed to nominate his own successor—and quickly chose his own son, Richard. At Oliver's death on 3 September 1658, Richard succeeded his father. Richard Cromwell, while not entirely without charisma and leadership, never quite gained the loyalty of either the Parliament or the Army, and was forced to resign in May 1659, ending the Protectorate. The Commonwealth was briefly reinstated as George Monck (1608–1670), the English governor of Scotland, marched on London at the head of a column of New Model Army regiments to restore to power the original Long Parliament, fearing the country would descend into anarchy otherwise.

The Long Parliament promptly dissolved itself and the first general election in nearly 20 years was held. Even though Royalist candidates were officially prohibited from standing for election, the rules were widely ignored and the resulting Commons was evenly split between Anglicans and Presbyterians, Royalists and Parliamentarians. This "Convention Parliament" assembled on 25 April 1660 and began passing constitutional reforms, under Monck's guidance, with the goal of enthroning Charles II as King of England under a restored monarchy. Word soon came that Charles had signed the Declaration of Breda, in which he agreed to pardon many of his father's enemies. The Parliament voted to proclaim Charles king and invited him to return to England, news that reached Charles on 8 May. He immediately began making preparations to return.

Charles' Triumphant Return to London

Charles set sail, landing in Dover on 25 May and entering London on 29 May, his 30th birthday. As the *Cambridge Historical Reader* (1911) put it,

> The people were mad with joy, to know that the king had got his own again. The church bells rang, bonfires blazed, the fountains ran with wine, and crowds cheered the king, as he drove in his fine coach to London. "Really," said king Charles, who was very witty, "I have been very foolish not to come back sooner, for everyone seems very glad to see me."

Samuel Pepys wrote in his diary entry for 1 June 1660 that

> Parliament had ordered the 29 of May, the King's birthday, to be for ever kept as a day of thanksgiving for our redemption from tyranny and the King's return to his Government, he entering London that day.

Charles was crowned at Westminster Abbey on 23 April 1661 and ruled until his death on 6 February 1685.

After the Restoration, Charles granted permanent annuities to the Penderels for their loyalty and assistance in preserving the King's life, which are still paid to their descendants. Charles also recognized Colonel Careless, making him a Gentleman of the Privy Chamber and, by letters patent, granting him the new surname "Carlos" (Spanish for "Charles") and a new coat of arms, shown in Fig. 22.10 and featuring the Royal Oak.

Charles and his new Parliament granted amnesty to fifty of Cromwell's associates through the Act of Indemnity and Oblivion.[6] Nine of the Regicides were executed for their past activities, while others received sentences of life imprisonment or were banned from holding public office for life. On 30 January 1661, twelve years to the day after Charles I's execution, Cromwell's body was exhumed from its resting place in Westminster Abbey and subjected to posthumous execution by decapitation, as were the bodies of the Commonwealth Naval Admiral Robert Blake, and Regicides John Bradshaw and Henry

---

[6] 12 Cha. II c. 11 (1660).

SUBDITUS FIDELIS REGIS ET SALUS REGNI

22.10 The coat of arms granted to William Carlos (Careless) in recognition of helping to save King Charles II by hiding in the Royal Oak. The Latin motto means "Faithful subjects of the King and of His safety". Rendered by Wikimedia Commons user Bvs-aca, licensed under CC BY-SA 3.0

Ireton. Cromwell's body was ceremonially hanged in chains on the Tyburn, and his head was skewered on a pike and so displayed outside Westminster Hall for the next 24 years.

Charles II himself died in the same year aged 54. On the last night of his life it is said he converted to Catholicism, effectively undoing 150 years' worth of effort to keep England Protestant. He was succeeded by his brother James II (also James VII of Scotland) whose brief reign ended when he was deposed in the Glorious Revolution of 1688. He was the last Roman Catholic monarch to rule over the United Kingdom.

The Royal Oak was not forgotten. Its role in Charles' story, sheltering him at his time of greatest need, was recounted widely after the Restoration, and the tree itself became a target of tourists desiring a memento. Over the next two centuries, so many sprigs and leaves were removed by visitors to Boscobel that the tree became largely denuded, and the owner of the field in which it sat fenced it off to keep the curious away. As a result, the tree standing today on the site is not the original oak but rather a tree called "Son of Royal Oak" thought to be a first-generation descendant of the original dating to the eighteenth century. This tree was badly damaged in a violent storm in 2000, losing many branches; in 2010 it was found to have developed several large cracks that threatened its health. As of 2011, the oak is again fenced in to protect it and the safety of visitors alike.

The anniversary of Charles' return to London was celebrated routinely for two hundred years. The official holiday was abolished in 1859, but a few rural areas in England continue to celebrate it. The tree remained current in popular culture through the nineteenth century based on a Restoration-era nursery rhyme intended to remind children of the letters of the alphabet. The letter 'O' stood for 'Oak':

> The Royal Oak, it was the tree,
> That saved his royal majesty.

There is a curious astronomical post-script to the story of the Royal Oak involving the previously unnamed star α Canem Venaticorum. Afterward this star took the name Cor Caroli (Charles' Heart), first appearing on a chart published in 1673 by Francis Lamb. Lamb labeled it with the longer title Cor Caroli Regis Martyris ("the heart of Charles, martyred king"), in distinct reference to the *first* King Charles. Some confusion over the reference developed since, according to Allen (1899), the name Cor Caroli was suggested by "the court physician, Sir Charles Scarborough, who said that it had shone with special brilliancy on the eve of the king's return to London on the 29th of May, 1660." Allen attributed the effort that led to the adoption of the name to Halley, but Burnham (1978) disputes this, writing[7] that "the attribution of the name to Halley appears in a report published by J. E. Bode at Berlin in 1801, but seems to have no other verification." Lamb and other cartographers marked the position of α with a small heart surmounted with a crown (Fig. 22.11) to cement the reference to Charles I.

DISAPPEARANCE

As a constellation, the fate of Charles' Oak was effectively sealed after the publication of Lacaille's *Coelum Australe Stelliferum* in 1763 in which he broke up Argo into its constituent pieces. When he made these divisions, he permanently solved the problem of Argo's missing front end by expanding the reach of Carina to encompass Robur's stars. We have Lacaille's own words[8] to understand his motivation, from 1752s *Memoire*:

> You will not find here the new constellation which Edmond Halley included in his planisphere in 1677, which he called Robur Carolinum because I have put the brightest stars in Argo, which that astronomer, aged 21 years, had detached to pay homage to the king of England. However praiseworthy his motives, I cannot approve of the way in which he went about delineating his constellation, since in order to isolate it, he so abbreviated Argo and left unnamed some quite bright stars between his tree and Argo: and in order to make out that the stars which formed his tree were new and never before observed, he did not compare their positions with those of the ancient catalogues, as he always did with the stars in other constellations: as it is, of

---

[7] Volume 1, page 359.

[8] Translated by Evans (1992).

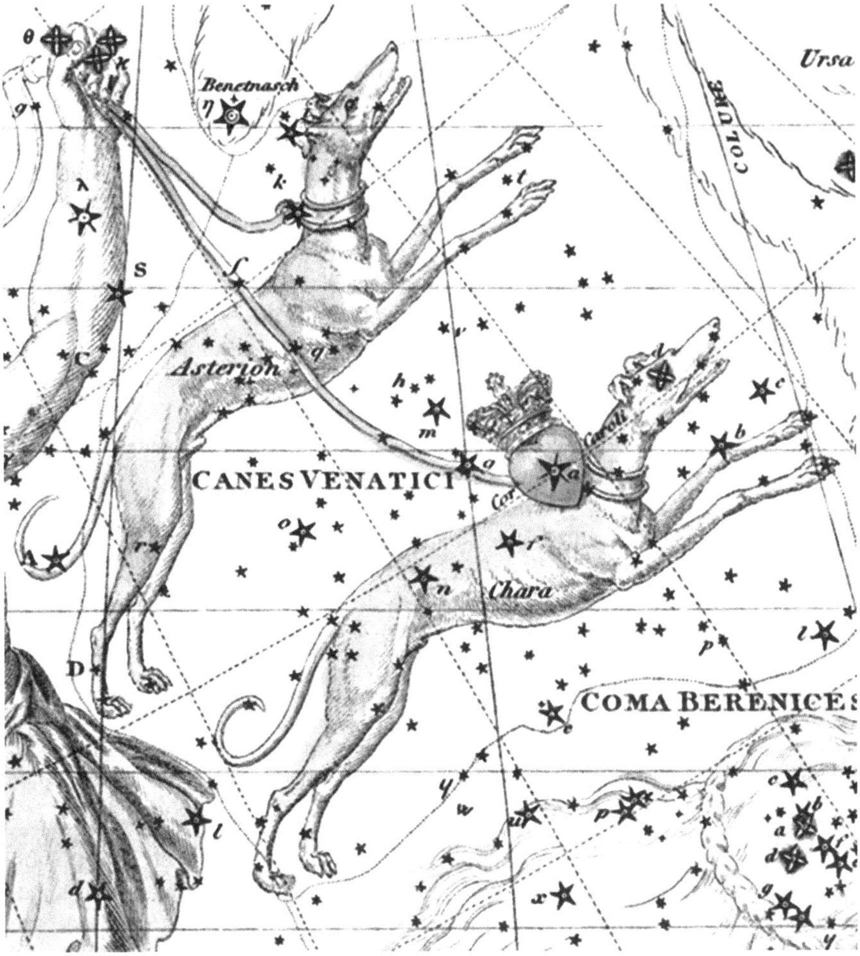

22.11 Canes Venatici, shown on Plate VII of Alexander Jamieson's *Celestial Atlas* (1822). α Canem Venaticorum, the brightest star in the figure of the hunting dog named Chara, is marked according to its popular name "Cor Caroli"

> the 12 stars of this tree, nine are in ancient catalogues and designated by particular letters on Bayer's planisphere in the constellation Argo.

The Frenchman respected Halley as an individual, but was less impressed by the English astronomer's appeal to honor his sovereign, whereas in his own invented constellations Lacaille evidently sought to avoid nationalistic symbols by instead placing in the firmament objects of the arts and of science. However, due to the figure's popularity on charts he couldn't entirely dismiss Halley's creation without offering some sort of justification as to why the southern skies were simply better off without it. "One cannot doubt," he wrote,

> that everybody who observed the southern stars in the fifteenth and sixteenth centuries, to place them in new constellations, always put the stars of Halley's tree in Argo. Otherwise it is not reasonable to believe that they would have formed the constellations of the flying fish [Volans] and Chamaeleon, which are next to Argo, and whose brightest stars are of the fifth magnitude, while leaving with no constellation, between Centaurus and Argo a large space filled with stars of the first, second, third, and fourth magnitude, so well grouped with Argo?

While undoubtedly a subtext of French patriotism threads through his words, Lacaille's objections to Halley's invention were superficially rooted in criticism of what he considered the Englishman's sloppy approach to surveying since the explorers of the centuries immediately preceding Halley "always put the stars of Halley's tree in Argo." In this, Lacaille does make a rather strong argument to differentiate Halley's approach from those who considered only faint stars left unformed by Ptolemy, such as Hevelius. While his point is well taken that Halley merely appropriated stars known to the ancients to his own ends, Lacaille wasn't about to allow a monument to an English king to remain in the southern skies.[9]

The great influence of Lacaille's work on the southern constellations meant that most cartographers after Lacaille followed his lead. An important exception was Bode, who kept the Oak in *Uranographia* (1801b). In the companion volume, *Allgemeine Beschreibung und Nachweisung der Gestirne*, he held[10] that it was the other way around and that Lacaille took stars from the "ancient" understanding of the tree in order to make a modern definition of Argo:

> [It was] placed in the heavens by Halley in memory of the tree where Charles II of England withdrew during his flight. de la Caille took the stars of the oak to create his ship [Argo], *but by rights I restored the ancient constellation.*

Allen summarized the situation thusly: "La Caille complained that the construction of the figure, from some of the finest stars in Ship, ruined that already incomplete constellation, 'and the Oak ceases to flourish after half a century of possession,' although Bode sought to restore it."

As a consequence of the reach of *Uranographia*, Robur held on through the first few decades of the nineteenth century; a notable appearance after Bode is on William Croswell's *Mercator map of the starry heavens* (1810; Fig. 22.4). Elijah Hinsdale Burritt cataloged 25 stars belonging to the Oak in *The geography of the heavens* (1833), but by mid-century the constellation rapidly became an anachronism. It received few mentions thereafter, in sources such as Ezra Otis Kendall's *Uranography* (1845), George Chambers' *Handbook of Descriptive Astronomy* (1877), and the *Poole Bros. Celestial Handbook*

---

[9] Evidently there is no record of what Lacaille thought of Hevelius in the latter's patent display of Polish patriotism by placing his patron's shield among the stars as Scutum (Sobieskii).

[10] "Hat Halley zum Andenken Carls II, Königs in England, der sinstens auf eine Eiche flüchtete, unter die südlichen Gestirne gesetz. *De la Caille* gebrauchte die Sterne dieser Eiche mit zur Formierung des Schiffs, ich habe solche aber wieder hergestellt."

(1892). While Allen was less forceful in his words to describe the currency of Robur compared to other constellations he considered defunct at the time of his writing, it is clear that by the beginning of the twentieth century the Oak had disappeared from maps. It was not counted as canonical by the International Astronomical Union when constellation abbreviations were published after the first General Assembly in 1922, nor did it appear on Eugène Delporte's maps in 1930 that established the modern constellation boundaries (Fig. 22.12).

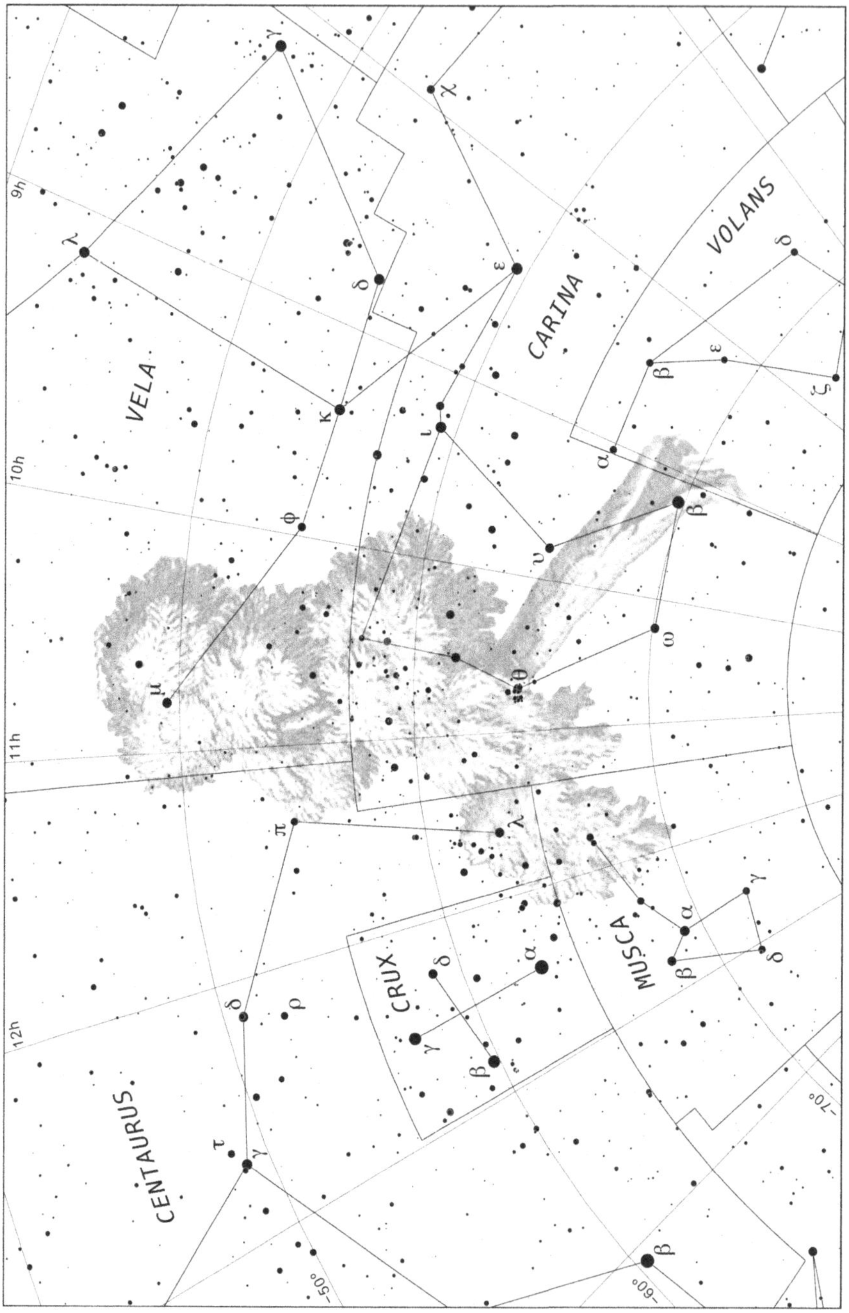

22.12 The figure of Robur Carolinum from Figure Eee of Hevelius' *Prodromus Astronomiae* (1690) overlaid on a modern chart

23

Sagitta Australis

The Southern Arrow

Genitive: Sagittae Australis
Abbreviation: SgA
Alternate names: "Sagitta Austr" (Plancius, 1612/3); "Sagitta Austra" (Cellarius, 1661); "Sagitta Australe" (Bakich, 1995)
Location: Between Sagittarius and Scorpius[1]

ORIGIN AND HISTORY

Petrus Plancius decided to fill the otherwise empty space between Sagittarius and Scorpius by creating this constellation from two unformed fourth-magnitude stars on a globe printed by Pieter van der Keere (1571–*c.* 1646) in Amsterdam in 1612 or 1613[2]. A partial reconstruction of this map from a 1649 reprint is shown in Fig. 23.1. Plancius' Arrow would be considered "single sourced"—and might have been completely forgotten—were it not for its inclusion in Andreas Cellarius' *Harmonia Macrocosmica* (1661; Fig. 23.2). Cellarius closely followed Plancius in showing the figure, although he labeled Antares in its correct position.

It is not entirely clear to which exact stars Plancius referred based on his depiction of the Arrow, and it does not appear that Cellarius followed Plancius particularly carefully in drawing it for his atlas. Following Cellarius strictly, one finds only fair alignment involving ϵ Scorpii (near the arrow point), 45 Ophiuchi (= HR 6492 = SAO 185412; V=+4.3—near,

[1]Bakich (1995) has it as "north of Aquila," which seems to erroneously refer to the Ptolemaic constellation Sagitta. Kanas (2007) uses similar language, evidently following Bakich.

[2]On this same map Plancius also introduced the new constellations Camelopardalis, Cancer Minor (Chap. 6), Gallus (Chap. 10), Jordanis (Chap. 13), Monoceros, and Tigris (Chap. 27). Of these, only Camelopardalis and Monoceros remain canonical.

J.C. Barentine, *The Lost Constellations*, Springer Praxis Books,
DOI 10.1007/978-3-319-22795-5_23

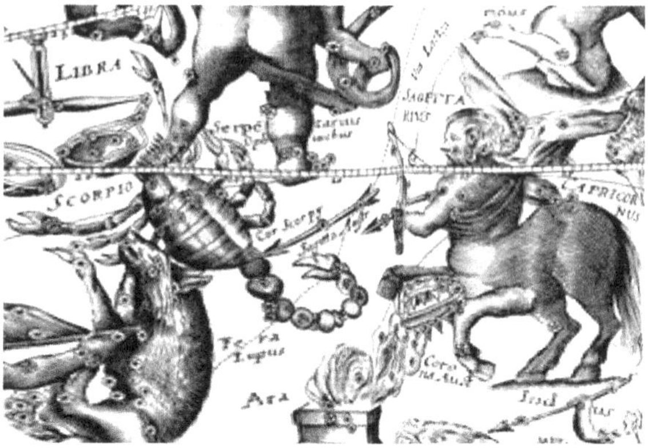

23.1 Petrus Plancius' original depiction of Sagitta Australis on a 1649 reprint of his *Globus coelestis in quo stellae fixae …* This scene has been digitally reconstructed from three adjacent globe gores

but distinctly off, the shaft), and 3 Sagittarii (= HR 6616 = SAO 185755; $V = +4.5$—on one of the arrow's flights). However, shifting the figure slightly north of its location in *Harmonia Macrocosmica* brings it into considerably better alignment with a plausible set of constituent stars (Fig. 23.5). The arrow's point remains ϵ Scorpii, while the rest of the arrowhead is defined by the stars HD 154090 (= HR 6334 = SAO 208377; $V = +4.9$), HD 152636 (= HR 6282 = SAO 208205; $V = +6.3$), HD 153613 (= HR 6316 = SAO 208324; $V = +5.0$) and 27 Scorpii (= HR 6288 = SAO 208232; $V = +5.5$). Its shaft is then better marked by 3 Sagittarii and 45 Ophiuchi, the latter of which lies closer to the figure than shown in Cellarius.

Contemporary atlases, such as Bayer's *Uranometria* (1603), often do not show these stars at all; the density of much fainter stars in the Milky Way clouds in this area renders it difficult to the visual observer to pick out individual stars below about the third magnitude. Jacob Bartsch included many of Plancius' new inventions of 1613 in *Usus Astronomicus* (1624), but Sagitta Australis was left off the list along with Cancer Minor (Chap. 6). Shortly thereafter in his attempted Christianization of the night sky in *Coelum Stellatum Christianum* (1627), Julius Schiller dealt with these stars in his rebranding of Ophiuchus as Saint Benedict (his 'Constellation XIII'; Fig. 23.3); his star '34,' shown as the tip of a crozier held by Benedict, appears to be 45 Ophiuchi.

ICONOGRAPHY

Plancius' Arrow draws an otherwise new connection between Sagittarius and Scorpius in that he positioned it such that it appears to have been fired toward the Scorpion by the archer. The orientation of the Arrow, based on the stars he chose, has it aimed toward the base of the Scorpion's tail, which would not make for an effective kill shot; as if to make up for this deficiency, Plancius wrote the words "Cor Scorpy" (a slight misspelling of Cor Scorpii, meaning "Scorpion's Heart") above the Arrow, evidently indicating the target.

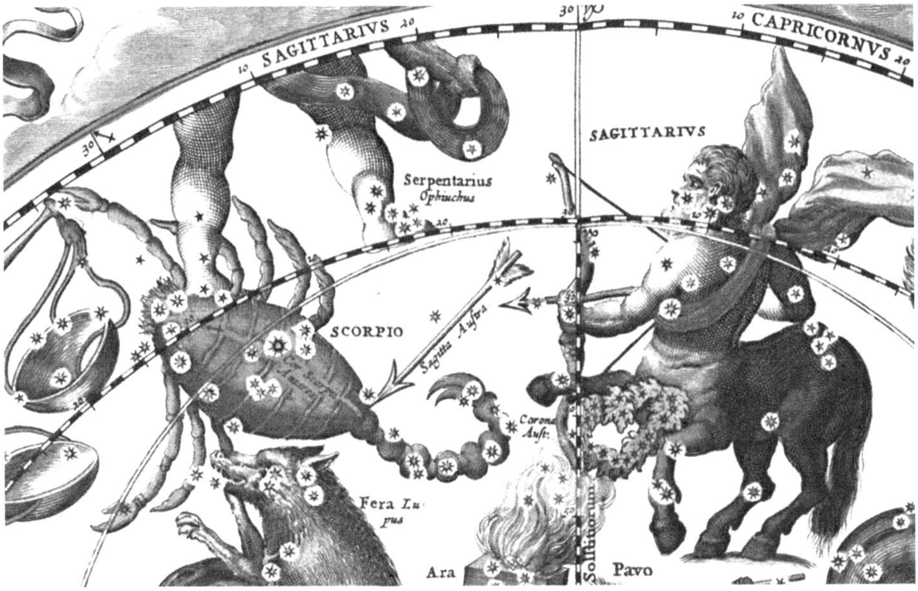

23.2 Sagitta Australis shown in a detail from Plate 27 of Andreas Cellarius' *Harmonia Macrocosmica* (Cellarius (1661))

Cor Scorpii is a traditional name given to the bright star Antares (α Scorpii), much further north along the Scorpion's body.

DISAPPEARANCE

The Southern Arrow had a brief flight across the first half of the seventeenth century, but its appearance on Cellarius' atlas was its most notable (and probably its very last). Augustin Royer did not include it in his *Cartes du Ciel Reduites en Quatre Tables* (1679a), instead making the constituent stars into part of Ophiuchus' left foot; Ignace-Gaston Pardies similarly treated it on Plate 5 of *Globi coelestis* (1674). None of its stars were shown in either volume of Stanislaus Lubieniecki's *Theatrum Cometicum* (1667; 1681). The Arrow's fate was probably sealed when not even its stars were shown in Johannes Hevelius' *Prodromus Astronomiae* (1690; Fig. 23.4).

By 1700, Sagitta Australis was certifiably extinct. When the modern constellation boundaries were formalized in 1928, the stars along the Arrow's path were divided between Scorpius, Ophiuchus and Sagittarius (Fig. 23.5).

23.3 Ophiuchus rebranded as Saint Benedict ('Constellation XIII') in Julius Schiller's *Coelum Stellatum Christianum* (1627). The star '34' in the tip of Benedict's crozier (*extreme bottom-center*) is 45 Ophiuchi, a star that once constituted part of Sagitta Australis

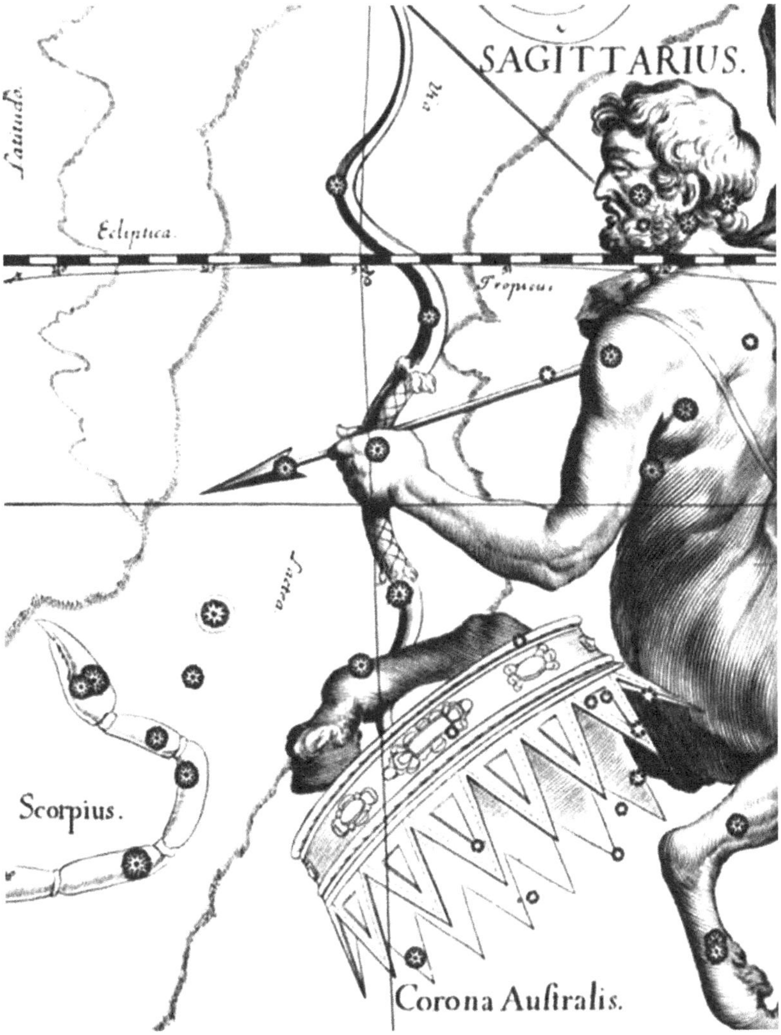

23.4 Sagittarius, Corona Australis, and the tail of Scorpius as shown in Figure Kk of Johannes Hevelius' *Prodromus Astronomiae* (1690)

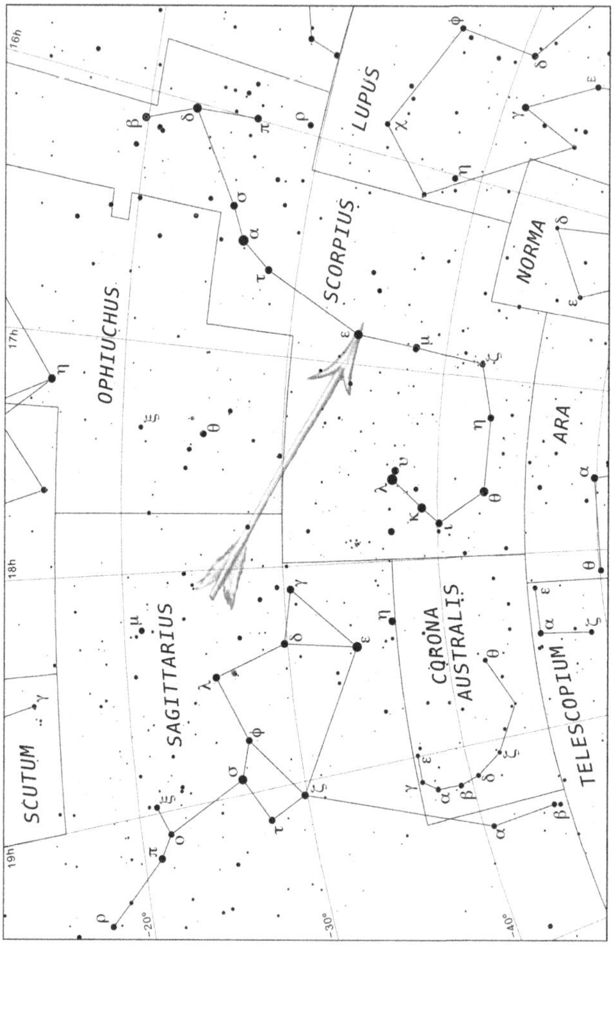

23.5 The figure of Sagitta Australis from Plate 27 of Andreas Cellarius' *Harmonia Macrocosmica* (1661) overlaid on a modern chart. The exact location is conjecture based on the description in the main text

24

Sceptrum Brandenburgicum

The Brandenburg Scepter

Genitive: Sceptri Brandenburgici
Abbreviation: ScB
Alternate names: "Sceptrum Brandenburg" (Hall, 1825); "Brandenburgium Sceptrum" (Ryan, 1827)
Location: West of Rigel, among the stars at a bend in Eridanus[1]

ORIGIN AND HISTORY

Gottfried Kirch introduced Sceptrum Brandenburgicum in *Acta Eruditorum* (1688), the first scientific journal in the German-speaking parts of Europe published between 1682 and 1782. It appeared in the same issue in which he also suggested Pomum Imperiale (see Volume 2), presumably to honor his home state of Brandenburg in Germany. Ridpath (1989) suggested that it was "more likely" intended to glorify Frederick I, Elector of Brandenburg (1688–1713) and Duke of Prussia. Kirch's shameless elevation of a symbol of Frederick's temporal authority to the heavens seems to have paid off: he was appointed by the Duke to the newly-created Brandenburg Society of Sciences in 1700 and made the

[1] *"Between the windings of Eridanus to the southwest of Orion"* (Bode, 1801a); *"Between Rigel in the foot of Orion and the quadrilateral figure made by the stars near the paws of Cetus, about one-fourth of the distance from Rigel"* (Green, 1824); *"Between Rigel and the parallelogram in the Whale, about one-fourth the distance from Rigel"* (Kendall, 1845); *"Between Lepus and Eridanus, and south of Orion and Taurus"* (Bouvier, 1858); *"Below the first bend in the River, west from Lepus"* (Allen, 1899); *"Just west of Lepus"* (Olcott, 1911); *"Near the foot of Orion in a large bend in the river Eridanus"* (Ridpath, 1989).

J.C. Barentine, *The Lost Constellations*, Springer Praxis Books,
DOI 10.1007/978-3-319-22795-5_24

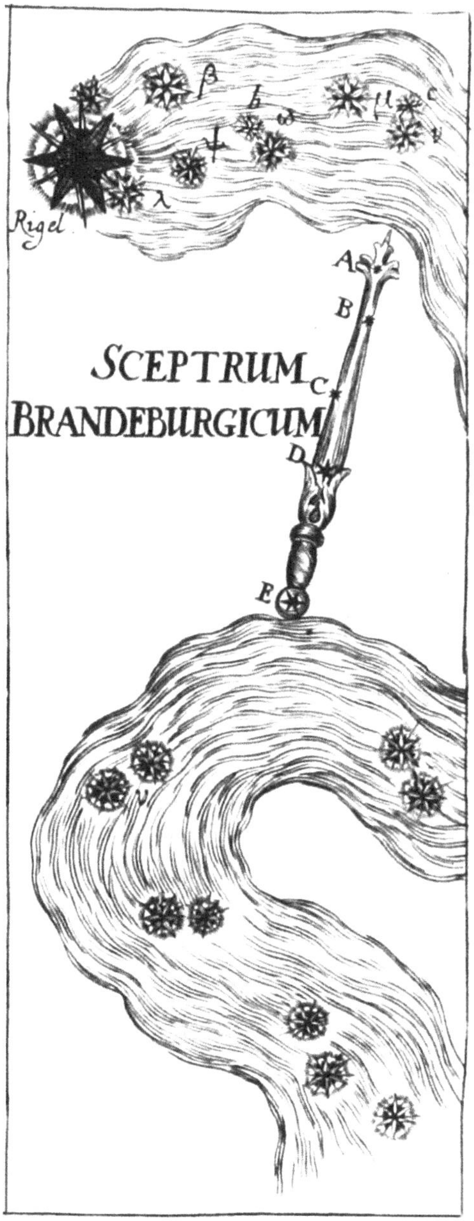

24.1 Sceptrum Brandenburgicum as shown in *Acta Eruditorum* (1688). Gottfried Kirch devised the constellation from stars situated between turns of the River Eridanus, below the bright star Rigel (β Orionis) at *upper left*. The label contains the misspelling "Brandeburgicum." Copperplate engraving, 16.5 × 6 cm

Table 24.1 Stars in Sceptrum Brandenburgicum according to Kirch (1688)

Kirch	*Designation*	*HR*	*SAO*	*V*
"A"	47 Eridani	1451	131315	+5.2
"B"	HD 29065	1452	131316	+5.2
"C"	HD 29573	1483	149789	+5.0
"D"	53 Eridani	1481	149781	+3.9
"E"	54 Eridani	1496	149818	+4.3

first director of the Society's observatory at Berlin. However, Kirch never formally took possession of the facility, dying before the new building was officially dedicated.

Kirch generally described[2] the location of his figure:

> *[T]he uppermost point of the scepter is situated below the highest bend in River Eridanus, immediately below Bayer's star ν* [Eridani]*; the four other stars (or more properly the five, because the third of these, consisting of two contiguous stars, appear as if one) stretch towards the extremes of Orion and beyond.*

The five stars identified by Kirch in Fig. 24.1 that define Sceptrum Brandenburgicum appear to be as indicated in Table 24.1 from north to south. Of these stars, 53 Eridani still carries the proper name "Sceptrum," befitting its former position as the *lucida* of Kirch's constellation.

While Kirch benefitted from creating this figure out of otherwise unremarkable, faint stars on the banks of Eridanus, other astronomers did not initially take note. In the century before and immediately after Kirch's invention was published, most cartographers dealt with these stars by either not showing them at all or adjusting the winding course of Eridanus to take them up. A notable exception is Edward Sherburne who, in the hemispheres published with his *Sphere of Marcus Manilius* (1675), assigned the stars to the long tail of the animal skin that Orion holds in his left hand (Fig. 24.2).

The Scepter is absent from key atlases of the century after Kirch such as Coronelli (1693), the 1693 second edition of Pardies (1674), the 1700 second edition of Seller (1680), de La Hire (1702), Allard (1706), de Broen (1709), the first edition of Flamsteed (1729), Thomas (1730), Semler (1731), Doppelmayr (1742), Bevis (1750), and Jean Nicolas Fortin's second edition of Flamsteed[3] (1776).

---

[2]"[Q]uarum superior apicem sceptri occupans, sub flexura Eridani suprema, proxime infra stellam nu Bayerianam conspicua est ; caeterae numero quatuor (aut verus quinque, nam tertia harum ex duabus contiguis, in unam velut coeuntibus, constat) versus extrema Orionis & ultra porriguntur." (p. 452)

[3]Fortin never accepted Kirch's constellation: it did not appear in the third edition of Flamsteed (1795), either.

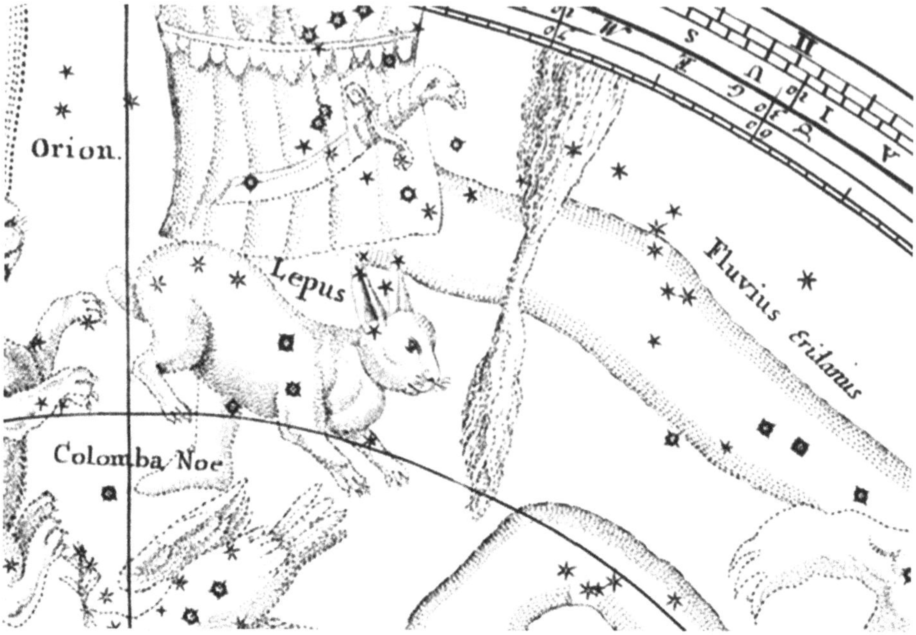

24.2 The stars later incorporated into Sceptrum Brandenburgicum by Gottfried Kirch are here shown here outlined by the long tail of the animal skin held aloft by Orion in Edward Sherburne's *Sphere of Marcus Manilius* (1675)

The Scepter might have truly disappeared were it not for the interest taken by a fellow Berliner. Kanas (2007) describes Sceptrum Brandenburgicum as "reintroduced" by Johann Elert Bode; it resurfaced on Plate XXIV of *Vorstellung der Gestirne* (1782), on which it is labeled as "Die Brandenbrugische Scepter" (Fig. 24.3). However, it seems that Bode was not quite ready to promote the Scepter to full constellation status: he encloses it inside a common boundary shared with Eridanus. It is unclear whether Bode, at that point, considered Sceptrum an independent constellation or merely an asterism of Eridanus. In *Uranographia* (1801b) it received its own set of boundaries (Fig. 24.4), which is taken to mean that Bode then considered it a fully-fledged constellation. In *Allgemeine Beschreibung und Nachweisung der Gestirne*, the descriptive companion volume to *Uranographia*, Bode described[4] the constellation as

[4]"Steht zwischen den Krümmungen des Eridans südwestwärts vom Orion, und macht sich an einigen Sternen vierter Grösse, die under einander stehen, kenntlich. Er ist 1688 von dem ersten Berlinschen Astronomen Gottfried Kirch eingenführt; ich nahm ihn 1782 in meine kleinen Himmelscharten auf, und habe ihn in den gegenwärtigen grossen Charten mit der Names-Chiffre unsers jetzt glorreich regierenden Königs bezeichnet." (p. 17)

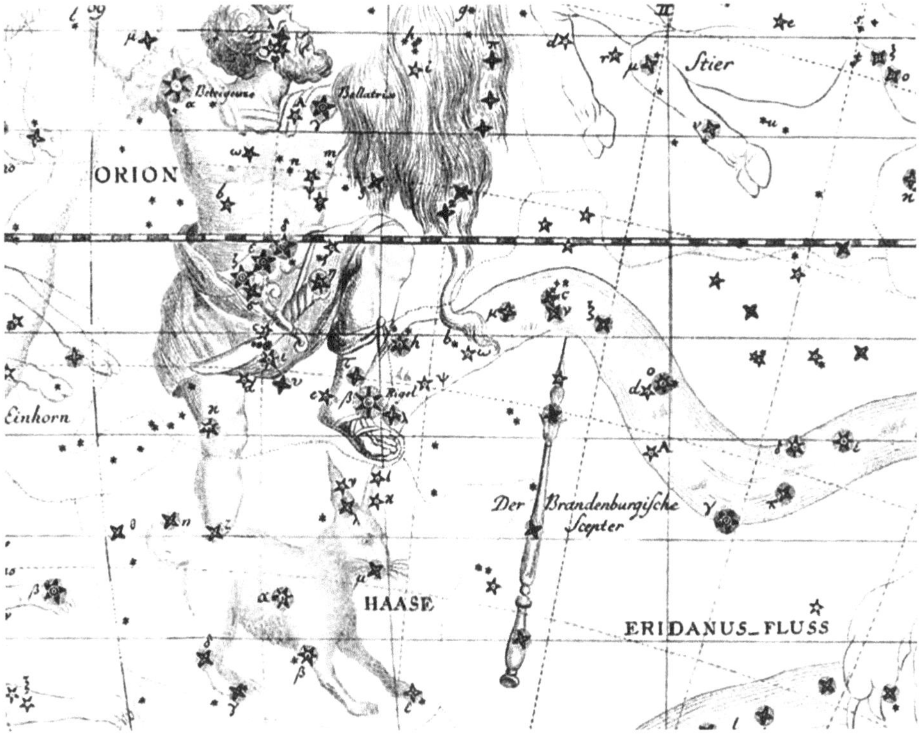

24.3 Sceptrum Brandenburgicum shown on Plate XXIV of Johann Elert Bode's *Vorstellung der Gestirne* (1782)

> situated between the windings of Eridanus to the southwest of Orion, and is characterized by some recognizable stars of the fourth magnitude located one atop another. This constellation was dedicated in 1688 by Gottfried Kirch, the foremost astronomer of Berlin; I adopted it in 1782 on my small star charts,[5] and in the present atlas[6] I added the figure of the King who at this moment gloriously rules Prussia.

Here Bode directly imitated Kirch by referencing another Frederick of the House of Hohenzollern: the Prussian King Frederick William III (1770–1840) who reigned at the time *Uranographia* was published. Bode unsubtly reinforced this in his figural depiction of the Scepter in *Uranographia*, encircling it with a ribbon-like device reading "F W III."

Bode's resurrection of the Scepter assured that it would grace a number of charts in the nineteenth century as many cartographers took their cues from the influential *Uranographia*. After its rebirth in the first edition of *Vorstellung der Gestirne*, Christian

[5] *Vorstellung der Gestirne*, 1st ed.

[6] *Uranographia*.

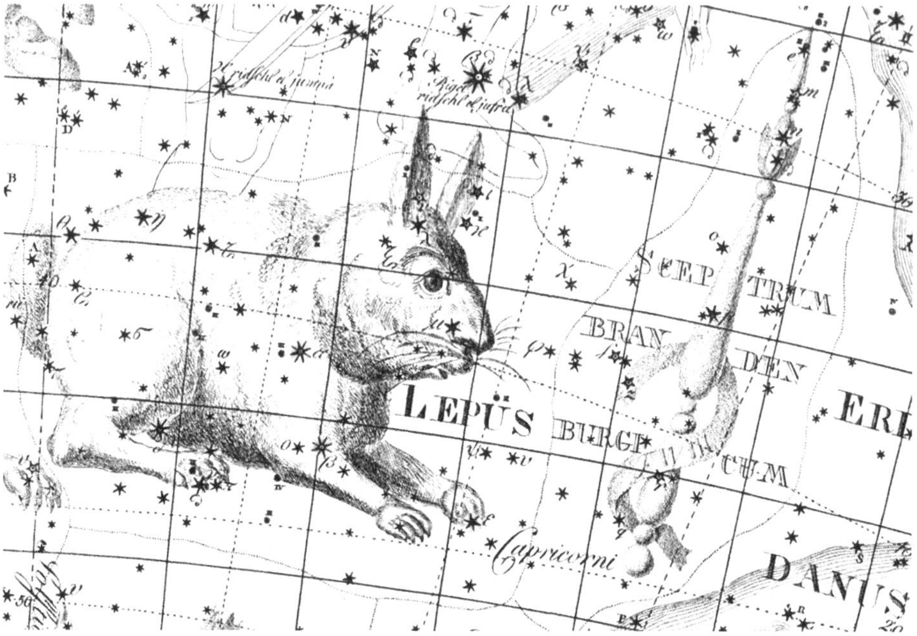

24.4 Sceptrum Brandenburgicum shown on Plate 18 of Johann Elert Bode's *Uranographia* (1801b)

Goldbach included the constellation as "Brandenburgisches Scepter" on Map 24 of his *Neuester Himmels* (1799). It appeared under the same name on Fig. XXXI, Table 29 of August Gottleib Meissner's *Astronomischer Hand-Atlas* (1805). Thomas Young showed the figure, labeled "Sceptre of Brandenburg," on Plate XXXVII, Fig. 518 of his *Course of Lectures on Natural Philosophy and the Mechanical Arts* (1807), while William Croswell depicted "Sceptrum Brandenburgium" on his *Mercator map of the starry heavens* three years later (Fig. 24.5). Alexander Jamieson gave the figure a very detailed rendering in his *Celestial Atlas* (1822; Fig. 24.6).

Jacob Green included it on Plate 16 of *Astronomical Recreations* (1824), in which he wrote

> The Sceptre of Brandenburg is a modern constellation, formed in 1688, by Geoffroi Kirch, a Prussian astronomer. This asterism occupies a small space in the heavens made by one of the bends in the river Eridanus. It may be easily found by its three principal stars, which are nearly in a straight line perpendicular to the equator. It lies between Rigel in the foot of Orion and the quadrilateral figure made by the stars near the paws of Cetus, about one-fourth the distance from Rigel.

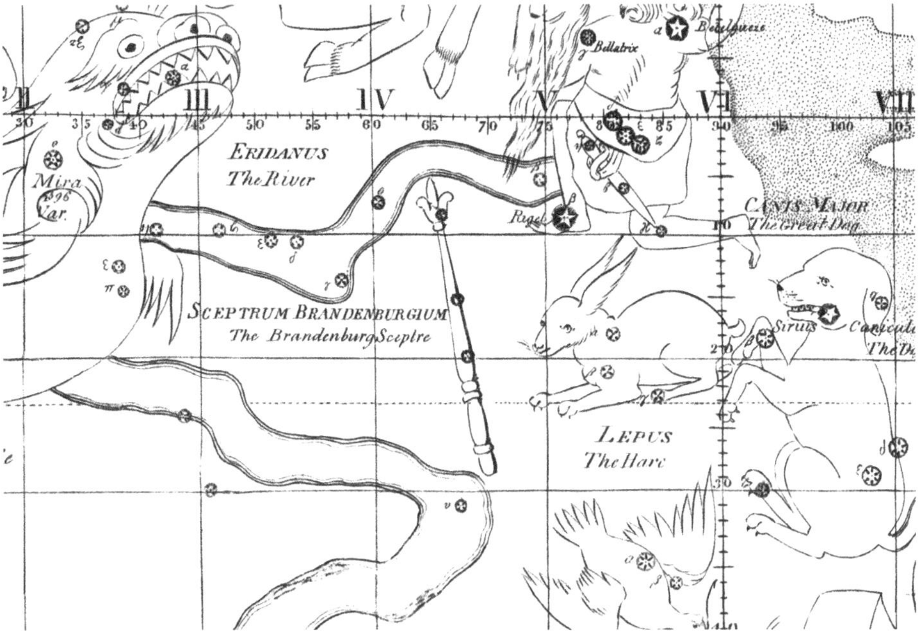

24.5 William Croswell's depiction of Sceptrum Brandeburgicum on his *Mercator map of the starry heavens* (1810)

Ezra Otis Kendall largely copied directly from Green's text in describing the constellation in *Uranography* (1845):

> *The Sceptre*, or the *Brandenburgh Sceptre*, is situated between the Hare and the Harp. It consists of three small stars in a vertical line between Rigel and the parallelogram in the Whale, about one-fourth the distance from Rigel. It was formed in 1688 by Kirch, a Prussian astronomer.

In her description of the Scepter in her *Familiar Astronomy* (1858), Hannah Bouvier capitalized on an opportunity to draw attention to the role of women in the history of astronomy:

> *Q*. Describe the constellation SCEPTRUM BRANDENBURGIUM.
> A. *It is a small new constellation, situated between* Lepus *and* Eridanus, *and south of* Orion *and* Taurus. *It contains no very conspicuous stars. The constellation was formed in 1688 by Kirch, a celebrated German astronomer. His wife, Mary Margaret, was his assistant, and the author of several astronomical works.*

Due to its position in the sky, the constellation was not properly shown in *Urania's Mirror* (1825), but the label "Sceptrum Brandenburg" appears west of Lepus on Plate 30. Elijah Hinsdale Burritt showed it on Plate 3 of his *Atlas Designed to Illustrate the*

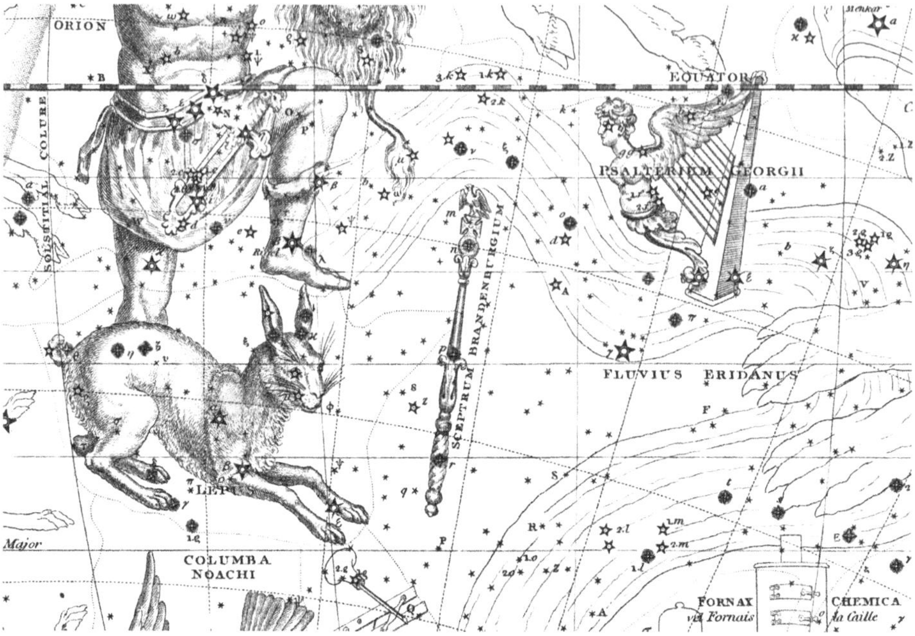

24.6 Sceptrum Brandenburgicum and Psalterium Georgianum (*upper right*; Chap. 18) shown on Plate 24 from Alexander Jamieson's *Celestial Atlas* (1822)

Geography of the Heavens (1835a) and it appeared on Plate 2 of James Middleton's *Celestial Atlas* (1842). Christian Gottlieb included Sceptrum on Plates 1 and 18 of his *Himmels-Atlas* (1849), but its popularity was already dwindling by midcentury. Ludwig Preyssinger showed it on Plate II of his *Astronomie Populaire ou Description des Corps Célestes*, and one of its last appearances is in the southern hemisphere on page 108 of Eliza A. Bowen's *Astronomy By Observation* (1888). The Scepter received mentions (but not figural depictions) in several other nineteenth century sources including Ryan (1827), Kendall (1845), Chambers (1877), and the *Poole Bros' Celestial Handbook* (1892).

ICONOGRAPHY

Brandenburg-Prussia

The Margraviate of Brandenburg in eastern Germany emerged from the Medieval period after its original founding as the "Northern March" in the tenth century. At the dawn of the early modern period of German history, around the beginning of the seventeenth century, Brandenburg achieved the status of one of the seven electoral states of the Holy Roman Empire. Along with its neighbor, Prussia, it formed the core of the nascent

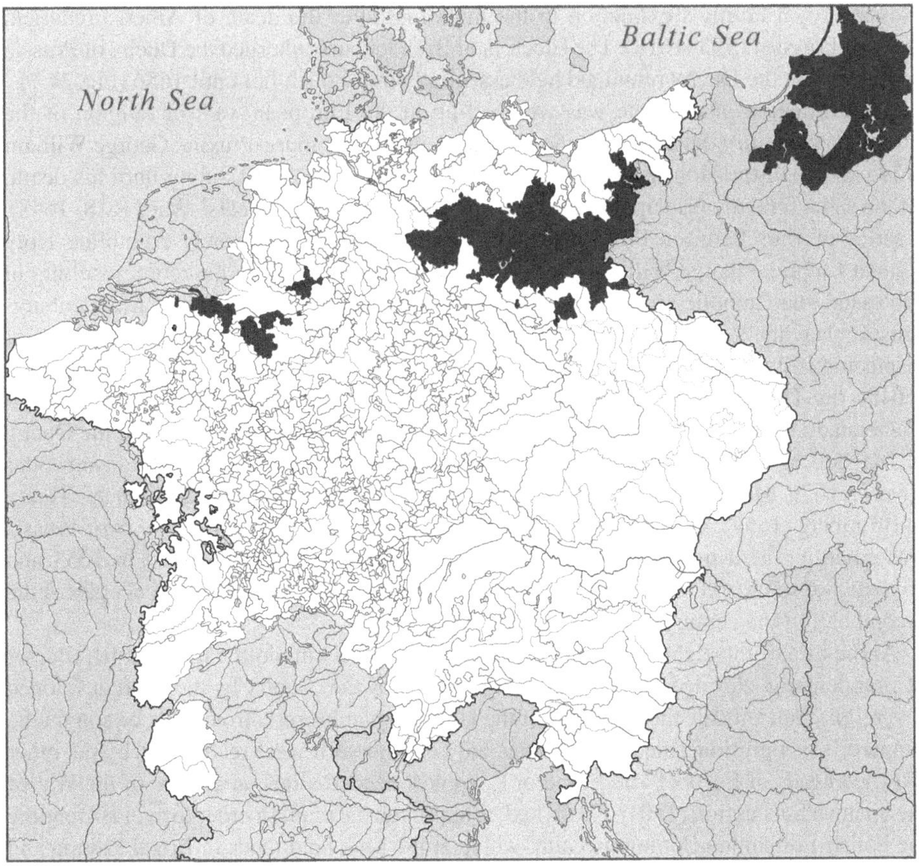

24.7 Map of north-central Europe in 1618 showing Brandenburg-Prussia (*dark gray*) relative to the territories controlled by the Holy Roman Empire (*white*). Principally drawn from "Deutschland: 1618–1648" (p. 122) in Josef Engel (ed.), *Großer Historischer Weltatlas, herausgegeben vom Bayerischen Schulbuch-Verlag: Dritter Teil, Neuzeit, Munich and Tübingen*. Munich: Bayerischer Schulbuch-Verlag (1967). Licensed under CC BY-SA 4.0 by Wikimedia Commons user Sir Iain

German Empire, which would become the first unified German state in the latter half of the nineteenth century. Berlin, the future German national capital, was embedded deeply within its territory.

Brandenburg became a principality of the Holy Roman Empire after Charles IV issued the Golden Bull of 1356, fixing for more than four centuries the main constitutional structure of the Empire. From 1415 the The Margraviate was governed by the Hohenzollern dynasty whose monarchs would lead the German Empire to its dissolution after the end of the First World War in 1918. Toward the end of the sixteenth century the main branch of the Hohenzollerns intermarried with the ruling family of the Duchy of Prussia, bringing the two realms into personal union and establishing the succession after the

Prussian royal family's extinction in the male line after the death of Albert Frederick, Duke of Prussia (1553–1618). The Electors of Brandenburg inherited the Duchy of Prussia thereafter, but the Duchy remained held essentially as a Polish fief until 1656 (Fig. 24.7).

The formation of the state was wrapped up in the European wars of religion of the seventeenth century. Shortly after the formation of Brandenburg-Prussia, George William (1595–1640) of the Hohenzollern dynasty came to power in 1619, ruling until his death. At the same time the principality became involved in the Thirty Years' War (1618–1648), a series of wars between Protestant and Catholic states in the slowly crumbling Holy Roman Empire. George William switches sides three times during his reign, resulting in Protestant and Catholic armies alternately sweeping back and forth across Brandenburg-Prussia, leaving a trail of destruction each time in their wake and either killing or displacing half the population. At the conclusion of the War with the Peace of Westphalia (1648), the The Margraviate expanded its territory as it gained control over Minden and Halberstadt; it further received the succession of Farther Pomerania (1653) and the Duchy of Magdeburg (1680), and its holdings in Pomerania were extended to the lower Oder River in 1679. Meanwhile, Prussia's time as a client state of Poland ended with the Treaty of Bromberg (1657). By the time Kirch was publishing in the 1680s, Brandenburg-Prussia was emerging as a major European power. A standing army was formed in 1653 and consisted of some 30,000 men by 1688. It saw a number of major military successes in the second half of the century.

At the turn of the eighteenth century, Prussia became a kingdom. Frederick III, Elector of Brandenburg, elevated his status to King in 1701 as the country by then had developed sovereign status outside the influence of the Holy Roman Empire. Ironically, he sought the Emperor's recognition, since a crown was only of value if it were recognized by the other crowned heads of Europe. The Emperor Leopold I needed allies on the eve of the War of the Spanish Succession (1701–1714), and he approved the plan. In order to avoid offending the Polish-Lithuanian Commonwealth, which at the time held lands in Royal Prussia and Ermland, Frederick styled himself "Frederick I, King *in* Prussia" rather than "King *of* Prussia." From this point on, Brandenburg essentially ceased to exist as a polity and the Hohenzollern domains in eastern Germany were often simply referred to as "Prussia."

The union of Brandenburg and Prussia persisted until the end of the Holy Roman Empire in 1806, although by that time the Emperor held only symbolic power over his realms. Frederick and his successors continued to consolidate power in Berlin by fundamentally changing the character of Brandenburg-Prussia, which traditionally had functioned as a series of diverse principalities under a weak central government. While the The Margraviate of Brandenburg continued to exist as a legal entity after the Kingdom was proclaimed, it came to an end with the Empire, replaced by the Prussian Province of Brandenburg in 1815. Its territory survived intact after the First World War, but its eastern third of Ostbrandenburg/Neumark was ceded to Poland at the conclusion of the Second World War in 1945. While Prussia no longer exists, Brandenburg remains one of sixteen federated states of Germany with its capital at Potsdam; Berlin retains its status as an independent, federal city-state surrounded on all sides by Brandenburg.

Frederick I of Prussia

The Elector of Brandenburg in Kirch's time, Frederick III (1657–1713; Fig. 24.8) was the first Hohenzollern king of Prussia as Frederick I after 1700. He was born in Königsberg to Frederick William, Elector of Brandenburg, and his first wife, Louise Henriette of Orange-Nassau; through his mother's family in the House of Orange he was a cousin to King William III of England. On the death of his father on 29 April 1688 he became Elector Frederick III of Brandenburg and Duke of Prussia.

Frederick's royal ambitions were evident soon after he came to power when he decided for himself that he wanted the more prestigious title of "king" and to raise the profile of Brandenburg-Prussia. Under German law of his time, with the exception of Bohemia, kingdoms within the Holy Roman Empire were prohibited. While Brandenburg was a principality of the Empire, the Duchy of Prussia was not; seeing an opportunity to form a political alliance against Louis XIV of France, Frederick suggested to Emperor Leopold I that he be allowed the title of King strictly within his Prussian holdings. Given that Prussia had no historical association with the Empire and he already ably governed it with full sovereignty, he argued there was no political or procedural reason he should not be able to rule as its king. Leopold gave his assent to Frederick's coronation on 16 November 1700, and Frederick crowned himself and his wife Sophia Charlotte of Hanover (1668–1705) at Königsberg Castle on 18 January 1701.

Frederick was slow to gain acceptance as a European monarch. Shortly after the coronation, Frederick Augustus I, Elector of Saxony (and later King August II of Poland, 1670–1733), was the first regional leader to congratulate the new king; most of the other Electors followed in short order. The kingdom of Denmark-Norway acknowledged his elevation in February 1701 in hopes he would become an ally in the Great Northern War (1700–1721). Sweden accepted Frederick's status as king in 1703, followed much later by Spain and France in 1713. But the knights of the Teutonic Order, who persisted in their Medieval claims to Prussian territory, refused to recognize Frederick. The Grand Master of the Order filed an official protest with Leopold's court and successfully petitioned Pope Clement XI to forbid Catholic monarchs from accepting Frederick as the rightful Prussian King; neither did the neighboring Polish-Lithuanian Commonwealth acknowledge a new royal figure on its doorstep until 1764.

The rule of the first Prussian King was largely peaceful and unremarkable, although Frederick slowly undermined the independent authority of his Brandenburg territories until they came under the hegemony of the Prussian monarchy. During this time, a Prussian national identity first emerged that persisted for some two centuries. Frederick died in Berlin on 25 February 1713. He was succeeded by his son Frederick William I, popularly known as the "Soldier King," as King in Prussia and Elector of Brandenburg. Frederick William I's son and successor, Frederick II "the Great" of Prussia, is profiled in Chap. 12 in connection with another lost constellation, Honores Frederici.

24.8 Portrait of Federick III of Brandenburg (Frederick I of Prussia) in a contemporary engraving housed at the Fürstlich Waldeckschen Hofbibliothek in Bad Arolson, Germany. Note the tiny scepter in the roundel above and to the left of the portrait, the royal symbol Gottfried Kirch chose to represent the temporal authority of the Elector of Brandenburg when forming his proposed constellation in 1688. The scepter is surrounded by the Old French "Honi Soi Qui Mal Y Pense" ("Evil be on him who thinks evil"), best known as the motto of the British chivalric Order of the Garter. The Latin motto on the ribbon below Frederick's arms reads FREDERICUS TERTIUS D(EI) G(RATIA) MARCHIO BRENDENB(URGICUM): S(ACRUM) R(OMANUM) I(MPERIUM) ARCHCAM(ERARIUS): ET ELECTOR SUPREMUS DUX BORUSSIAE, etc. ("Frederick the Third by the grace of God Marquis of Brandenburg, Arch-Chamberlain of the Holy Roman Empire and Supreme Elector, Duke of Prussia, etc."

24.9 Frederick William III in a portrait from page 10 of L.D. Eberhard's *Wegweiser durch die preußischen Staaten*, Berlin: Verlagsort (1831)

Frederick William III of Prussia

At the time that Bode revived Kirch's constellation, the Prussian King was Frederick William III (1770–1840; Fig. 24.9) who led the country through the turbulent period of the Napoleonic Wars and the fall of the first German Empire. His reign saw the Congress of Vienna (1814–1815) during which ambassadors of European states drafted a plan for the long-term peace of Europe by wiping out French territorial gains and redrawing the boundaries of kingdoms and empires to better balance the influence of the main Continental powers.

Frederick William was born in Potsdam on 3 August 1770, the son of King Frederick William II (1744–1797) and Frederika Louisa of Hesse-Darmstadt (1751–1805). As was the custom of the time, his care and upbringing were handed to a set of tutors in his early childhood, and he spent several happy years in the royal palace at Paretz in Havelland, west of Berlin. He was a shy and often melancholy boy who spoke little and avoided personal pronouns in his speech, a characteristic that was thought to indicate a natural tendency toward military leadership. In his youth he trained to be an officer, expected of a Prussian prince; he became a lieutenant in 1784 and a colonel in 1790, seeing action against France during 1792–1794.

On Christmas Eve 1793 he married Duchess Louise of Mecklenburg-Strelitz (1776–1810) at the Kronprinzenpalais in Berlin. In the 16 years until her death the pair enjoyed a happy marriage that produced nine children, two of which later became monarchs: King Frederick William IV of Prussia (1795–1861), who succeeded his father in 1840, and the first German Emperor William I (1797–1888). Queen Louise was especially loved by the Prussian people, helping to improve the popularity of the House of Hohenzollern.

Frederick William acceded to the Prussian throne upon the death of his father on 16 November 1797, making clear early on that he intended to put his own distinct stamp on the monarchy. He began by dismissing Frederick William II's ministers, reducing expenditures of the royal household and instituting some government reforms aimed at tempering the excesses of the former regime. In particular, he wanted his reign to project an image of sobriety and morality, which was only somewhat reluctantly followed by his subjects. He did not especially trust the advice of his ministers, but at the same time avoided actions that established clear policies and priorities.

His lack of strong political will led to difficulties for Prussia in the Napoleonic Wars (Fig. 24.10). Frederick William's instinct was to stay out of the wars completely, pursuing a policy of neutrality through the period of the Third Coalition in 1805. However, the Queen was strongly in favor of Prussia's participation in the war and influenced Frederick William's decision to enter it in October 1806. He personally led his army into battle against the French at Jena-Auerstädt on 14 October but suffered a defeat that left the army badly damaged. Fearing that support for the monarchy would follow, the royal family fled for far eastern Prussia in pursuit of the favor and protection of the Russian Emperor Alexander I (1777–1825).

However, Alexander had his own problems with the French, his army having been defeated by Napoleon. He sued for peace with France; Prussia, swept up in the current of capitulation, signed on to the agreement. As a consequence, it lost much of its Polish land holdings along with all native Prussian territory west of the River Elbe. France intended to punish Prussia for its role in the war, forcing the government to pay a large indemnity and to finance the cost of a French military occupation of certain Prussian cities it felt represented the places where anti-French sentiment was most likely to flare up. Consistent with his weak leadership, Frederick William accepted French dominion while his ministers tried to rebuild with new reforms in both government and the military. Shortly thereafter, Queen Louise died, striking a further blow to the King.

Frederick William reluctantly joined the coalition against France in 1813 following Napoleon's defeat in Russia. This was a dangerous move, with Berlin still under French occupation; the King secretly signed an alliance with Moscow and slipped out of his capital. With the military backing of Russia, his army contributed to the allied victories of 1813–1814. As a consequence Prussia won back some of the territories lost under the terms of the peace treaty with France, although his ministers failed to expand further through their desired annexation of Saxony. However, emboldened by Prussian ascendancy, he reneged on certain promises made before the war in the way of political reforms, in particular his assent to a constitutional system limiting his powers.

As peace and stability returned to his kingdom Frederick William married a second time to Countess Auguste von Harrach, Princess of Liegnitz (1800–1873). The marriage was

24.10 Map of the Kingdom of Prussia (*grey shaded area*) at the time Johann Elert Bode's *Uranographia* was published. This map covers the years 1797–1805. From T. Schade, *Atlas zur Geschichte des Preußischen Staates*. Głogów: Carl Fleming (1881)

morganatic, bringing happiness to Frederick William but depriving Auguste of the title of Queen. He died in Berlin on 7 June 1840, succeeded by his eldest son, Frederick William IV. His remains were laid to rest in the Mausoleum at Schlosspark Charlottenburg, Berlin; Auguste, ranked below all the living princes and princesses at court, was denied the right to attend his funeral.

The Scepter as a Symbol of Royal Power

The scepter is an ancient part of royal regalia in much of the Old World, representing both the temporal power of a monarch as well as the indication of good and just government. It takes the form of a short rod or staff held by the ruling sovereign. It is related to, but distinct from, pastoral staffs and croziers borne by clerics in the sense that the two would never be confused.

In Western culture, the earliest known representations of the scepter are seen in Egyptian art. A symbol called the *was* is often found in association with both gods and the pharaoh, indicating "power" and "dominion" in both the temporal and spiritual senses. Depictions of the *was* typically show a rod roughly as long as human and divine figures are tall, forked at one end and with a stylized animal head at the other (Fig. 24.11).

The use of staffs as signs of royal authority seems to have emerged independently in Greece, with instances in literature as early as the Homeric era; at least, the Greek practice dates to a time before the Hellenistic dominion over Egypt. The modern word *scepter* derives from the Ancient Greek σκῆπτρον, meaning a stick or baton. In Greek iconography, the σκῆπτρον is shown as a long staff topped by the metal figurine of an animal representative of some aspect of the holder; if, for example, when Zeus is depicted holding a scepter it is topped by the figure of a bird. Such scepters were the instruments of the κήρυκες, or heralds, whose possession of the objects indicated the protection of Zeus on foreign missions as an early form of diplomatic immunity.

The Greek-influenced Etruscans depicted scepters in the hands of both their kings as well as senior religious figures in scenes from tomb paintings across Etruria. Some Etruscan scepters in gold survive, often decorated with fine elements of metalwork. In Republican Rome, consuls were recognizable by their ivory scepters. Later Emperors incorporated the scepter into the imperial regalia in reference to the derivation of their temporal powers; the *sceptrum Augusti*, made of ivory, was topped by a small eagle in gold. The more recent art historical device of a Christian monarch holding both a scepter and orb also originated in this period, except that the orb was topped by a small figurine of Victory rather than a Christian cross.

The Cross replaced pagan symbols on European scepters after the end of Antiquity, although during the Medieval period the decorations drew from various secular designs and national symbols. The United Kingdom retains in its coronation ceremony the use of two scepters, one each topped with a cross and a dove. One French royal scepter was topped with a fleur de lys and the other with the *Main de Justice* (the Hand of Justice), a motif repeated in the figure of another lost constellation, Sceptrum et Manus Iustitiae (see Volume 2). To the end of the Holy Roman Empire, the scepter remained a highly visible reminder of the Emperor's authority over much of central and northern Europe (Fig. 24.12).

24.11 The Egyptian goddess Hathor (*right*), bearing a *was* staff, leading the 19th dynasty Queen Nefertari (*left*). Painting from Valley of the Queens tomb QV66, Thebes

24.12 Portrait of the last Holy Roman Emperor Francis II in his coronation robes (1805), by Leopold Kupelwieser (1796–1862). In his right hand Francis holds the Imperial Scepter

The Scepter as a Symbol of the Electorate of Brandenburg

The Scepter became associated specifically with Brandenburg during the twelfth century in the figure of Albert I ("the Bear"; *c.* 1100–1170), the first Margrave of Brandenburg, who also served as the Duke of Saxony from 1138 to 1142. Around the time that he renounced the Duchy of Saxony he was made *Erzkämmerer* (Archchamberlain) of the Holy Roman Empire, a ceremonial office that gave his successors the rights of prince-electors. The "High Officers" of the Empire discharged their duties only during coronations of new Emperors, during which they functioned as the bearers of the Imperial crown and regalia. Since the traditional office of chamberlain (Latin *camerarius*) was the head of a royal or

24.13 Arms of the Electorate of Brandenburg and the later Prussian province of Brandenburg. The Scepter is shown in the eagle's right talon and is also emblazoned on the escutcheon at *center*. Rendered by David Liuzzo; licensed under CC BY-SA 3.0

imperial household, the scepter became the officeholder's symbol as it symbolized the king or emperor's temporal authority.

The Holy Roman High Officers were entitled to augment their coats of arms with the symbols of their offices. The augmentations appeared as inescutcheons at the center of the Electors' shields, above all other charges. The Archchamberlain's augmentation was a gold scepter against a blue field (*azure a sceptre palewise Or*), and was formally added to the Brandenburg arms under Frederick II (1440–1470). The result is shown in Fig. 24.13.

At the time Kirch formed his constellation honoring Frederick I, the Scepter was a natural choice of figure as the most well-known symbol of the Brandenburg Electorate. But it was introduced long after the union of royal houses that resulted in the merging of the Electorate with the Duchy of Prussia. Kirch gave no clear indication as to why he chose the symbol most closely associated with Brandenburg, other than perhaps it was the older and more securely-established feature of the ruling Elector's symbolism. More likely it was intended to specifically honor Kirch's home state and thus reflects some degree of patriotic fervor.

DISAPPEARANCE

The Scepter rapidly fell out of favor and disappeared from charts as the nineteenth century drew to a close and cartographers gradually consigned it to the dustbin of history. Writing in 1899, Richard Hinckley Allen noted its revival under Bode, but also that it had again fallen into disuse:

> Sceptrum Brandenburgicum, the Brandenburg Sceptre was charted in 1688 by Gottfried Kirch, the first astronomer of the Prussian Royal Society of Sciences, and, more than a century thereafter, was published by Bode, who thus rescued it for a time from the oblivion into which, however, it seems to have lapsed again.

Like most other obsolete constellations, its apparent visual obscurity, simple design and situation in an otherwise sparse part of the sky seems to have doomed it. Hannah Bouvier (1858) summed up the situation well: "It contains no very conspicuous stars." Among its last mentions in popular texts is this description in William Tyler Olcott's *Star Lore of All Ages* (1911):

> Just west of Lepus is the little asterism known as "the Brandenburg Scepte," designed by Kirch in 1688. It contains but four stars of the 4th and 5th magnitudes, and the scepter is represented in Burritt's Atlas as standing upright in the sky.

At the first IAU General Assembly in 1922, Sceptrum Brandenburgicum was dropped from the list of constellations for which three-letter abbreviations were devised, and its fate was finally sealed when Eugène Delporte incorporated its stars within the boundaries of Eridanus in *Délimitation scientifique des constellations* (1930b) (Fig. 24.14).

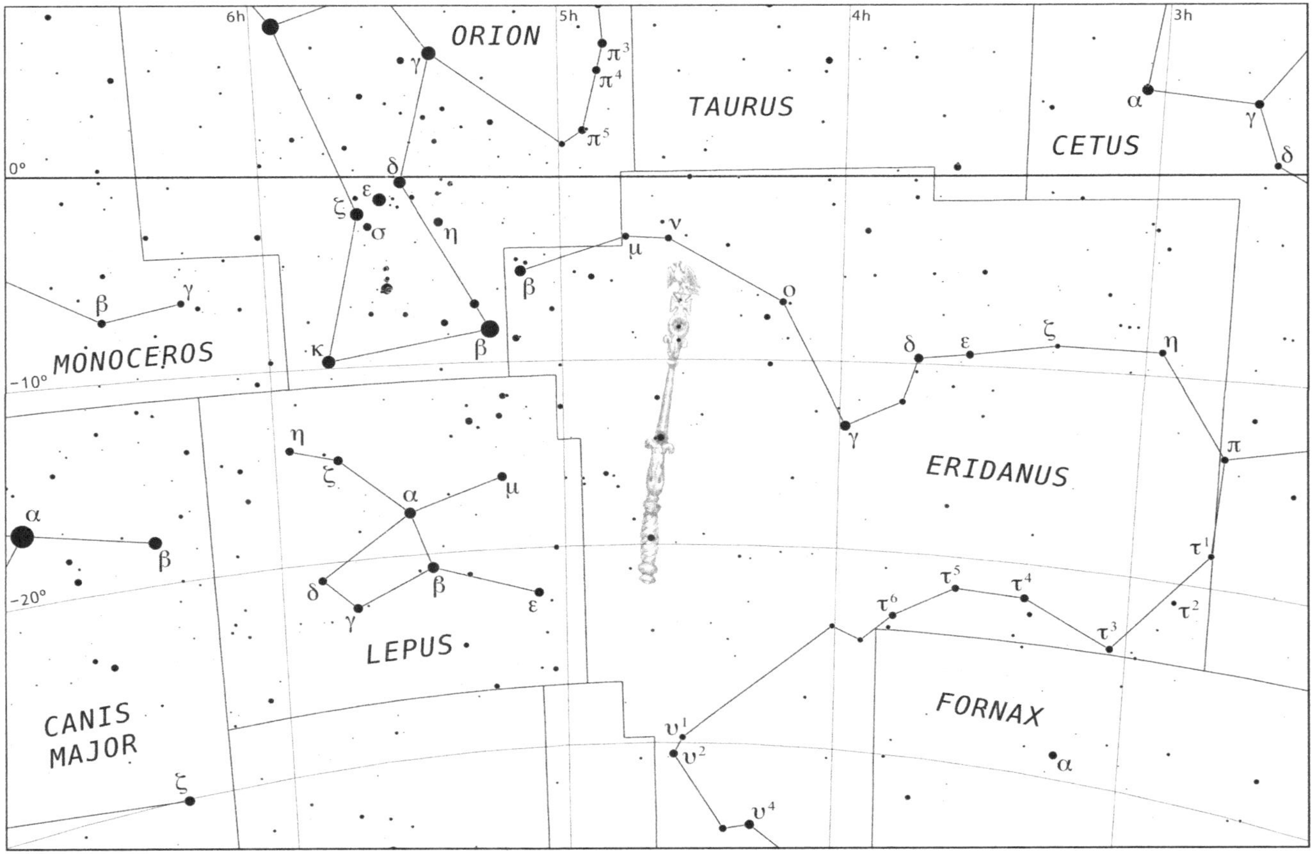

24.14 The figure of Sceptrum Brandenburgicum from Plate 24 of Alexander Jamieson's *Celestial Atlas* (1822) overlaid on a modern chart. Note that the figure is slightly distorted due to projection errors in Jamieson's original map

25

Taurus Poniatovii

Poniatowski's Bull

Genitive: Tauri Poniatovii
Abbreviation: TaP
Alternate names: "Taurus Poniatowski" (Ryan, 1827; Smyth, 1844b; Kendall, 1845; Chambers, 1877; Cottam, 1891); "Taurus Poniatowii" (Colas, 1892); "Taurus Poniatowskii" (Steele, 1899); "Bull of Poniatowskia" (Olcott, 1911); "Poniatowsky's Bull" (Young, 1807)
Location: Immediately east of the upper body of Ophiuchus and north of Serpens Cauda[1]

ORIGIN AND HISTORY

Poniatowski's Bull (Fig. 25.1) was introduced by Marcin Odlanicki Poczobut (1728–1810), a Jesuit astronomer educated at Vilnius University (then in Poland; now in Lithuania) and Charles University at Prague, with additional stints in France, Italy, and Germany from 1754 to 1764. For a time he was based at Marseille Observatory, studying under the French Jesuit astronomer Esprit Pézenas (1692–1776); it was this experience that inspired him to pursue astronomy as a career. After completing his doctorate, he became a professor at Vilnius and director of its Observatory in 1764. Only recently established, the Observatory was fairly lacking in modern instrumentation and Poczobut went to great

[1] *"Contiguous to the shoulder of Ophiuchus"* (Green, 1824); *"The* face *of the animal is near the point where the* solstitial colure *crosses the* equinoctial*" (Bouvier, 1858); "Midway between Altair in Aquila, and Ras Algethi in Hercules."* (Rosser, 1879); *"Between Aquila and Ophiuchus on the borders of Hercules"* (Cottam, 1891); *"About fifteen minutes* [of arc] *east of the star γ Ophiuchi"* (Olcott, 1911).

J.C. Barentine, *The Lost Constellations*, Springer Praxis Books,
DOI 10.1007/978-3-319-22795-5_25

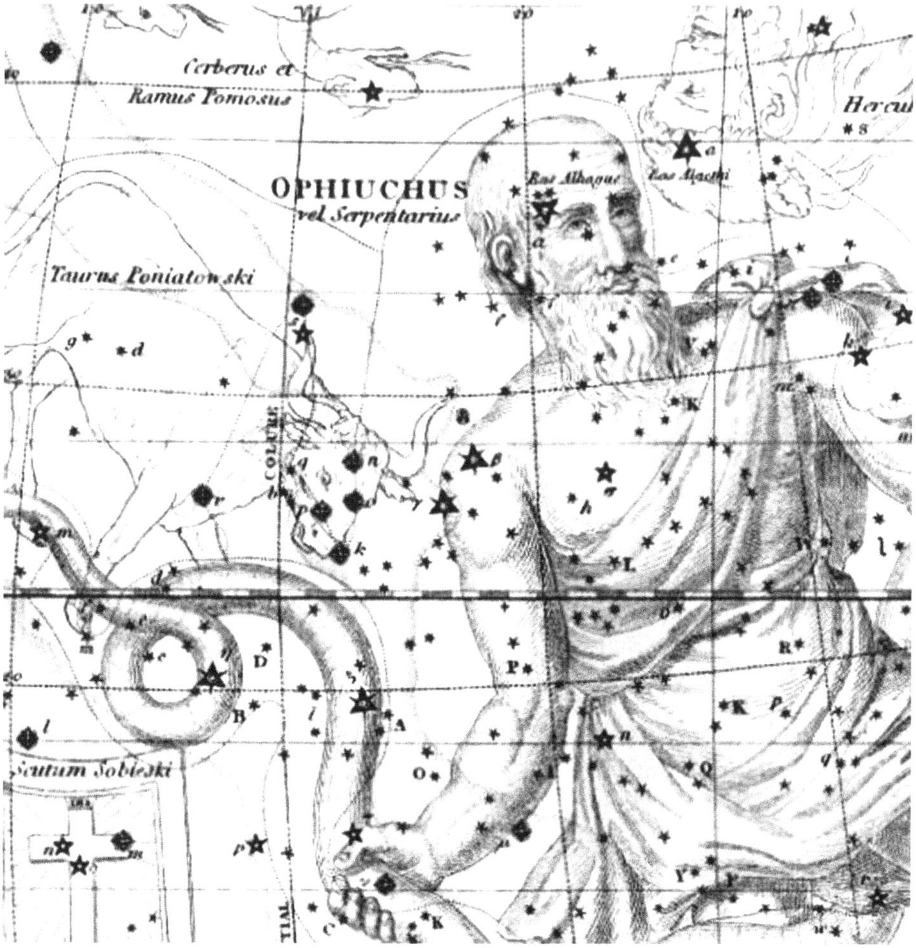

25.1 Taurus Poniatovii (labeled *Taurus Poniatowski*, *center-left*) as depicted on Plate 9 of Alexander Jamieson's *Celestial Atlas* (1822). Note the extinct constellation Cerberus (as "Cerberus et Ramus Pomosus"; Chap. 7) at *upper left*

lengths to equip it as well as possible. Despite official suppression of the Jesuits in the 1770s, the Observatory won the patronage of the Polish king, King Stanisław II August Poniatowski; it was renamed the Royal Observatory and Poczobut became the Astronomer Royal.

In 1777, Poczobut published the results of his initial investigations at the Observatory in *Cahiers des observations astronomiques faites à l'observatoire royal de Vilna en 1773*. In it he cataloged a V-shaped group of 16 stars east of Ophiuchus that Ptolemy did not consider part of Ophiuchus itself. Richard Hinckley Allen (1899) noted that the same stars were earlier identified by Jacob Bartsch as part of his constellation Tigris (Chap. 27):

"A century and a half before Poczobut's time these stars, with those of our Vulpecula, had been introduced by Bartsch into his plates as the River Tigris, although this probably had previously been a recognized constellation." Tigris seems to have been extinct by the mid-eighteenth century, and Poczobut evidently felt entitled to re-form its stars as he saw fit. The shape comprising those stars reminded Poczobut of the Hyades star cluster that outlines the face of the ancient constellation Taurus, suggesting to him a form for his new constellation. He attached the name of his royal patron to the new figure, "a formal permission to that effect having been obtained from the French Academy," according to Admiral Smyth (1844b). However, the dedication was evidently not especially well received by Stanisław himself. According to a description of Poczobut's book in the contemporary British literary review *The Critical review, or, Annals of literature* (1778):

> This collection contains a great number of accurate astronomical observations made by Abbé Poczobut and M. Streki, with very good instruments, under the auspices and patronage of his Polish majesty, to whom the astronomer intended to dedicate the new constellation of Poniatowski's Royal Bull, (Taureau Royal de Poniatowski), already mentioned in the supplement to the Parisian Cylopedia; but the king's modesty declined the intended honour, and obliged the astronomer to expunge that denomination; while his munificence rewarded the astronomer's merit with a medal struck in honour of him, with the legend on one side: 'Martin Poczobut, Astron. Reg. Pol. S. R. Lond. n. 1728.' On the other[2] 'Sic itur et astra. Bene merentis Laudi dedit Stan. Aug. Rex. 1775,' with the attributes of astronomy. A medal that will, among many other proofs, evince the king's generosity, as well as the subject's merits.

If Poczobut ever formally retracted the constellation, he failed to prevent it from attracting notice elsewhere in Europe. Poniatowski's Bull first appeared in charts as "le Taureau Royal de Poniatowski" in a 1778 edition of Jean Fortin's *Atlas Céleste*, and was included by Joseph Jérôme Lefranois de Lalande on a celestial globe he produced in 1779. Fortin again published it in his 1795 edition of Flamsteed's atlas, but its appearance in Bode's *Uranographia* (1801b) assured its survival well into the nineteenth century. Bode first adopted the figure in *Vorstellung der Gestirne* (1782), in which he labeled it in German "Der Königliche Stier von Poniatowski" ("Poniatowski's Royal Bull;" Fig. 25.2, top) but later Latinized its name as "Taurus Poniatovii" (Fig. 25.2, bottom). He described[3] it in *Allgemeine Beschreibung und Nachweisung der Gestirne*, complete with a reference to its canonical namesake:

> Poczobut, abbot of Wilna, placed this bull among the constellations in honor of Stanisław II August Poniatowski, King of Poland. It is adjacent to the shoulder of

[2]"Thus you shall go and the stars. Well deserving of praise given by the August King Stanislaw, 1775." The first sentence is probably an erroneous rendering of Virgil's famous line from *Aeneid* IX, 641, "Sic itur *ad* astra" ("thus you shall go to the stars"), said by Apollo to Aeneas's son Iulus.

[3]Poczobut, abbé de Wilna, plaça ce taureau au nobre des constellations en l'honeur de Stanislas Poniatowsky, roi de Pologne. Il est contigu à l'épaule du serpentaire en tirant vers l'orient, & situé dans la voie lactée; on le reconnoit à cinq étoiles situées a la tête de taureau, & qui auparavant apprtenoient à Ophiuchus. Ces étoiles forment un V, comme celles situées à la tête du taureau dans le zodiaque. (p. 10)

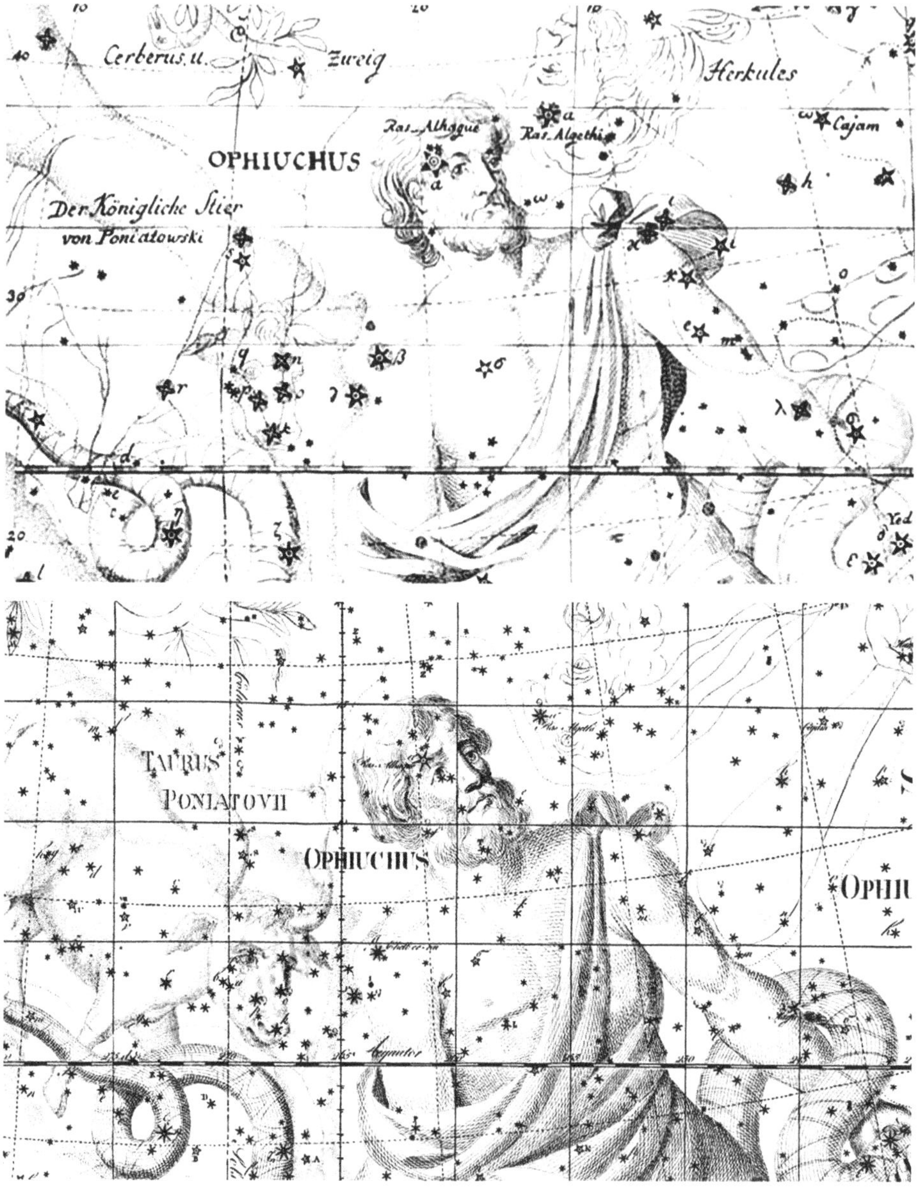

25.2 Two views of northern Ophiuchus in atlases published by Johann Elert Bode: *Vorstellung der Gestirne* (1782; *top*) and *Uranographia* (1801b, *bottom*), both numbered Plate 9

> the serpent-bearer stretching eastward, and is in the Milky Way; we recognize five stars located at the head of the bull formerly belonging to Ophiuchus. These stars form a "V," such as those located at the head of the bull in the zodiac [Taurus].

Bode also greatly expanded the number of stars cataloged as belonging to the Bull; Admiral Smyth noted that "Poczobut was content with seven component stars, but Bode has scraped together no fewer than eighty."

The influence of *Uranographia* on both sides of the Atlantic ensured wide recognition of Poczobut's constellation, although the details of its backstory sometimes became lost in translation. For example, the American author James Ryan attributed to the Bull to the wrong Poniatowski in *The New American Grammar of the Elements of Astronomy* (1827):

> The Bull of Poniatowski was so called in honour of Count Poniatowski, a Polish officer of great merit, who saved the life of Charles XII, king of Sweden, at the battle of Pultowa, a town in Russia, and capital of the government of the same.

The constellation remained in fairly wide circulation past midcentury, appearing in *Urania's Mirror* (1825; Fig. 25.3), Ezra Otis Kendall's *Uranography* (1845, as "Taurus Poniatowski—The Polish Bull"), and George Chambers' *Handbook of Descriptive Astronomy* (1877).

Considering his usual dislike for contrived constellations of recent design, Jacob Green gave Taurus Poniatovii a comparatively favorable treatment in *Astronomical Recreations* (1824), noting it

> is a small constellation in the Milky Way, contiguous to the shoulder of Ophiuchus, and appears to form part of that group. It was first arranged by the Abbé Poczobut, in 1778, in honour of Stanislaus Poniatowski, king of Poland. It may be known by four stars of the fourth magnitude, arranged in the figure of the letter V, very much in the same manner as the stars which form the head of Taurus in the Zodiac. It was, no doubt, this resemblance that suggested the name which has been applied to this group.

He included it among his charts on Plate 7 (Fig. 25.4); unusually, relative to the depictions of other cartographers, Green showed only the Bull's head and a vague suggestion of its neck and upper back. Since he did not draw the rest of the Bull's body, it is reasonable to conclude that he felt the constellation as proposed by Poczobut only applied to the "V" asterism around 70 Ophiuchi. Similarly, Hannah Bouvier identified in her *Familiar astronomy* (1858) only the asterism as the constellation itself:

> *Q.Are there any* REMARKABLE STARS *in this constellation?*
> *A.There are* not. *A few small stars form the letter V, which are thought to resemble the Hyades and Aldebaran in the zodiacal constellation Taurus.*

From the end of the eighteenth century through much of the nineteenth, Taurus Poniatovii appeared in many charts under a surprisingly large variation of names depending on the source language and preferences of the individual cartographer. A sample of these instances is collected in Table 25.1.

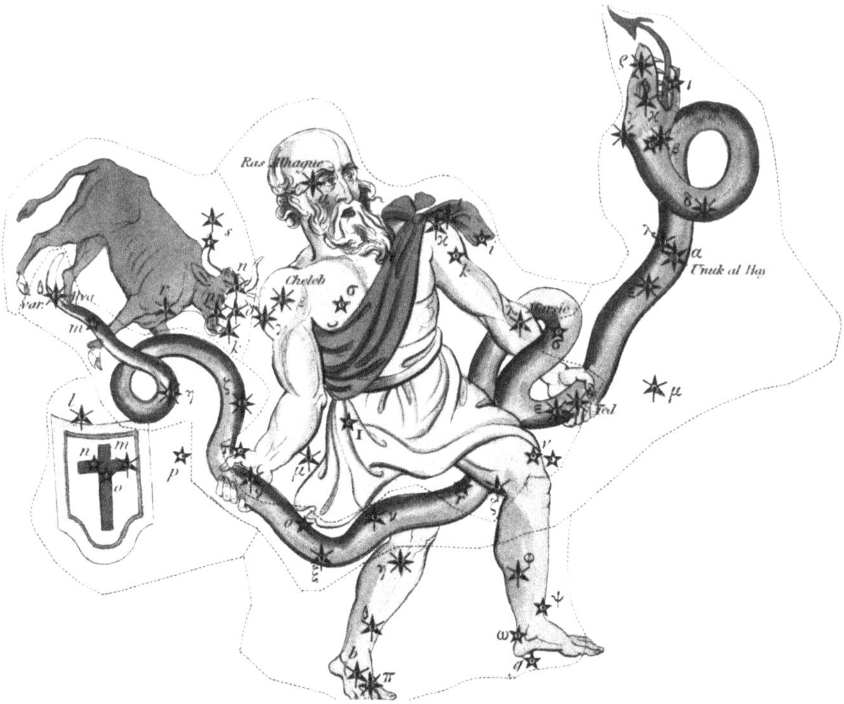

25.3 Taurus Poniatovii (*left*) shown on Plate 12 of *Urania's Mirror* (1825). Another Polish namesake, Scutum Sobeskii, is shown below Serpens' tail at *lower left*

ICONOGRAPHY

Stanisław II Poniatowski of Poland

A controversial figure in Polish history to this day, King Stanisław II August was born Stanisław Antoni Poniatowski (Fig. 25.5) on 17 January 1732 to Stanisław and Konstancja (Czartoryska) Poniatowski at Wołczyn, in the Polish-Lithuanian Commonwealth. He was the last elected King and Grand Duke of the Commonwealth, known as a reformer and patron of the arts and sciences. He is mainly remembered as the individual who failed to prevent the disintegration of the Polish nation after having come to power under irregular circumstances.

Poniatowski had a sheltered upbringing within the *szlachta*, the highest echelon of the Polish nobility. His childhood was spent in Gdańsk and later in Warsaw, where his introverted personality left him with few friends and a voracious appetite for reading books. At the age of sixteen he left home for the first time, following the Russian army's advance into Germany; he visited territory as far west as the Netherlands on the trip and returned home before the end of the year.

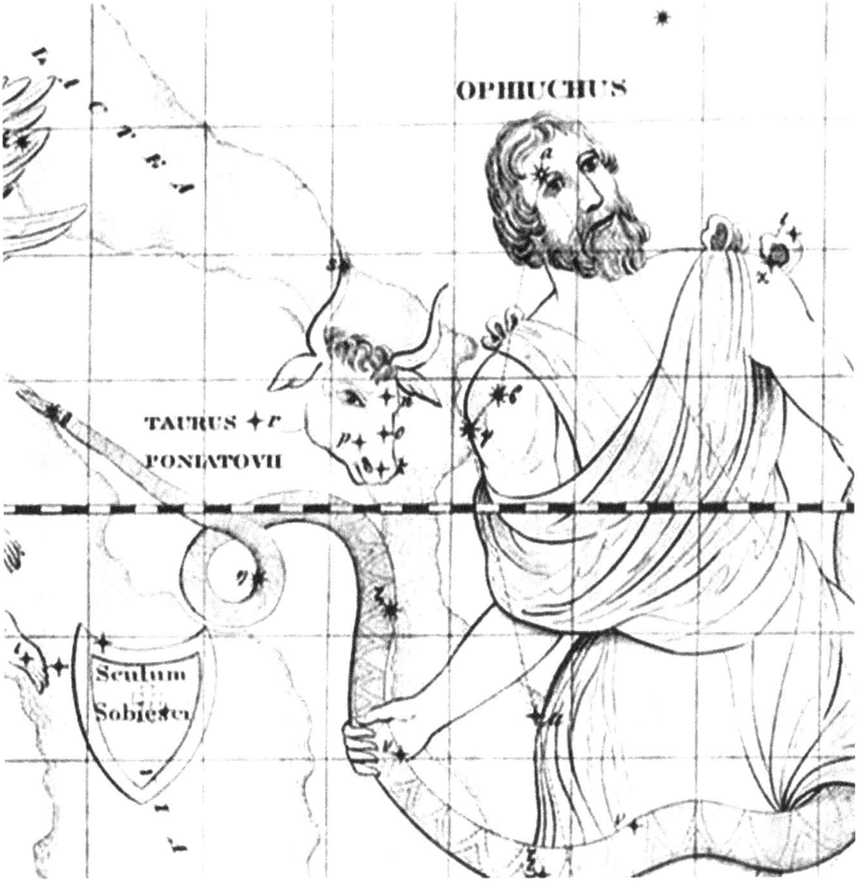

25.4 Taurus Poniatovii shown only as a disembodied head on Plate 7 of Jacob Green's *Astronomical Recreations* (1824)

His political career began early. In 1749 he apprenticed in the chancellery of Michał Fryderyk Czartoryski, then-Deputy Chancellor of Lithuania and traveled to Berlin the following year where he became fast friends with the British diplomat Charles Hanbury Williams. His travels continued to the Austrian court in Vienna during 1752. In the interim he was elected to the Treasury Tribunal in Radom and later as a deputy in the Sejm, or Polish parliament. In 1753–1754 he traveled trough Hungary to Vienna again before heading onward to Holland, Paris, and finally England where he met Charles Yorke, who would become Lord Chancellor of Great Britain in 1770. He returned to Poland toward the end of 1754 but did not participate in the Sejm, held back at his parents' requests in order to insulate him from the political intrigue of the time swirling around the Ostrogski family.

Table 25.1 Appearances of Taurus Poniatovii on various maps and the labels used to identify it

Source	*Map*	*Label*
Fortin (1795)	Map 9	"Taureau Royal de Poniatowski"
Goldbach (1799)	Map 8	"Poniatowsk. Stier"
	Map 9	"Konigl. Stier des Poniatowski"
	Map 10	"d. Konigliche Stier von Poniatowski"
Bode (1801b)	Plate 9	"Taurus Poniatovii"
Meissner (1805)	Figure XXVI, Table 24	"Stier Poniatowsky"
	Figure XXVIII, Table 26	"Koniglicher Stier von Poniatowsky"
	Figure XXIX, Table 27	"Koniglicher Stier von Poniatowsky"
Young (1807)	Plate XXXVI, Fig. 517	"Poniatowsky's Bull"
Brooke (1820)	Plate 4	"Taurus Poniatowski"
Jamieson (1822)	Plate 9	"Taurus Poniatowski"
Burritt (1835a)	Plate 5	"Taurus or BULL of PONIATOWSKIA"
Littrow (1839)	Tables 1–2	"Stier de Poniatowski"
Middleton (1842)	Plate 4	"Taurus Poniatowski"
Riedig (1849)	Plate 1	"Stier Poniatomski"
	Plate 9	"Poniatowsk. Stier"
Preyssinger (1862)	Plate II	"Taureau de Poniatowski"
Bowen (1888)	Map, page 107	"Taurus Ioniatowski"

A fateful meeting with the future Empress of Russia that led to his election as King of Poland. His connections with the powerful Czartoryski family and its political wing, the 'Familia,' led to his being sent to St. Petersburg in 1755 in the service of Williams, recently named the British ambassador to the Russian Empire. There he met young Catherine Alexeievna, who would later become Catherine II the Great. Catherine took young Poniatowski as her lover, while he saw the connection as his ticket to career advancement, but plans were cut short in July 1756 when Poniatowski was abruptly forced out of St. Petersburg on account of intrigue at court. Catherine pressed Empress Elizabeth and Grand Chancellor Count Alexey Petrovich Bestuzhev-Ryumin to secure his appointment as the ambassador of Saxony to the Russian court by early 1757. This did not sit well with various European governments, some of which demanded his withdrawal. Poniatowski left St. Petersburg again in August 1758.

Back in the Commonwealth, he attended the Sejms of 1758, 1760 and 1762, advocating a political stance favoring Russia and opposing Prussia. In 1762 his father died, leaving him a reasonable inheritance. During the same year, Catherine acceded to the Russian throne; Poniatowski held out hope that she would consider marrying him in order that he could exert direct control over some aspects of Russian politics, but no proposal was forthcoming. Meanwhile, the Familia planned a coup to overthrow King Augustus III of Poland and Poniatowski became peripherally involved, and in August 1763 Catherine made clear to him that she would not back the effort to depose Augustus.

25.5 Portrait of King Stanisław II Poniatowski of Poland (1788) by Johann Baptist von Lampi the Elder (1751–1830). Oil on canvas, 55 × 69 cm; Hermitage Museum, St. Petersburg, Russia

August died that October, and negotiations opened for the election of his successor. With Augustus out of the way, Catherine gave Poniatowski her blessing and support as a candidate. Russia spent some 2.5 million rubles backing Poniatowski's election, and after his supporters and opponents engaged in physical violence Catherine called out her army, which took up camp mere miles from the meeting site of the Sejm at Wola. No serious opposition candidates emerged, and Poniatowski was elected king on 7 September 1764. He was crowned in Warsaw on 25 November, taking the regnal name "Stanisław August" that combined the names of his two most immediate royal predecessors.

While Poniatowski was a capable administrator his conciliatory nature and support for progressive reforms were his ultimate undoing. Highly popular at the start of his reign, he attempted to reform the ineffective central government. That involved changing the structure of the government itself, reducing the power of the military and making its commanders part of commissions elected by the Sejm and reporting directly to him. The Familia assumed he would adopt its goals and function more or less as its puppet, becoming alarmed when he began making peace with various opponents of their policy priorities. Tolerance of religious minorities was a major sticking point in relations between the King and the Familia, as Poniatowski supported tolerance and the Familia opposed it.

For the first decade of his reign, Poniatowski maintained an uncomfortable alliance with the Familia, but Russia sensed strain in their relationship and intended to exploit that weakness to its advantage by meddling in the Commonwealth's affairs and destabilizing Poniatowski's government. Catherine herself opposed the King's reform efforts; she supported his tenure on the Polish throne only insofar as the country remained a weak Russian client state. The King's reforms enabled people who trusted Russia the least, and his effort gradually came to be perceived as a threat to Russian hegemony over the Polish state.

As the Familia pulled away from Poniatowski it found itself set up in opposition to Russia, who supported the cause of Polish dissidents favoring Russian influence over the culture and politics of their homeland. A political faction known as the Bar Confederation emerged that aligned itself against Russia, the dissidents and the King. Poniatowski and the Familia reached out unsuccessfully to western European powers to build political and military alliances but found themselves rebuffed at every turn. Facing growing domestic discontent, they were left with little choice but to accept the status of a de facto Russian protectorate in order to retain power.

The Bar Confederation declared Poniatowski dethroned in 1770 and kidnapped him the following year with the intent of deposing him and putting their own candidate on the throne. While they briefly held the King prisoner near Warsaw, he managed to escape. Rapidly losing popular support, Poniatowski's government was in danger of imminent collapse. Russia, Prussia and Austria allied to launch a military intervention in the Commonwealth, but in return unilaterally demanded major territorial concessions. Seeing no other way out, the King and his Sejm capitulated to the allies and consented to the partition treaty. In it, the Polish King's power was significantly curtailed, leaving him subservient to Russian suzerainty (Fig. 25.6).

Nevertheless, Poniatowski continued to press for political reforms within the scope of the power he retained. His efforts culminated in the reformist Constitution of 1791, which was a major step forward but well short of recent efforts in Revolutionary France and America. It made Poniatowski a constitutional monarch in the English model with a federal system, in the King's words,[4] "founded principally on those of England and the United States of America, but avoiding the faults and errors of both, and adapted as much as possible to the local and particular circumstances of the country."

The new constitution made the absolute monarchies of Europe nervous. Frederick William II of Prussia officially congratulated the Poles on their achievement, but behind the scenes worried about high-level contacts between Polish reformers and their revolutionary counterparts in the French National Assembly. Prussia feared that a newly-empowered Poland would demand the return of lands acquired during the First Partition.

The Polish nobility didn't like it either. In short order the nobles formed the Targowica Confederation and took aim at the new constitution, which they claimed too severely restricted the privileges they traditionally enjoyed in Polish society. The confederates turned toward St. Petersburg, precipitating the Polish-Russian War of 1792. Poniatowski's Sejm acted to increase the size of the standing army to 100,000 men, but failed to allocate

[4] Recorded in *The Scots Magazine*, Edinburgh, Vol. 53, page 291 (June 1791).

25.6 An allegory of the First Partition of Poland (1772) by Noël Le Mire (1724–1801). Catherine II of Russia is seated at left, while Joseph II of Austria and Frederick II ("the Great") of Prussia argue at right over their respective territorial gains

the resources to assemble, train and deploy such a force; instead, Poniatowski was forced to field an army of only 37,000 ill-prepared recruits. Nevertheless, the army commanded by King's nephew, Józef Poniatowski, and Tadeusz Kościuszko put up an impressive resistance to the Russian advance, repeatedly fighting them to a draw and even defeating the Russian army in a few cases.

The Polish force eventually buckled under the numerical superiority of the Russians, and Poniatowski's kingdom quickly began to unravel. The King attempted in vain to negotiate with the Russian court, and as its army closed in on Warsaw in July 1792 he began to consider surrender. Russian ambassador Yakov Bulgakov assured Poniatowski that his country would assert no new territorial claims as a consequence of a Polish surrender, and a majority of the King's ministers voted in favor of proceeding with the plan. Shortly thereafter Poniatowski joined the Targowica Confederation and his army dissolved. The end of the Commonwealth seemed near.

Poniatowski and the reformist faction had lost almost all of their influence, both domestically and with the Russian court. A Second Partition of Poland soon followed, despite previous Russian promises. The so-called Grodno Sejm was convened and on 23 November 1793 it annulled all the acts of its predecessor, including the Constitution of 1791. The King considered abdication. In early 1794, Tadeusz Kościuszko led an uprising against Russia and Prussia in opposition of the Second Partition, but it failed in its goal of liberating Poland, Lithuania and Belarus from Russian influence. The suppression of the Kościuszko Uprising marked the end of the Commonwealth in fact.

The King held on to power for a few more months in the face of his nation's collapse, but Catherine lost what remaining patience she had with him and in early December 1794 demanded that he leave Warsaw. Shortly after the new year, Poniatowski consented and relocated to Grodno under a Russian military escort. The formal end of the Commonwealth followed the signing of the treaty affecting the Third Partition of Poland on 24 October 1795, and Poniatowski signed an instrument of abdication on 25 November, 31 years to the day after his coronation.

The former king lived out his final years in Russia. Catherine died on 17 November 1796 and was succeeded by Paul I. A few months later Poniatowski left for St. Petersburg, hoping the new Emperor would allow him to travel abroad. Paul's government refused the request, and he lived out the remainder of his days a virtual prisoner in the Marble Palace. Catherine had left him a small pension, but his old age was spent in varying degrees of financial stress. He worked on his memoirs and intermittently pleaded the Polish case to the Russian court before dying of a stroke on 12 February 1798.

Paul furnished a royal state funeral, after which Poniatowski was buried at the Catholic Church of St. Catherine in St. Petersburg. The Soviet Union planned to demolish the church in 1938, allowing his remains to be transferred secretly to a church in his birthplace of Wołczyn. That church, in the Belarussian Soviet Socialist Republic, fell into disrepair under the Soviet regime, and his body was moved again to St. John's Cathedral in Warsaw where a final funeral ceremony was held on 14 February 1995. Poniatowski remains interred there to this day.

While Poniatowski supported progressive reforms during a turbulent period of European history and was perhaps the most important patron of the arts of the Polish Enlightenment, he presided over the functional end of his country. At the time Poczobut formed a constellation in his honor, after the First Partition, Poland achieved some degree of security and territorial integrity. It was in that time that Poczobut noticed the little "V" asterism of stars around 70 Ophiuchi, which reminded him of Aldebaran and the Hyades. From these stars and those further east he formed a figure in recognition of his embattled sovereign, whose rule and life he would see draw to a close some two decades hence.

DISAPPEARANCE

From the mid-nineteenth century Poniatowski's Bull gradually disappeared from charts, its stars increasingly identified as outermost constituents of Ophiuchus. It was shown neither in Otto Möllinger's *Himmels-Atlas* (1851) nor Alexander Keith Johnston's *Atlas of Astronomy* (1855). Jean Charles Houzeau included the Bull among a list of constellations described as "not adopted and ... not in use" in *Vade-mecum de l'astronome* (1882). Arthur Cottam showed the constellation it on Chart 32 of *Charts of the Constellations From the*

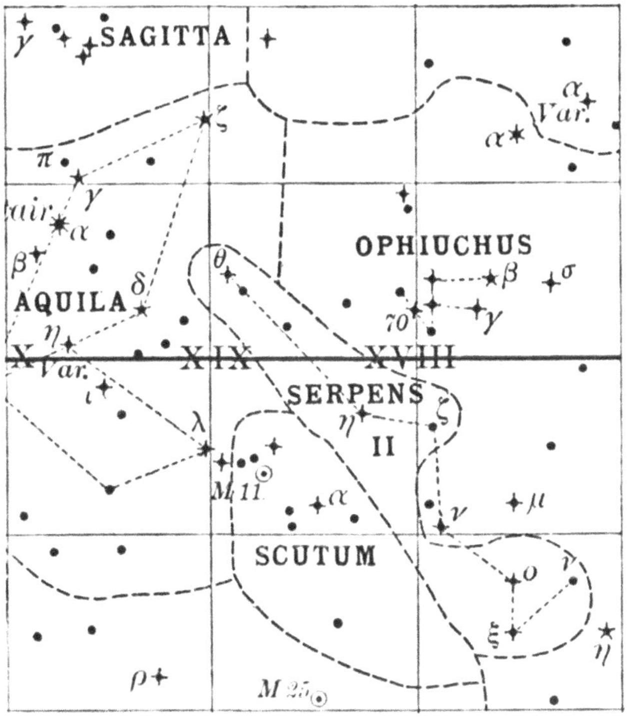

25.7 An excerpt of Map IV in Charles Augustus Young's *Lessons in Astronomy* (1903). The V-shape of stars representing the face of the former constellation Taurus Poniatovii is shown right of center, including the *star marked* "70" (70 Ophiuchi)

North Pole to between 35 & 40 Degrees of South Declination (1891), but seems to have relegated it to the status of an historical oddity:

> Poniatowski's Bull. A small modern asterism added by Poczobut in 1777. ... It is not shown as a separate constellation on the chart, but all the objects given by Webb as included in it are shown. It lies in a very rich part of the Galaxy.

This sense became permanent around the turn of the twentieth century. Joel Dorman Steele seemed to think of it as an asterism within Ophiuchus, writing in the second edition of *Popular Astronomy* (1899) "There is a small cluster near *β*, called *Taurus Poniatowskii*." In the same year, Allen wrote "as a distinct constellation it is not generally recognized by astronomers, and its stars have been returned to Ophiuchus," and in *Star Lore of All Ages* (1911), William Tyler Olcott referred to the Bull as a "discarded asterism" though he included it on a chart of Ophiuchus.

But by that time other authors were simply drawing the borders of Ophiuchus and Serpens around its stars and rejecting Poczobut's identification of the Bull. It was not accorded constellation status by the International Astronomical Union at the First General Assembly in 1922, and its stars fully absorbed into Ophiuchus when the modern constellation boundaries were drawn by Eugène Delporte (1930b) (Fig. 25.8).

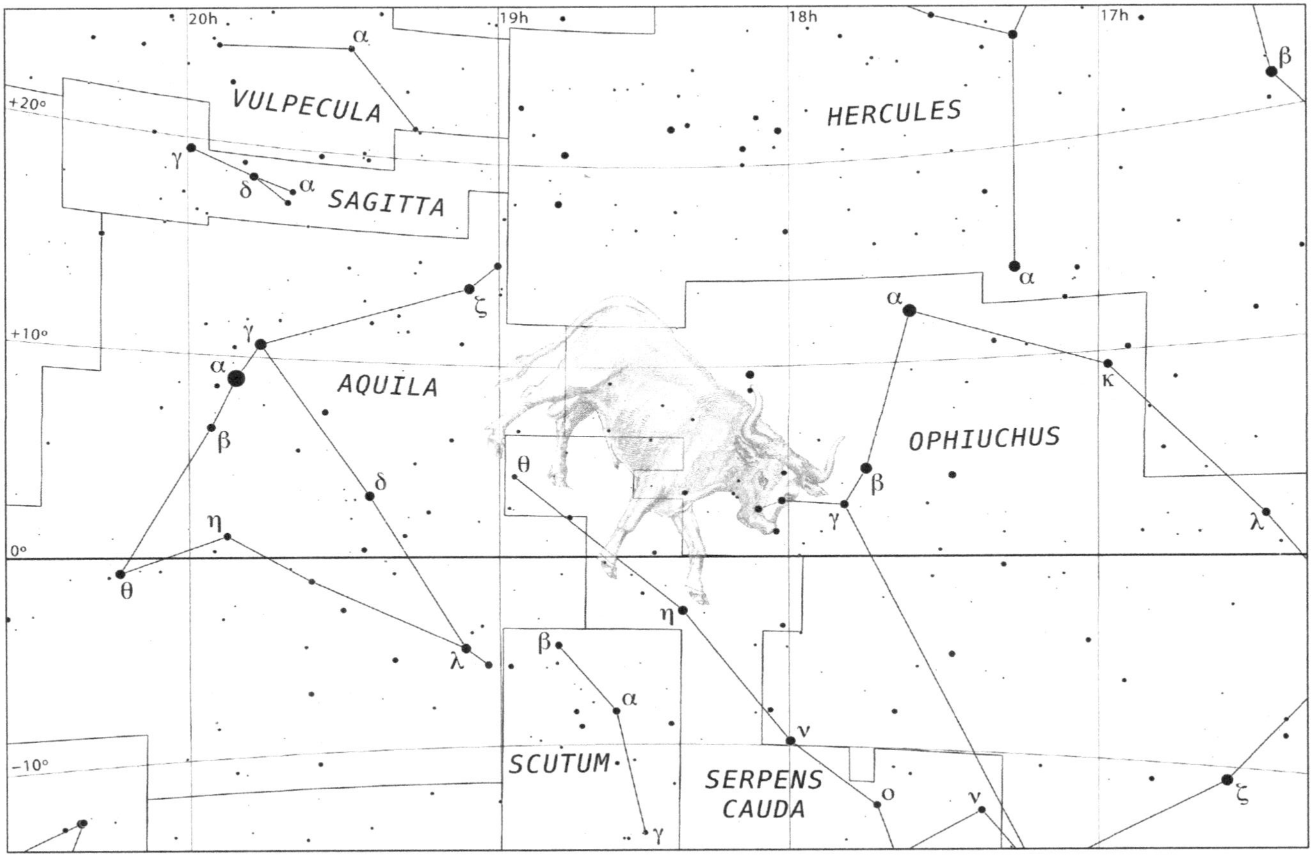

25.8 The figure of Taurus Poniatovii from Plate 10 of Alexander Jamieson's *Celestial Atlas* (1822) overlaid on a modern chart. Note that the forelegs of the Bull in this figure are a hypothetical reconstruction, given that on Jamieson's chart they are almost entirely obscured by the body of Serpens

26

Telescopia Herschelii

Telescopium Herschelii Major (Herchel's 20-Foot Telescope)

Genitive: Telescopii Herschelli Majoris
Abbreviation: THMa
Alternate names: "Tubus Hershelii Maior" (Hell, 1789); "Telescopium Herschelii" (Bode, 1801b); "Telescopium Herscelii" (Jamieson, 1822); "Telescopium Herschilii" (Hall, 1825)
Location: Between Lynx and Gemini[1]

Telescopium Herschelii Minor (Herchel's 7-Foot Telescope)

Genitive: Telescopii Herschelii Minoris
Abbreviation: THMi
Alternate names: "Tubus Hershelii Minor" (Hell, 1789)
Location: East of the Hyades and north of Orion[2]

---

[1] *"Between the bridle* [of Auriga] *and Gemini"* (Green, 1824) *"Lies between the Lynx and the Wagoner"* (Kendall, 1845); *"Situated between the Lynx and Gemini"* (Bouvier, 1858); *"It lay between the Lynx and Gemini... The star π of Gemini marks its former location, the western end having been among the ψ stars of Auriga, not far from the latter's β."* (Allen, 1899)

[2] *"Crammed awkwardly between Orion and the head of Taurus"* (Ridpath, 1989); *"Just to the east of the Hyades in Taurus"* (Bakich, 1995).

J.C. Barentine, *The Lost Constellations*, Springer Praxis Books,
DOI 10.1007/978-3-319-22795-5_26

ORIGIN AND HISTORY

Maximilian Hell introduced two new constellations in 1789 commemorating the discovery of the planet Uranus by Sir William Herschel (1738–1822) 8 years earlier. Both took the form of telescopes belonging to Herschel, and first appeared in an annual almanac published at Milan called *Ephemerides astronomicae* for the year 1790, subtitled *Anni 1790 ad meridianum mediolanensem* ("For the year 1790 at the meridian of Milan"). To underscore the Uranian connection, the two telescopes nearly bracketed the region of sky where Herschel found the first planet discovered since antiquity in March 1781, marked between the horns of Taurus on Hell's map reproduced here in Fig. 26.2. Hell's "Tubus Hershelii Maior" and "Tubus Hershelii Minor" were intended to represent, respectively, a 20-foot reflector and the 7-foot reflector actually used in the Uranus discovery. Based on the depictions in *Ephemerides astronomicae* (Figs. 26.1 and 26.2) it is clear that, as Ridpath (1989) put it, "Hell had not seen either telescope."

Hell struggled mightily to find a group of stars in the vicinity immediately east of the Hyades from which to form something looking vaguely like the tube of a telescope, settling on a handful of faint stars that formed the de facto boundary between Orion and Taurus on some eighteenth century charts. To these few stars he ascribed the figure of a refracting telescope, a completely different optical design than the reflecting instruments used exclusively by Herschel. Hell's "Tubus Hershelii Minor" even has a smaller "finder

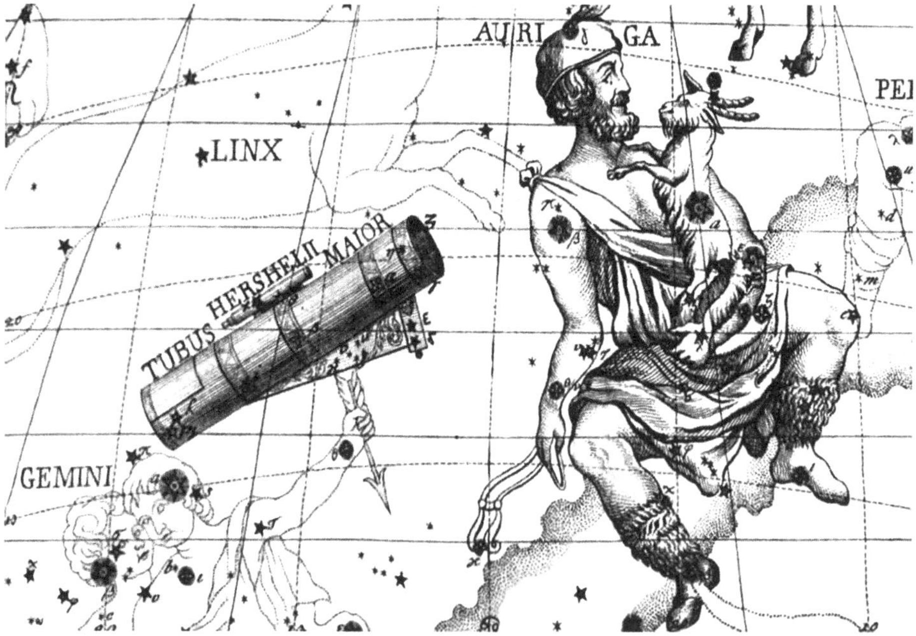

26.1 "Tubus Hershelii Maior" as suggested by Maximilian Hell in *Ephemerides astronomicae: Anni 1790 ad meridianum mediolanensem*

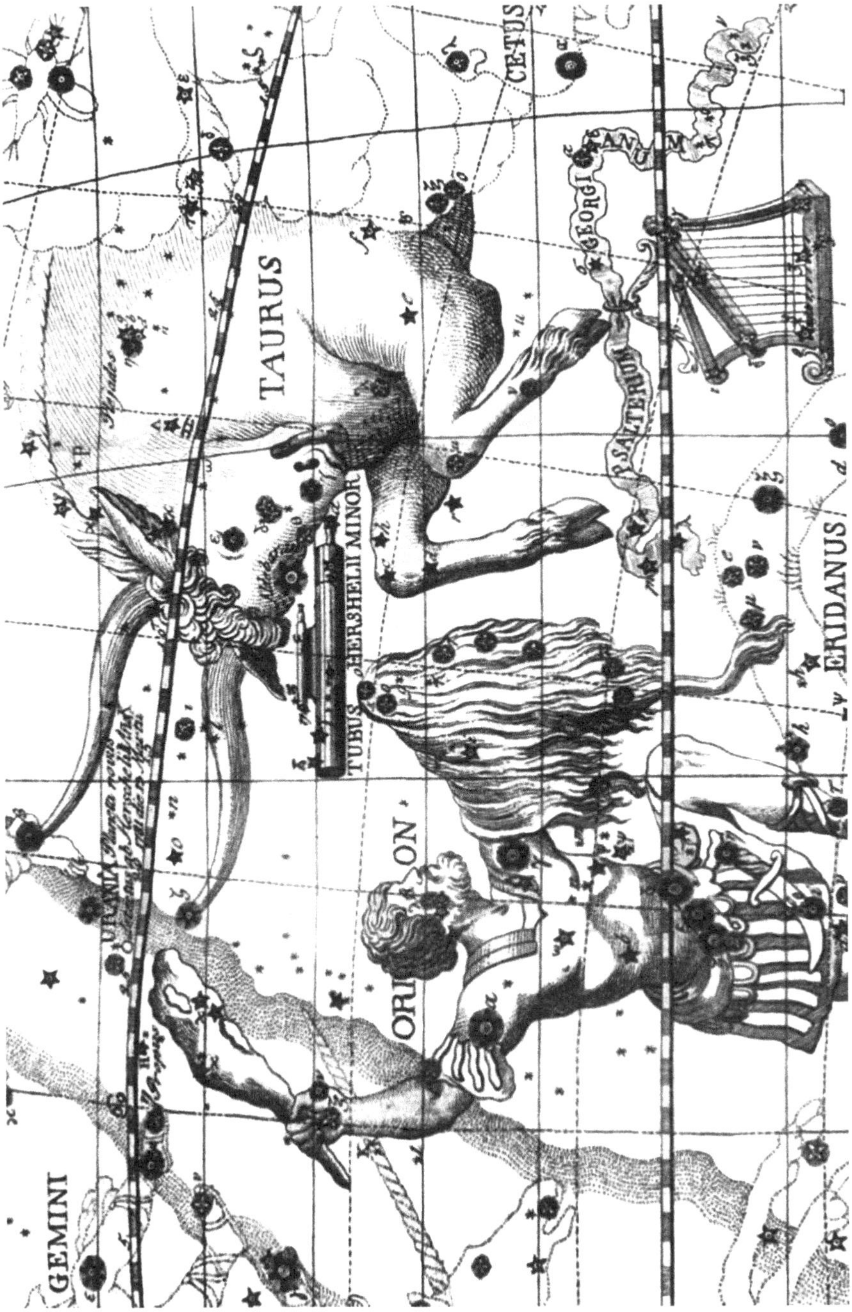

26.2 "Tubus Hershelii Minor" as suggested by Maximilian Hell in the 1790 edition of *Ephemerides astronomicae*. The Latin motto between the Bull's horns reads "URANIA planeta novus detectus ab Herscheli Anno 1781 die 13 Martii" (*"The new planet Uranus discovered by Herschel on 13 March 1781"*). Psalterium Georgianum (Chap. 18) appears at *lower right*

scope" attached to it, providing the low-magnification context of a region of sky for use in guiding the pointing of the more powerful telescope. The placement of Hell's invention on his map borders on the whimsical, as it appears Taurus is trying to peer through the tiny eyepiece of the finder scope. In the creation and placement of this constellation, the idea of astronomers suggesting figures to fill in spaces on their charts seemed to have reached its absurd nadir.

There were few early adopters of Hell's creations. Prior to Johann Elert Bode's landmark *Uranographia* (1801b), the most prominent inclusion of either telescope was Jean Nicolas Fortin's depiction of the smaller telescope ("Petit Telescope de Herschel"), in his third edition of John Flamsteed's *Atlas Coelestis* (1795); it is unclear why Fortin kept this figure and not the other. Six years later, Bode dropped the smaller telescope but kept the larger, giving it for the first time a reliable depiction (Fig. 26.3) as Bode knew Herschel personally and purchased telescopes from him. In turn, Bode retroactively inserted it into the second edition of his *Vorstellung der Gestirne* (1805), seen in the bottom panel of Fig. 26.4. His taste for the smaller instrument would render further references to the "20-foot" telescope Hell associated with these stars obsolete. He described[3] the relatively new invention in *Allgemeine Beschreibung und Nachweisung der Gestirne*, the catalog supplement to *Uranographia*:

> Abbé Hell suggested this constellation a few years ago, resulting in its formation. The delineation which I give it in these charts, to the east of the constellations within the box, is to commemorate the 7-foot telescope with which the famous Dr. Herschel saw the planet Uranus for the first time on 13 March 1781, in the same part of the sky between the Bull's horns and the feet of the Twins. The main stars included in this new constellation belonged previously to Auriga.

Due to the widespread influence of *Uranographia*, it was assured that at least a few mapmakers would emulate Bode and include the Telescope, such as Young (1807), Jamieson (1822; Fig. 26.5), and Green (1824). The *Urania's Mirror* artist drew a very generic refracting telescope on a three-legged tripod in 1825 (Fig. 26.6), seemingly unaware of Bode's accurate depiction a generation earlier. Elijah Hinsdale Burritt showed the Telescope in *The geography of the heavens* (1833), and Ezra Otis Kendall described the constellation in *Uranography* (1845) as having been "proposed by P. Hell in honour of Herschel and his discoveries in the heavens." It was evidently still current through the mid-nineteenth century; Hanna Bouvier wrote of it in her *Familiar astronomy* (1858), although she was evidently unaware of its actual inventor: "This constellation was formed by Bode, in honor of the discoveries of Sir William Herschel." The attribution to "Bode's maps" was shared by George Chambers in his *Handbook of Descriptive Astronomy* (1877).

---

[3] *"Der Abt Hell gab vor einigen Jahren zur Einführung desselben die Veranlassung. So wie ich es nun in diesen neuen Himmelscharten ostwärts beym Furhmann abbilde, soll es die Gestalt uns Aufstellungsart des 7füssigen* [sic] *im Andenken erhalten, womit der berühmte D. Herschel im Jahr 1781 den 13 Marz in deffen Nachbarschaft, nehmlich zwischen den Hörnern des Siters und den Füssen der Zwillinge, den Uranus zuerst also Planet erkannte. Die vornehmsten Sterne deffelben wurden sonst zum Furhmann gerechnet."* (p. 7)

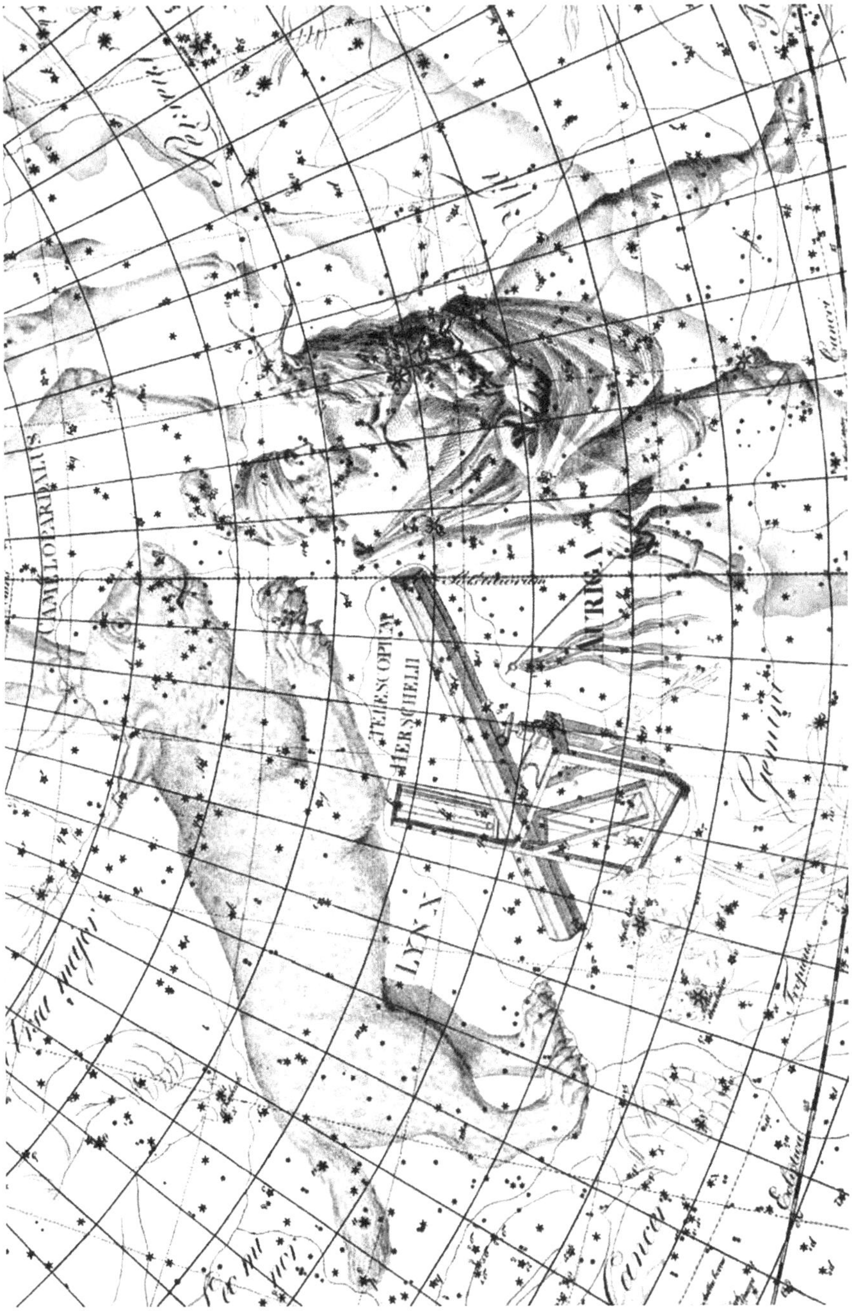

26.3 Telescopium Herschelii Major shown on Plate 5 of Johan Elert Bode's *Uranographia* (1801b) with its own constellation boundaries

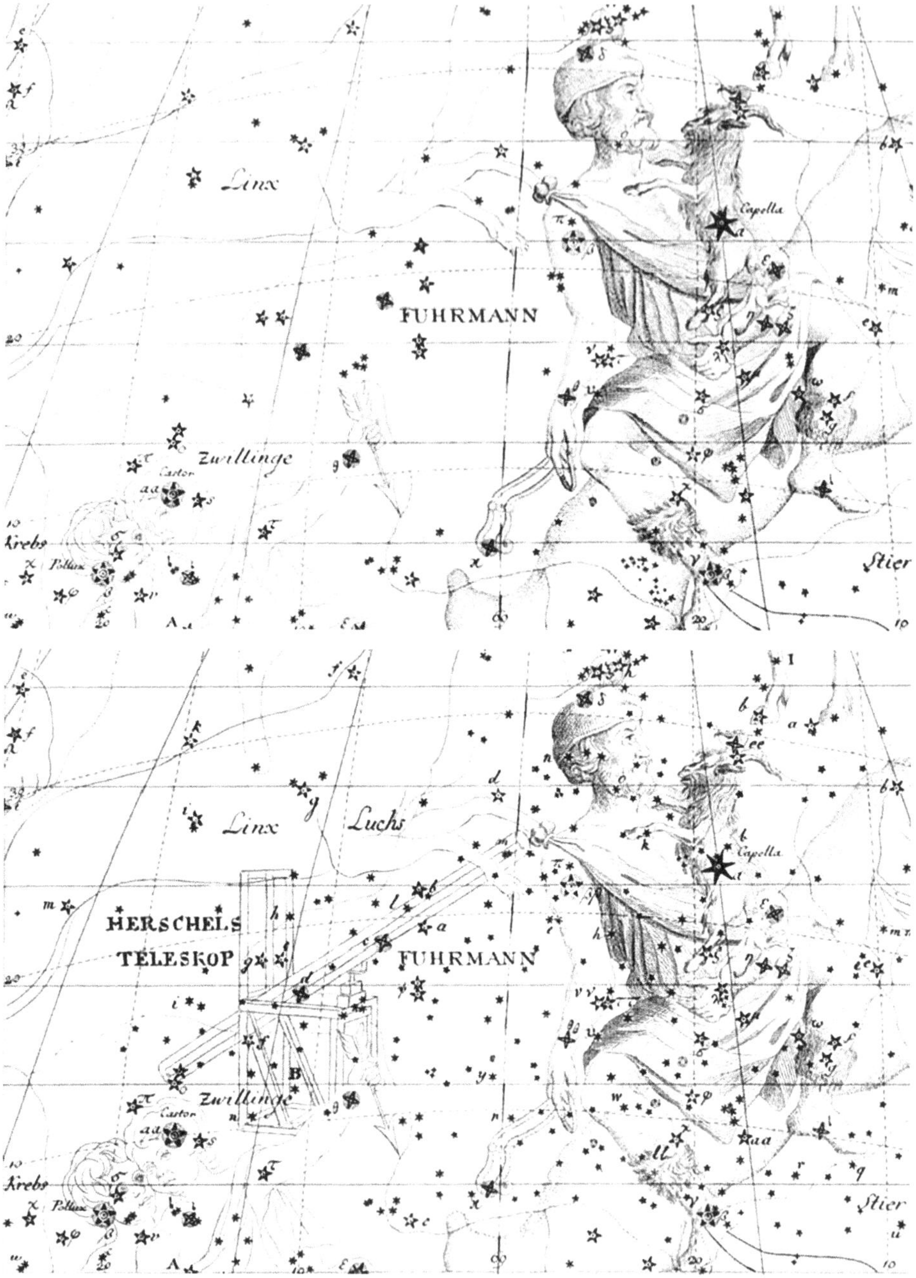

26.4 Before and after. Excerpts of Plate 4 from two editions of Johan Elert Bode's *Vorstellung der Gestirne*: 1782 (*top*) and 1805 (*bottom*). In the interim between the two editions, Bode decided to include Telescopium Herschelii Major ("Herschels Teleskop") in *Uranographia* (1801b), taking care to include the figure in the second edition of his earlier charts

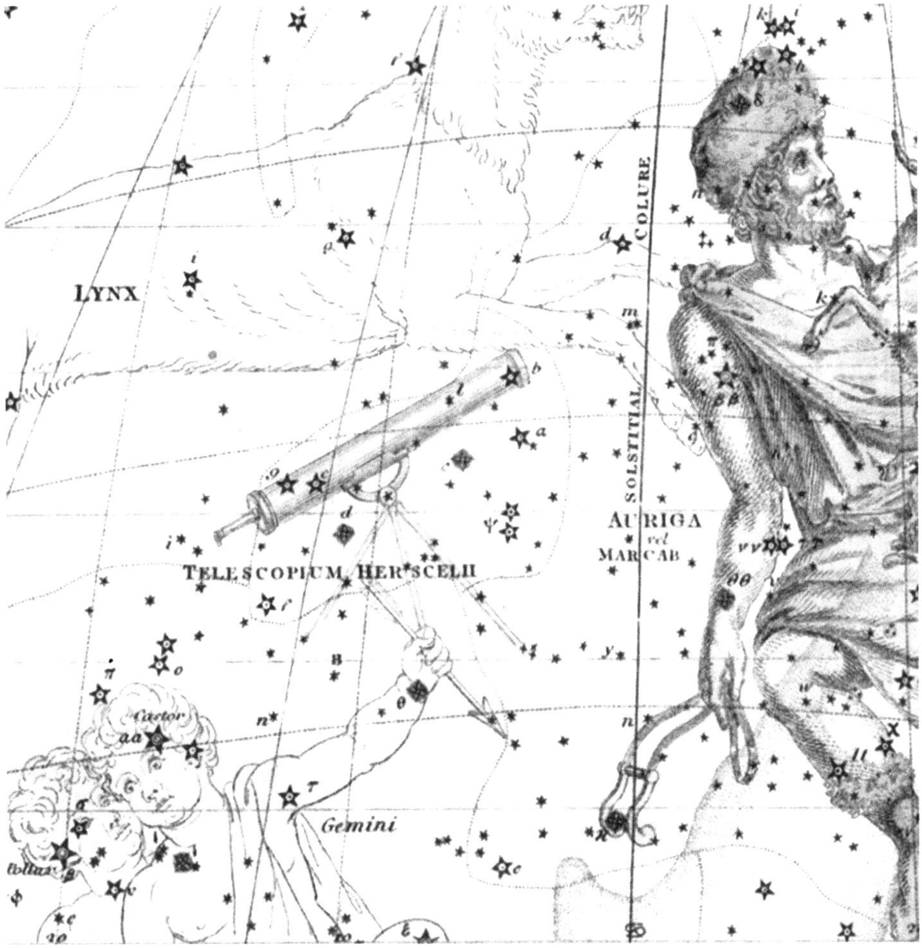

26.5 Telescopium Herschelii Major shown as "Telescopium Herscelii" in Jamieson's *Celestial Atlas* (1822). Note that the telescope shown is a refractor on a tripod, utterly unlike the instrument actually used by Herschel

The smaller Telescope might otherwise be considered a "single-sourced" constellation (see Volume 2) but for one significant appearance between the time of Hell and Bode's *Uranographia*. It is shown on Plate 14 in Jean Nicolas Fortin's third edition of John Flamsteed's *Atlas Coelestis* (1795; Fig. 26.7) as "Petit Telescope de Herschel," presumably referring to the 7-footer. In this rendition, with the telescope still a refractor with a finder scope, Fortin turned the instrument around so that it appears to be pointing directly into the Bull's right eyeball (the bright star Aldebaran, α Tauri).

26.6 Plate 8 from *Urania's Mirror* (1825) showing a fearsome-looking Lynx and "Telescopium Herschilii"

Fortin did not show the larger Telescope, which would otherwise have appeared on Map 4; curiously, in the forward to *Atlas Céleste de Flamstéed*, he explicitly mentioned[4] the plural "Telescopes of Herschel:"

> In this new edition we added the Solitaire [Turdus Solitarius], *according to the Citizen Lemonnier, Mém*[oires] *de l'Academie, 1776; Messier* [Custos Messium], *according to Lalande's globe, engraved in 1779; Taurus Poniatowski, according to Poczobut, Polish astronomer; the Telescopes of Herschel and George's Harp* [Psalterium Georgianum], *after Hell in the Almanac of Vienna; the Trophy of Frederick* [Honores Frederici], *according to Bode, who engraved it in 1787.*

Fortin clearly recognized that Hell introduced *two* telescopes, but for reasons that are not clear he showed only one of them on his charts. It is possible that the failure to include the larger telescope on Map 4 was a simple oversight. The map appears virtually unchanged between the 1776 and 1795 editions, leading one to conclude that the 1776 plate was

[4] *"Dans cette nouvelle édition nous avons ajouté le Solitaire, d'après le citoyen Lemonnier, Mém. de l'Academie, 1776; le Messier, d'après le Globe de Lalande, gravé en 1779; le Taureau de Poniatowski, d'après Poczobut, Astronome de Pologne; les Télescopes de Herschel & la Harpe de Georges, d'après Hell dans les Ephémérides de Vienne, 1790; le Tropheé de Frédéric, d'après Bode, qui l'a fait graver en 1787."* (page v.)

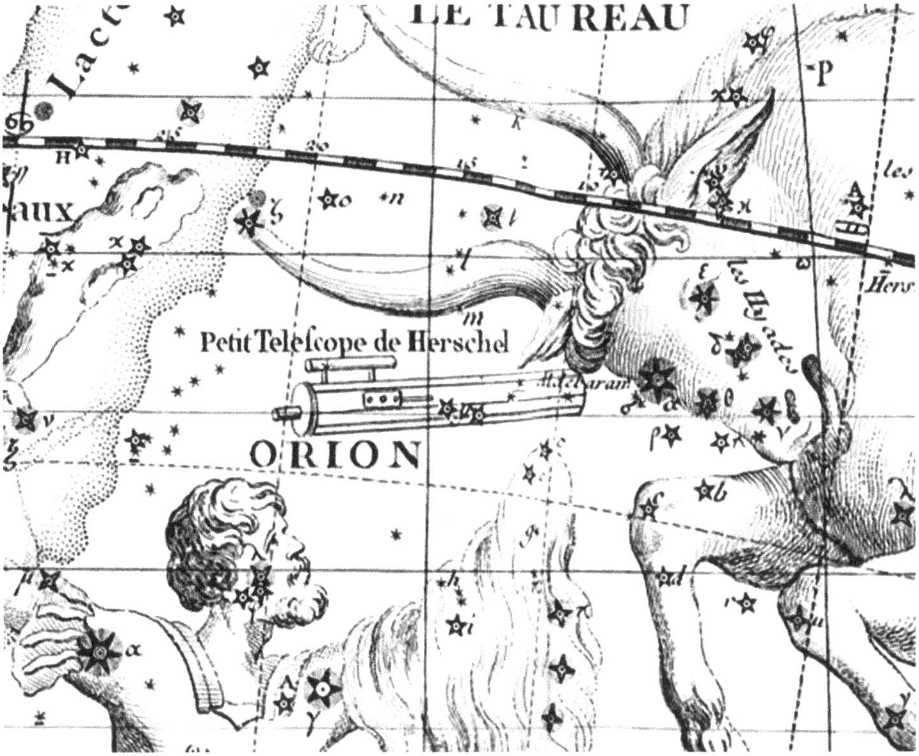

26.7 A representation of Telescopium Herschelii Minor ("Petit Telescope de Herschel") on Plate 14 of Fortin's third edition of Flamsteed's *Atlas Coelestis* (1795)

simply reused. Whatever the case, Herschel's smaller Telescope would never grace the pages of an atlas after Fortin. It was ignored by Bode in *Uranography* (1801b), who drew the boundary between Taurus and Orion in such as a way as to roughly evenly distribute the stars identified by Hell between the two and setting the stage for its complete extinguishment at the turn of the nineteenth century.

ICONOGRAPHY

William Herschel

The maker of both telescopes commemorated by Hell was born Friedrich Wilhelm Herschel (Fig. 26.8) on 15 November 1738 in the Electorate of Brunswick-Lüneburg, then part of the Holy Roman Empire and more commonly referred to as Hanover. He was one of ten children born to Isaak and Anna Ilse (Moritzen) Herschel, whose surname betrays

26.8 Portrait of William Herschel by Lemuel Francis Abbott (1760–1802). Oil on canvas, 76.2 × 63.5 cm, 1785. National Portrait Gallery, London: NPG 98

the family's Jewish origins; according to Herschel's biographer, Edward Singleton Holden (1846–1914), they were descended from Jewish Moravians who converted to Lutheranism in the seventeenth century.

William was born into a family of musicians, and he began life as a working performer and composer. Isaak Herschel was an oboist in the Hanover Military Band, and sons Wilhelm and Jakob played oboe in the same regiment of the Hanoverian Guards. In 1755, their regiment was sent to England during a period in which Hanover and Great Britain were ruled as a personal union of George II Augustus (1683–1760). A threatened war with France resulted in the brothers Herschel being sent back to defend Hanover; within months of the Guards' defeat at the Battle of Hastenbeck on 26 July 1757, Isaak sent both sons back to England. There he anglicized his name to "Frederick William" and at age nineteen began a residency that would take up the remainder of his life. Jakob was there formally discharged from the Guards while William was accused of desertion, an alleged crime for which he was finally pardoned by George III in 1782.

During 1760–1761 he led the Durham Militia band, and in the latter year he played violin for one season with Charles Avison's (1709–1770) Newcastle Orchestra. In addition to his accomplishment with the oboe, William played violin, harpsichord, and later, the organ. After his military service ended, he composed dozens of symphonies and concertos.

After a few years in and around Newcastle, he moved first to Leeds and then to Halifax, where he served as the first organist at the Church of Saint John the Baptist, now Halifax Minster. He continued to follow musical work south, landing in Bath in the mid-1760s. There he was appointed the organist at the Octagon Chapel in 1766 and was Director of Public Concerts for the city. His first concert at the chapel was given on 1 January 1767 and featured a then-incomplete organ; Herschel instead performed some of his own compositions on violin, harpsichord, and oboe. Brothers Jakob, Dietrich and Alexander variously performed stints as musicians at Bath with William in the 1770s. By 1780, William was named director of the Bath Orchestra.

Music led him to astronomy, by way of mathematics and optics. In his early twenties, he met the fifth English Astronomer Royal, Nevil Maskelyne (1732–1811), who encouraged his studies. He learned enough to begin experimenting with making telescope mirrors out of speculum metal, a highly reflective (but brittle) alloy of about two parts copper to one part tin, with traces of arsenic and other metals. These mirrors were relatively easy to figure and polish but tarnished quickly, requiring frequent repolishing to retain sufficiently high reflectivity to make them useful in astronomical telescopes. Herschel worked relentlessly to perfect his mirror-making technique, sometimes spending as many as 16 h a day in the workshop.

His first good look at the heavens through his telescopes came in mid-1773, and he recorded the first observations in what became decades of astronomical journals with details of Saturn's rings and the Great Nebula in Orion on 1 March 1774. Holden quotes "the writer in *European Magazine*" on this phase of Herschel's life[5]:

> All this time he continued his astronomical observations, and nothing now seemed wanting to complete his felicity, but sufficient leisure to enjoy his telescopes, to which he was so much attached, that at the theatre he used frequently to run from the harpsichord to look at the stars, during the time between the acts.

His sister, Caroline (1750–1848), emigrated to England in 1772 after the death of their father and settled with William in 19 New King Street, Bath. According to Holden, she was[6]

> a perfectly untried girl, of very small accomplishments and outwardly with but little to attract. The basis of her character was the possibility of an unchanging devotion to one object; for the best years of her life this object was the happiness and success of her brother William, whom she profoundly loved.

A great deal of William's success as an observer can be attributed to the attention, care and encouragement he received from Caroline and her "almost spaniel-like allegiance" to him. However, she became an important astronomer in her own right, and the success of one of the Herschel siblings cannot easily be separated from that of the other.

---

[5]Chapter II, page 38.

[6]Chapter II, page 33.

26.9 A reproduction of William Herschel's 7-foot telescope with which he discovered the planet Uranus on 13 March 1781, on display at the Herschel Museum in Bath, England. Compare with Bode's depiction of the same in Fig. 26.4 (*bottom*) and Fig. 26.3

His early astronomical interest was in double stars. From the back garden of the house on New King Street (now the site of the Herschel Museum), he set out in October 1779 to observe "every star in the Heavens" with a 6.2-in. (160 mm) aperture, 7-foot (2.1 m) focal length Newtonian telescope (Fig. 26.9). The telescope enabled him to discover a great number of binary and multiple stars, which he published in two catalogs presented to the Royal Society in 1782 and 1784; together with a third catalog published in 1821 of systems discovered after 1783, he discovered a total of 848 new multiple-star systems. In 1797, he re-observed many of the same systems he previously published, finding that the apparent relative positions of the constituent stars had changed in a way that could not be explained by parallax induced by the Earth's motion around the Sun. Five years later, in 1802s *Catalogue of 500 new Nebulae, nebulous Stars, planetary Nebulae, and Clusters of Stars; with Remarks on the Construction of the Heavens*, he advanced the idea that some of these changes must represent true physical motion of stars around each other under their mutual gravitational pull. This work became the basis for modern binary star studies, and his catalogs stood alone until the later efforts of James South (1785–1867), founder of the Royal Astronomical Society), Friedrich Georg Wilhelm von Struve (1793–1864), of the famous Struve astronomical dynasty, and John Herschel (1792–1871), William's own son.

The Discovery of Uranus

On 13 March 1781, Herschel was carrying out his usual sweeps for multiple stars when he stumbled upon an object that defied his initial recognition. "In the quartile near ζ Tauri," he wrote[7] in his journal, "[is] either [a] Nebulous star or perhaps a comet." Being the most likely explanation, Herschel quickly concluded[8] it was a new comet:

> On Tuesday the 13th of March, between ten and eleven in the evening, while I was examining the small stars in the neighbourhood of H Geminorum, I perceived one that appeared visibly larger than the rest: being struck with its uncommon magnitude, I compared it to H Geminorum and the small star in the quartile between Auriga and Gemini, and finding it so much larger than either of them, suspected it to be a comet."

Whatever the object was, its disc was decidedly nonstellar and subject to a notable increase in apparent size under progressively stronger magnification:

> From experience I knew that the diameters of the fixed stars are not proportionally magnified with higher power, as the planets are; therefore I now put on the powers of 460 and 932, and found the diameter of the comet increased in proportion to the power, as it ought to be, on a supposition of its not being a fixed star, while the diameters of the stars to which I compared it were not increased in the same ratio.

The disc "appeared hazy and ill-defined with these great powers," which was sufficient to firmly distinguish it from a point source.

On subsequent nights, when he returned to the same field, he found the object missing from its previous position and, instead, in a new one. Each subsequent night, it showed a regular motion suggestive of a comet; on 19 March, for example, Herschel noted its motion as "$2\frac{1}{4}$ seconds per hour." He wrote to his old friend, Nevil Maskelyne, about his discovery; on 23 April, Maskelyne sent a befuddled reply[9]: "I don't know what to call it. It is as likely to be a regular planet moving in an orbit nearly circular to the sun as a Comet moving in a very eccentric ellipsis. I have not yet seen any coma or tail to it."

Based on Herschel's careful observations, the Russian astronomer Anders Lexell (1740–1784) computed a low-eccentricity orbit for the object that was far more planetary

---

[7]Royal Astronomical Society MSS W.2/1.2, 23, quoted in Ellis D. Miner, *Uranus: The Planet, Rings and Satellites* (1998), p. 8.

[8]*Philosophical Transactions of the Royal Society of London*, 1 January 1781, Vol. 71, pp. 492–501.

[9]RAS MSS Herschel W1/13.M, 14, quoted in Miner (1998), p. 8.

than cometary, with an orbital radius placing it beyond Saturn. Maskelyne asked[10] Herschel to "do the astronomical world the faver [sic] to give a name to your planet, which is entirely your own, [and] which we are so much obliged to you for the discovery of." In hopes of catching the attention of the King, Herschel began referring[11] to the new planet as 'Georgium Sidus' (George's Star):

> In the fabulous ages of ancient times the appellations of Mercury, Venus, Mars, Jupiter and Saturn were given to the Planets, as being the names of their principal heroes and divinities. In the present more philosophical era it would hardly be allowable to have recourse to the same method and call it Juno, Pallas, Apollo or Minerva, for a name to our new heavenly body. The first consideration of any particular event, or remarkable incident, seems to be its chronology: if in any future age it should be asked, when this last-found Planet was discovered? It would be a very satisfactory answer to say, 'In the reign of King George the Third'.

The name failed to gain traction outside of Britain. The French, preferring to avoid any reference to the monarch who still claimed the French throne in pretense, called the object "Herschel" at the suggestion of Joseph Jérôme Lefrançois de Lalande, who wished to honor the discoverer. The Swedish astronomer Erik Prosperin (1739–1803) proposed "Neptune," which saw some popular support as a means of celebrating victories of the British Royal Navy in the American Revolutionary War. But the winner turned out to be Johann Elert Bode, who argued that because Saturn was the father of Jupiter, the next most distant planet yet should be named Uranus after the father of Saturn. The suggestion was amplified by Bode's colleague at the Royal Academy, the German chemist Martin Heinrich Klaproth (1743–1817), who in 1789 named the newly-discovered chemical element uranium after the same. The last holdout still referring to the planet as Georgium Sidus, the Royal Nautical Almanac Office, finally adopted Bode's suggestion in 1850.

The 20-Foot Telescope

While it was not the largest instrument he built, the 20-foot telescope was certainly the most useful. However, there was no single "20-foot telescope," but rather there were two. Since the descriptor refers to the focal length of the telescope and not its aperture, Herschel distinguished the two telescopes he made with this common focal length as the "large 20-foot" (18.75-in. aperture, Fig. 26.10) and the earlier "small 20-foot" (12-in. aperture). Much of the survey work resulting in his later published catalogs began with the same 7-foot telescope used to discover Uranus, but by that time Herschel had reached the limits of the aperture of that telescope and knew he needed more light-gathering capacity to see further.

[10]RAS MSS Herschel W.1/12.M, 20, quoted in Miner (1998), p. 12.

[11]Letter to Joseph Banks, quoted in J.L.E. Dreyer, *The Scientific Papers of Sir William Herschel* (1912).

26.10 "The 20-Foot Telescope, From a drawing made either at Datchet or at Clay Hall" in *The Scientific Papers of Sir William Herschel* (1912)

The small 20-foot was inadequate for the job, both due to its relatively small aperture and the awkward means of its use. The focal length of the telescope was sufficiently long that it could not be operated using a normal mount. Instead, as shown in the figure, the upper end of its tube was suspended from a mast by ropes, and the observer reached its focus only by standing high on a ladder. Once an observation was made, he had to descend the ladder and, once safely again on solid ground, record what he saw. Ultimately, Herschel decided that the mounting issue for these long-focus instruments was a serious problem, and he completely reconsidered his design for the large 20-foot. He had constructed a large wooden frame, about two stories in height, which sat upon a circular track. The telescope focus was accessed from a stable platform attached to the mount which could be lifted and lowered according to the attitude of the telescope. Helpers on the ground were needed to rotate the entire contraption in azimuth and to raise and lower the tube by means of a hand crank. Two 18.75-in. speculum blanks were cast in the fall of 1783, and one of them was sufficiently figured and polished to see first light on 23 October.

The large 20-foot was originally operated as a Newtonian instrument, but this led to very inconvenient positioning of the focus. After 1786, Herschel reconfigured the optics by tilting the primary mirror, leading to an off-axis arrangement wherein focus could be viewed with an eyepiece mounted in a fixed position at the mouth of the tube. Still, Herschel noted that about 15 min of effort were required to point the telescope at a new object, so his observing strategy for nebulae changed. Rather than laboriously changing the telescope pointing in azimuth, Herschel allowed the apparent motion of the night sky to change pointings of the telescope in right ascension. In this mode, the telescope was fixed in azimuth, usually along the meridian, and his helpers moved the tube up and down in altitude through a 2° angle. Over the course of a night, Herschel surveyed a strip of sky 2° wide in declination and several hours of right ascension in length.

The Portuguese astronomer João Jacinto de Magalhães (also known as Jean Hyacinthe de Magellan, 1723–1790) visited Herschel in 1785, afterward writing a firsthand account[12] of the large 20-foot in operation:

> I spent the night of the 6th of January at Herschel's, in Datchet, near Windsor, and had the good luck to hit on a fine evening. He has his twenty-foot Newtonian telescope in the open air and mounted in his garden very simply and conveniently. It is moved by an assistant, who stands below it ... Near the instrument is a clock regulated to sidereal time ... In the room near it sits Herschel's sister, and she has Flamsteed's Atlas open before her. As he gives her the word, she writes down the declination and right ascension and the other circumstances of the observation. In this way Herschel examines the whole sky without omitting the least part. He commonly observes with a magnifying power of one hundred and fifty, and is sure that after four or five years he will have passed in review every object above our horizon. He showed me the book in which his observations up to this time are written, and I am astonished at the great number of them. Each sweep covers 2° 15′ in declination, and he lets each star pass at least three times through the field of his telescope, so that it is impossible that anything can escape him. He has already found about 900 double stars and almost as many nebulæ.
>
> I went to bed about one o'clock, and up to that time, he had found that night four or five new nebulæ. The thermometer in the garden stood at 13° Fahrenheit; but, in spite of this, Herschel observes the whole night through, except that he stops every three or four hours and goes in the room for a few moments. For some years Herschel has observed the heavens every hour when the weather is clear, and this always in the open air, because he says that the telescope only performs well when it is at the same temperature as the air. He protects himself against the weather by putting on more clothing. He has an excellent constitution, and thinks about nothing else in the world but the celestial bodies.

After the Herschels moved to Slough, Berkshire, in April 1786, the large 20-foot was set up in their garden near the house and observations continued. He continued discovering new objects with it and completed a set of star counts in different regions of the sky intended to locate the limits of the Milky Way as a stellar system. He made two key assumptions in his analysis: (1) all stars have the same intrinsic brightness, so the differences in the apparent brightness indicate only distance; and (2) his telescope was

[12]Quoted in Johann Elert Bode's *Astronomisches Jahrbuch* for 1788, p. 161.

capable of showing stars all the way to the edges of the system. From his observations, he devised a model of the Galaxy (Fig. 1.2) published in 1785 with a flattened geometry in which the Sun was located very near to the center.

Herschel made his last sweeps with the large 20-foot in 1802, and its use became infrequent after that time. By 1817, his son John was preparing to take up his father's work, and had two new mirrors cast for it that June; one was figured and polished by John under William's direction, while the other was later completed entirely on his own. The deteriorating tube was restored, salvaging everything possible from the original, and by 1820 the instrument was back on the sky. John took this rebuilt 20-foot telescope with him to South Africa during his expedition of 1834–1838; with it he expanded the number of nebulae and clusters he observed to over 4000 and the number of double stars to nearly 5500.

Herschel's Later Life and Death

In 1783, William gave Caroline a telescope and she began to make her own observations her own discoveries, including five comets entirely on her own and three which were co-discoveries with other observers. At William's suggestion, she took up the work of correcting some of John Flamsteed's stellar positions, the result of which was eventually published as an updated edition of Flamsteed's *British Catalog of Stars*. Anticipating better conditions outside Bath, William and Caroline moved first to Clay Hall at Windsor in 1785 and then to a new home on Windsor Road in Slough the following year. The year after that, George III granted her an annual salary of £50 for her working assisting William, making her the first woman known to have been paid for her scientific labor.

In May 1788, William married a wealthy widow, Mary Baldwin Pitt, at St. Laurence's Church, Upton, in Slough. The marriage introduced a new presence and obligation into William's life that strained his relationship with his sister. Mary displaced Caroline in the managerial functions of the household, demoting her to a lower standing that necessitated her relocation out of the main house; she was also deprived of the keys to the observatory and attached workroom, cutting her off from her own work space. Whatever her feelings on the developments of this period, they were lost later when she destroyed her own personal journals from the decade 1788–1798. William and Mary's son, John, was born in 1792, and Caroline became immediately attached to him; she would go on to serve him as an assistant after William's death and even eventually reconciled with Mary.

William continued to make observations for nearly four decades, showing a dogged determination few of his era matched. "Every serene dark night was to him a precious opportunity," wrote[13] Miss Agnes Clerke in 1895, "availed of to the last minute."

> The thermometer might descend below zero, ink might freeze, mirrors might crack; but, provided the stars shone, he and his sister worked from dusk to dawn.... 'Many

[13] *The Herschels and Modern Astronomy*, pp. 34–35.

> a night,' he states, 'in the course of eleven or twelve hours of observation, I have carefully and singly examined not less than 400 celestial objects, besides taking measures, and sometimes viewing a particular star for half an hour together, with all the various powers.' ... On one occasion he is said to have worked without intermission at the telescope and the desk for seventy-two hours, and then slept unbrokenly for twenty-six hours. His instruments were never allowed to remain disabled. They were kept, like himself, on the alert. Relays of specula were provided, and one was in no case removed from the tube for re-polishing, unless another was ready to take its place. Even the meetings of the Royal Society were attended only when moonlight effaced the delicate objects of his particular search.

The results of untold thousands of hours spent sweeping the night sky were a catalog of star clusters, nebulae and galaxies—collectively called "deep-sky objects"—without comparison in the history of astronomy to that point, even though no one had any inkling into their true physical nature. In the period of 1782–1802, and most intensively during the 7 years spanning 1783–1790, the Herschels used both the 7-foot telescope and the two 20-foot telescopes to observe and carefully catalog over 2400 diffuse, non-stellar objects collectively termed "nebulae" by William. While the instruments were capable of revealing the true shapes of objects (especially galaxies) and resolving into stars some of the rich open and globular clusters he saw, it did not yield any particular insight into the space beyond the Solar System.

William took a very practical approach toward surveying the contents of a literally brand-new field still taken by scientists today: he formed a taxonomy of the things he saw. The objects were grouped morphologically into eight "classes:" (I) bright nebulae; (II) faint nebulae; (III) very faint nebulae; (IV) planetary nebulae; (V) very large nebulae; (VI) very compressed and rich clusters of stars; (VII) compressed clusters of "small and large" (i.e., faint and bright) stars; and (VIII) coarsely scattered clusters of stars. Caroline herself contributed the discovery of eleven objects. In addition to the *Catalogue of 500 New Nebulae ...* (1802), he published the discoveries in the *Catalogue of One Thousand New Nebulae and Clusters of Stars* (1786) and *Catalogue of a Second Thousand New Nebulae and Clusters of Stars* (1789). Later, his son John added 1754 discoveries of his own, largely made from an observing site in South Africa, which he published as the *General Catalogue of Nebulae and Clusters* in 1864. The combined published output of the Herschels was later edited by the Danish-Irish astronomer John Louis Emil Dreyer (1852–1926), who added various other nineteenth century discoveries to the lists and published the entire product as the *New General Catalog* (1888), containing a total of 7840 objects. The "NGC" identifier lives on to this day as perhaps the most common catalog designation of brighter deep-sky objects.

Herschel's last observations were made with the large 20-foot. Miss Clerke (1895) relates:

> On June 1st, 1821, he inserted into the tube with thin and trembling hands the mirror of the twenty-foot telescope, and took his final look at the heavens. All his old instincts were still alive, only the bodily power to carry out their behests was gone.

An unparalleled career of achievement left him unsatisfied with what he had done. Old age brought him no Sabbath rest, but only an enforced and wearisome cessation from activity.

William died at Observatory House on Windsor Road, Slough, on 25 August 1822. He is buried at nearby St. Laurence's Church, Upton; John Herschel later erected a monument to his father containing this eloquent epitaph[14]:

H. S. E. (Hic situs est)
Gulielmus Herschel Eques Guelphicus
Hanoviæ natus Angliam elegit patriam
Astronomis ætatis suæ præstantissimis
Merito annumeratus
Ut leviora sileantur inventa
Planetam ille extra Saturni orbitam
Primus detexit
Novis artis adjumentis innixus
Quæ ipse excogitavit et perfecit
Coelorum perrupit claustra
Et remotiora penetrans et explorans spatia
Incognitos astrorum ignes
Astronomorum oculis et intellectui subjecit
Qua sedulitate qua solertia
Corporum et phantasmatum
Extra systematis nostri fines lucentium
Naturam indagaverit
Quidquid paulo audacius conjecit
Ingenita temperans verecundia
Ultro testantur hodie æquales
Vera esse quæ docuit pleraque
Siquidem certiora futuris ingeniis subsidia
Debitura est astronomia
Agnoscent forte posteri
Vitam utilem innocuam amabilem
Non minus felici laborum exitu quam virtutibus
Ornatam et vere eximiam
Morte suis et bonis omnibus deflenda
Nec tamen immatura clausit
Die XXV Augusti A. D. CIƆIƆCCCXXII
Ætatis vero suæ LXXXIV

[14]Quoted in Holden, Chap. III, page 117.

Rendered in English:

Here lies
William Herschel, Guelphic knight
Born of Hanover, England his chosen country
The outstanding astronomer of his age
Rightly numbered
And lighter than silence found
Outside the orbit of the planet Saturn
First discovered
A new art upon which to lean for support
Which he devised and perfected
He broke through the barriers of the heavens
And further penetrating and exploring spaces
The unknown fires of the stars
And under the eyes and understanding of astronomers
Which assiduous skill requires
Of bodies and of spectres
Casting light beyond our system of boundaries
Tracing out nature
Whatever slight, more boldly cast
The vast tempered by awe
Besides serving as witness that today are equals
True are most of the things they taught
Since future talents more certainly aid
Are owed to astronomy
They will, by luck, follow
The useful life of the innocent and lovable
Is no less happy than the result of virtuous effort
Elegant and truly exceptional
Laments his death and all the good
And yet prematurely ended
On the 25th day of August, 1822
In the 84th year of his age

He left behind more than four hundred telescopes built during his lifetime, of which two were proposed as his memorials in the firmament.

DISAPPEARANCE

As is typical for many minor constellations in the nineteenth century, there is no definitive expiration date for Herschel's Telescopes. Telescopium Herschelii Minor disappeared within a span of less than 10 years and was firmly discarded when Bode's *Uranographia* displaced Fortin's 1795 edition of Flamsteed's atlas as the reference of choice among most astronomers. Bode was the tastemaker for a century to come, and he evidently found nothing memorable about a small telescope awkwardly wedged between Orion and Taurus. Despite the exhaustive nature of Richard Hinckley Allen's research, Telescopium Herschelii Minor didn't even merit a mention in *Star Names* (1899), and was entirely forgotten at the time the modern constellation boundaries were set by Eugéne Delporte in 1930 (Fig. 26.12).

Telescopium Herschellii Major persisted for over a century in one form or another, retained by some authors and discarded by others. Among those who did not show it on charts or describe it in their books were Argelander (1843), Proctor (1876), Rosser (1879), and Pritchard (1885). It was named as "Telescopium Herschelii" in the table "Old And New Constellations In Chronological Order" in *Poole Bros. Celestial Handbook* (1892) and attributed to "Hell, 1789, AD" One of its last appearances on charts of the day with apparent full constellation status was in Ernest Lebon's *Histoire Abrégée de l'astronomie* (1899) as "Telescope Herschel." In the same year, Allen referred to its appearance in "Burritt's Atlas," but noted that the Telescope "since [Burritt's] day has passed away from the maps and catalogues." William Tyler Olcott retained it as something akin to an asterism in his *Star Lore of All Ages* (1911), labeling its stars with the name "Herschel's Telescope" while not connecting the stars with lines as in the case of other figures. The Telescope did not receive an International Astronomical Union three-letter abbreviation at the first General Assembly in Rome, essentially sealing its fate. Upon publication of *Délimitation scientifique des constellations* by Eugéne Delporte (1930b), its stars were divided by new boundaries defining Lynx, Gemini, and Auriga (Fig. 26.11).

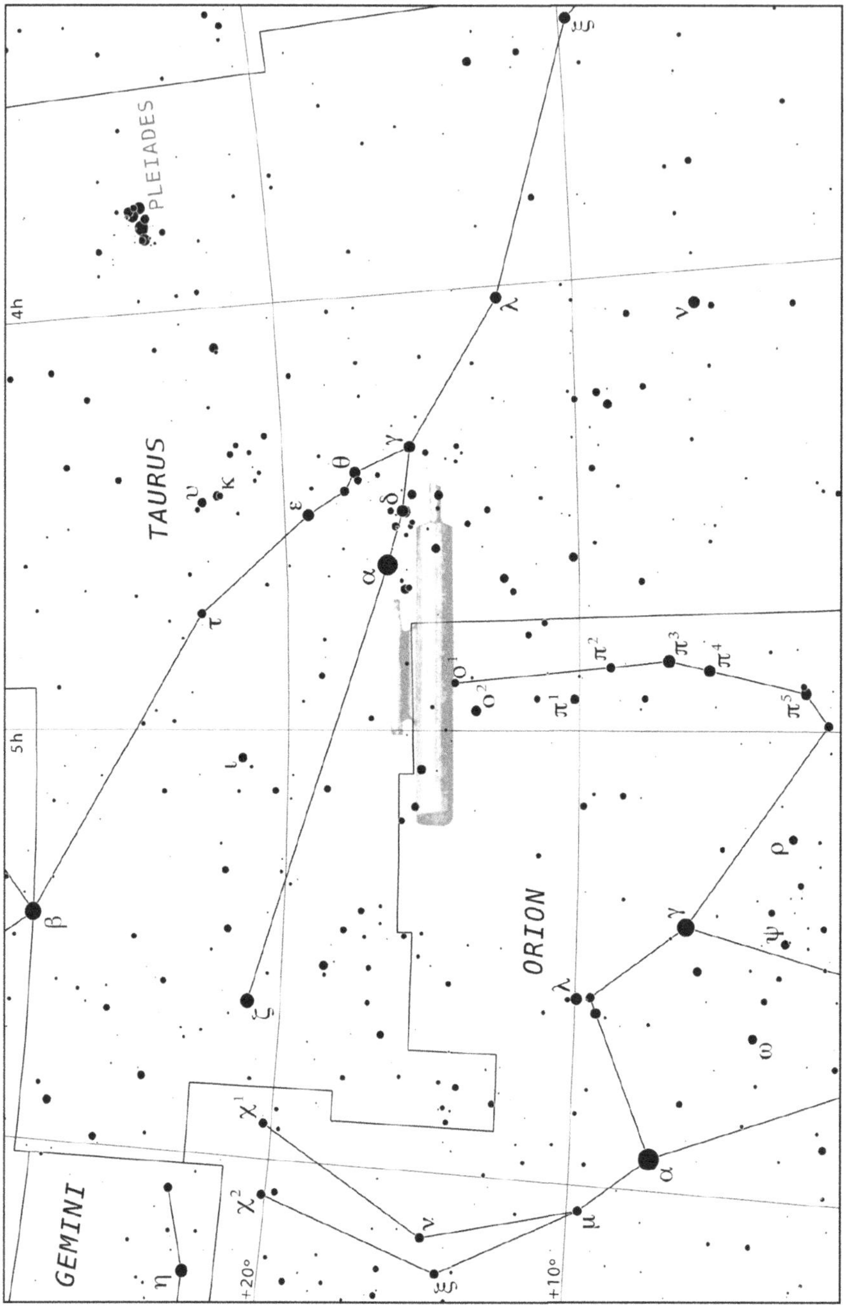

26.11 The figure of Telescopium Herschelii Major from Plate 5 of Johann Elert Bode's *Uranographia* (1801b) overlaid on a modern chart

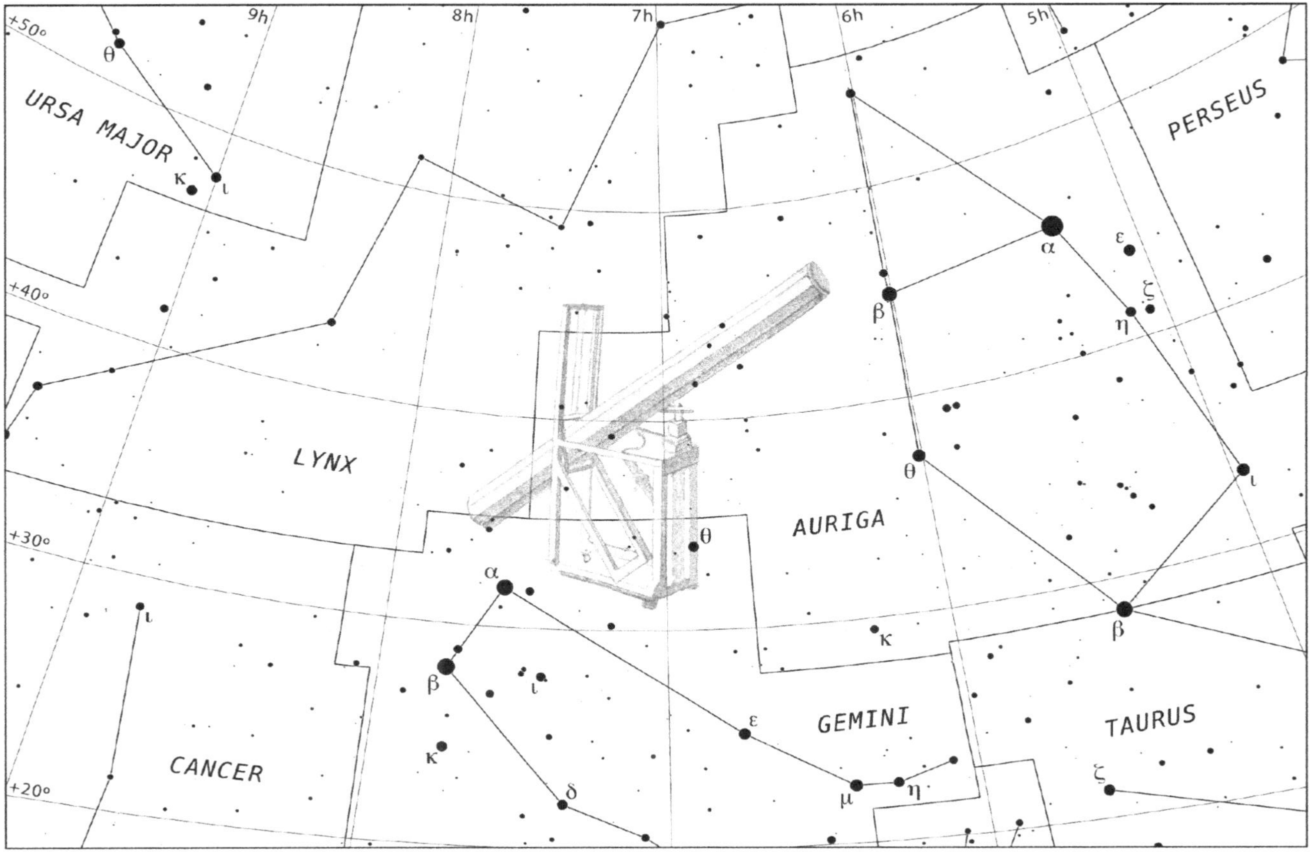

26.12 The figure of Telescopium Herschelii Minor from the 1790 edition of *Ephemerides astronomicae* overlaid on a modern chart

27

Tigris

The River Tigris

Genitive: Tigris
Abbreviation: Tig
Alternate names: "Tigris et Euphrates Fluvii" (Cellarius, 1661); "Tygris Fluvius" (Royer, 1679a)
Location: From near the eastern shoulder of Ophiuchus to near Pegasus, flowing between Cygnus and Aquila, including parts of modern Vulpecula[1]

ORIGIN AND HISTORY

The River Tigris as a constellation (Fig. 27.1) was introduced sometime in the first quarter of the seventeenth century. Ptolemy left the stars comprising this constellation unformed, although mostly acknowledged, and this tradition was kept as late as the time of Bayer (1603). The main body of the river in part paralleled the path of the "Great Rift" in the Milky Way, a layer of obscuring interstellar dust in the mid plane of the Galaxy stretching roughly from Cygnus to Centaurus. In this sense the identification of a constellation here is opposite the usual sense in which stars outline a figure; rather, Tigris followed a path suggested by the relative *absence* of stars. It is sometimes shown on maps as two-branched, reflecting the proper geographic sense of the rivers Tigris and Euphrates as they flow through lands referred to as the "Fertile Crescent."

[1] *"From β and γ, in the right shoulder of Ophiuchus, onwards between Aquila and the left hand of Hercules; thence between Albireo (β Cygni) and Sagitta to Equuleus and the front parts of Pegasus, ending at the latter's neck"* (Allen, 1899); *"Began at Pegasus and flowed between Cygnus and Aquila (an area now occupied by Vulpecula), ending by the right shoulder of Ophiuchus"* (Ridpath, 1989).

J.C. Barentine, *The Lost Constellations*, Springer Praxis Books,
DOI 10.1007/978-3-319-22795-5_27

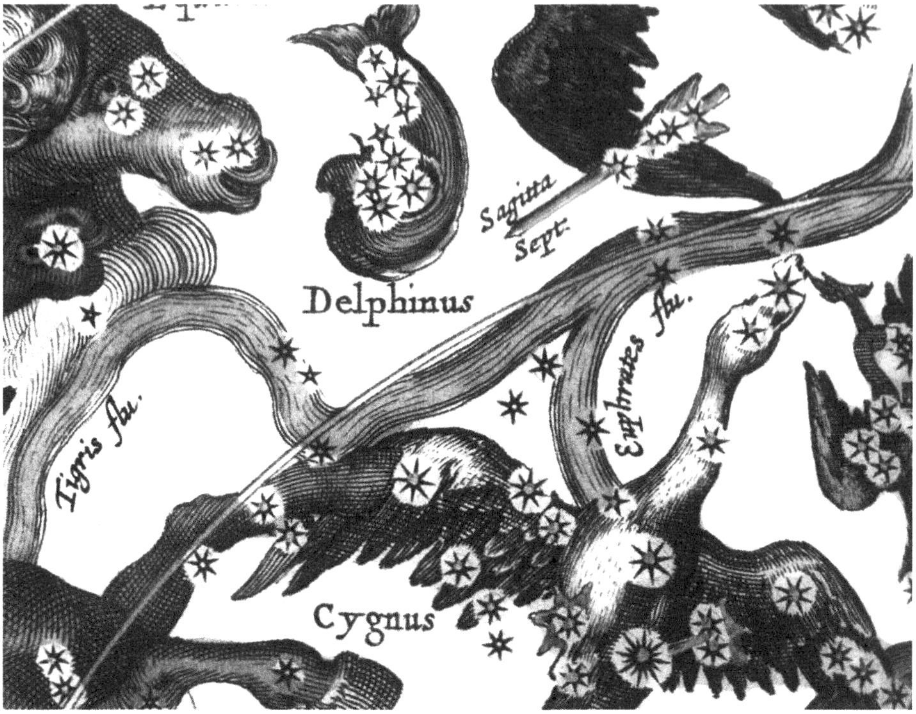

27.1 Detail of the confluence of Tigris and Euphrates shown on Plate 24 of Andreas Cellarius' *Harmonia Macrocosmica* Cellarius (1661)

There is some confusion among previous authors as to exactly who proposed it and when it made its first appearance on charts. Ridpath (1989) wrote that it was "introduced in 1612 by Petrus Plancius on the same globe as the river Jordan made its first appearance."[2] Jordanis (Chap. 13) also has very similarly confused origins, as Richard Hinckley Allen (1899) identified Jacob Bartsch as the person responsible for its introduction. Allen also claims Bartsch as the originator of Tigris, in *Usus Astronomicus* (1624). As in the case of Jordanis, Bartsch did indeed mention[3] Tigris:

[2]Both rivers also appear on a 1649 reprint of the *Globus coelestis in quo stellae fixae* ... gores, which seems to have been the stylistic model for Andreas Cellarius' *Harmonia Macrocosmica* (1661).

[3]"TIGRIS fluvius Asiae insignis, Mesopotamiam ab ortu alluens, quem ex Paradiso oriri facrae literae testantur. Gen. 2. Noviter ex variis Pegasi, Equulei, Cygni, & Ophiuchi informibus formatus & adjectus, iisdem alluit." (p. 55)

> The famous river Tigris of Asia and Mesopotamia, flowing from the east, which the sacred scriptures testify arises from Paradise (Genesis 2).[4] Newly shaped and added from various unformed [stars of] Pegasus, Equuleus, Cygnus, and Ophiuchus, flowing near the same.

The river also appeared as "Tigris Fluvius" on Isaac Habrecht's contemporaneous *Planiglobium Coeleste, et Terrestre* (1628; Fig. 27.2, top), probably directly influenced by Bartsch.

The first instance in which the river was split into two branches corresponding to the Tigris and Euphrates appears to be in *Harmonia Macrocosmica*, the grand 1661 work of Andreas Cellarius. In his Plate 27 (Fig. 27.1) the confluence of the rivers is indicated just above Sagitta in the space between Cygnus and Delphinus, in a region now enclosed by the modern boundaries of Vulpecula. In *Tabula Universalis Longitudinum et Latitudinum Stellarum* (1679b), the set of tables accompanying his maps in *Cartes du Ciel Reduites en Quatre Tables*, Augustin Royer counted[5] six stars "at the bend, the Swan's wing and Dolphin's head," marking the confluence of the rivers.[6] Royer claimed a total of 46 stars in Tigris including four taken from Ophiuchus.[7]

As the sixteenth century gave way to the seventeenth, some cartographers continued to include the figure on their maps. Tigris appears on Plate 17 of Vincenzo Coronelli's *Epitome Cosmographica* (1693) as "Fiume Tigre", and as "Tigris Fluvius" on Carel Allard's 1706 map *Hemisphaerium meridionale et septentrionale planisphaerii coelestis* (Fig. 27.3). Corbinianus Thomas plotted the course of the river on a chart in *Mercurii philosophici firmamentum firmianum* (1730; Fig. 27.4), writing[8] that it "issues forth from Paradise, which the Scriptures call Pison,[9] today flows through Mesopotamia, whose name recent astronomers [have placed], in [otherwise] empty parts of the sky, curving between the stars of Pegasus and Ophiuchus."

---

[4]Genesis 2:10–14: "A river watering the garden flowed from Eden; from there it was separated into four headwaters. The name of the first is the Pishon; it winds through the entire land of Havilah, where there is gold. (The gold of that land is good; aromatic resin and onyx are also there.) The name of the second river is the Gihon; it winds through the entire land of Cush. The name of the third river is the Tigris; it runs along the east side of Ashur. And the fourth river is the Euphrates." (NIV).

[5]"Ex 6. pr. inter anconem alae cygni & caput delphini" (p. 42).

[6]Royer may have meant Brocchi's Cluster (Collinder 399), a small asterism commonly known as the "Coathanger" from its distinctive shape. At least six of its stars are brighter than seventh magnitude, and may be visible to observers with sufficient visual acuity under superb conditions.

[7]"Ex 4. pr. in origine vers serpentarii" (p. 38).

[8]"*Tigris. Fluvius* ex Paradiso derivatus, quem sacra pagina *Pison* vocat, hodie Mesopotamiam interluit, cujus nomen *Recentiores* Astronomi, in vacuas coeli sedes, sidera nempe *Pegasum* inter & *Ophiuchum* sinuata, transmisére." (p. 184)

[9]Thomas appears to confuse the Tigris with the Pishon, mentioned in Genesis but whose course has never been clearly identified. Alternately, by "sacra pagina" he may have referred to "The Book of the All-Virtuous Wisdom of Joshua ben Sira," also known as the "Wisdom of Sirach," a work on ethics written in the first quarter of the second century BC by the Jewish scribe Shimon ben Yeshua ben Eliezer ben Sira. In Sirach 24:25 the Law of Moses "fills men with wisdom, like the Pishon, and like the Tigris at the time of the first fruits." (Revised Standard Version)

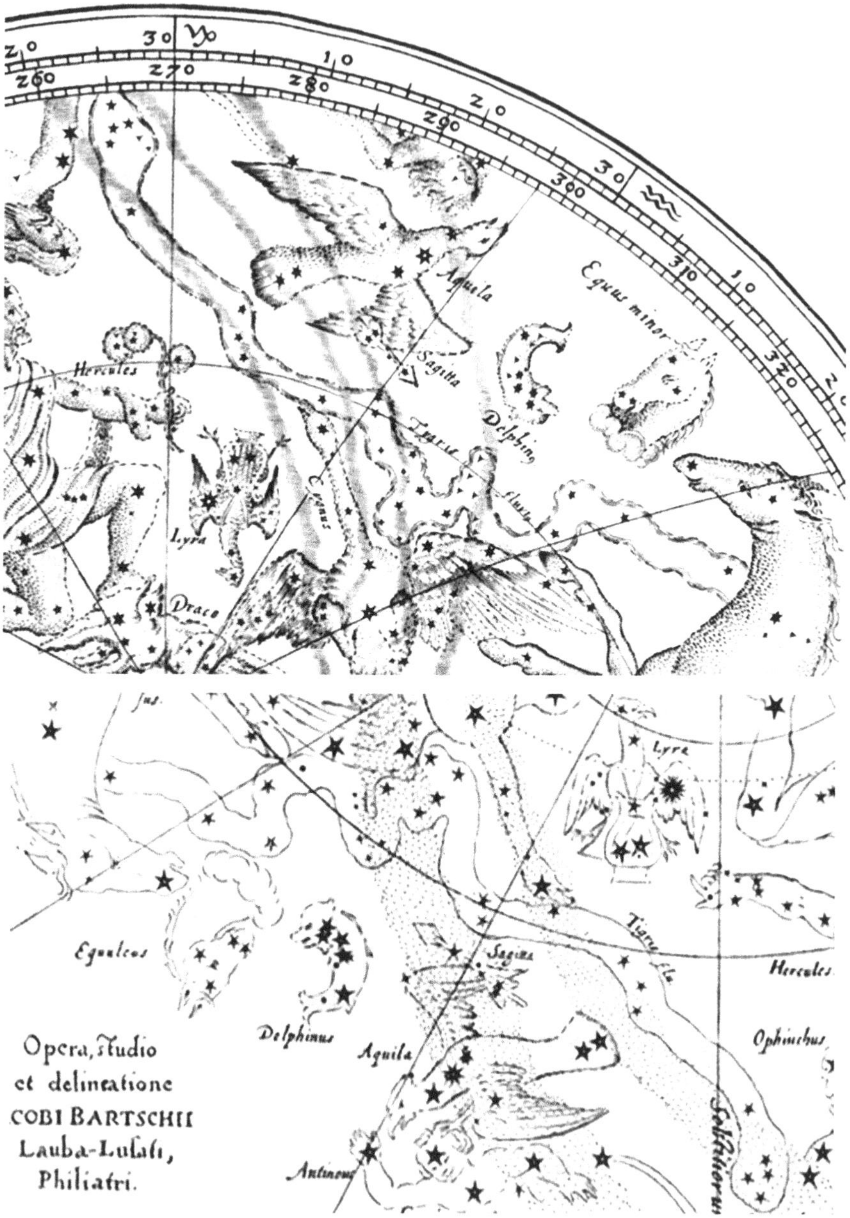

27.2 Two depictions of Tigris in early seventeenth century charts: Isaac Habrecht's *Planiglobium Coeleste, et Terrestre* (1628; *top*) and Jacob Bartsch's *Planisphaerium stellatum* (1661; *bottom*). The charts are reversed relative to one another because Habrecht plotted his charts from the perspective of an observer inside the celestial sphere while Bartsch placed the observer outside the sphere

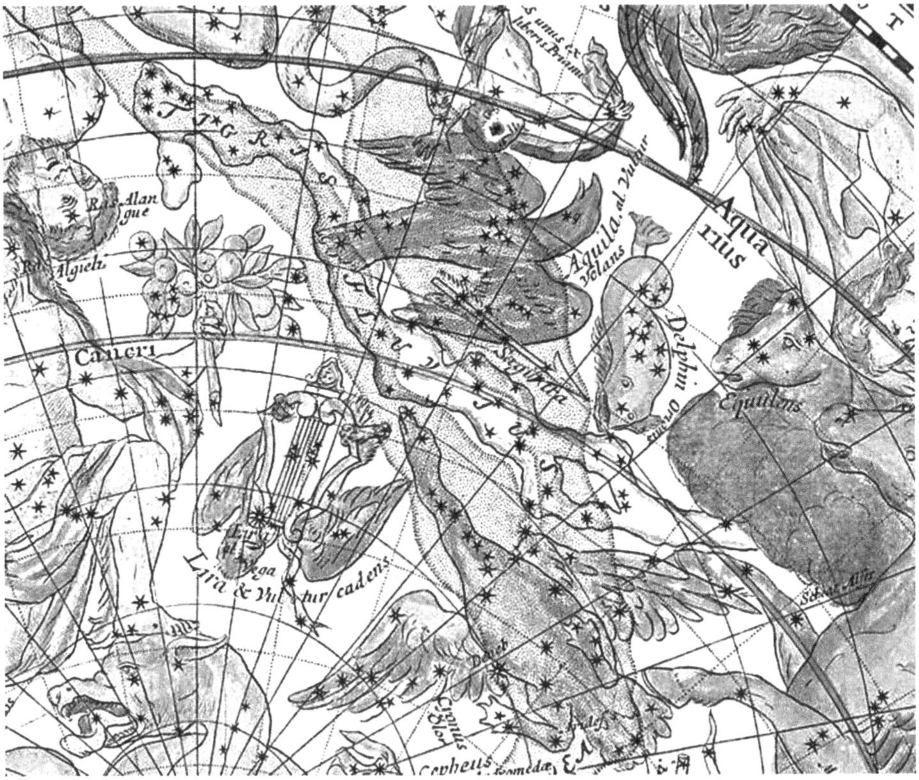

27.3 The path of Tigris shown across Carel Allard's *Hemisphaerium meridionale et septentrionale planisphaerii coelestis* (1706)

ICONOGRAPHY

The Tigris-Euphrates River System

From sources in the Taurus Mountains of eastern Turkey, the Tigris, Euphrates and their tributaries (Fig. 27.5) form one of the principal river systems of Western Asia. The waters emerge from the Anatolian Plateau and descend through mountain valleys to the uplands of the northern Levant then flow toward the southeast across the plains of central Iraq, discharging into the Persian Gulf below their confluence at Al-Qurnah.

Much of the rivers' course is through lands classified as subtropical to arid desert, and they are important sources of water for irrigating farmland. The silt washed down the rivers has made the surrounding lands very fertile, and the rivers themselves teem with life. Several small tributaries flow into the primary rivers from shallow freshwater lakes and springs emerging from the surrounding desert, forming important marshlands that contribute to the ecology of the region. Marshes and adjacent lands once covering

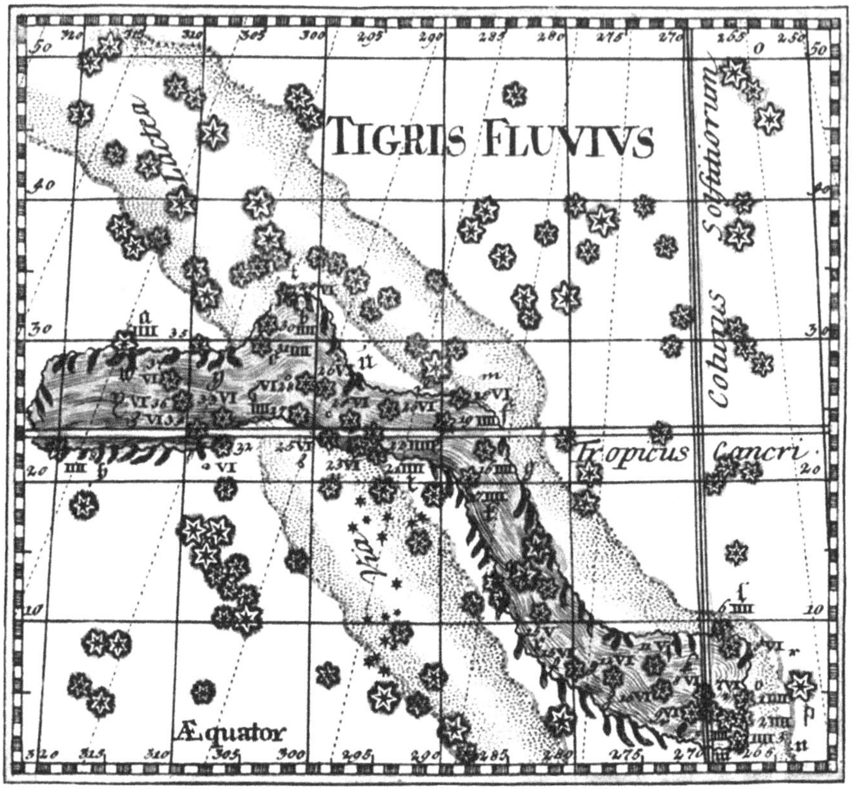

27.4 Tigris received its own treatment as a constellation in Corbinianus Thomas' *Mercurii philosophici firmamentum firmianum ...* (1730). Thomas also depicts the path of the Milky Way ("Via Lactea") illustrating how part of the river followed the path of the Great Rift, shown as a bifurcation in the Galaxy running diagonally from *center to lower right*

up to 20,000 km^2 host migrating and resident water birds, small mammals, fish, snakes, frogs, and lizards. Over 90 % of marshlands in the river system have been destroyed since the 1970s due to various drainage projects carried out by the Saddam Hussein regime to achieve political control of Shi'a Muslims living in the area. Millennia of agriculture have resulted in desertification and increase soil salt levels, threatening local wildlife populations and access to potable water supplies.

Water rights on the rivers remain a volatile political subject among the nations in the system's watershed. Turkey began implementation of a public works water project in the 1960s involving the construction of 22 dams along the upper reaches of the Tigris and Euphrates for both irrigation and hydropower production. Downstream, Iraq saw water levels drop and considered the Turkish policy a threat, magnified in the 1990s when Syria and Turkey participated in the economic embargo following the Gulf War. Iraq experienced

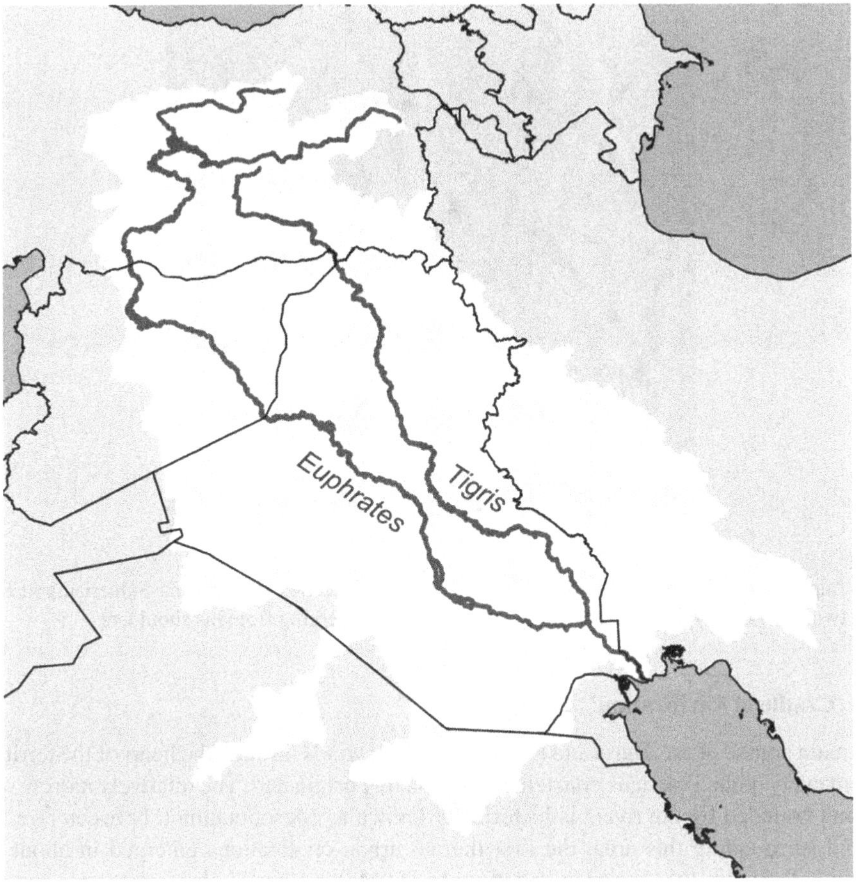

27.5 Map based on United States Geological Survey data showing the Tigris-Euphrates river system. The combined drainage of both rivers is indicated by the *lightest gray shading*. Drawn by Karl Musser and licensed under CC-BY-SA-2.5

a drought in 2008, protesting Turkish restrictions on water flows released from its dams. As a consequence of the drought, which affected all countries in the region, Iraq signed a memorandum of understanding in 2009 to improve flow monitoring on the rivers and improve communications between their governments on water issues. In exchange for Turkey releasing more water from its dams, Iraq agreed to trade more of its oil and crack down on the activities of Kurdish separatists along the countries' shared border. As of this writing, the water infrastructure of the upper Tigris-Euphrates drainage is under threat by the militant group calling itself the Islamic State. The future of water supplies in the region is tied to climate change, and is certain to remain an important obstacle to political and social stability.

27.6 Impression of an Akkadian cylinder seal (*c.* 2350–2150 BC) showing the Sumerian god Enki with twin rivers, interpreted as the Tigris and Euphrates, emerging from his shoulders

The 'Cradle of Civilization'

The main course of the Tigris and Euphrates in Iraq winds through the heart of the territory where many of the Polemaic constellations probably originated. The relatively narrow strip of land bounded by the rivers is historically known as Mesopotamia ("between rivers"); in and surrounding this area, the first literate urban civilizations emerged in about the fourth millennium BC. A narrower definition of "Mesopotamia" places that area north of Baghdad, whereas to the south lies Babylonia, named for the ancient city of Babylon.

The Sumerians believed that the Tigris was created by the god Enki (Fig. 27.6), who initially filled the river with water and caused it to flow.[10] In Hurrian mythology the river was a divinity called Aranzahas, son of Kumarbi; spat out of Kumarbi's mouth onto Mount Kanzuras, he later plots to destroy his father with the sky-god Anu and the storm-god Adad.

Babylonian versions of Sumerian myths assign a different origin to the rivers. In the seventh generation of gods, the Igigi, children of Enlil and Ninlil, quit their job of continuing the working of creation. The god of fresh water, Abzu, threatened to destroy the world with a flood in response, causing panic among the gods. They solicited help from Enki, who put Abzu to sleep in waters confined within the earth; this enraged Tiamat, the primordial ocean goddess and Abzu's consort. Tiamat resolved to take back creation, again

[10] Carlos A. Benito, "Enki and Ninmah and Enki and the world order." Ph.D. dissertation, University of Pennsylvania (1969).

27.7 Byzantine mosaic showing a personification of the Tigris as one of the Four Rivers of Paradise. East Church, Qasr-el-Lebia, Libya, AD 539–540

panicking the gods who turned once more to Enki. But Enki refused to involve himself further, leaving the gods to petition their father, Enlil, who offered to end the crisis on the condition that the others make him King of the Gods. Enlil destroyed Tiamat by forcefully blowing his winds down her throat. From her body he plucked several ribs, forming the arch of the night sky, and her tail to form the Milky Way. The tears from both of Tiamat's eyes became the source of the waters of the Tigris and Euphrates (Fig. 27.7).

The Assyrians similarly regarded the Tigris in association with a river god, which may have influenced the authors of the Jewish Bible. The Tigris is named twice in the Bible,

27.8 Victorian personifications of the Tigris and Euphrates rivers from a set of badges representing the Four Rivers of Paradise. Mathilde Claßen-Schmid, *Musterbuch für Frauenarbeiten mit erklärendem Text—Erster Band.* Leipzig: Hoffmann & Ohnstein (1893)

first in the Book of Genesis[11] wherein it is described as the third of four rivers branching out from the river exiting Eden, and then when Daniel[12] received a vision of a great coming war (Fig. 27.8).

Tigris is a Greek word (Τίγρις) descending through the Old Persian *Tigrā* and Elamite *ti-ig-ra* from the Sumerian word **id igna* describing the fast flow of a river's waters; the Tigris historically flowed faster then the Euphrates, whose slower flow rate causes it to build up silt more quickly. The original sense of the Sumerian word was preserved in the Akkadian *Idiqlat*, from which are derived equivalents in Hebrew, Arabic and Syriac. While the Middle Persian *Arvand Rud* referred specifically to the "swift" Tigris, in modern Persian the term is identified with the confluence of the Tigris and Euphrates. The Kurds refer to the Tigris simply as *Ava Mezin*, the 'Great Water.' "According to Pliny," wrote[13] Sir William Smith, "the river in the upper part of its course, where it flowed gently, was called Diglito; but lower down, where it moved with more rapidity, it bore the name of Tigris, which, in the Median language, signified an arrow."[14]

---

[11]2:10-14; see Footnote 4.

[12]10:4–6, "On the twenty-fourth day of the first month, as I was standing on the bank of the great river, the Tigris, I looked up and there before me was a man dressed in linen, with a belt of fine gold from Uphaz around his waist. His body was like topaz, his face like lightning, his eyes like flaming torches, his arms and legs like the gleam of burnished bronze, and his voice like the sound of a multitude."

[13]*A dictionary of Greek and Roman geography, Volume 2*, London: John Murray (1872), p. 1208.

[14]"The Tigris flows through this lake after issuing from the mountainous country near the Niphates; and because of its swiftness it keeps its current unmixed with the lake; whence the name Tigris, since the Median word for 'arrow' is 'tigris.'" Strabo, *Geography* Book 11, trans. H.C. Hamilton and W. Falconer.

The Sumerians began to harness the power of the Tigris and Euphrates as shipping channels early in their history, founding their cities in places that afforded them easy river access. They dug canals to irrigate their crops and bring water to cities further away on the plains, but poor management of the water use coupled with the soil erosion brought by seasonal flooding seems to have rendered the water of the lower Tigris too salty for agricultural use by the end of the third millennium BC. Nevertheless, the Babylonians and later the Assyrians maintained a network of masonry dams on the river, which were "cut through by Alexander, in order to improve the navigation."[15] The Tigris remains vital today as a source of water and transportation along much of its course (Fig. 27.9).

DISAPPEARANCE

Tigris the constellation persisted on maps into the first few decades of the eighteenth century, but it is difficult to pinpoint exactly when it fell out of favor. It appeared in neither volume of Stanislaus Lubieniecki's *Theatrum Cometicum* (1667; 1681), even though Lubieniecki showed other inventions tracing back to the time of Plancius, such as Gallus (Chap. 10). Allen wrote that recognition of Tigris "continued until as late as 1679 with Royer, but has long since disappeared from the maps, and indeed from the memory of most observers." However, Allen must have been unaware of the maps of Carel Allard, who depicted it on *Hemisphaerium meridionale et septentrionale planisphaerii coelestis* (1706; Fig. 27.4) and Corbinianus Thomas in *Mercurii philosophici firmamentum firmianum* (1730; Fig. 27.4).

One of the last eighteenth century mentions of Tigris in print is in *Urania: or, A Compleat View of the Heavens* (1754), the cyclopedic astronomical dictionary of John Hill. Hill wrote generically of the "River," mentioning both the classical Eridanus and the later creation:

> FLUVIUS, the River. A name by which one of the constellations is called by some of the old writers; it is that which we characterise by the name of Eridanus. There was no impropriety in this at the time when they wrote, nor will there be any confusion about it when we know that they always mean the Eridanus by it; but it would be an occasion of perplexity to denominate the constellation simply the river now, because Royer has since exalted the Tigris into the sides; so that it is necessary to say now which river we mean, and to particularise this by the name Eridanus.

To say "which river we mean" suggests that Tigris was still in some practical use in Hill's time. But the preponderance of evidence, in the sense of charts in which it does *not* appear (e.g., Lubieniecki 1667; Pardies 1674; Sherburne 1675; Lubieniecki 1681; Hevelius 1690; de La Hire 1702; de Broen 1709; Flamsteed 1729; Doppelmayr 1742), strongly suggests that most cartographers simply never considered it legitimate in the first place.

---

[15] Smith, p. 1209.

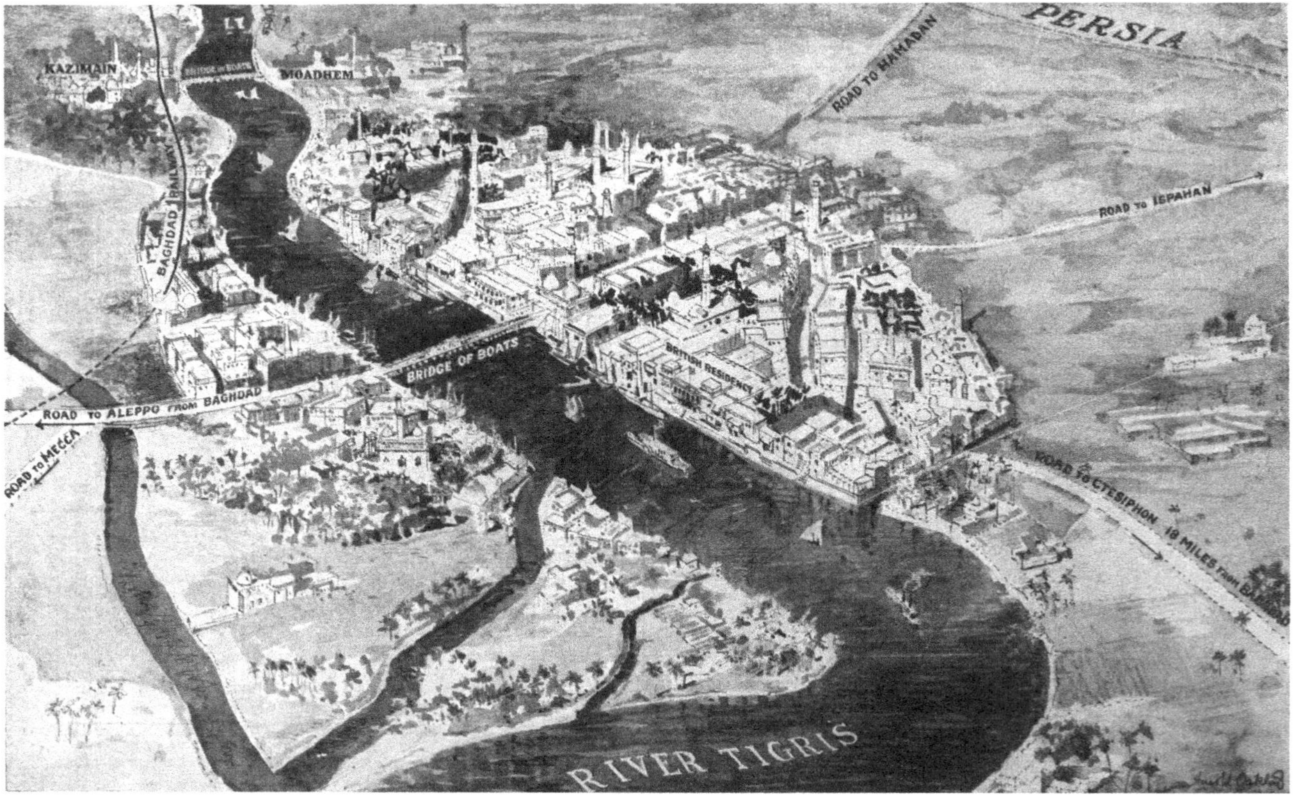

27.9 “Baghdad—a bird’s-eye view of the historic city on the Tigris” from *The Illustrated London News* for Saturday, 17 March 1917, Vol. CL, Issue 4065, page 306

That its stars were still considered unformed, and thus fair game for other creations, is evident in that when Marcin Poczobut introduced his own Taurus Poniatovii (Chap. 25) in 1777, he took for it the stars that previously marked the headwaters of Tigris. It is clear that by the time of Johann Elert Bode's influential charts *Vorstellung der Gestirne* (1782; 1805) and *Uranographia* (1801b), Tigris was no longer fashionable even as an asterism. That it was not adopted by Bode in *Uranographia* demonstrates it was functionally obsolete by 1800. At the time the modern constellation boundaries were established in the early twentieth century, Tigris was split across four[16] constellations (Ophiuchus, Hercules, Vulpecula and Pegasus; Fig. 27.10).

[16]If the spur representing the Euphrates on some maps is included, the result would be five (adding Cygnus).

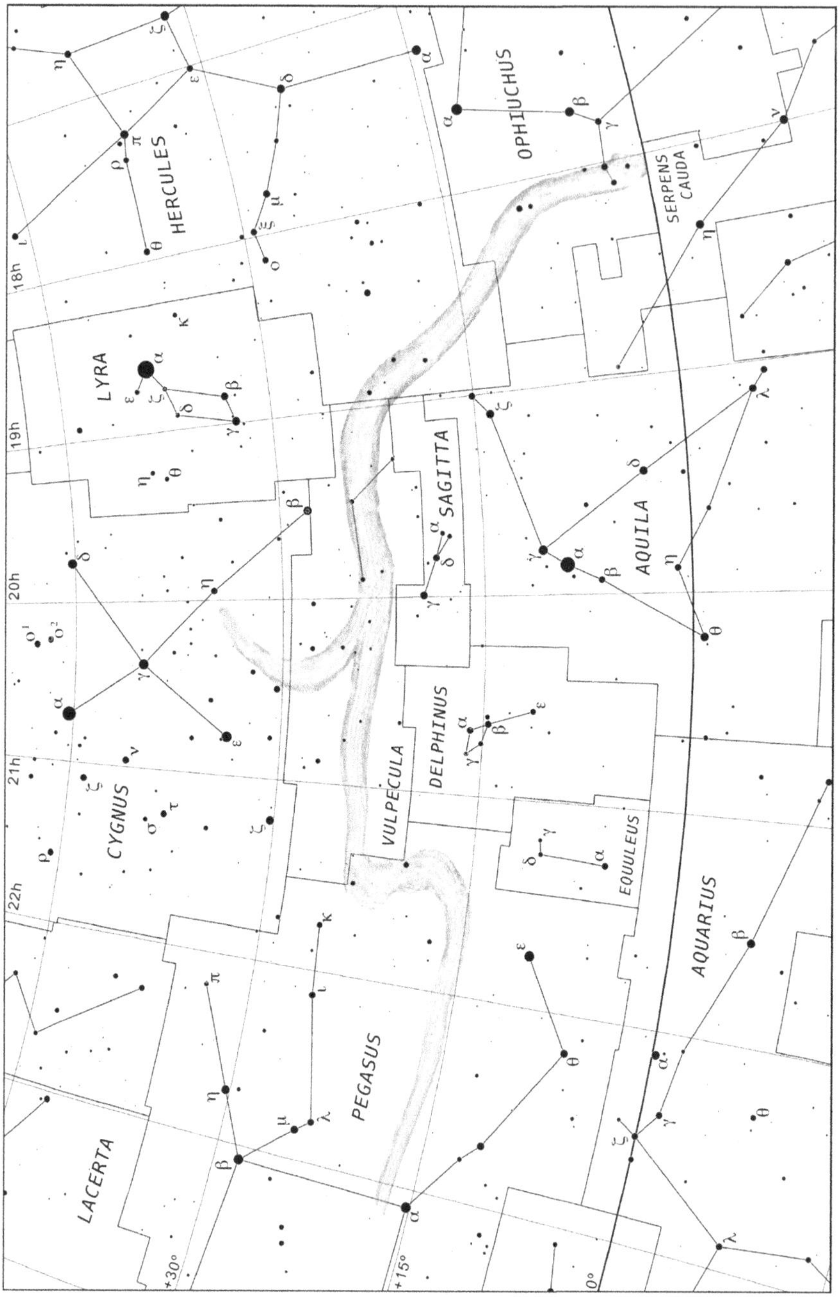

27.10 The figure of Tigris from Plate 24 of Andreas Cellarius' *Harmonia Macrocosmica* (1661) overlaid on a modern chart. The tributary north of Sagitta represents the river Euphrates

28

Triangulum Minus

The Lesser Triangle

Genitive: Trianguli Minoris
Abbreviation: TrM
Alternate names: "Triangulum Minor" (Chambers, 1877); "Triangulum Minora" (Allen, 1899)
Location: Bounded by the stars 6, 10, and 12 Trianguli[1]

ORIGIN AND HISTORY

Described rightly by Ridpath (1989) as "one of the least imaginative constellations," Triangulum Minus was introduced by Johannes Hevelius and published posthumously in *Prodromus Astronomiae* (1690). To form the new figure, he appropriated three faint stars later cataloged by Flamsteed as 6, 10, and 12 Trianguli; the remaining brighter stars traditionally associated with the existing constellation were labeled "Triangulum Majus" to distinguish the two (Fig. 28.1).

Earlier mapmakers who plotted such faint stars generally associated these three with the ancient constellation; two of the three, for example, were shown by Johan Bayer on Plate W of *Uranometria* (1603; Fig. 28.2). Bayer only placed five stars, α through ϵ, properly within the figure of Triangulum. Flamsteed added six others (η, ι, and four receiving Roman letters), but numbered a total of 16. As Wagman (2003) points out, Flamsteed's

[1] *"Three … stars between it* [Triangulum] *and the head of Aries"* (Smyth, 1844b); *"Lies between the head of Aries and the southern foot of Andromeda"* (Kendall 1845, referring to both this and the ancient Triangulum); *"A little to the south-east of Andromeda"* (Bouvier, 1858); *"Three small stars immediately to the south of the major constellation"* (Allen, 1899); *"Just south of the existing celestial triangle, Triangulum"* (Ridpath, 1989).

J.C. Barentine, *The Lost Constellations*, Springer Praxis Books,
DOI 10.1007/978-3-319-22795-5_28

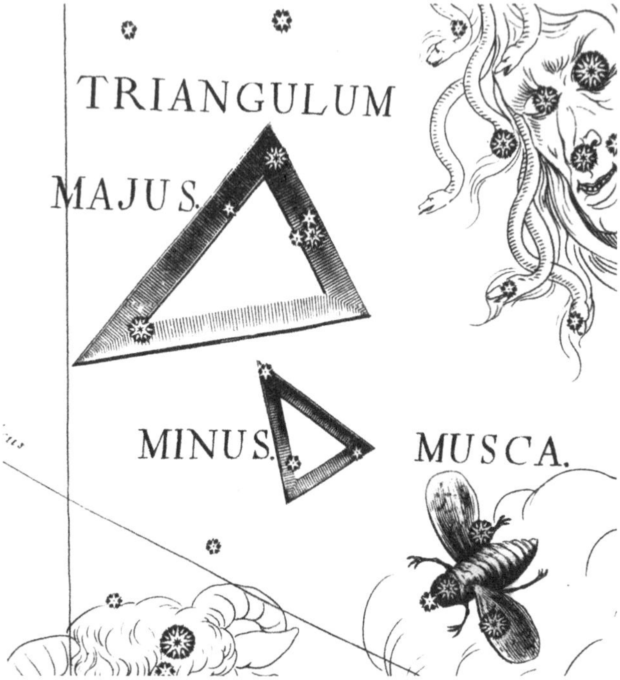

28.1 Detail of Triangulum Minus depicted on Figure Aa of Hevelius' *Prodromus Astronomiae* (1690) with Musca Borealis (Chap. 16). The face of Caput Medusae (see Volume 2) looks on at *upper right*

numbers 1 and 16 are no longer in use, because 1 Trianguli's coordinates were evidently in error relative to what Flamsteed published in *Catalogus Britannicus* (1725) and 16 Trianguli was later absorbed into nearby Aries.

Prior to Hevelius' time, most cartographers simply left the stars of Triangulum Minus off their charts altogether; seventeenth century examples of this include Plate 27 of Andreas Cellarius' *Harmonia Macrocosmica* (1661) and on various maps in Stanislaw Lubieniecki's two-volume *Theatrum Cometicum* (1667 and 1681). Those who showed its stars inevitably incorporated them into existing constellations. Julius Schiller added them to the figure of St. Peter, Constellation XXI in his *Coelum Stellatum Christianum* (1627; Fig. 28.3). Schiller remade the classical Triangulum into the papal tiara, being gently delivered to the apostle in a not-so-subtle endorsement of Catholicism at a time that Europe was being torn apart by religious warfare between Catholic and Protestant monarchs and their armies. The stars of Triangulum Minus are shared between Peter's face and the angel's ample drapery.

The adoption of the new figure at Hevelius' suggestion was initially slow. It does not appear in a number of contemporaneous charts, including Coronelli (1693), the 1693 second edition of Ignace-Gaston Pardies' *Globi coelestis* (1674), de La Hire (1702), Allard (1706) and de Broen (1709). It became popular beginning with Johann Leonhardt Rost's

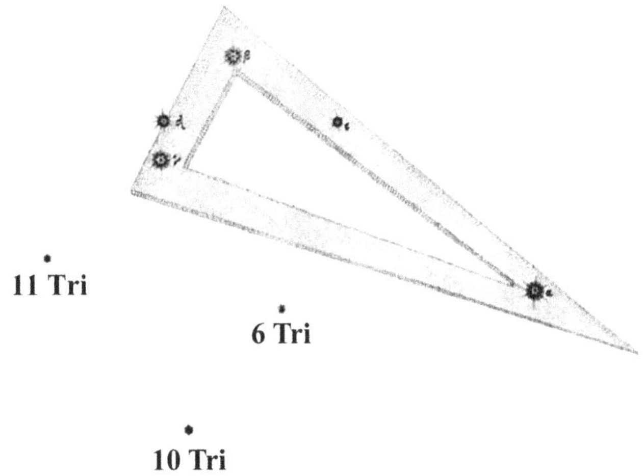

28.2 The classical constellation Triangulum shown on Plate W of Johannes Bayer's *Uranometria* (1603). *Labels* are overplotted showing the Flamsteed designations of the stars outside the figure of Triangulum proper that Bayer charted. Of these, two stars (6 and 10 Trianguli) were later used by Hevelius to form Triangulum Minus; the *third star marking* a vertex of Hevelius' figure, 12 Trianguli, was not shown by Bayer. Instead, he plotted 11 Trianguli, a star that was not used by Hevelius

Atlas Portatilis Coelestis (1723). Johann Gabriel Doppelmayr showed both Triangles on Plate XVII ("Hemisphaerium Coeli Boreale") of *Atlas Coelestis* (1742). Christoph Semler showed the smaller Triangle in *Coelum Stellatum in Quo Asterismi* (1731), as did John Bevis in Table 11 of *Uranographia Britannica* (*c.* 1750) and Christian F. Goldbach in *Neuester Himmels* (1799). John Flamsteed (1729) elaborated on the design of both Triangles in their depiction on his Plate XIII, placing many more stars within the bodies of each figure than appeared in historical atlases, an approach kept by most subsequent cartographers. Jean Nicolas Fortin kept Flamsteed's figures as "les Triangles" on Map 13 of both 1776 and 1795 editions of Flamsteed's *Atlas Coelestis*.

It was its inclusion in Johann Elert Bode's landmark atlases that ensured its survival into the nineteenth century. Bode followed Flamsteed's lead and showed both Triangles (as "Die Triangel") on Plate 13 of *Vorstellung der Gestirne* (1782) and Plate 11 of *Uranographia* (1801b; Fig. 28.4). In the accompanying explanatory volume *Allgemeine Beschreibung und Nachweisung der Gestirne* Bode wrote[2]

> The former figure [Triangulum] *was previously introduced in the ancient world, and the poets say, Ceres asked Jupiter to put the triangular shape of the fertile island of*

[2]Page 6: *"Jenes ist bereits seiner Figur wegen im Alterthum eingeführt, und die Dichter fagen, Ceres habe den Jupiter gebeten, die dreyeckige Gestalt der fruchtbaren Insel Sicilien an den Himmel zu versetzen. Letzteres ist erst von Hevel hinzugefügt worden. Beyde stehen zwischen Almak am Fuss der Andromeda, und dem Kopf des Widders."*

28.3 Constellation XXI (St. Peter's Tiara, aka Triangulum) in Julius Schiller's *Coelum Stellatum Christianum* (1627). An angel bears the first Pope's crown to the apostle (Constellation XXII). The stars of Triangulum Minus are split between Peter's face and forehead, and the robes of the angel. The ends of the crossed keys Peter holds in his left hand were alternately imagined as the constellation Musca Borealis (Chap. 16)

> *Sicily in the heavens. The latter* [Triangulum Minus] *was added by Hevelius. Both are between Almak at the foot of Andromeda and the head of Aries.*

As a result, both Triangles regularly appeared in atlases in the first half of the nineteenth century, including Gottleib (1805), Young (1807), and as "Triangula" in Jamieson (1822).

Jacob Green recounted what he thought to be 'certain' history of the Triangles in *Astronomical Recreations* (1824):

> According to the Poets, Jupiter assigned the island of Sicily a place in the heavens, under the figure of a Triangle. Other say, that the large Triangle owes its origin to

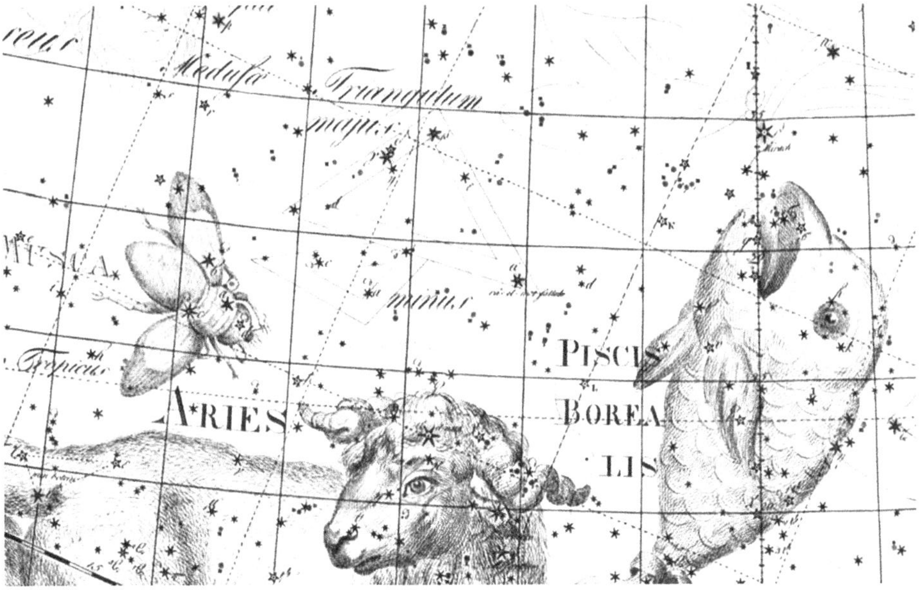

28.4 Triangulum Minus and Musca Borealis (Chap. 16) as depicted on Plate 11 of Bode's *Uranographia* (1801b) along with Triangulum ("Majus"), Pisces ("Piscis Borealis"), and Aries. Bode took clear liberties in his depiction of both Triangles relative to most other sources

> some one who wishes to represent the figure of the Delta in Egypt, among the stars. Although uncertainty hangs over the history of the large Triangle, we know that Hevelius introduced the small one, which is contiguous to it. The stars which form the small figure can be distinguished under favorable circumstances.

James Ryan similarly described its origin among "the unformed stars between the Triangulum Borealis and the Head of Aires" in *The New American Grammar of the Elements of Astronomy* (1827). Elijah Hinsdale Burritt extended Bode's influence in America by illustrating both Triangles on Plate 2 of *Atlas Designed to Illustrate the Geography of the Heavens* (1835). Later, some atlases stopped noting the Triangles separately, labeling both with names like "Triangulum" (e.g., Johnston 1855) and "Triangel" (e.g., Riedig 1849).

ICONOGRAPHY

Triangulum is among the smallest constellations in the Ptolemaic canon, and is often shown on historic charts as the type of device that might be used by a carpenter or draughtsman. The earliest instance of this figure have decidedly agrarian origin. In the Mesopotamian MUL.APIN star catalog, the stars of Triangulum were called "the plow"; Schaefer (2006) argued that it was the communication of knowledge of geometry from Egypt to Greece in the sixth century BC that led to the introduction of the constellation, for "only with this transformation would anyone seek to commemorate the triangle, as the basis of geometry, in the sky." He put the origin of the Greek constellation at somewhere between the sixth and fourth centuries BC.

Another version of its origin story references the island of Sicily. Triangulum was alternately known to the Greeks as Deltoton (Δελτωτόν) because of its resemblance to the upper-case letter delta. Eratosthenes associated it with the shape of the Nile River delta in his native Egypt. The name was at first transliterated by the Romans; Manilius described it in the first century AD as having "two equal sides parted by one unequal, a sign seen flashing with three stars and named Deltoton, called after its likeness." Around the same time, an unknown Latin poet imitating the style of Ovid referenced a figure "which the Constellation in the heavens, and the fourth letter among the Greeks have."[3] Later Roman authors Latinized it as "Deltotum." Hyginus agreed that the Triangle represented the island of Sicily, which was previously known as *Trinacria* on account of its shape.

Admiral Smyth, writing in *A Cycle of Celestial Objects* (1844b), thought that the figure had changed shape over historical time: "Several very old illustrations delineate Deltoton as an equilateral triangle, with a star at each angle – 'in unoquoque angulo unum[4];' but it has latterly been drawn as a scalene figure." Richard Hinckley Allen echoed this conclusion: "[The ancients] drew it as equilateral, but now it is a scalene figure, β, δ, γ at the base and α at the vertex." He further traced its history:

> Τϱίγωνον, used by Hipparchos and Ptolemy, became Trigonum *with Vitruvius, and* Trigonus *with Manilius, translated* Trigon *by Creech.* Tricuspis, *Three-pointed, and* Triquetrum, *the Trinal Aspect of astrology, are found for it; while Bayer had* Triplicitas *and* Orbis terrarum tripertitus *as representing the three parts of the earth, Europe, Asia, and Africa; and* Triangulus Septentrionalis, *to distinguish it from his own Southern Triangle. Pious people of his day said that it showed the Trinity, its shape resembling the Greek initial letter of Διος; while others of the same sort likened it to the Mitre of Saint Peter.*

Fundamentally we are left with the question of *why* were the three bright stars in the classical Triangulum called out as a constellation in their own right in the first place? The answer, Robert Brown wrote in *Researches into the Origin of the Primitive Constellations of the Greeks, Phoenicians and Babylonians* (1899), may simply be 'because it is there.'

> This little constellation supplies a very good illustration of the principles which obtained in the formation of the Signs. The school of O. Müller and the modern 'untutored anthropologist' would deal with its origin in the same futile manner with which Müller treats the constellation of the Arrow. They would say that someone noticed these stars, saw they resembled a triangle, called them the *Triangle*, as

[3]"Nux" verse 82, trans. H.T. Riley in *The Heroides or Epistles of the Heroines, the Amours, Art of Love, Remedy of Love, and Minor works of Ovid*, London: H.G. Bohn (1852).

[4]'One in each corner'.

everyone else followed suit; a pretended explanation *which merely repeats the fact that such a constellation exists.*

But suppose we ask, As there are hundreds of stars which might have been combined in triangles, how comes it that these particular stars, which, moreover, form a perfect isosceles triangle, were selected? To this Ignorance would answer that the stars *chanced* to be selected, and that the circumstance that the figure is an isosceles triangle was also accidental and devoid of any significance. But, rejecting this vain repetition of the facts of the case, in the first place we observe that Aratos says[5]: –

> Below Andromeda, in three sides measured
> Like-to-a-Delta; equal two of them
> As it has, less the third, yet good to find
> The sign, than many better stored with stars.

DISAPPEARANCE

Triangulum Minus began to vanish from charts in the first half of the nineteenth century. It made one of its last appearances in Alexander Jamieson's *Celestial Atlas* (1822; Fig. 28.5). Shortly thereafter, Jacob Green referred to the figure in name only, instead drawing a figure only for Triangulum (Majus). "Two triangles," Green wrote, "are commonly marked on celestial globes; but as the stars which compose the smaller one, are not of a sufficient magnitude to be embraced by our plan, we have not sketched it." Toward midcentury it was increasingly left off of popular charts; examples include Littrow (1839), Middleton (1842), Argelander (1843) and Mölinger (1851).

In 1844b, Admiral Smyth described Triangulum Minus as "discontinued;" the following year, however, it appeared in a list of constellations in Ezra Otis Kendall's *Uranography*. Kendall referred to both figures as "Triangula" and described Triangulum Minus as a "recent constellation" without detailing any of its history. Hannah Bouvier listed the smaller triangle and ascribed its creation to Hevelius in her *Familiar Astronomy* (1858), but noted that "it is no longer continued in some maps." George Chambers' *Handbook of Descriptive Astronomy* (1877) labeled it "Triangulum Minor."

Probably its last appearance on a chart in wide circulation was on the northern hemisphere all-sky map in Eliza A. Bowen's *Astronomy By Observation* (1888), in which the two figures were distinguished by the "Majus"/"Minus" appellation. The second edition of Joel Dorman Steele's *Popular Astronomy* (1899) made a strange passing reference to the existence of Triangulum Minus in describing Aries: "A line drawn from Almach [γ Andromedae] to Arietis [α Arietis] will pass through a beautiful figure of three

[5] *Phainomena*, lines 233–237.

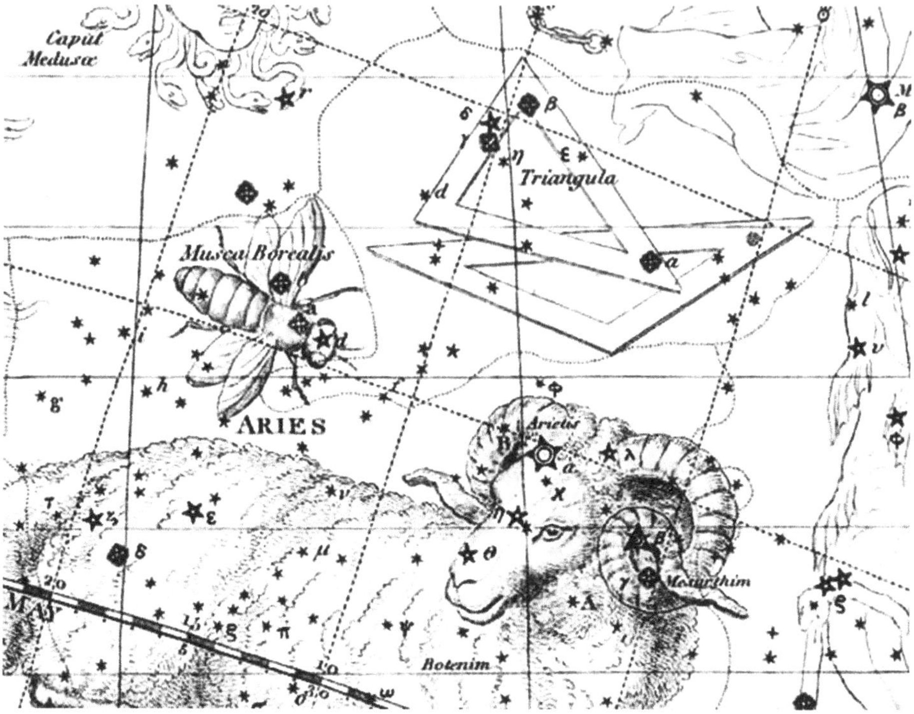

28.5 Detail of Triangulum Minus shown on Plate XIII of Alexander Jamieson's *Celestial Atlas* (1822), along with Musca Borealis (Chap. 16). Jamieson followed the design lead of Bode (1801b) in how he drew the figures

stars called *The Triangles*." Allen himself thought it was defunct nearly a century and a half before his time, although he refers to a late renaissance of sorts:

> Triangulum Minora was formed, and thus named, by Hevelius, from three small stars immediately to the south of the major constellation, towards Hamal of Aries; but it has been discontinued by astronomers since Flamsteed's day. Still Gore has recently revived it in the title Triangula on the planisphere in his translation of l'Astronomie Populaire, *as did Proctor in his reformed list.*

Only Triangulum (Majus) was recognized as canonical by the International Astronomical Union at Rome in 1922. Eugéne Delporte followed the same convention in *Délimitation scientifique des constellations* (1930b), carefully defining a polygon consisting of 14 segments to enclose both the classical figure as well as Hevelius' invention (Fig. 28.6).

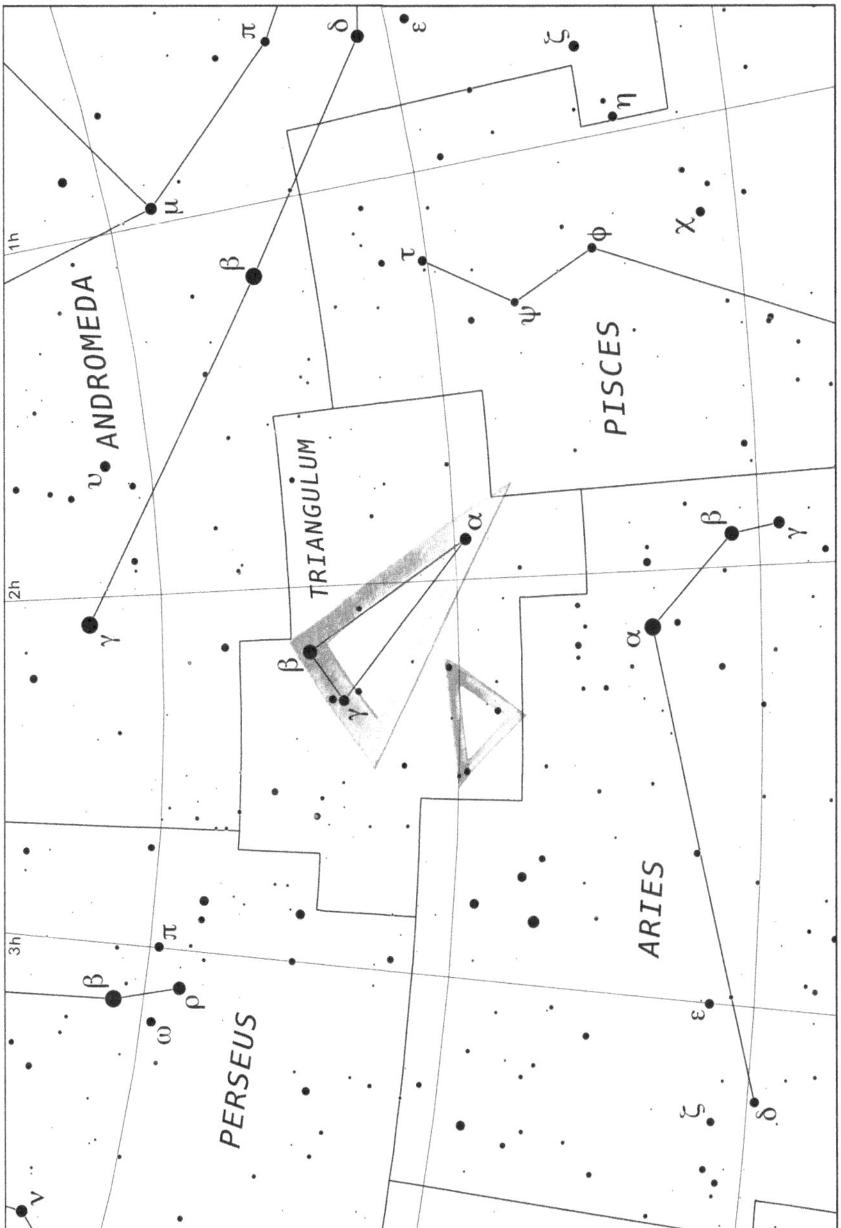

28.6 The figures of Triangulum (center, upper) and Triangulum Minus (center, lower) from Figure Aa of Hevelius' *Prodromus Astronomiae* (1690) overlaid on a modern chart

29

Turdus Solitarius/Noctua

The Solitary Thrush/the Night Owl

Genitive: Turdi Solitarii/Noctuae
Abbreviation: TuS
Alternate names: "Solitaire" (Le Monner 1776); "Solitarius" (Chambers, 1877; Rosser, 1879); "Avis Solitarius" (Bouvier, 1858); "Sitarius" (Chambers, 1877)
Location: Perched above the end of the tail of Hydra[1]

ORIGIN AND HISTORY

"Solitaire"

Pierre-Charles Le Monnier (1715–1799) introduced this constellation in a 1776 issue of *Mémoires de l'Académie Royale des Sciences*, intending it to represent "a bird of the Indies and the Philippines" that he believed already extinct by his time. Le Monnier included a table of stars (p. 561) and a map depicting the bird, reproduced here as Fig. 29.1. In addition, he described his motivation for defining the new figure on page 562:

> Le grand vide si apparent sous les bassins de la Balance, & qu'on aperçoit dans les zodiaques de Senex & d'Heulland, m'avoit déterminé il y a long-temps a vérifier dans les crépuscules, & à chaque mois lunaire, la position des plus petites Etoiles

[1] *"Between the southern pan of the balance [Libra], and the tail of Hydra"* (Bode, 1801a); *"North of Libra"* (Kendall, 1845); *"Over the tail-tip of the Hydra"* (Allen, 1899); *"On the extreme tail-tip of Hydra"* (Bakich, 1995).

J.C. Barentine, *The Lost Constellations*, Springer Praxis Books,
DOI 10.1007/978-3-319-22795-5_29

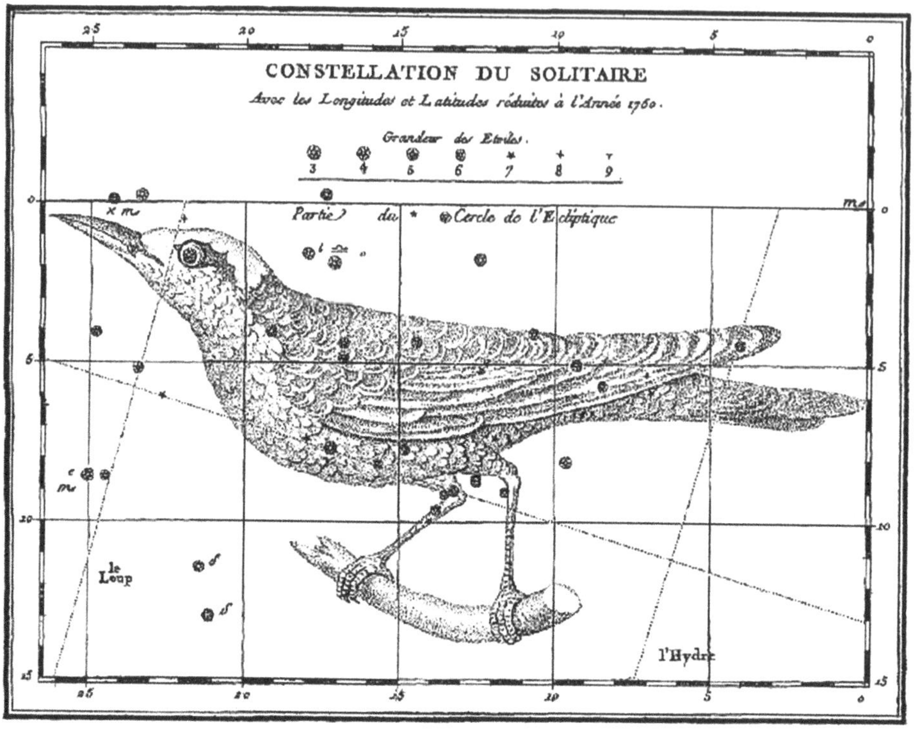

29.1 "Solitaire" as envisioned by Pierre-Charles Le Monnier in *Mémoires de l'Académie Royale des Sciences* (1776)

de la sixième à la neuvième grandeur, puisqu'on les aperçoit dans le ciel avec les nouvelles lunettes achromatiques ordinaires, quoiqu'on les ait omises dans les Catalogues.

La Lune, comme je viens de la dire, rencontre à chaque mois ces Étoiles; en sorte que leurs occultations nous donnent aussi sûrement que les autres Étoiles du zodiaque éclipsées par la Lune, les positions géographiques terrestres, puisque les latitudes des Etoiles aident à déterminer la longitude de cet Astre, aux instans des occultations instantanées, comme j'en ai averti dès l'année 1741.

J'ai observé la plus grand partie de ces Étoiles à mes deux quarts-de-cercles muraux, & la figure de la constellation du Solitaire (oiseau des Indes & des Philippines) a été préférée, en mémoire du voyage en l'île Rodrigue, m'ayant été fournie par Mrs Pingré & Brisson: voyez le *tome II de l'Ornithologie*: cette constellation sera voisine du Corbeau & de l'Hydre sur nos planisphères & globes célestes.

Rendered in English:

> The great void apparent in the pans of the Balance, and that one sees in the zodiacs of Senex[2] and d'Heulland,[3] I had long ago determined to check in the twilight, and each lunar month, the position of the smaller stars of the sixth to the ninth magnitude, since they are seen in the sky with new ordinary achromatic telescopes, although they have been omitted from Catalogs.
>
> The Moon, as I say it, meets these stars each month; ensuring that their occultations also certainly give us that the other Stars of the zodiac are eclipsed by the Moon, the terrestrial geographic positions, since the latitudes of the Stars help to determine the longitude of this Star, to the instantaneous moments of occultations, as I have noted since 1741.
>
> I observed most of these stars with my two mural quadrants, and the figure of the constellation of the Solitaire (bird of the Indies & the Philippines) was chosen in memory of the voyage to the island of Rodrigues, I have been provided by Messrs. Pingré[4] & Brisson[5]: see Volume II of *Ornithology*: this constellation appears near the Crow and the Hydra on our planispheres and celestial globes.

The "voyage to the island of Rodrigues" and "Messr. Pingré" refers to the 1761 expedition led by Alexandre Guy Pingré to observe that year's transit of Venus across the face of the Sun. The 1761 transit is among the earliest examples of international scientific collaboration, as astronomers planned expeditions across the eastern hemisphere to observe the transit from locations as widely separated by geography as possible. It had been theorized as early as 1691 by Edmond Halley that such observations could be used to accurately determine the solar parallax, and therefore the distance between the Earth and the Sun.

There has been some historical confusion as to the identification of the bird species Le Monnier meant by "Solitaire." Richard Hinckley Allen wrote that the constellation's name "is said to be that of the Solitaire, formerly peculiar to the little island Rodriguez in the Indian Ocean ... although the bird has been extinct for two centuries," implying that it was not found on the island at the time of Pingré's visit. In a note to the entry in his book, Allen further explains the background of the name: "The generic word Turdus, however, is erroneous; for the bird was not a thrush, but, as its correct name, Pezophaps solitaria, denotes, an extremely modified form of flightless pigeon allied to the dodos, yet larger and taller than a turkey." Ridpath (1989) decided that Allen's characterization of Le Monnier's

[2] The English cartographer and engraver John Senex (1678–1740).

[3] The French engraver Guillaume d'Heulland (1700–1770).

[4] Alexandre Guy Pingré (1711–1796) was a French astronomer and canon regular best known for publishing an extensive history of comet observations.

[5] Mathurin Jacques Brisson (1723–1806) was a French zoologist best known for his comprehensive *Ornithologie* (1760–1763).

figure as an extinct flightless bird "seems to be a misunderstanding" and suggested that "the bird shown on Le Monnier's diagram of the constellation resembles a female blue rock thrush (*Monticola solitarius*, family *Turdidae*)." As a result of this resemblance, he argued, the figure was labeled "Turdus Solitarius" on most later maps.

Le Monnier's figure was not readily adopted by other cartographers at first. Johann Elert Bode, for example, left Turdus out of the first edition of *Vorstellung der Gestirne* (1782; top of Fig. 29.2), published 6 years after Le Monnier's paper. The stars identified by Le Monnier were assigned to Libra, according to the constellation boundaries Bode arbitrarily drew. However, he included Turdus on Plate 14 of *Uranographia* (1801b; Fig. 29.3), perhaps as a result of the fact that Jean Nicolas Fortin adopted it as "le Solitaire" for his revised, third edition of Flamsteed's *Atlas Coelestis* in 1795 (Fig. 29.4). Of the Thrush, Bode wrote in *Allgemeine Beschreibung und Nachweisung der Gestirne*,[6] "This bird of India was placed in the firmament by Le Monnier in 1776, between the southern pan of the balance [Libra], and the tail of Hydra, and of stars that were once part of these constellations." Fortin created a perch for the bird by drawing in a fragment of a branch unattached to anything, and including only the star 52 Hydrae. Since he did not include constellations boundaries, he was not confronted with the practical concerns that later faced Bode.

The rendering of the bird in these last two sources is sufficiently similar that it is no reach to conclude that Bode was influenced by Fortin's atlas; he then retroactively placed Turdus ("Der Vogel Einsiedler", or The Hermit Bird) in the second edition of *Vorstellung der Gestirne* (1805; bottom of top of Fig. 29.2). However, he did not adjust the boundaries of Libra, Hydra, or Virgo in order to accommodate the new figure, suggesting that Bode thought of Turdus more like an asterism and less as an independent constellation. Also evident in Fig. 29.2, Bode extended the tail of Hydra in order to give the bird something on which to perch; while the figure of the tail terminates at π Hydrae in the 1782 edition, in 1805 Bode drew it out as far west as 58 Hydrae, a star he labeled with the Roman letter "k" and placed within the boundaries of Libra. Similarly, the star 50 Hydrae (a star he labeled with the proper name "Veninderlich") was also incorporated into the figure of the tail, whereas Bode's own boundaries specifically assigned it to Libra.

In the end, Bode must have felt obligated to incorporate Le Monnier's constellation as he had included it in *Uranographia*, but its placement in the 1805 edition of *Vorstellung der Gestirne* seems like an afterthought as boundaries were not changed to accommodate the new figure. This treatment is at strikingly at odds with his depiction of Turdus in *Uranographia* just 4 years earlier, when he drew a clear constellation boundary around the bird (Fig. 29.3). It is unclear why he kept the 1782 boundaries for the later edition of *Vorstellung der Gestirne* after having promoted Turdus as a full-fledged constellation at the turn of the new century. In each case from 1795 to 1805, Turdus was drawn oriented eastward, following Le Monnier, with its head completely overlapping the southern pan of Libra.

---

[6]"Cet oiseau de l'Inde a été placé par Monnier en 1776 au firmament, entre le bassin méridional de la balance, & la queue de l'hydre, & formé d'étoiles qui autrefois faisoient partie de ces constellations." (p. 13)

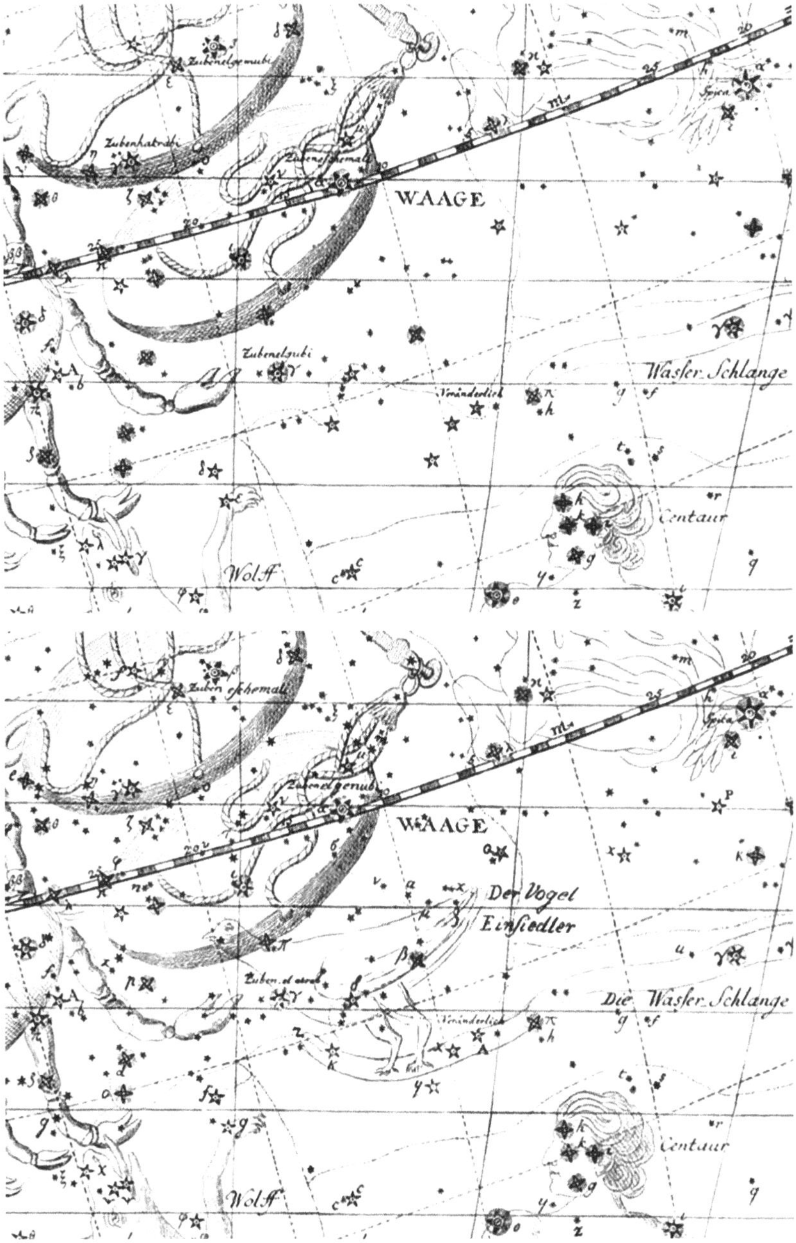

29.2 Before and after. Excerpts of Plate 19 from two editions of Johann Elert Bode's *Vorstellung der Gestirne*: 1782 (*top*) and 1805 (*bottom*). In the interim between the two editions Bode decided to include Turdus Solitarius in *Uranographia* (1801b), taking care to add the figure in the second edition of his earlier charts. In the 1805 edition of *Vorstellung der Gestirne* he labeled the bird "Der Vogel Einsiedler" ("The Hermit Bird") in German

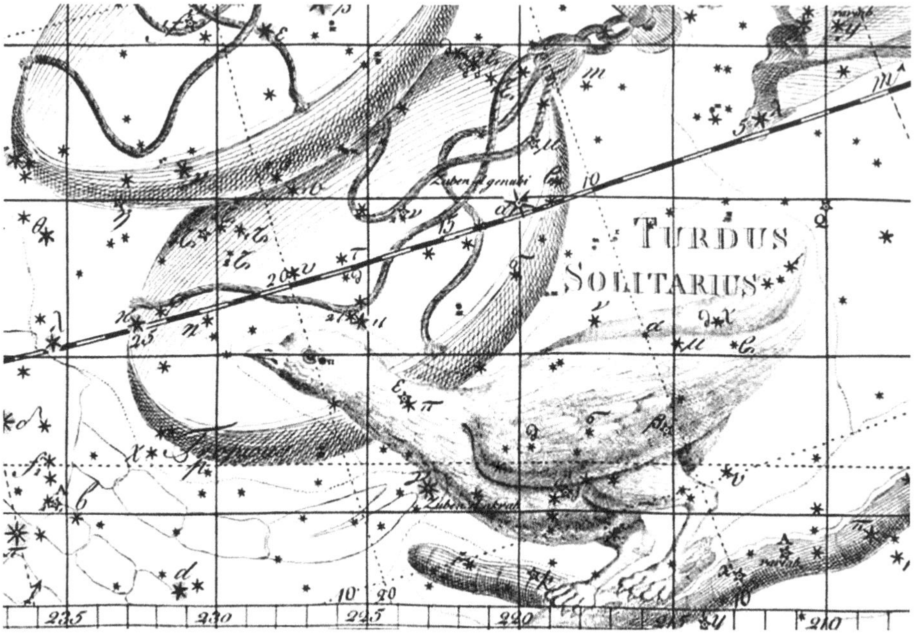

29.3 Turdus Solitarius shown on Plate 14 of Johann Elert Bode's *Uranographia* (1801b) with its own constellation boundaries

Noctua, the Night Owl

Despite the influence of Bode, this avian constellation experienced an identity crisis within a few years of the publication of *Uranographia*. The English polymath Thomas Young (1773–1829) remade the stars into the "Mockingbird" on a star chart he published in *A Course of Lectures on Natural Philosophy and the Mechanical Arts* (1807), the same work in which he attempted to introduce the Battery of Volta to the heavens (see Volume 2). As with the Battery, the rebranding of Turdus did not persist and his figure is found on no other chart. Young's depiction (Fig. 29.5) imagined the bird as a different passerine species, leaning toward the ground in order to prevent its head from colliding with Libra.

Alexander Jamieson took Young's lead and substituted a completely different species of bird, an owl he called "Noctua" (Fig. 29.6, top). Jamieson introduced a further innovation which solved a problem and inspired other mapmakers: he literally turned the bird around. Still perched upon Hydra's tail, Jamieson's version faces west, thereby avoiding overlap with Libra entirely. In explaining his de facto rebranding of Le Monnier's bird, Jamieson explained that he felt it was odd there was no owl in the heavens "considering the frequency it is met with on all Egyptian monuments." A similar owl appeared on Plate 32 of *Urania's Mirror* (1825; Fig. 29.6, bottom), Map IV of Elijah Hinsdale Burritt's *Atlas Designed to Illustrate the Geography of the Heavens* (1835a), and Plate 4 of James Middleton's *Celestial Atlas* (1842). Authors such as Allen (1899) and Burnham (1978) have written

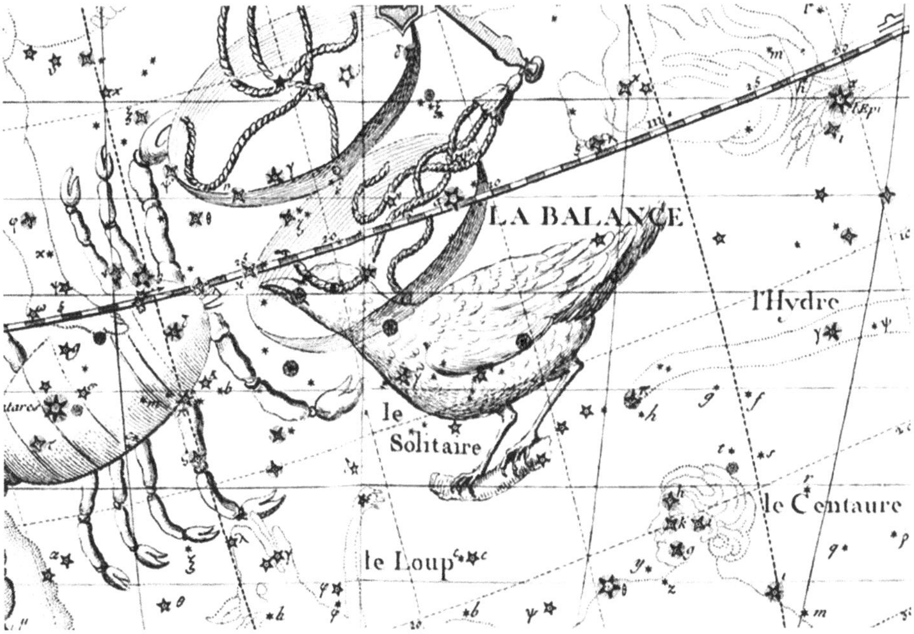

29.4 Turdus Solitarius depicted as "le Solitaire" on Map 19 in Jean Nicolas Fortin's 1795 edition of Flamsteed's *Atlas Coelestis*. The stars were left unformed in the 1776, the year Le Monnier's constellation was published

that Noctua's first appearance was in Burritt's *Atlas of the Heavens* (1835), but it clearly predates Burritt's works by more than a decade.

Burnham mentioned Noctua in the context of Hydra on page 1031 of his magnum opus, the eponymous *Celestial Handbook* (1978): "The small group [of stars] consisting of 54, 55, 56, and 57 Hydrae... is the site of the obsolete asterism Noctua, the 'Night-Owl', which appears on the star charts of E. H. Burritt and on other maps of the early 19th century."

ICONOGRAPHY

As "Turdus Solitarius"

The name of the bird identified by Le Monnier refers to the former designation of the blue rock thrush (*Monticola solitarius*; Fig. 29.7) when it was considered part of the family *Turdidae*.[7] The thrush is a species of chat, a group of small, insectivorous birds native to

[7]Carl Linnaeus introduced the binomial *Turdus solitarius* in the tenth edition of *Systema Naturae* (1758).

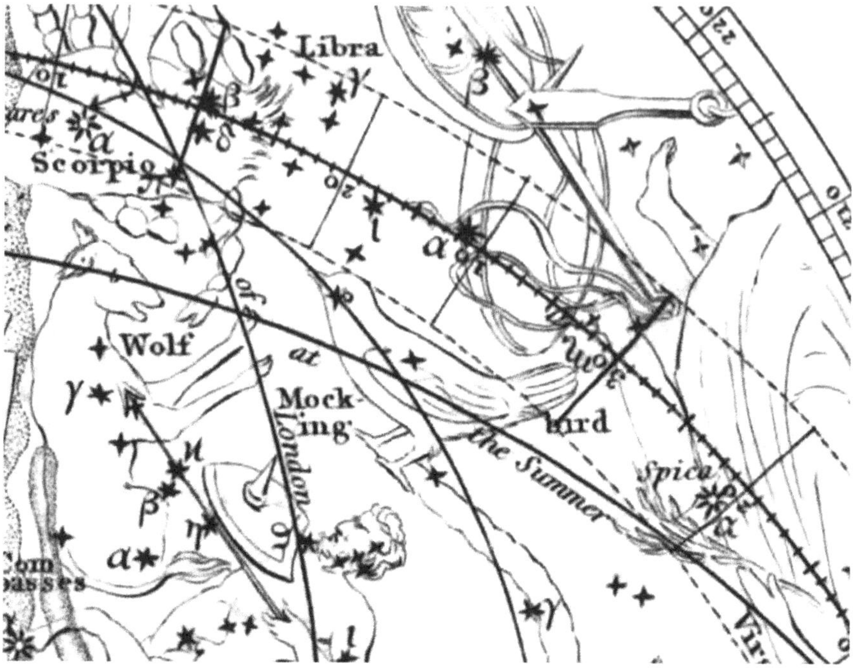

29.5 An attempted rebranding of Turdus Solitarius as "Mockingbird" on Plate XXXVII, Fig. 518 in Thomas Young's *A Course of Lectures on Natural Philosophy and the Mechanical Arts* (1807)

the Old World. It is found predominantly from southern Europe to northern Africa, across central Asia and as far east as China and Malaysia.

The adult birds are roughly 20 cm in length, comparable to the size of the typical starling. Their name is derived from the deep blue color of the adult male on top, slightly brighter around the head and darker on the wings and tail. In comparison, the adult female is dull in appearance, a dusty blue-grey on top and streaked buff/brown on the lower face and upper chest. Juveniles are dark brown, absent any blue coloration, and more strongly spotted and scaled than adult females. The blue rock thrush is differentiated from other types of rock thrush by the lack of red outer tail feathers. The adult male's song is clear and noticeably louder than that of the common rock thrush.

Blue rock thrushes are largely resident in Europe, north Africa and southeast Asia, while other Asian populations tent to be migratory. The latter move from central Asia to their winter homes in sub-Saharan Africa, India and parts of southeast Asia. They are rarely seen in northern and western Europe. They are adapted to a variety of terrains and climates, from African deserts to the high mountains of the central Himalayas and at elevations between sea level and about 4000 m.

The birds feed on a variety of prey during much of the year. In summer they favor various types of invertebrates including insects, earthworms and snails, and some small

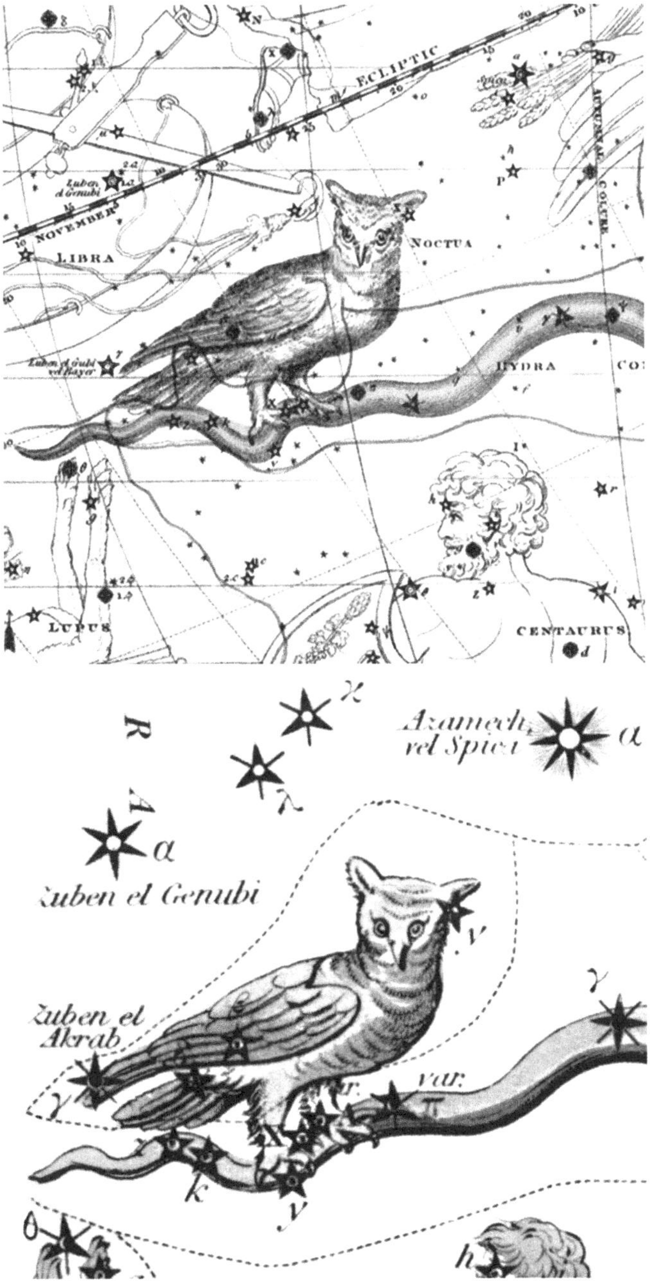

29.6 The stars of Turdus rendered as Noctua (the Night Owl), perched on the tail of Hydra. *Top*: Plate XXVII of Alexander Jamieson's *Celestial Atlas* (1822). *Bottom*: Plate 32 of *Urania's Mirror* (1825)

29.7 Three specimens of rock thrush (genus *Monticola*) from Johann Friedrich Naumann's *Naturgeschichte der Vögel Mitteleuropas* ("Natural history of the birds of Central Europe"), published at Gera in 1897

29.8 Obverse of the Maltese 1 Lira coin from the series of 1986–2007. The design features the blue rock thrush, Malta's national bird

vertebrates such as snakes and mice. In winter they supplement their diet with berries and seeds. Its long, thin bill is well adapted to manipulating small objects and picking up prey. The birds forage by scanning the ground from an elevated perch, fluttering to the ground when a prey item is identified; otherwise, they spend little of their time on the ground. They have even been observed catching small insects on the wing (Fig. 29.8).

The wide range of blue rock thrushes means that different populations and subspecies breed at different times through the year. They prefer open mountainous areas for breeding, generally higher in elevation than those favored by the common rock thrush. The breeding period begins as early as January and may extend as late as July. Females nest in cavities among rocks and cliffs up to 5 m above the ground, building loose, shallow nests of coarse plant material that is lined with feathers and plant down. The typical clutch size is 3–6 eggs; the eggs hatch after an incubation period of 12–15 days. Chicks fledge around 2 weeks after hatching, but remain dependent on adults for feeding up to 2 weeks beyond fledging.

Le Monnier referred to a "bird of the Indies & the Philippines," which may indicate the *Monticola solitarius philippensis* subspecies recently suggested as a full species in its own right.[8] The *philippensis* birds are distinguished from other rock thrushes by a stronger blue color on top and a dark red-brown breast and undertail. These birds are distributed from southeast Siberia to Japan and winter in Indonesia.

[8]D. Zuccon and P.G. Ericson, "The Monticola rock-thrushes: phylogeny and biogeography revisited". *Molecular Phylogenetics and Evolution* 55(3), 901–910 (2010).

29.9 Nineteenth-century illustrations of two owl species. *Left*: *Athene noctua*, the Little Owl (date and artist unknown). *Right*: *Asio otus*, the Long-eared Owl (Chromolithograph signed "E. Kohler," 1880)

As "Noctua"

Understanding which owl species Noctua represents is not as straightforward as identifying the blue rock thrush as the model for Turdus Solitarius. The name "Noctua" is suggestive of the Little Owl (*Athene noctua*; Fig. 29.9, left), an Old World owl species whose name is derived from the Greek goddess to whom it was sacred. However, close inspection of the owl drawn by Alexander Jamieson for his *Celestial Atlas* (1822) reveals a key trait not shared by *Athene noctua*: ear tufts. These feather-covered skin projections common to members of several owl genera superficially resemble mammalian ears, but they do not seem to be involved in the owls' sense of hearing and their true function remains unknown.

Jamieson, who spent the entirety of his life in Britain up to the point of the publication of the *Celestial Atlas*, would likely have been familiar only with the five owl species resident on the island: the Little Owl (*Athene noctua*), the Barn Owl (*Tyto alba*), the Long-eared Owl (*Asio otus*), the Short-eared Owl (*Asio flammeus*) and the Tawny Owl (*Strix aluco*). Of these only *A. otus* (Fig. 29.9, right) has prominent ear tufts and a similar pattern of coloration as Jamieson's owl, making it the most probable candidate for its prototype. It fails to fully fit Jamieson's drawing only in that the owl in the drawing lacks the clear facial disc of *A. otus*; forgiving some amount of artistic license the resemblance is much more clear than that with *Athene noctua*. It is unclear why he would have applied the term "Noctua" to an owl that clearly does not resemble *Athene noctua*; the answer may be that he simply applied a name evoking the night to an unrelated owl he liked.

A. otus is part of a family of owls known as *Strigidae*, which is a larger group of owls than the barn owls (*Tytonidae*). Four subspecies have been identified: *A. o. otus*, occupying a large range throughout the Old World from western Europe to east-central China; *A. o. canariensis*, isolated on the Canary Islands; *A. o. tuftsi*, found in western North America; and *A. o. wilsonianus*, the eastern North America counterpart. All *A. otus* owls require a combination of open space such as grasslands for foraging and tall trees or shrubs for nesting. The birds are semi-migratory, wintering in the southern extent of their temperate range.

Long-eared owls are medium-sized owls averaging 30–40 cm in length with a wingspan of 85–100 cm. Adults have a body mass of between about 200 and 400 g. The ear tufts are usually dark in color and stand up above the center of the head, making the animal appear larger to other owls while it is perched. Males and females show clear sexual dimorphism, as females are almost always larger and darker in color than males.

Long-eared owls overlap in much of their geographic range with *A. flammeus*, and the two can be difficult to discern from a distance. While *A. otus* has clearly longer ear tufts, they can appear artificially foreshortened when the animal holds them relatively flat to its head. The better discriminator is the color of the iris, which is orange in Long-eareds and more yellow in Short-eareds. The coloration of the two also varies; Long-eared owls tend to be darker than their Short-eared relatives.

A. otus is fully nocturnal, spending its days roosting in the dense interiors of trees and shrubs where it is best camouflaged. For further protection, the owl will stretch out its body to appear flattened like a tree branch. They become active at dusk and hunt rodents and small mammals in clearings at night, although it has also been known to eat other birds. The owls are highly skilled flyers with an acute sense of hearing enabling them to successfully hunt in total darkness. They fly relatively low to the ground at heights averaging 1–2 m, with their heads tilted to one side as the listen for the sounds of their prey. On locating suitable prey, the owls attack with their powerful talons and either swallow small items whole or carry larger ones away with their talons or bills for convenient feeding.

Long-eared owls breed between February and July, nesting in trees. Males move into nesting territories first and begin territorial calls in winter. They perform erratic display flights among nesting trees; females respond with a nest call. After mating, adults will roost close together but the females prefer roosting on nests almost exclusively. Females avail themselves of nests abandoned by other birds when possible, laying between 4 and 6 eggs and incubating them for 25–30 days. The young have a distinct call whose sound has been compared to that of a rusty metal hinge. They begin walking out of the nest onto nearby tree branches at around three weeks but do not fledge until five weeks, and become independent of the parents by two months.

DISAPPEARANCE

James Middleton's 1842 atlas is among the last that featured a bird of any kind at this place in the sky. A few other contemporary authors, such as Ezra Otis Kendall (1845), mentioned the constellation, but by midcentury it had effectively vanished from charts; no mention is made of it by authors such as Johnston (1855), Proctor (1876) or Pritchard (1885). It appeared as "Solitarius" in William Henry Rosser's list of "Modern Constellations" in *The Stars and Constellations* (1879). George Chambers wrote of it in his *Handbook of Descriptive Astronomy* (1877), corruptly rendering its name as "Sitarius," while in the *Poole Bros' Celestial Handbook* (1892) it somehow came out as "Tartus Solitarius." By this time its origins were confused, as the *Handbook* attributed its formation to "Lalande, 1774, A.D." Still, it appeared in charts as late as Eliza Bowen's *Astronomy By Observation* (1888) where it simply labeled "Solitarius."

Writing at the close of the nineteenth century, Allen pointed out that the bird originally identified by Le Monnier in 1776 "has been extinct for two centuries—as indeed now is the constellation." When the modern constellation boundaries were set in 1928, the stars formerly comprising its figure were distributed among Libra, Virgo and Hydra (Figs. 29.10 and 29.11).

It is fitting to quote Burnham (1978) in the context of Noctua at the conclusion to the last chapter in this book on a widely-recognized yet extinct constellation: "The loss of this little asterism might seem unfortunate to astronomers, since no other celestial creature so appropriately honored their profession."

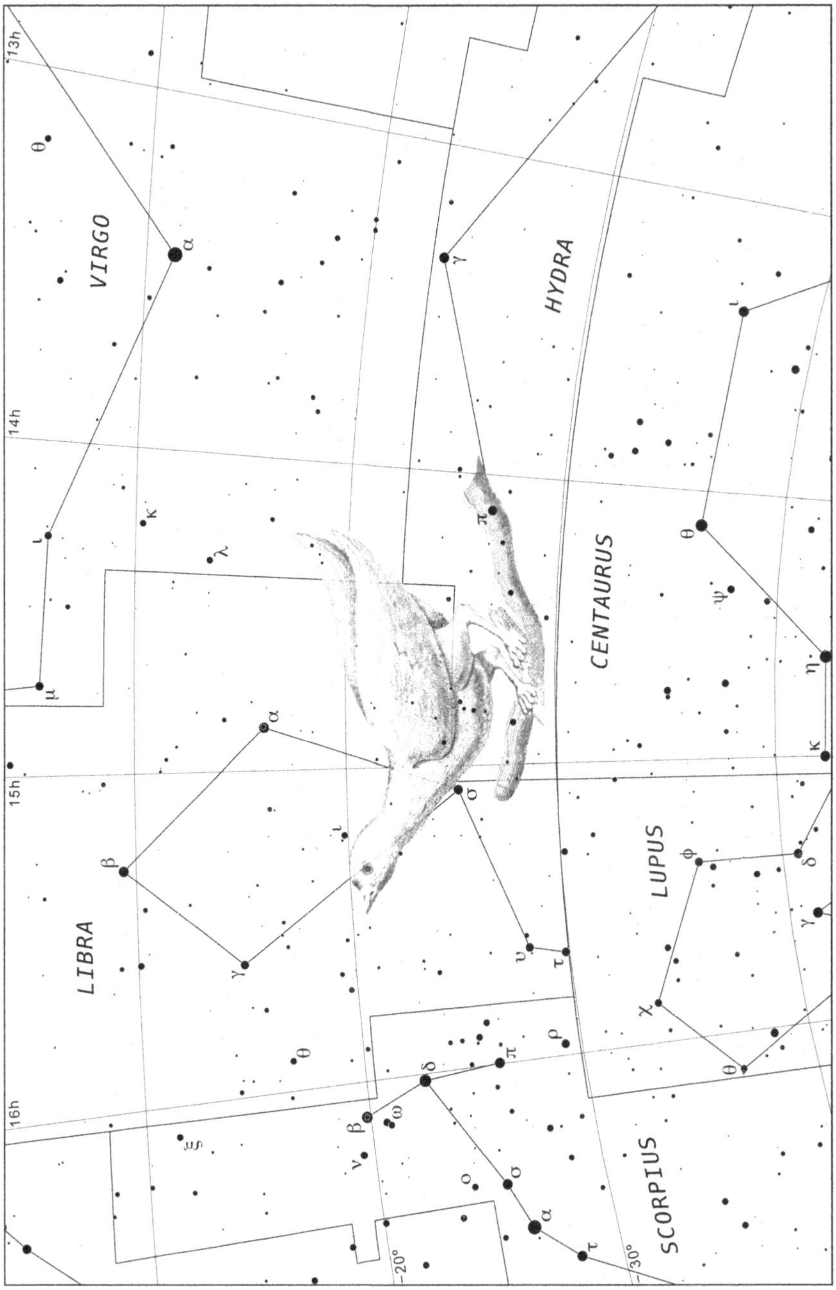

29.10 The figure of Turdus Solitarius from Plate 14 of Bode's *Uranographia* (1801b) overlaid on a modern chart

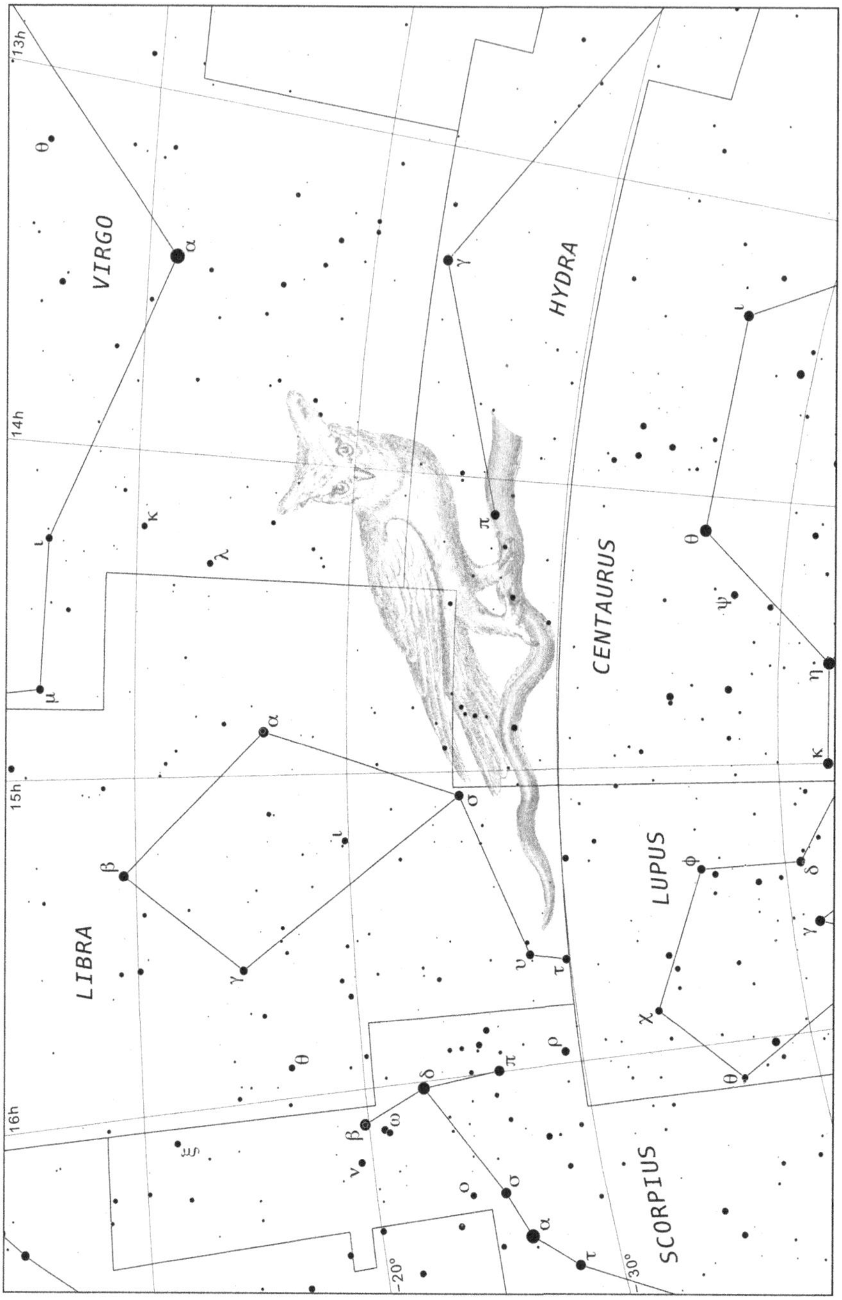

29.11 The figure of Noctua from Plate 27 of Alexander Jamieson's *Celestial Atlas* (1822) overlaid on a modern chart

Part III
Conclusion

30

Lost and Found

> With a few exceptions, the likeness is purely fanciful. Not only are the figures uncouth, and the origin often frivolous, but the boundaries are not distinct. Stars occur under different names ; while one constellation encroaches upon another. Though, however the constellations are thus rude and imperfect, there seems little hope of any change. Age gives them a dignity that insures their perpetuation.
>
> – J. D. Steele, *Popular Astronomy, Being The New Decriptive Astronomy* (1899)

This book is the result of an effort to identify all of the constellations in published Western sources introduced after Ptolemy that achieved at least some currency in the area of celestial cartography, but were left out of the set of 88 constellations considered "official" by the world professional astronomy community. Along the way, we have gained some insight into the mind of pre-twentieth century astronomy in Europe and America, and the transition of astronomy to a fully scientific discipline in the modern sense. Other than as a convenience, the constellations serve no practical purpose for modern astronomers; rather, the exploration of historical astronomy tells us more about the human condition than how our cosmos is constructed.

The notion of what makes for a *constellation*, as opposed to an asterism, was never clearly defined by scientists. Mapmakers tended to defer first to the tradition established by Ptolemy, then to the southern explorers, and finally to everyone else. The production and dissemination of standalone atlases and globes diminished in importance through the eighteenth century; in the century that followed, star charts evolved away from a means of storytelling rather than a substantial inventory of the night sky and found a new home in popular works on astronomy. By 1900 the role of charts was mostly to introduce students and laypersons to the folklore and history of the night sky. Compared to previous eras, relatively few new constellations were introduced in the nineteenth century, and the transition of general atlases from primarily scholarly to lay audiences inhibited the adoption of new figures by astronomers.

J.C. Barentine, *The Lost Constellations*, Springer Praxis Books,
DOI 10.1007/978-3-319-22795-5_30

Before about 1700, the makers of maps and globes uniquely determined what a constellation was, and their influence was proportional to their ability to design, print and sell attractive charts. Camelopardalis, Cancer Minor, Columba, Gallus, Jordanus, Musca Borealis/Vespa; see also Gallus and Polophylax in Volume 2. In the eighteenth century, the reorganization of the night sky had much to do with scientific societies putting their stamp of approval on expeditions, such as the highly successful efforts of Nicolas Louis de Lacaille. A brief renaissance of artistic cartography in the first four decades of the nineteenth century led to a few additional suggestions by mapmakers like Alexander Jamieson and Elijah Hinsdale Burritt, but their promotion of newly-invented figures ultimately amounted to nothing.

One goal of this book is to understand something about the process by which constellations became canonical over historical time, and to determine if any particular factors made it more or less likely that a new constellation survived. Tracing the history of the lost constellations leads to a few useful conclusions.

Attempts to flatter patrons (Sceptrum Brandenburgicum and Taurus Poniatowski in this volume; Gladii Electorales Saxonici, Leo Palatinus and Pomum Imperiale in Volume 2) or to commemorate nationalist symbols (Robur Carolinum) or honor significant individuals (Custos Messium and the Telescopia Herschelii here; Corona Firmiana and Marmor Sculptile in Volume 2) did not work well in most cases, but one made it through the International Astronomical Union culling of the 1920s: Johannes Hevelius' constellation Scutum. A few such figures persisted for a while (Honores Fredrici, Psalterium Georgianum), but did not ultimately gain wide acceptance.

Many new constellations, particularly in the northern sky, were created from "unformed stars" not clearly assigned to any particular classical constellation by Ptolemy in the *Almagest*, but these stars often turned out to be too faint to make the figures at all distinct (Sagitta Australis). Their popularity seems to have suffered as a result. Some new constellations extended the practice of referencing ancient mythology directly (Cerberus et Ramus Pomifer) or more obliquely (Mons Maenalus), although they seem like small of well-established stories. Some might plausibly still exist as asterisms within canonical constellations, but they do not merit labels on modern star charts. There is one instance in which a southern constellation (Polophylax), was created as something of the mirror image of a northern figure (Arctophylax, better known now as Bootes), but it, too, had only a very brief heyday.

Remaking the celestial sphere to reflect Biblical themes was a flop (Jordanis, Tigris), not to mention Julius Schiller's attempt to rebrand the entire firmament in *Coelum Stellatum Christianum* (1627). Only one constellation introduced after Ptolemy referencing a Biblical story—Columba Noae—survives on modern maps. Certain asterims referencing Biblical subjects (e.g., the 'Job's Coffin' figure in Delphinus) endured informally into the nineteenth century.

Placing images of devices representing technological achievements among the stars was fashionable and relatively fruitful from the mid-eighteenth to the mid-nineteenth centuries. Lacaille was spectacularly triumphant (Antlia, Caelum, Circinus, Fornax, Horologium, Mensa, Microscopium, Norma, Octans, Pictor, Pyxis, Reticulum, Sculptor, Telescopium). Other such efforts (Globus Aerostaticus, Machina Electrica, Quadrans

Muralis, Quadratum, Solarium, and the Battery of Volta in Volume 2) did not fare as well. Many of those constellations were strictly composed of faint stars, predisposing them to unpopularity.

Ironically, the effort to name constellations after living astronomers (Custos Messium, for Charles Messier) or their effects (the Telescopia Herschelii, for William Herschel) failed. With the exception of Antinoüs, which was still found on charts as late as the turn of the twentieth century, all human figures incorporated into widely circulating constellation designs were mythical. Even the astronomers whose inventions were intended to catch the attention of potential royal patrons shied away from depicting living (or recently deceased) people and instead tended toward heraldic or other symbols suggesting the monarchs' institutions. Political sensitivities, particularly in the tumultuous seventeenth century, were enough to give pause in the consideration of any sense of overt nationalism or sectarian preference.

Not even sentimentality guaranteed a constellation's permanent inclusion in the heavenly pantheon. Early efforts at inventing figures of exotic animals from faraway southern lands (Apus, Chamaeleon, Dorado, Pavo, Tucana, Volans) are still found on modern star charts, but other animal-themed figures (Felis, Noctua/Turdus Solitarius, Rangifer and Sciurus Volans here; Gallus and Sciurus Volans in Volume 2) have long since fallen away. The constellations invented by John Hill and published in his *Uranua* (1754; see Appendices in Volume 2)—if not an elaborate in-joke—represent fifteen further names added to this list.

Some figures were reconfigured internally to become asterisms within associated constellations: Anser was subsumed into Vulpecula; Antinoüs became part of Aquila; Cerberus (et Ramus Pomifer) was always seen connected to Hercules; and Caput Medusae walked the fine line between asterism and constellation. While many are still drawn on modern maps, none carries the distinction of a separate label it once held. In some instances, attempts were made to break parts away from existing constellations and give them their own identities (Malus from Argo Navis; Norma Nilocta from Aquarius, Volume 2). None is found on any modern map of the night sky.

Trying to understand why some constellations became famous while others did not is like something like trying to develop a predictive theory of art history: it simply does not work. Why certain post-Ptolemaic figures survived to the twentieth century and others didn't is analogous to asking why Michelangelo and Van Gogh are known around the world in the twenty-first century but most of their artist contemporaries are not. The reason why is not just that they "made better art". Similarly, it's impossible to say which contemporary artists will be world-famous many centuries from now—or even whether the "modern" constellations will be found on star maps of the distant future.

For many minor constellations, the nineteenth century was one long, drawn-out goodbye. There are few instances of constellations abruptly disappearing from maps everywhere; such examples tend to be the results of displacement from existing maps by the works of exceptionally influential cartographers. Nearly all instances are before the time of Johann Elert Bode's seminal *Uranographia* (1801). Rather, many constellations, and in particular those proposed by Bode and Joseph Jérôme Lefrançois de Lalande, experienced a period of decades in the nineteenth century during which some cartographers plotted them and others simply didn't, each according to his personal taste. Some authors described them in terms implying current, widespread recognition among astronomers, while others wrote that they had been by that time abandoned, if not earlier. As a result, the canon of constellations we know today was already essentially complete by 1900. In the quarter-century before Eugéne Delporte's work for the International Astronomical Union (Chap. 2), there was virtually no change in the contents of that canon, which made it easier to avoid disputes resulting from eliminating constellations then still in common circulation.

Why did these constellations not "make it' into the twentieth century canon? Fundamentally, there are three reasons.

(a) Most were made of relatively faint and undistinguished stars. Visibility seems to have been something of a factor: If people couldn't readily find the figures on maps, then the new constellations were less likely to have circulated as a matter of popular communication.
(b) Nearly all later inventions were increasingly contrived figures. The human brain prefers clear patterns that are easily identified (and therefore committed to memory). The oldest constellations—Ptolemy's set of 48 figures named in the *Almagest*—feature the brightest stars in some of the most obvious patterns.
(c) There may have been a sense that humanity didn't need figures for every single group of stars. To leave room for constellation boundaries, each figure required a well-defined "edge."

This book synthesizes a great deal of previously-published material into a single reference, resolves some apparent mysteries (or misunderstandings) among those sources, and illuminates a time in the history of astronomy when visual observations gradually gave way to more scientifically robust means of sensing light using both analog and digital media at wavelengths well beyond the rainbow of colors to which the human eye is sensitive. It highlights a time when artistic cartography of the sky, steeped in historical tradition, was still relevant. But it leaves unanswered another question: Will our own "modern" constellations survive for centuries hence, or will they, too, become "lost?" That is a question only the history of the future will answer.

Appendix 1

The Modern Constellations

Nominative	*Genitive*	*Abbreviation*	*Origin*
Andromeda	Andromedae	And	Ancient (Ptolemy)
Antlia	Antliae	Ant	Lacaille (1754)
Apus	Apodus	Aps	Keyser & de Houtman
Aquarius	Aquarii	Aqr	Ancient (Ptolemy)
Aquila	Aquilae	Aql	Ancient (Ptolemy)
Ara	Arae	Ara	Ancient (Ptolemy)
Aries	Arietis	Ari	Ancient (Ptolemy)
Auriga	Aurigae	Aur	Ancient (Ptolemy)
Boötes	Böotis	Boo	Ancient (Ptolemy)
Caelum	Caeli	Cae	Lacaille (1754)
Camelopardalis	Camelopardalis	Cam	Plancius (1613)
Cancer	Cancri	Cnc	Ancient (Ptolemy)
Canes Venatici	Canum Venaticorum	CVn	Hevelius (1690)
Canis Major	Canis Majoris	CMa	Ancient (Ptolemy)
Canis Minor	Canis Minoris	CMi	Ancient (Ptolemy)
Capricornus	Capricorni	Cap	Ancient (Ptolemy)
Carina	Carinae	Car	Lacaille (1754)
Cassiopeia	Cassiopeiae	Cas	Ancient (Ptolemy)
Centaurus	Centauri	Cen	Ancient (Ptolemy)
Cepheus	Cephei	Cep	Ancient (Ptolemy)
Cetus	Ceti	Cet	Ancient (Ptolemy)
Chamaeleon	Chamaeleontis	Cha	Keyser & de Houtman
Circinus	Circini	Cir	Lacaille (1754)
Columba	Columbae	Col	Plancius (1592)
Coma Berenices	Comae	Com	Vopel (1536)

J.C. Barentine, *The Lost Constellations*, Springer Praxis Books,
DOI 10.1007/978-3-319-22795-5

The Modern Constellations (continued)

Nominative	*Genitive*	*Abbreviation*	*Origin*
Corona Australis	Corona	CrA	Ancient (Ptolemy)
Corona Borealis	Corona	CrB	Ancient (Ptolemy)
Corvus	Corvi	Crv	Ancient (Ptolemy)
Crater	Crateris	Crt	Ancient (Ptolemy)
Crux	Crucis	Cru	Plancius (1598)
Cygnus	Cygni	Cyg	Ancient (Ptolemy)
Delphinus	Delphini	Del	Ancient (Ptolemy)
Dorado	Doradus	Dor	Keyser & de Houtman
Draco	Draconis	Dra	Ancient (Ptolemy)
Equuleus	Equulei	Equ	Ancient (Ptolemy)
Eridanus	Eridani	Eri	Ancient (Ptolemy)
Fornax	Fornacis	For	Lacaille (1754)
Gemini	Geminorum	Gem	Ancient (Ptolemy)
Grus	Gruis	Gru	Keyser & de Houtman
Hercules	Herculis	Her	Ancient (Ptolemy)
Horologium	Horologii	Hor	Lacaille (1754)
Hydra	Hydrae	Hya	Ancient (Ptolemy)
Hydrus	Hydri	Hyi	Keyser & de Houtman
Indus	Indi	Ind	Keyser & de Houtman
Lacerta	Lacertae	Lac	Hevelius (1690)
Leo	Leonis	Leo	Ancient (Ptolemy)
Leo Minor	Leo Minoris	Lmi	Hevelius (1690)
Lepus	Leporis	Lep	Ancient (Ptolemy)
Libra	Librae	Lib	Ancient (Ptolemy)
Lupus	Lupi	Lup	Ancient (Ptolemy)
Lynx	Lyncis	Lyn	Hevelius (1690)
Lyra	Lyrae	Lyr	Ancient (Ptolemy)
Mensa	Mensae	Men	Lacaille (1754)
Microscopium	Microscopii	Mic	Lacaille (1754)
Monoceros	Monocerotis	Mon	Plancius (1613)
Musca	Muscae	Mus	Keyser & de Houtman
Norma	Normae	Nor	Lacaille (1754)
Octans	Octantis	Oct	Lacaille (1754)
Ophiuchus	Ophiuchi	Oph	Ancient (Ptolemy)
Orion	Orionis	Ori	Ancient (Ptolemy)
Pavo	Pavonis	Pav	Keyser & de Houtman
Pegasus	Pegasi	Peg	Ancient (Ptolemy)
Perseus	Persei	Per	Ancient (Ptolemy)
Phoenix	Phoenicis	Phe	Keyser & de Houtman
Pictor	Pictoris	Pic	Lacaille (1754)
Pisces	Piscium	Psc	Ancient (Ptolemy)
Piscis Austrinus	Piscis Austrini	PsA	Ancient (Ptolemy)
Puppis	Puppis	Pup	Lacaille (1754)
Pyxis	Pyxidis	Pyx	Lacaille (1754)

The Modern Constellations (continued)

Nominative	*Genitive*	*Abbreviation*	*Origin*
Reticulum	Reticuli	Ret	Lacaille (1754)
Sagitta	Sagittae	Sge	Ancient (Ptolemy)
Sagittarius	Sagittarii	Sgr	Ancient (Ptolemy)
Scorpius	Scorpii	Sco	Ancient (Ptolemy)
Sculptor	Sculptoris	Scl	Lacaille (1754)
Scutum	Scuti	Sct	Hevelius (1690)
Serpens	Serpentis	Ser	Ancient (Ptolemy)
Sextans	Sextantis	Sex	Hevelius (1690)
Taurus	Tauri	Tau	Ancient (Ptolemy)
Telescopium	Telescopii	Tel	Lacaille (1754)
Triangulum	Trianguli	Tri	Ancient (Ptolemy)
Triangulum Australe	Trianguli Australis	TrA	Keyser & de Houtman
Tucana	Tucanae	Tuc	Keyser & de Houtman
Ursa Major	Ursae Majoris	UMa	Ancient (Ptolemy)
Ursa Minor	Ursae Minoris	UMi	Ancient (Ptolemy)
Vela	Velorum	Vel	Lacaille (1754)
Virgo	Virginis	Vir	Ancient (Ptolemy)
Volans	Volantis	Vol	Keyser & de Houtman
Vulpecula	Vulpeculae	Vul	Hevelius (1690)

Bibliography

Académie des inscriptions et belles-lettres (France). 1776. "Review of 'Explication des nouveaux Globes céleste et terrestre, d'un pied de diamètre le céleste"'. *Journal des Sçavans*, 762–763.

Aldrovandi, Ulisse. 1621. *Quadrupedum omnium bisulcorum historia ....* Bologna: Sebastianum Bonhommium.

Allard, Carel. 1706. *Hemisphaerium meridionale et septentrionale planisphaerii coelestis*. Amsterdam: Covens et Mortier.

Allen, Richard Hinckley. 1899. *Star Names: Their Lore And Meaning*. New York: Dover.

Apianus, Petrus. 1536. *Imagines syderum coelestium ...* Ingolstad: Self-published.

Argelander, Friedrich W. A. 1843. *Neue Uranometrie*. Berlin: Simon Schropp.

Bachman, Frank P. 1918. *Great Inventors and Their Inventions*. New York: American Book Company.

Baily, Francis. 1845. *The Catalogue Of Stars Of The British Association For The Advancement Of Science*. London: Richard & John E. Taylor.

Bakich, Michael E. 1995. *The Cambridge guide to the constellations*. Cambridge, UK: Cambridge University Press.

Barritt, L. 1916. *no title*.

Bartsch, Jacob. 1624. *Usus astronomicus planisphaerii stellati seu vice-globi in plano*. Strasbourg: Heribert Rosweyde.

Bartsch, Jacob. 1661. *Planisphaerium stellatum seu vive-glocbus coelestis in plano delineatus*. Nuremburg: Christophorus Gerhardus.

Bayer, Johann. 1603. *Uranometria, omnium asterismorum continens schemata, nova methodo delineata, aereis laminis expressa*. Augsburg, Germany: Christophorus Mangus.

Berezkin, Yuri. 2005. The cosmic hunt: variants of a Siberian – North-American myth. *Folklore*, **31**, 79–100.

Bevis, John. 1750. *Uranographia Britannica*. London: John Neale.

Blaauw, Adriaan. 1994. *History of the IAU. The Birth and First Half-Century of the International Astronomical Union*. Dordrecht, Netherlands: Kluwer.

Black, Jeremy. 2006. *George III: America's Last King*. New Haven: Yale University Press.

Bode, Johann Elert. 1787. Friedrichs Sternen-Denkmal. *Astronomisches Jarhbuch fur das Jahr 1790*.

Bode, Johann Elert. 1792. Monument astronomique consacré à Frédéric II. *Mémoires de l'Académie Royale des Sciences et Belles Lettres 1786-87*, 57–60.

Bode, Johnann Elert. 1782. *Vorstellung der Gestirne auf XXXIV Kupfertafeln nach der Pariser Ausgabe des Flamsteadschen Himmelsatlas*. Berlin: Gottlieb August Lange.

Bode, Johnann Elert. 1801a. *Allgemeine Beschreibung und Nachweisung der Gestirne*. Berlin: Bode, Johnann Elert.

J.C. Barentine, *The Lost Constellations*, Springer Praxis Books,
DOI 10.1007/978-3-319-22795-5

Bode, Johnann Elert. 1801b. *Uranographia, sive astrorum descriptio*. Berlin: Frederico de Haan.
Bode, Johnann Elert. 1805. *Vorstellung der Gestirne auf vier und dreyssig Kupfertafeln nebst einer Anweisung zum Gebrauch und einem Verzeichnisse von 5877 Sternen, Nebelflecken und Sternhaufen*. Second ed. edn. Berlin: Gottlieb August Lange.
Bouvier, Hannah M. 1858. *Bouvier's Familiar astronomy; or, An introduction to the study of the heavens*. Philadelphia: Sower, Barnes, & Potts.
Bowen, Eliza A. 1888. *Astronomy By Observation: An Elementary Text-Book for ook for High-Schools and Academies*. New York: D. Appleton & Co.
Brahe, Tycho. 1602. *Astronomiae Instauratæ Progymnasmata*. Uraniborg, Hven, Denmark: Publisher.
Brooke, Henry. 1820. *A Guide To The Stars*. London: Taylor and Hessey.
Brooke, John. 1972. *King George III*. London: Constable.
Brown, Robert. 1885. *The Phainomena or 'Heavenly Display' of Aratos*. London: Longmans, Green, & Co.
Brown, Robert. 1899. *Researches into the Origin of the Primitive Constellations of the Greeks, Phoenicians and Babylonians*. London: Williams and Norgate.
Bryant, Jacob. 1807. *A New System: Or, an Analysis of Ancient Mythology*. London: Self published.
Burnham, Robert Jr. 1978. *Burnham's Celestial Handbook*. New York: Dover.
Burritt, Elijah Hinsdale. 1833. *The geography of the heavens; or, Familiar instructions for finding the visible stars and constellations accompanied by a celestial atlas*. Hartford, Conn.: F.J. Huntington.
Burritt, Elijah Hinsdale. 1835a. *Atlas designed to illustrate the geography of the heavens*. New York: Huntington and Savage.
Burritt, Elijah Hinsdale. 1835b. *Atlas of the Heavens*. Hartford, Connecticut: F.J. Huntington.
Butterwick, Richard. 1998. *Poland's Last King and English Culture: Stanislaw August Poniatowski, 1732–1798*. Oxford: Oxford University Press.
Case, Linda P. 2003. *The Cat: Its Behavior, Nutrition, and Health*. Ames, Iowa: Iowa State University Press.
Cellarius, Andreas. 1661. *Harmonia Macrocosmica Seu Atlas Universalis Et Novus: Totius Universi Creati Cosmographiam Generalem, Et Novam Exhibens*. Amsterdam: Jan Janssonius.
Chambers, Ephraim (ed). 1728. *Cyclopædia, or, An universal dictionary of arts and sciences*. London: James and John Knapton, et al.
Chambers, G. F. 1913. *Astronomy*. New York: D. Van Nostrand Company.
Chambers, George Frederick. 1877. *A Handbook of Descriptive Astronomy*. 3rd edn. Oxford: Clarendon Press.
Clark, Christopher. 2009. *Iron Kingdom: The Rise and Downfall of Prussia, 1600-1947*. Cambridge, Mass.: Harvard University Press.
Clerke, Agnes Mary. 1895. *The Herschels and Modern Astronomy*. London: Cassell and Company, Ltd.
Colas, Jules A. (ed). 1892. *Poole Bros' Celestial Handbook, companion to their Celestial Planisphere*. Chicago: Poole Bros.
Collins, Arthur H. 1913. *Symbolism of animals and birds represented in English architecture*. London: Sir I. Pitman & Sons, Ltd.
Cooley, Jeffrey L. 2011. "An OB Prayer to the Gods of the Night". Pages 77–83 of: Lenzi, Alan (ed), *Reading Akkadian Prayers and Hymns*. Ancient Near East Monographs, no. 3. Society of Biblical Literature.
Coronelli, Vincenzo M. 1693. *Epitome cosmografica, o Compendiosa introduttione all'astronomia, geografia, & idrografia...* Venice: Andrea Poletti.
Cottam, Arthur. 1891. *Charts of the Constellations From the North Pole to between 35 & 40 Degrees of South Declination*. London: Edward Stanford.
Croswell, William. 1810. *A Mercator map of the starry heavens : comprehending the whole equinoctial and terminated by the polar circles*. Boston: T. Wightman.
de Broen, Johannes. 1709. *Hemelskaart voor de noordelijke en zuidelijke sterrenhemel uitgevoerd in Mercatorprojectie*. Find out.
de La Hire, Philippe. 1702. *Planisphère céléste septentrionale*. Paris: N. de Fer.

de Lalande, Jérôme Lefrançois. 1803. *Histoire abrégée de l'astronomie, depuis 1781 jusqu'à 1802*. Paris: Imprimerie de la République.
de Mapertuis, Pierre-Louis Moreau. 1738. *La Figure de la Terre, déterminée par les Observations de Messieurs de Maupertuis, Clairaut, Camus, Le Monnier, de l'Académie Royale des Sciences, & de M. l'Abbé Outhier, Accompanés de M. Celsius, Faltes par Ordre du Rois au Cercle Polaire*. Paris: L'Imprinere Royale.
de Maupertuis, Pierre-Louis Moreau. 1752. *Les oeuvres de Mr. de Maupertuis*. Desden: Walther, George Conrad.
Delporte, Eugène J. 1930a. *Atlas Céleste*. Cambridge University Press.
Delporte, Eugène J. 1930b. *Délimitation scientifique des constellations*. Cambridge University Press.
Doppelmayr, Christian. 1742. *Atlas Coelestis*. Homännische Erben.
Duffy, Christopher. 1985. *Frederick the Great: A Military Life*. London: Routledge & Kegan.
Dürer, Albrecht. 1515. *Imagines coeli Septentrionales cum duodecim imaginibus zodiaci*. Nuremburg: Self published.
Engels, Donald W. 2001. *Classical Cats: The Rise and Fall of the Sacred Cat*. London: Routledge.
Evans, David S. 1992. *Lacaille: astronomer, traveller; with a new translation of his journal*. Tucson: Pachart.
Firmicus Maternus, Julius. 1499. *Matheseos Liber*. Venice: Aldus Manutius.
Flammarion, Camille. 1882. *Les Etoiles et les Curiosités du Ciel*. Paris: C. Marpon and E. Flammarion.
Flammarion, Camille, and Gore, J. E. 1894. *Popular Astronomy*. English translation edn. New York: D. Appleton & Co.
Flammarion, G. C. 1956. "Eugène Delporte (1882–1955)". *L'Astronomie*, **70**(Oct.), 379.
Flamsteed, J., Crosthwait, J., and Sharp, A. 1725. *Historiae coelestis britannicae volumen primum[–tertium]* ... London: Edmund Halley.
Flamsteed, John. 1725. *Stellarum inerrantium Catalogus Britannicus*. Historia Coelestis Britannica, vol. 3. London: Self published.
Flamsteed, John. 1729. *Atlas Coelestis*. London: Self published.
Fortin, Jean Nicolas. 1776. *Atlas Céleste de Flamstéed*. Second edn. Paris: F. G. Deschamps.
Fortin, Jean Nicolas. 1795. *Atlas Céleste de Flamstéed*. Third edn. Paris: Lamarche.
Gillespie, Charles Coulston. 1983. *The Montgolfier Brothers and the Invention of Aviation 1783–1784*. Princeton, NJ: Princeton University Press.
Goldbach, Christian F. 1799. *Neuester Himmels-Atlas*. Weimar: Verlage des Industrie-Comptoirs.
Gould, Benjamin Apthorp. 1879. *Uranometria Argentina*. Resultados del Observatorio Nacional Argentino en Cordoba, vol. 1. Buenos Aires: Impr. de P.E. Coni.
Green, Jacob. 1824. *Astronomical Recreations, or Sketches of the Relative Position and Mythological History of the Constellations*. Philadelphia: Anthony Finley.
Guthnick, P., and Prager, R. 1928. Benennung von veränderlichen Sternen. *Astronomische Nachrichten*, **232**(June), 353.
Habrecht, Isaac. 1621. *Globus Coelestis. Omnes Stellas fixas a Nobilis Tychone Brahe* .... Strasbourg: Self published.
Hall, Sidney. 1825. *Urania's Mirror, or A View of the Heavens; Consisting of Thirty-Two Cards on Which are Represented all the Constellations Visible in Great Britain; on a Plan Perfectly Original, Designed by a Lady*. London: Samuel Leigh.
Halley, Edmond. 1679. *Catalogus Stellarum Australium sive Supplementum Catalogi Tychoni* ... London: R. Harford.
Harbord, J. B. 1883. *Glossary of Navigation*. 2nd edn. Portsmouth, UK: Griffin & Co.
Heis, Eduard. 1872. *Atlas Coelestis Novus*. Cologne: M. Dumont-Schauberg.
Hell, Maximilian. 1789. *Ephemerides astronomicae: Anni 1790 ad meridianum mediolanensem ; Acc. appendix cum observ. et opusc*. Milan: Joseph Galeatium.
Herschel, John F. W. 1843. "Farther Remarks on the Revision of the Southern Constellations". *Monthly Notices of the Royal Astronomical Society of London*, **6**(1), 60–62.

Hevelius, Johannes. 1687. *Firmamentum Sobiescianum sive Uranographia*. Danzig: Johann Zacharias Stoll.
Hevelius, Johannes. 1690. *Prodromus Astronomiae*. Danzig: Johann Zacharias Stoll.
Hewitt, J. F. 1895. *The ruling races of prehistoric times in India, South-western Asia and Southern Europe*. Vol. 2. Edinburgh: T. and A. Constable.
Hill, John. 1754. *Urania: or, A Compleat View of the Heavens*. London: T. Gardner.
Holden, Edward Singleton. 1881. *Sir William Herschel: His Life And Works*. New York: Charles Scribner's Sons.
Houzeau , Jean Charles. 1882. *Vade-mecum de l'astronome*. Brussels: F. Hayez, L'Académie Royale de Belgique.
Howey, M. Oldfield. 1930. *The Cat in Magic and Myth*. Rider & Co.
Hubatsch, Walther. 1975. *Frederick the Great of Prussia: Absolutism and Administration*. London: Thames and Hudson.
Hunger, Hermann, and Pingree, David. 1999. *Astral Sciences in Mesopotamia*. Handbook of Oriental Studies, vol. 1 (The Near and Middle East). Leiden, Netherlands: Brill.
Jamieson, Alexander. 1822. *A celestial atlas : comprising a systematic display of the heavens in a series of thirty maps : illustrated by scientific description of their contents and accompanied by catalogues of the stars and astronomical exercises*. London: G. & W.B. Whittaker.
Johnston, Alexander Keith. 1855. *Atlas of astronomy : comprising, in eighteen plates a complete series of illustrations of the heavenly bodies, drawn with the greatest care, from original and authentic documents*. London: William Blackwood and Sons.
Kanas, Nick. 2007. *Star Maps: History, Artistry, and Cartography*. Chichester, UK: Praxis.
Kendall, Ezra Otis. 1845. *Uranography: or, a description of the heavens; designed for academies and schools; accompanied by an atlas of the heavens, showing the places of the principal stars, clusters and nebulae*. Philadelphia: E. H. Butler & Co.
Kepler, Johannes. 1606. *De stella nova in pede serpentarii*. Prague: Paul Sessius.
Kepler, Johannes. 1627. *Tabulae Rudolphinæ(Rudolphine Tables)*. Ulm, Germany: Self published.
Kirch, Gottfried. 1688. Godofredi Kirchii Pomum Imperiale et Sceptrum Brandeburgicum. *Acta Eruditorum*, 452.
Kopff, A. 1956. "Eugène J. Delporte". *Astronomische Nachrichten*, **283**(May), 140.
Lacaille, Nicolas Louis de. 1752. Table des ascensions droites et des déclinaisons apparentes des Etoiles australes renfermées dans le tropique du Capricorne, observées au cap de Bonne-espérance, dans l'intervalle du 6 Août 1751, au 18 Juillet 1752. *Histoire de l'Académie Royale des Sciences*, 539–592.
Lacaille, Nicolas Louis de. 1756. Planisphere contenant les Constellations Celestes. *Memoires Academie Royale des Sciences pour* 1752. Paris.
Lacaille, Nicolas Louis de. 1763. *Coelum Australe Stelliferum; seu Observationed ad Construendum Stellarum Australium Catalogum Institutae*. Paris: H. L. Guerin & L.F. Delatour.
Le Monnier, Pierre-Charles. 1743. *La Théorie des Comètes*. Paris: G. Martin, J.B. Coignard, & les Freres Guerin.
Le Monnier, Pierre-Charles. 1776. Constellation du Solitaire. *Mémoires de l'Académie Royale des Sciences*.
Lebon, Ernest. 1899. *Histoire Abrégée de l'astronomie*. Paris: Gauthier-Villars.
Lewis, Isabel M. 1948. Constellation Boundaries. *Nature Magazine*, October.
Littrow, Joseph Johann. 1839. *Atlas des Gestirnten Himmels*. Stuttgart: Hoffmann'sche Verlags-Buchhandlung.
Longomontanus, Christen Sorensen. 1622. *Astronomica Danica*. Amsterdam: Willem Janszoon Blaeu.
Lubieniecki, Stanislaus. 1667. *Theatrum cometicum, duabus partibus constans ...* Vol. 1. Leiden: Franciscum Cuyperum.
Lubieniecki, Stanislaus. 1681. *Theatrum cometicum, duabus partibus constans ...* Vol. 2. Leiden: P. van der Meersche.
Luvaas, Jay. 1999. *Frederick The Great On The Art Of War*. New York: Da Capo Press.

Mallory, J. P., and Adams, D. Q. 2006. *The Oxford Introduction to Proto-Indo-European and the Proto-Indo-European World*. Oxford Linguistics. Oxford: Oxford University Press.
Marion, Fulgence. 1870. *Wonderful Balloon Ascents; or, The Conquest of the Skies*. London: Cassel Petter & Galpin.
Martin, Henri-Jean. 1995. *The History and Power of Writing*. Chicago: University of Chicago Press.
Meissner, August Gottlieb. 1805. *Astronomischer Hand-Atlas zu Rüdigers Kentniss des Himmels*. Leipzig: Siegfried Lebrecht Crusius.
Middleton, James. 1842. *A Celestial Atlas*. London: Whittaker and Co.
Miner, Ellis D. 1998. *Uranus: The Planet, Rings and Satellites*. New York: John Wiley and Sons.
Möllinger, Otto. 1851. *Himmels-atlas mit transparenten sternen*. Solothurn, Switzerland: Self-published.
Moore, Patrick, and Rees, Robin. 2011. *Patrick Moore's Data Book of Astronomy*. Cambridge, UK: Cambridge University Press.
Murdin, Paul. 2008. *Full Meridian of Glory: Perilous Adventures in the Competition to Measure the Earth*. New York: Copernicus.
Newcomb, Simon. 1878. *Popular Astronomy*. London: Macmillan and Co.
Oberlies, Thomas. 1998. *Die Religion des Rigveda*. Publications of the De Nobili Research Library, vol. 26. Vienna: Vienna University.
"A Society of Gentlemen" 1778. Foreign Literary Intelligence. *The Critical review, or, Annals of literature*, **46**, 146.
Ogilvie, Sheilagh (ed). 1996. *Germany: A New Social and Economic History Volume 2: 1630–1800*. Vol. 2. London: Arnold.
Olbers, Heinrich W. M. 1841. On a Reformation of the Constellations, and a Revision of the Nomenclature of the Stars. *Monthly Notices of the Astronomical Society of London*, **V**(13), 101–105.
Olcott, William Tyler. 1911. *Star Lore of All Ages*. New York: G. P. Putnam's Sons.
Pardies, Ignace-Gaston. 1674. *Globi coelestis in tabulas planas redacti descriptio*. Paris: Sebastien Mabre-Cramoisy.
Pecker, J.-C. (ed). 1964. *Transactions of the International Astronomical Union (XIIth General Assembly, Hamburg, Germany, August 25 - September 3, 1964)*. Vol. 12. Cambridge, UK: Blackwell Scientific Publications.
Peltier, Leslie C. 1965. *Starlight Nights: The Adventures of a Star-Gazer*. New York: Harper & Row.
Peltier, Leslie. 1967. *Starlight Nights*, London: Macmillan.
Philippot, H., and Delporte, E. 1914. Catalogue de 3553 étoiles de repère de la zone 21° – 22° pour la carte photographique du ciel. *Annales de l'Observatoire Royal de Belgique Nouvelle serie*, **13**, 1.
Pierach, C. A., and Jennewein, E. 1999. Friedrich Wilhelm I and porphyria. *Sudhoffs Archiv*, **83**(1), 50–66.
Poczobut, Martin O. 1777. *Cahiers des observations astronomiques faites à l'observatoire royal de Vilna en 1773*. Royal publisher.
Prager, M. R. 1928. *Katalog und Ephemeriden Veränderlicher Sterne für 1928*. Catalog 3. Universitäts-sternwarte zu Berlin.
Preyssinger, Ludwig. 1862. *Astronomie Populaire ou Description des Corps Célestes*. Brussels: Kiessling & Co.
Pritchard, Charles. 1885. *Uranometria Nova Oxoniensis*. Oxford: Clarendon Press.
Proctor, Richard Anthony. 1872. *A new star atlas for the library, the school, and the observatory*. London: Longmans, Green, and Company.
Proctor, Richard Anthony. 1876. *The Constellation-Seasons*. London: Longmans, Green, and Company.
Pugh, Philip. 2012. *Observing the Messier Objects with a Small Telescope*. The Patrick Moore Practical Astronomy Series. New York: Springer.
Rawlins, Dennis. 1993. Tycho's 1004-Star Catalog. *DIO, the International Journal of Scientific History*, **3**(October), 1–106.
Ridpath, Ian. 1989. *Star Tales*. Cambridge, UK: Lutterworth Press.
Riedig, Christian Gottlieb. 1849. *Himmels-Atlas in 20 Blattern nach den grossen Bodenschen Sternkarten...* Leipzig: bei Schreibers Erben ("by the writer's heirs").
Rogers, J. H. 1998. "Origins of the ancient constellations: I. The Mesopotamian traditions". *Journal of the British Astronomical Association*, **108**(Feb.), 9–28.

Rosser, William Henry. 1879. *The Stars and Constellations; How and When to Find and Tell Them.* London: Charles Wilson.

Rost, Johann Leonhardt. 1723. *Atlas Portatilis Coelestis.* Nuremburg: Johann Ernst Udelbulner.

Royer, Augustin. 1679a. *Cartes du Ciel Reduites en Quatre Tables, Contenant Toutes les Constellations.* Paris: Jean Baptiste Coignard.

Royer, Augustin. 1679b. *Tabula Universalis Longitudinum et Latitudinum Stellarum.* Paris: Jean Baptiste Coignard.

Ryan, James. 1827. *The New American Grammar of the Elements of Astronomy: On an Improved Plan.* New York: Collins and Hannay.

Schaefer, B. E. 2006. "The Origin of the Greek Constellations". *Scientific American,* **295**(5), 050000–101.

Schiller, Julius. 1627. *Coelum Stellatum Christianum.* Augsburg: Andrea Apergeri.

Schlesinger, F. 1932. "Commission des Notations, des Unités, et de L'Économic des Publications". Page 19 of: Stratton, F. J. M. (ed), *Transactions of the International Astronomical Union: Fourth General Assembly Held at Cambridge, Massachusetts, September 2 to September 9, 1932,* vol. IV. Cambridge University Press, for International Astronomical Union.

Seller, John. 1680. *Atlas cœlestis : containing the systems and theoryes of the planets, the constellations of the starrs, and other phenomina's of the heavens.* London: Self published.

Semler, Christoph. 1731. *Coelum Stellatum in Quo Asterismi.* Halle: Self published.

Sharpless, I., and Philips, G. M. 1882. *Astronomy for Schools and General Readers.* 4th edn. J. B. Lippincott Company.

Sherburne, Edward. 1675. *The Sphere of Marcus Manilius made an English Poem, with annotations and an astronomical appendix.* London: Nathanael Brooke.

Smyth, W. H. 1844a. *A cycle of celestial objects for the use of naval, military, and private astronomers: The Bedford Catalogue.* Vol. 2. London: John W. Parker.

Smyth, William Henry. 1844b. *A cycle of celestial objects for the use of naval, military, and private astronomers: Prolegomena.* Vol. 1. London: John W. Parker.

Steele, Joel Dorman. 1899. *Popular Astronomy, Being The New Decriptive Astronomy.* 2nd edn. New York: American Book Company.

Stratton, F. J. M. (ed). 1928. *Transactions of the International Astronomical Union (IIIrd General Assembly, Leiden, The Netherlands, July 5–13, 1928).* Vol. 3. Cambridge, UK: Cambridge University Press.

Stratton, F. J. M. (ed). 1932. *Transactions of the International Astronomical Union (IVth General Assembly, Cambridge, Massachusetts, USA, September 2–9, 1932).* Vol. 4. Cambridge, UK: Cambridge University Press.

Stroobant, P., Delvosal, J., H., Philippot, Delporte, E., and Merlin, E. 1907. *Les observatoires astronomiques et les astronomes.* Brussels: Hayez.

Terrall, Mary. 2002. *The Man Who Flattened the Earth: Maupertuis and the Sciences in the Enlightenment.* Chicago: University of Chicago Press.

Thomas, Corbinianus. 1730. *Mercurii philosophici firmamentum firmianum descriptionem et vum globi artificialis coelestis.* 1st edn. Frankfurt: Prostat Franckofurti & Lipsiae.

Thoren, V. E. 1989. Tycho Brahe. Pages 3–21 of: Taton, R., Wilson, C., and Hoskin, M. (eds), *Planetary Astronomy from the Renaissance to the Rise of Astrophysics. Part A: Tycho Brahe to Newton.*

Wagman, Morton. 2003. *Lost Stars: Lost, Missing and Troublesome Stars from the Catalogues of Johannes Bayer, Nicholas Louis de Lacaille, John Flamsteed, and Sundry Others.* Blacksburg, Virginia: McDonald & Woodward.

Watson, Rita, and Horowitz, Wayne. 2011. *Writing Science Before the Greeks: A Naturalistic Analysis of the Babylonian Astronomical Treatise MUL.APIN.* Culture and History of the Ancient Near East. Leiden, Netherlands: Brill.

Wesley, W. G. 1978. The Accuracy of Tycho Brahe's Instruments. *Journal for the History of Astronomy,* **9**, 42.

White, Gavin. 2008. *Babylonian Star-Lore: an Illustrated Guide to the Star-Lore and Constellations of Ancient Babylonia*. 2nd edn. London: Solaria Publications.

Vopel, Caspar. 1536. CASPAR. VOPEL. MEDEBACH HANC. COSMOGRA: faciebat sphæram. Coloniæ. Å. 1536. Globe. Medebach, Germany.

Young, Charles Augustus. 1903. *Lessons in Astronomy, Including Uranography: A Brief Introductory Course Without Mathematics*. 2nd edn. Boston: Ginn & Company.

Young, Thomas. 1807. *A Course of Lectures on Natural Philosophy and the Mechanical Arts*. Vol. 1. London: Joseph Johnson.

Zamoyski, Adam. 1992. *The Last King of Poland*. London: Jonathan Cape.

Index

Symbols

J.C. Barentine, *The Lost Constellations*, Springer Praxis Books,
DOI 10.1007/978-3-319-22795-5

B

F

H

I

M

Printed in the United States
By Bookmasters